INTERNATIONAL SYMPOSIUM ON VECTOR BOSON SELF-INTERACTIONS

AIP CONFERENCE PROCEEDINGS 350

INTERNATIONAL SYMPOSIUM ON VECTOR BOSON SELF-INTERACTIONS

LOS ANGELES, CA FEBRUARY 1995

EDITORS: **ULRICH BAUR**
UNIVERSITY OF BUFFALO
STEVEN ERREDE
UNIVERSITY OF ILLINOIS
THOMAS MÜLLER
UNIVERSITY OF CALIFORNIA, LOS ANGELES

American Institute of Physics   Woodbury, New York

Authorization to photocopy items for internal or personal use, beyond the free copying permitted under the 1978 U.S. Copyright Law (see statement below), is granted by the American Institute of Physics for users registered with the Copyright Clearance Center (CCC) Transactional Reporting Service, provided that the base fee of $6.00 per copy is paid directly to CCC, 222 Rosewood Drive, Danvers, MA 01923. For those organizations that have been granted a photocopy license by CCC, a separate system of payment has been arranged. The fee code for users of the Transactional Reporting Service is: 1-56396-520-8/ 96 /$6.00.

© 1996 American Institute of Physics.

Individual readers of this volume and nonprofit libraries, acting for them, are permitted to make fair use of the material in it, such as copying an article for use in teaching or research. Permission is granted to quote from this volume in scientific work with the customary acknowledgment of the source. To reprint a figure, table, or other excerpt requires the consent of one of the original authors and notification to AIP. Republication or systematic or multiple reproduction of any material in this volume is permitted only under license from AIP. Address inquiries to Office of Rights and Permissions, 500 Sunnyside Boulevard, Woodbury, NY 11797-2999; phone 516-576-2268; fax: 516-576-2499; e-mail: rights@aip.org.

L.C. Catalog Card No. 95-79865
ISBN 1-56396-520-8
DOE CONF-950287

Printed in the United States of America.

CONTENTS

Preface vii
International Advisory Committee viii

A) PLENARY TALKS

Gauge Symmetries and Vector Boson Self-Interactions 3
    S. Willenbrock
Status of the Electroweak Sector of the Standard Model 20
    T. Takeuchi
Precision Data from LEP 33
    D. Schaile
Probing Trilinear Gauge Boson Couplings at Colliders 46
    D. Zeppenfeld
QCD Corrections to Diboson Production 60
    J. Ohnemus
$W\gamma$ and $Z\gamma$ Production at the Tevatron 72
    H. Aihara
WW and WZ Production at the Tevatron 84
    T. A. Fuess
Standard Model Higher Order Corrections to the $WW\gamma/WWZ$ Vertex 98
    J. Papavassiliou
Static Quantities of the W Bosons in the MSSM 110
    A. B. Lahanas
Rare b Decays and Anomalous Couplings 124
    S. M. Playfer
Theory of Rare B Decays 136
    A. F. Falk
Single Photon and Radiative Events at LEP 148
    P. Mättig
Experimental Signatures of a Parity Violating Anomalous Coupling g_5^Z 160
    G. Valencia
Effective Lagrangians and Anomalous Couplings 171
    J. Wudka
Low Energy Constraints on Electroweak Vector Boson Self-Interactions 182
    K. Hagiwara
W and Z Boson Production at HERA 198
    R. Walczak
Quartic Gauge Boson Couplings 209
    S. Godfrey
Exact and Approximate Radiation Amplitude Zeros—Phenomenological Aspects 224
    T. Han
Strong $W_L W_L$ Scattering 239
    R. S. Chivukula

Triple Gauge Couplings: Does LEP-1 Obviate LEP-2? 250
  P. Hernández
Understanding Something About Nothing: Radiation Zeros 261
  R. W. Brown
Measuring WWZ and $WW\gamma$ Couplings at LEP II 273
  J. Busenitz
Testing Vector Boson Self-Interactions in Future Tevatron Experiments 285
  C. Wendt
Probing Tri-linear Couplings at the LHC 299
  J. Womersley
Studies of $WW\gamma$ and WWZ Couplings at Future e^+e^- Linear Colliders 307
  T. L. Barklow
Anomalous Gluon Self-Interactions and $t\bar{t}$ Production 323
  E. H. Simmons and P. Cho
Summary Talk: Gauge Boson Self-Interactions 335
  I. Hinchliffe

B) PARALLEL TALKS

Rare Z^O Decays 347
  P. Giacomelli
The R_b Excess at LEP: Clue to New Physics at the TEVATRON? 352
  E. Ma and D. Ng
The Measurement of Tri-Linear Gauge Boson Couplings at e^+e^- Colliders 357
  G. Couture, M. Gintner, and S. Godfrey
The Ward Identities of the Gauge Invariant Three Boson Vertices 362
  K. Philippides
One-Loop Effects of a Heavy Higgs Boson: A Functional Approach 367
  S. Dittmaier and C. Grosse-Knetter
Search for W Boson Pair Production in Dilepton Decay Modes at DØ 372
  H. Johari
Search for $WW \to l^+l^- + X$ at $\sqrt{s} = 1.8\ TeV$ 377
  D. L. Carlsmith and L. Zhang
Search for Anomalous $ZZ\gamma$ and $Z\gamma\gamma$ Couplings with DØ 382
  G. Landsberg
CDF Results on $Z\gamma$ Production 388
  R. G. Wagner for the CDF Collaboration
Results on $W\gamma$ Production at CDF 393
  D. Neuberger
Towards Probing the $WW\gamma$ Vertex at HERA 399
  A. Schöning
Limits on Rare B Decays $B \to \mu^+\mu^-K^\pm$ and $B \to \mu^+\mu^-K^*$ 404
  C. Anway-Wiese
List of Participants 407
Author Index 409

PREFACE

The International Symposium on Vector Boson Self-Interactions was held at UCLA, February 1–3, 1995.

The purpose of the Symposium was to provide an overview of our current theoretical and experimental understanding of trilinear and quartic vector boson couplings ($WW\gamma$, WWZ, $ZZ\gamma$, etc.), and to discuss the prospects for measuring these couplings in future experiments (high precision experiments, Tevatron, LEP II, HERA, LHC, NLC). Over the last five years experiments at LEP, the SLAC linear collider and the Tevatron have beautifully confirmed many of the predictions of the Standard Model of Electroweak Interactions. Nevertheless the most direct consequences of the $SU(2) \times U(1)$ symmetry upon which the Standard Model is based, the non-abelian self-couplings of the W, Z and photon, remain poorly measured to date.

As the first Symposium focused on this important theme, the meeting was designed to bring together experimentalists and theorists working on a wide range of related processes and problems. Plenary sessions concentrated on overviews of the major theoretical issues and summaries of new experimental results; parallel sessions provided opportunities for specialized (and frequently lively) discussions among experts. The active participation of 85 registered physicists from North America, Europe, and Japan, representing most major laboratories and universities, and their enthusiasm at the Symposium, are testimony to the timeliness of the meeting. The organizers would like to thank all speakers and participants for contributing to the success of the Symposium.

The Symposium was sponsored by Fermilab, the National Science Foundation, the Department of Energy, and the University of California at Los Angeles. The extraordinary efforts of An-Chi Kao, Jim Kolonko, and Melinda Laraneta were instrumental to the success of the Symposium. Thanks also go to the staff of the UCLA Faculty Center, where the meeting was held, for their smooth and efficient help.

Ulrich Baur
Steven Errede
Thomas Müller
June 1995

INTERNATIONAL ADVISORY COMMITTEE

W. Buchmüller	DESY
J. Busenitz	University of Alabama
D. Cline	UCLA
K. Einsweiler	LBL
J. Ellison	UC Riverside
S. Eno	University of Maryland
H. Fritzsch	University of Munich
P. Grannis	SUNY Stony Brook
F. Halzen	University of Wisconsin
J. Hauser	UCLA
P. Mättig	University of Bonn
A. Miyamoto	KEK
R. Peccei	UCLA
C. Quigg	Fermilab

A) PLENARY TALKS

Gauge Symmetries and Vector-Boson Self Interactions

S. Willenbrock

Department of Physics
University of Illinois
1110 West Green Street
Urbana, IL 61801

I review why we believe the electromagnetic, strong, and weak interactions are gauge theories, and what this implies for the self interactions of the gauge bosons. The modern point of view regarding non-renormalizable effective field theories is emphasized.

INTRODUCTION

The discovery of the W and Z bosons at the CERN $Sp\bar{p}S$ in 1983 (1) began the era of the weak vector boson. It opened with a bang, earning a Nobel prize for the discovery of the particles and the development of the machine that made it possible (2). That era is now in its maturity, with precision studies of Z bosons taking place at the CERN LEP and SLAC SLC e^+e^- colliders, and studies of both Z and W bosons taking place at the Fermilab Tevatron $p\bar{p}$ collider.

The era of weak boson pair production began more quietly about two years ago with the first WZ event at the Tevatron, shown in Fig. 1. We will hear at this meeting of the first direct evidence for the WWZ interaction from the CDF Collaboration (3,4). Soon we will see the production of W^+W^- pairs at the CERN LEP II e^+e^- collider (5), and large numbers of weak boson pairs will be provided by the CERN LHC (6). Future e^+e^- colliders will further contribute to the study of W^+W^- and ZZ pairs at high energy (7).

Given the present situation, this is an appropriate time to ask two questions:

- What have we learned from the weak-boson era?

- What can we learn from the era of weak-boson pair production?

The language for this discussion will be quantum field theory. As far as we know, quantum field theory is the only possible way to wed quantum mechanics and special relativity.[1] More precisely, it is the only formalism capable of simultaneously implementing the constraints of Lorentz invariance,

---

[1] There is also string theory, but at low energies this reduces to quantum field theory.

© 1996 American Institute of Physics

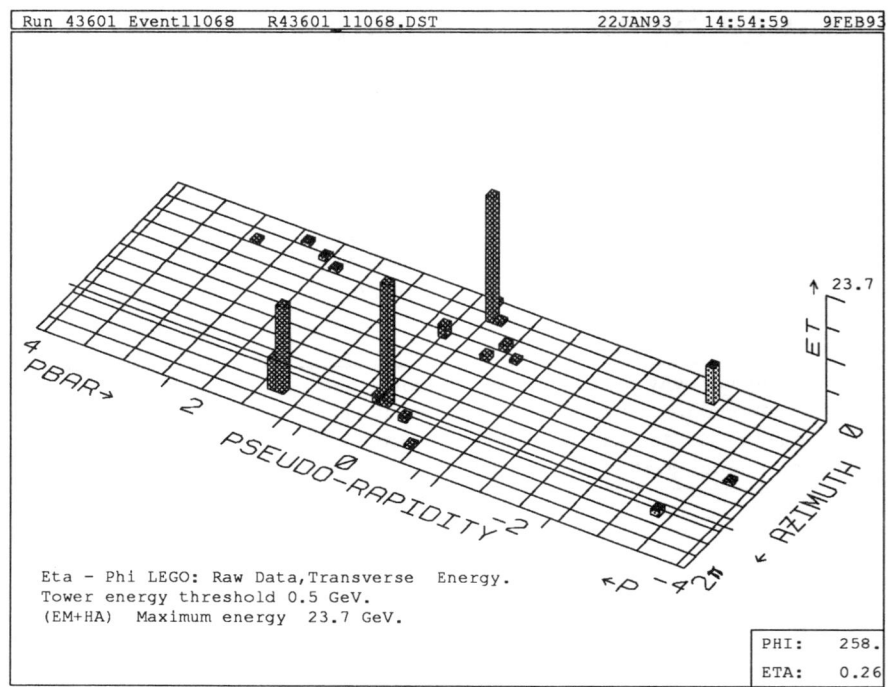

FIG. 1. The first WZ event at the Fermilab Tevatron, with leptonic decay of both weak bosons. The two tallest towers are the e^+e^- decay products of the Z boson, and the third tallest tower is an e^+ from W^+ decay.

unitarity, analyticity, and cluster decomposition (8).[2] Due to the well-known ultraviolet divergences of quantum field theory, it is unlikely that it is a valid description of nature to arbitrarily high energies. Thus we believe that at the energies currently available to us, nature must be described by an "effective" quantum field theory, even though we do not believe that quantum field theory is truly fundamental (8,9).

In this talk I present a discussion of vector-boson self interactions from a modern point of view. The presentation is ahistorical, although I occasionally make remarks pertinent to the historical development of the theory. In particular, the modern point of view regarding non-renormalizable effective field theories figures prominently in the discussion. My goal is to present a theoretical overview of the subject, and to point to subsequent speakers who will develop various subtopics in more detail. In keeping with this style, I will

---

[2]Cluster decomposition is the requirement that scattering amplitudes factorize when two particles are separated by a large spacelike distance.

often leave the citation of the literature to these speakers.[3]

Although the subject of this talk is mostly of interest for the weak interaction, it is instructive to also consider the electromagnetic and strong interactions. The order of presentation is as follows:

- Quantum Electrodynamics

- Quantum Chromodynamics

- Weak Interaction:

  o Higgs model

  o No-Higgs model

In a final section I reflect upon what we have learned from our deliberations.

QUANTUM ELECTRODYNAMICS

Let us begin by building the theory of quantum electrodynamics from two experimental facts:

1. The photon is massless.[4]

2. The photon has spin one.

From these experimental facts, the challenge is to construct a consistent quantum field theory of photons and electrically-charged fermions. The simplest field which contains spin one is the vector field,[5] so we begin by associating the photon with a field $A^\mu(x)$. In so doing, we immediately encounter two difficulties:

1. The photon has only two degrees of freedom, corresponding to helicity ± 1, while the vector field A^μ has four degrees of freedom.

2. The temporal component of the vector field has negative energy.

To see the latter point, consider the following Lagrangian for a vector field,

$$\mathcal{L} = -\frac{1}{2}\partial^\mu A^\nu \partial_\mu A_\nu \qquad (1)$$
$$= -\frac{1}{2}\left[\left(\frac{\partial A^0}{\partial t}\right)^2 - \left(\frac{\partial \mathbf{A}}{\partial t}\right)^2 + \cdots\right]$$

---

[3]Some of the observations made in this talk are also made in Ref. (10).

[4]The experimental upper bound on the photon mass is 3×10^{-27} eV. For the sake of argument, let us regard the photon as being exactly massless.

[5]Tensor fields also contain spin one, but do not reproduce Maxwell's equations in the classical limit (11).

which shows that in order for the spatial components of the vector field to have positive energy, the temporal component must have negative energy.

The resolution of these difficulties is well known. To eliminate the negative-energy component, we add an additional term to the Lagrangian which cancels the offending term above,

$$\mathcal{L} = -\frac{1}{2}(\partial^\mu A^\nu \partial_\mu A_\nu - \partial^\mu A^\nu \partial_\nu A_\mu) \tag{2}$$

$$= -\frac{1}{2}\left[\left(\frac{\partial A^0}{\partial t}\right)^2 + \cdots - \left(\frac{\partial A^0}{\partial t}\right)^2 + \cdots\right]$$

$$= -\frac{1}{4}F^{\mu\nu}F_{\mu\nu}$$

where the last line casts the Lagrangian in the familiar form in terms of the electromagnetic field-strength tensor

$$F^{\mu\nu} = \partial^\mu A^\nu - \partial^\nu A^\mu . \tag{3}$$

The field A^0 has been eliminated as a dynamical degree of freedom. We now notice that this Lagrangian is invariant under the transformation

$$A^\mu \to A^\mu - \partial^\mu \lambda \tag{4}$$

which allows us to eliminate another degree of freedom from the theory, bringing us down to the desired two degrees of freedom (12).

We recognize Eq. 4 as the familiar gauge invariance of QED. What the above argument shows in a heuristic way, and has been proven rigorously (11,13), is that gauge invariance is *mandatory*; it can be derived from the assumption of a massless spin one particle.[6] Gauge invariance is necessary to reconcile Lorentz invariance (the four-vector field A^μ) and unitarity (two degrees of freedom).[7]

The necessity of gauge invariance in the formulation of QED implies that photon self interactions of the form

$$\mathcal{L}_{int} = c_1 A^\mu A_\mu A^\nu A_\nu + c_2 \partial^\mu A^\nu A_\mu A_\nu \tag{5}$$

are strictly forbidden. Such terms are not gauge invariant, and their presence would destroy the consistency of the theory.

This does not mean that there cannot be photon self interactions, however. Let's write down the most general Lagrangian for the interaction of photons and fermions allowed by Lorentz invariance and gauge invariance:

---

[6] We now regard gauge invariance as fundamental, and use it to *explain* the masslessness of the photon, the reverse of the above logic. This point of view is largely a consequence of our realization that the strong and weak interactions are also gauge theories.

[7] An alternative point of view is that A^μ is not a four vector, because under Lorentz transformations it undergoes a gauge transformation as well. Again, gauge invariance is mandatory to ensure Lorentz invariance (11,14).

$$\mathcal{L} = -\frac{1}{4}F^{\mu\nu}F_{\mu\nu} + i\bar{\psi}\,\rlap{/}{D}\psi - m\bar{\psi}\psi \qquad (6)$$
$$+ \frac{c_1}{M^2}m\bar{\psi}\sigma^{\mu\nu}\psi F_{\mu\nu} + \frac{c_2}{M^2}\bar{\psi}\gamma^\mu\psi\bar{\psi}\gamma_\mu\psi$$
$$+ \frac{c_3}{M^4}(F^{\mu\nu}F_{\mu\nu})^2 + \cdots$$

where the terms are arranged in increasing powers of dimension,[8] and M is a mass scale introduced to make the constants c_i dimensionless. The first line above is the familiar Lagrangian of QED, and it describes the interaction of photons and fermions with remarkable success. The (infinite number of) additional terms are unnecessary; there is no experimental observation which requires any of them. In the past, such terms would have been dismissed on the grounds that they are non-renormalizable; they have coefficients with inverse powers of mass, the hallmark of non-renormalizable interactions. However, we no longer regard renormalizability as a fundamental requirement of a field theory, since we do not demand that a given field theory (or even field theory itself) be valid to arbitrarily high energy. Instead, we recognize that these additional terms are suppressed by inverse powers of M, which we regard as the energy scale at which ordinary QED ceases to be a valid description of the interaction of photons and fermions. The presence of such terms would be revealed to us by performing experiments at sufficiently high energy or with sufficient accuracy. The success of QED implies that M is a very large mass, at least 1 TeV. The renormalizability of ordinary QED ensures that these terms are not needed to cancel divergences, to all orders in perturbation theory, so the scale M can be arbitrarily large. However, the renormalizability of QED is just a consequence of the fact that M is much larger than the currently accessible energy and accuracy.

The last term in Eq. 6 represents a gauge-invariant four-photon interaction.[9] The observation of such an interaction would be evidence for new physics beyond QED, but would be consistent with what we already know about QED.

QUANTUM CHROMODYNAMICS

Let us now approach QCD in a manner analogous to our approach to QED. We again begin with a list of "experimental facts":

1. The gluon is massless.[10]

---

[8]QED possesses a global chiral symmetry, $\psi \to \exp[i\theta\gamma_5]\psi$, in the limit $m \to 0$, so we expect the coefficient of the term $\bar{\psi}\sigma^{\mu\nu}\psi F_{\mu\nu}$, which violates this symmetry, to contain an explicit power of the fermion mass.

[9]The term $F_\nu^\mu F_\rho^\nu F_\mu^\rho$ vanishes since $F^{\mu\nu}$ is antisymmetric.

[10]Since gluons (and quarks) are confined, their masses cannot be measured directly. There is no evidence for a bare gluon mass, so let us assume it is exactly massless. Gluons behave as if they have a dynamically-generated mass of order 300 MeV, in

2. The gluon has spin one.[11]

3. The gluon interacts with itself.[12]

Of course, these facts cannot be gleaned directly from experiment, which is the reason it took so many years to realize that QCD is the theory of the strong interaction. Let's construct a consistent theory which incorporates the above facts.

As with QED, we attempt to construct a theory based on the vector field $G^\mu(x)$. We encounter the same difficulties as in QED (too many degrees of freedom, one of which has negative energy), with the same resolution (gauge invariance). However, we argued that gauge invariance forbids vector-field self interactions, such as those in Eq. 5, so we run into a new problem: how do we allow the gluon to interact with itself and not spoil gauge invariance?

The resolution of this problem is also well known. Instead of a single gluon, we introduce a multiplet of gluons, eight to be exact. We expand the gauge transformation of QED, Eq. 4, to include a rotation of the eight gluons into each other under the group SU(3). The result is the familiar Yang-Mills theory of QCD, with eight self-interacting gluons. The essential point is that, as in QED, gauge invariance is *mandatory* for the consistency of the theory.[13]

The gluon self interaction is believed to be responsible for much of the physics of QCD which sets it apart from QED, such as confinement and asymptotic freedom. The gluon self interaction has been tested via $Z \to 4j$ events at LEP, where the decay $Z \to q\bar{q}gg$ involves the three-gluon interaction. Since gauge invariance is mandatory for the consistency of the theory, it is not acceptable to arbitrarily vary the three-gluon interaction when comparing theory with experiment. Instead, the analysis by the LEP experiments leaves the Yang-Mills gauge symmetry intact, but varies the gauge group (leaving the fermions in the fundamental representation, as in QCD) (20). Fig. 2 shows the result of an analysis of $Z \to 4j$, comparing the expectation of various gauge groups (boxes) and QCD (circle) with the data (star); the axes identify the gauge group, and are explained in the figure caption. The agreement of the data with the SU(3) prediction is impressive.

As with QED, gauge symmetry does not mean there cannot be anomalous vector-boson self interactions. The most general Lagrangian for gluons and quarks, consistent with Lorentz invariance and SU(3) gauge symmetry, is

---

the same sense that quarks have a dynamically-generated "constituent" mass of the same order; this should not be confused with the bare mass.

[11] As evidenced, for example, by the angular distribution of three-jet events in e^+e^- collisions (15).

[12] This is necessary to explain confinement, asymptotic freedom, and other phenomena.

[13] To the best of my knowledge, it has never been rigorously shown that Yang-Mills gauge theory is the unique theory of massless, interacting, spin-one particles, based on vector fields. The necessity of gauge symmetry is suggested by the Weinberg-Witten theorem on massless charged particles (16). Of course, we now regard the masslessness of the gluons to be a *consequence* of gauge invariance, just as in QED.

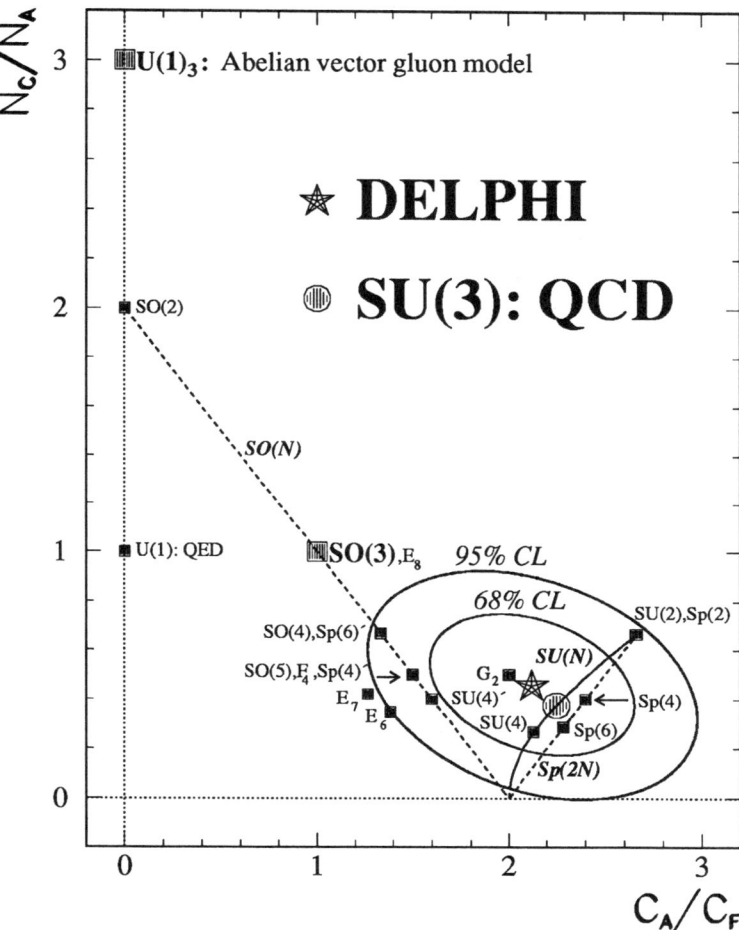

FIG. 2. Comparison of $Z \to 4j$ ($\star$) with the expectation of a Yang-Mills gauge theory based on various gauge groups ($\square$) and SU(3) ($\bigcirc$). N_C/N_A is the ratio of the number of quark colors to the number of gluons, and C_A/C_F is the ratio of the strength of the three-gluon interaction to that of gluon bremsstrahlung from quarks. Figure from Ref. (20).

$$\mathcal{L} = -\frac{1}{2}\text{Tr}\, G^{\mu\nu}G_{\mu\nu} + i\bar\psi\,\slashed{D}\psi - m\bar\psi\psi \qquad (7)$$
$$+ \frac{c_1}{M^2}\bar\psi\gamma^\mu\psi\bar\psi\gamma_\mu\psi + \frac{c_2}{M^2}\text{Tr}\, G^\mu_\nu G^\nu_\rho G^\rho_\mu + \cdots$$

where $G^{\mu\nu}$ is the non-Abelian field-strength tensor. The first line is the Lagrangian of ordinary QCD, and the (infinite number of) additional terms correspond to new physics associated with a mass scale M, as in QED. The first such term corresponds to a four-quark contact interaction, and is searched for in high-p_T jet events at the Tevatron, resulting in a lower bound on M of about 1 TeV (17). The second such term yields an anomalous three-gluon interaction,[14] and is best sought in top-quark production at the LHC (18).[15] It also yields an anomalous four-, five-, and six-gluon interaction.

As with QED, there is no reason not to expect these additional terms in the Lagrangian to be present, but there is also nothing which tells us at what mass scale, M, we should expect them to manifest themselves. The renormalizability of ordinary QCD ensures that these terms are not necessary to cancel divergences, to all orders in perturbation theory. However, as with QED, the renormalizability of the theory is simply a consequence of the fact that M is much greater than the currently accessible energy and accuracy.

WEAK INTERACTION

Let's move on to the weak interaction. As with QED and QCD, we begin with experimental facts:

1. The weak bosons are massive.

2. The weak bosons have spin one.

The big difference between the weak interaction and both QED and QCD is that the vector bosons are massive. Let's construct a consistent theory of massive vector bosons, associated with a field $W^\mu(x)$. We'll add more experimental facts later.

As with QED and QCD, we immediately run into the problem that the vector field W^μ has too many degrees of freedom.[16] This problem is less severe for a massive spin-one particle because it has three degrees of freedom, corresponding to helicities $\pm 1, 0$, rather than the two degrees of freedom of the massless case. As with QED and QCD, the temporal component of the vector field corresponds to a state of negative energy, so we must eliminate it as a dynamical degree of freedom.

Consider the following Lagrangian for a non-interacting vector field W^μ of mass M_W:

---

[14] As mentioned in a previous footnote, the analogous term vanishes in QED. It does not vanish in QCD because $G^{\mu\nu}$ is an SU(3) matrix.

[15] This term may also be sought in $Z \to 4j$, which yields a weak bound (19).

[16] At this point, I use W^μ to denote a generic massive vector field.

$$\mathcal{L} = -\frac{1}{4}W^{\mu\nu}W_{\mu\nu} + \frac{1}{2}M_W^2 W^\mu W_\mu \tag{8}$$

where

$$W^{\mu\nu} = \partial^\mu W^\nu - \partial^\nu W^\mu . \tag{9}$$

The kinetic part of the Lagrangian is written in terms of the field-strength tensor $W^{\mu\nu}$ in order to remove W^0 as a dynamical field, just as we did for QED and QCD. However, for the case of a massive vector field one can do even better; the extra degree of freedom can be removed in a manifestly Lorentz-invariant manner (21–24). Consider the equation of motion of the field W^μ, derived from the Lagrangian in Eq. 8:

$$\partial_\nu W^{\nu\mu} + M_W^2 W^\mu = 0 . \tag{10}$$

Now apply ∂_μ to this equation. The first term vanishes since $W^{\mu\nu}$ is antisymmetric, so we find

$$\partial_\nu W^\nu = 0 . \tag{11}$$

This is a constraint equation on the field W^μ, and allows us to remove one degree of freedom. Since it is a Lorentz-invariant condition, Lorentz invariance remains manifest.[17]

Although Eq. 11 is reminiscent of the familiar Lorentz gauge condition of QED and QCD, it is not a gauge condition at all. The massive vector theory has no gauge invariance whatsoever; gauge invariance is *non-existent* and *unnecessary*. This is in striking contrast to QED and QCD. Because we are so used to working with gauge theories, this simple point is sometimes forgotten. The quantization of a massless vector theory such as QED or QCD is a difficult task, and one tends to forget how easy it is to quantize a massive vector theory.[18]

This construction is not upset by the introduction of interactions with other fields, or even self interactions. One way to see this is to consider the propagator of the vector field. The free field equation, Eq. 10, written in terms of the vector field, is

$$\Box W^\mu - \partial_\nu \partial^\mu W^\nu + M_W^2 W^\mu = 0 . \tag{12}$$

This yields the momentum-space propagator

$$D^{\mu\nu}(p) = i\frac{-g^{\mu\nu} + \frac{p^\mu p^\nu}{M_W^2}}{p^2 - M_W^2} . \tag{13}$$

---

[17] Alternatively, one can simply impose this constraint on the field as an auxiliary condition (22).

[18] If the massive vector theory is a spontaneously-broken gauge theory, quantization is as complicated as in QED and QCD, of course.

The numerator of the propagator contains the sum over the three polarization states corresponding to the three helicity states of a massive spin-one particle, and nothing more.[19] Hence there is no concern about interactions potentially coupling to unphysical polarization states, as there is in QED and QCD (14,22,23).[20]

Given that gauge invariance has nothing to do with a generic massive vector boson theory, one must wonder why we believe the weak interaction is described by a gauge theory. The answer lies in a third experimental fact:

3. The couplings of the weak bosons to the three generations of quarks and leptons are, to high precision, those of an $SU(2)_L \times U(1)_Y$ gauge theory.

But if the weak interaction is a gauge theory, why aren't the weak bosons massless, as appears to be required of gauge bosons? The well-known solution to this puzzle is that the gauge symmetry is spontaneously broken (25). This means that while the Lagrangian *is* invariant under $SU(2)_L \times U(1)_Y$ gauge transformations, the solution to the Lagrangian is not.

A skeptic might ask if the (local) gauge symmetry is really necessary. Wouldn't it be enough to impose *global* $SU(2)_L \times U(1)_Y$ symmetry on the Lagrangian to reproduce the observed couplings of the weak bosons to fermions? The answer is no; one needs the local gauge symmetry to explain the universality of the weak interaction, i.e., to explain why the weak bosons couple the same to quarks as to leptons, and to all three generations (as far as we know).[21] To see this, consider the Lagrangian for the coupling of the weak bosons to fermions,[22]

$$\mathcal{L} = i\bar{\psi}_L \gamma^\mu (\partial_\mu + ig\frac{1}{2}\tau \cdot W_\mu)\psi_L \ . \tag{14}$$

Each term is separately invariant under global $SU(2)_L$ transformations, regardless of the value of g. However, both terms are needed to ensure invariance under *local* $SU(2)_L$ transformations $U(x)$,

$$\psi_L \to U\psi_L \tag{15}$$

$$\tau \cdot W^\mu \to U\tau \cdot W^\mu U^\dagger + \frac{2i}{g}(\partial^\mu U)U^\dagger \tag{16}$$

---

[19]This is the familiar "unitary gauge" propagator of a spontaneously-broken gauge theory, meaning it contains only the physical polarization states. Since we are not (yet) treating the massive vector field as a gauge field, we avoid this language.

[20]In QED and QCD, the propagator must couple to a conserved current, or gauge invariance (and hence Lorentz invariance) is lost (12).

[21]It is particularly important that we test the universality of the weak interaction with respect to the recently-discovered top quark (26).

[22]Here and throughout I consider only the $SU(2)_L$ part of the weak interaction, and ignore the hypercharge interaction. The field W^μ represents an $SU(2)_L$ triplet of gauge fields, and τ are the usual Pauli matrices. As usual, $\psi_L \equiv \frac{1}{2}(1-\gamma_5)\psi$ denotes the left-chiral fermion field.

and they must be present exactly as shown in Eq. 14, with the same coupling g for all fermions (27).

The skeptic might counter that, while willing to accept local gauge invariance as the explanation of the universality of the coupling of weak bosons to fermions, this universality need not extend to the gauge-boson self interactions. Couldn't one imagine that the gauge symmetry is present only in the fermionic sector of the theory? The answer to this is also negative. The one-loop correction to the coupling of weak bosons to fermions involves the weak-boson self interaction, and unless this interaction is of the Yang-Mills form, it will generally destroy the gauge-theory form of the fermionic coupling. While the couplings may be "readjusted" to their experimentally-observed values, the explanation of universality is lost. This problem is especially severe in light of the fact that the quarks experience the strong interaction, while the leptons do not, so the amount of "readjustment" necessary will generally differ for the two types of fermions. Thus we conclude that in order for gauge symmetry to explain the universality of the weak interaction, it must be a symmetry of the full Lagrangian, not just part of it.

Just as in QED and QCD, anomalous vector-boson self interactions may be introduced via higher-dimension terms in the Lagangian, suppressed by inverse powers of some mass scale, M. However, in the weak interaction, the implementation of this differs depending on whether or not a fundamental Higgs field is introduced in the Lagrangian. Below we pursue these possibilities separately.

Higgs model

Consider including the Higgs-doublet field, ϕ, to break the electroweak symmetry in the standard way. The Lagrangian is

$$\mathcal{L} = -\frac{1}{8}\text{Tr}\, W^{\mu\nu}W_{\mu\nu} + i\bar{\psi}_L \not{D}\psi_L \qquad (17)$$
$$+ (D^\mu \phi)^\dagger D_\mu \phi - V(\phi^\dagger \phi)$$
$$+ \frac{c_1}{M^2}(D^\mu\phi)^\dagger W_{\mu\nu} D^\nu \phi + \frac{c_2}{M^2}\text{Tr}\, W^\mu_\nu W^\nu_\rho W^\rho_\mu + \cdots$$

where $W^{\mu\nu}$ is the full non-Abelian field-strength tensor,

$$W^{\mu\nu} = \tau \cdot (\partial^\mu W^\nu - \partial^\nu W^\mu - g W^\mu \times W^\nu) . \qquad (18)$$

The first two lines are the standard electroweak Lagrangian. When the Higgs field ϕ acquires a vacuum-expectation value, the first term in the third line produces additional three- and four-W interactions. The last term contributes additional three-, four-, five-, and six-W interactions (27–29). These anomalous vector-boson self interactions are suppressed by inverse powers of some mass scale, M, which is the scale at which the ordinary electroweak theory ceases to be a valid description of nature. As with QED and QCD, we have no reason not to expect that such terms are there. Since the standard electroweak theory is renormalizable, these terms are not necessary to cancel divergences, to all orders in perturbation theory, so M can be arbitrarily large. However, radiative corrections to the Higgs vacuum-expectation value diverge quadratically,[23] so the value $v \approx 250$ GeV is natural only if there is new physics which cuts off the divergence at or below 1 TeV (30). Thus naturalness of the Higgs model suggests that M should not be greater than about 1 TeV.[24]

No-Higgs model

Although we believe that the electroweak interaction is a spontaneously-broken gauge theory, we do not know if the spontaneous symmetry breaking is the result of the vacuum-expectation value of a fundamental Higgs field. If we insist that the theory be renormalizable and perturbative (weak coupling), then the only option is indeed the standard Higgs model (31) and generalizations thereof, such as a two-Higgs-doublet model as employed in the supersymmetric standard model (32). However, we have no guarantee that nature is so kind as to provide us with a symmetry-breaking mechanism that can be analyzed perturbatively.

Whatever the symmetry-breaking mechanism, it must provide the three Goldstone bosons which are absorbed by the $W^\pm$ and Z bosons to become massive. A generic approach to the symmetry-breaking physics is then to introduce only these three Goldstone bosons into the Lagrangian, but no other fields (27–29,33,34). Although the resulting theory is non-renormalizable, it should be a valid effective field theory at energies below the mass scale of the symmetry-breaking physics responsible for the Goldstone bosons.

Let us introduce the three Goldstone-boson fields π via the field[25]

$$\Sigma \equiv \exp[i\tau \cdot \pi/v] \qquad (19)$$

where $v = 2M_W/g$. The Lagrangian is

$$\mathcal{L} = -\frac{1}{8}\text{Tr}\, W^{\mu\nu}W_{\mu\nu} + i\bar{\psi}_L \not{D}\psi_L \qquad (20)$$
$$+ \frac{v^2}{4}\text{Tr}\,(D^\mu\Sigma)^\dagger D_\mu\Sigma$$
$$+ c_1\frac{v^2}{M^2}(\text{Tr}\,(D^\mu\Sigma)^\dagger D_\mu\Sigma)^2 + c_2\frac{v^2}{M^2}\text{Tr}\,W^{\mu\nu}(D_\mu\Sigma)^\dagger D_\nu\Sigma + \cdots$$

where

---

[23] An equivalent argument is usually presented in terms of the Higgs mass.

[24] One possibility for new physics which cuts off the quadratic divergence is supersymmetry.

[25] The choice of the symbol π to denote the Goldstone bosons is by analogy with the physical pion field, which is an (approximate) Goldstone boson of spontaneously-broken chiral symmetry in QCD.

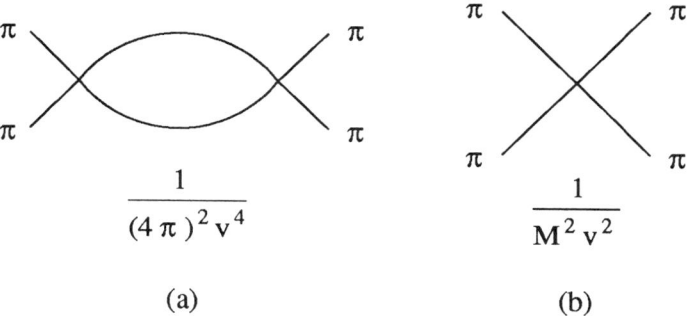

FIG. 3. (a) One-loop amplitude for four Goldstone bosons. (b) Tree-level interaction of four Goldstone bosons required to cancel the divergence from the one-loop amplitude.

$$D^\mu \Sigma = (\partial^\mu + i\frac{g}{2}\tau \cdot W^\mu)\Sigma \quad (21)$$

is the gauge-covariant derivative. The Σ field transforms under SU(2)$_L$ as

$$\Sigma \to U\Sigma . \quad (22)$$

The first line in Eq. 20 is the usual weak-interaction Lagrangian. The second line is is responsible for the W mass, which is evident when it is expanded in terms of the π fields:

$$\frac{v^2}{4}\text{Tr}\,(D^\mu\Sigma)^\dagger D_\mu\Sigma = \frac{1}{2}\frac{g^2}{4}v^2 W^\mu W_\mu + \frac{1}{2}\partial^\mu\pi \cdot \partial_\mu\pi + \frac{1}{2v^2}(\pi \cdot \partial^\mu\pi)(\pi \cdot \partial_\mu\pi) + \cdots \quad (23)$$

The physical content of the theory is manifest in the unitary gauge, $\Sigma = 1$, in which the Goldstone bosons are completely absorbed by the weak vector bosons, and disappear from the Lagrangian. However, it is convenient for our discussion (and for calculational purposes) to consider a gauge in which the Goldstone bosons are present.

The first term in the third line contributes an anomalous four-W interaction, and the second term an anomalous three- and four-W interaction. These terms are suppressed by inverse powers of a mass scale, M, which is the scale at which the physics responsible for spontaneous symmetry breaking resides. At first sight this mass scale can be made arbitrarily large, as in QED, QCD, and the Higgs model. However, the term responsible for the vector-boson masses, Eq. 23, contains a non-renormalizable four-π interaction with a coefficient proportional to $1/v^2$, as shown. This coefficient sets the scale for the other non-renormalizable terms. A one-loop four-π amplitude constructed from two of these four-π interactions, shown in Fig. 3(a), is of order $1/(4\pi)^2 v^4$, where the factor $1/(4\pi)^2$ arises from the loop integration.

This has the same dimensions as the contribution to the four-π amplitude from the terms in the last line of Eq. 20, shown in Fig. 3(b), of order $1/M^2v^2$. Since the one-loop amplitude is divergent, these terms must be there to cancel the divergence. Thus M must be of order $4\pi v \approx 3$ TeV or less. The physics responsible for electroweak symmetry breaking must therefore manifest itself by at least 3 TeV.

There is one last experimental fact we can add to our discussion:

4. $M_W \approx M_Z \cos\theta_W$

This is embodied in the ρ parameter,

$$\rho \equiv M_W^2/(M_Z^2 \cos^2\theta_W) \approx 1 \ . \tag{24}$$

In other words, not only are the W and Z bosons massive, but their masses are related. This can be explained by hypothesizing that the symmetry-breaking sector of the theory possesses a global SU(2) symmetry, called a "custodial" symmetry (30,34–36). Although it is not always made explicit, the standard Higgs model contains this symmetry. Models of dynamical electroweak symmetry breaking, such as Technicolor (30,35), must contain this symmetry even if Eq. 24 is satisfied at tree level, since the corrections are potentially large (strong coupling). The custodial symmetry is further evidence that the properties of the weak interaction are dictated by symmetry.

OUTLOOK

Of the electromagnetic, strong, and weak interactions, only the last guarantees new physics at an accessible mass scale. This is the physics associated with electroweak symmetry breaking. All we know for sure about this physics is that it must manifest itself by at least $4\pi v \approx 3$ TeV. Furthermore, the fact that the W and Z masses are related by $M_W \approx M_Z \cos\theta_W$ suggests that the symmetry-breaking sector contains a "custodial" global SU(2) symmetry.

The electroweak-symmetry-breaking sector could be the source of anomalous weak-vector-boson self interactions. We have considered two scenarios for the electroweak-symmetry-breaking physics:

1. Higgs model

2. No-Higgs model: only Goldstone bosons up to $4\pi v \sim 3$ TeV

These two models are so commonly studied that one begins to think they are the only possibilities. This is not the case. The symmetry-breaking physics could be very rich, containing resonances, new fermions, new gauge bosons, etc. The fact that nature makes use of gauge theories for the three known low-energy forces leads one to guess that the symmetry-breaking sector is also a gauge theory. Examples which implement this idea are fixed-point Technicolor (37), walking Technicolor (38), two-scale Technicolor (39), fermions in large representations of the gauge group (40,41), etc. From a theoretical point of

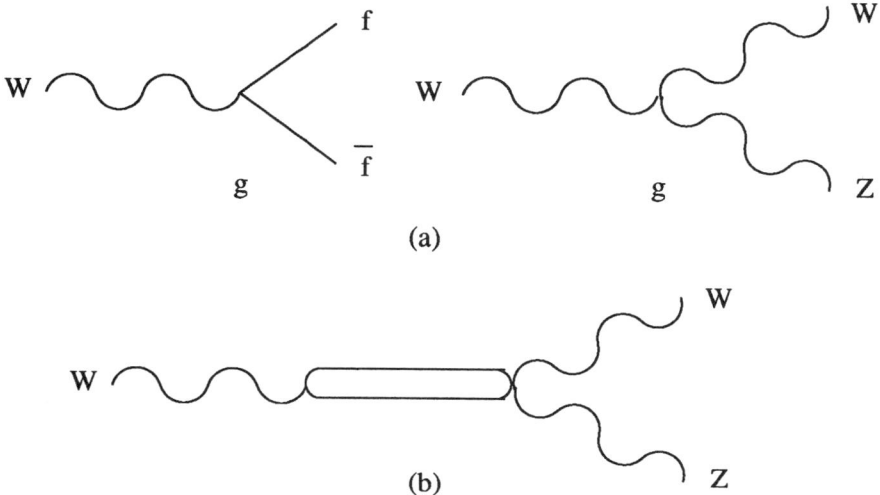

FIG. 4. (a) Schematic illustration of the universality of the coupling of weak bosons to fermions and to themselves. (b) An anomalous weak-boson self interaction produced by a $J = 1$ resonance.

view, these models receive less attention because they are neither amenable to a perturbative analysis (like the Higgs model) nor to a "model-independent" analysis (like the No-Higgs model). They also generically run into difficulty with precision electroweak experiments (42–44), something we have learned from the vector-boson era. Nature may not care about any of these objections. We should probe higher energies and keep an open mind regarding the manner in which the physics of electroweak symmetry breaking reveals itself.

After this long discussion, we are now prepared to go back and answer a question we posed at the beginning: What can we learn from the era of weak boson pair production? At the very least, we will see a confirmation of the universality of the weak interaction, extended to the weak-boson self interaction, as depicted schematically in Fig. 4(a). It is the universality of the fermionic couplings of the weak bosons which led us to the electroweak theory in the first place, so this confirmation will be a crowning achievement. However, we hope for much more from this era; we anticipate at least the first signs of the physics responsible for electroweak symmetry breaking, and, at best, the complete revelation of this physics. One possible manifestation of this physics is a $J = 1$ resonance which couples to the weak bosons, as depicted in Fig. 4(b). Although we would not usually regard this as an "anomalous vector-boson self interaction", there is no reason why we should not. If we observe such a resonance, it would be a *very* anomalous vector-boson self interaction.

ACKNOWLEDGMENTS

I am grateful for conversations and correspondence with T. Appelquist, W. Bardeen, C. Burgess, D. Dicus, E. Eichten, A. El-Khadra, W. Marciano, D. Morris, C. Quigg, J. Stack, G. Valencia, S. Weinberg, J. Wudka, D. Zeppenfeld, and D. Zwanziger. This work was supported in part by Department of Energy grant DE-FG02-91ER40677.

REFERENCES

1. UA1 Collaboration, G. Arnison et al., Phys. Lett. **122B**, 103 (1983); **126B**, 398 (1983); **129B**, 273 (1983); UA2 Collaboration, M. Banner et al., Phys. Lett. **122B**, 476 (1983); P. Bagnaia et al., Phys. Lett. **129B**, 130 (1983).
2. S. van der Meer, Rev. Mod. Phys. **57**, 689 (1985); C. Rubbia, Rev. Mod. Phys. **57**, 699 (1985).
3. T. Fuess, these proceedings.
4. CDF Collaboration, F. Abe et al., CDF/ANAL/ELECTROWEAK/CDFR/ 2951 (1995).
5. J. Busenitz, these proceedings.
6. J. Womersley, these proceedings.
7. T. Barklow, these proceedings.
8. S. Weinberg, Physica **96A**, 327 (1979).
9. The concept of "effective" quantum field theory has become widespread, and there are many excellent introductions. Among them are G. P. Lepage, in *From Actions to Answers, Proceedings of the 1989 Theoretical Advanced Study Institute (TASI)*, eds. T. DeGrand and D. Toussaint (World Scientific, Singapore, 1990), p. 483; S. Weinberg, in *Proceedings of the XXVI International Conference on High-Energy Physics*, ed. J. Sanford (American Institute of Physics, New York, 1993), p. 346; J. Polchinski, in *Recent Directions in Particle Theory, Proceedings of the 1992 Theoretical Advanced Study Institute (TASI)*, eds. J. Harvey and J. Polchinski (World Scientific, Singapore, 1993), p. 235; H. Georgi, Ann. Rev. Nucl. Part. Sci. **43**, 209 (1993).
10. M. Einhorn, in *Workshop on Physics and Experiments with Linear e^+e^- Colliders*, Hawaii, 1993, eds. F. Harris, S. Olsen, S. Pakvasa, and X. Tata (World Scientific, Singapore, 1993), p. 122.
11. S. Weinberg, Phys. Rev. **138**, B988 (1965); in *1964 Brandeis Summer Institute in Theoretical Physics, Lectures on Particles and Field Theory* (Prentice-Hall, Englewood Cliffs, 1965), Vol. 2, p. 405.
12. J. Bjorken and S. Drell, *Relativistic Quantum Fields* (McGraw-Hill, New York, 1965).
13. D. Zwanziger, Phys. Rev. **133**, B1036 (1964).
14. An engaging and non-technical discussion is given in M. Veltman, *Diagrammatica* (Cambridge University Press, Cambridge, 1994).
15. DELPHI Collaboration, P. Abreu et al., Phys. Lett. **274**, 498 (1992); L3 Collaboration, B. Adeva et al., Phys. Lett. **B263**, 551 (1991); OPAL Collaboration, G. Alexander et al., Z. Phys. C **52**, 543 (1991).
16. S. Weinberg and E. Witten, Phys. Lett. **96B**, 59 (1980).

17. CDF Collaboration, F. Abe et al., Phys. Rev. Lett. **71**, 2542 (1993); **74**, 3538 (1995).
18. E. Simmons, these proceedings.
19. A. Duff and D. Zeppenfeld, Z. Phys. C **53**, 529 (1992).
20. ALEPH Collaboration, D. Decamp et al., Phys. Lett. **B284**, 151 (1992); DELPHI Collaboration, P. Abreu et al., Z. Phys. C **59**, 357 (1993); L3 Collaboration, B. Adeva et al., Phys. Lett. **B248**, 227 (1990); OPAL Collaboration, R. Akers et al., Z. Phys. C **65**, 367 (1995).
21. P. Roman, *Introduction to Quantum Field Theory* (Wiley, New York, 1969).
22. N. Bogoliubov and D. Shirkov, *Introduction to the Theory of Quantized Fields* (Interscience, New York, 1959).
23. C. Itzykson and J.-B. Zuber, *Quantum Field Theory* (McGraw-Hill, New York, 1980).
24. L. Ryder, *Quantum Field Theory* (Cambridge University Press, Cambridge, 1985).
25. S. Weinberg, Phys. Rev. Lett. **19**, 1264 (1967); A. Salam, in *Elementary Particle Theory: Relativistic Groups and Analyticity (Nobel Symposium No. 8)*, ed. N. Svartholm (Almqvist and Wiksell, Stockhom, 1968), p. 367.
26. CDF Collaboration, F. Abe et al., Phys. Rev. Lett. **74**, 2626 (1995); D0 Collaboration, S. Abachi et al., Phys. Rev. Lett. **74**, 2632 (1995).
27. K. Hagiwara, these proceedings.
28. P. Hernandez, these proceedings.
29. J. Wudka, these proceedings.
30. S. Chivukula, these proceedings.
31. J. M. Cornwall, D. Levin, and G. Tiktopoulos, Phys. Rev. Lett. **30**, 1268 (1973); Phys. Rev. D **10**, 1145 (1974); C. Llewellyn-Smith, Phys. Lett. **B46**, 233 (1973).
32. A. Lahanas, these proceedings.
33. T. Appelquist and C. Bernard, Phys. Rev. D **22**, 200 (1980).
34. G. Valencia, these proceedings.
35. S. Weinberg, Phys. Rev. D **19**, 1277 (1979); L. Susskind, Phys. Rev. D **20**, 2619 (1979).
36. P. Sikivie, L. Susskind, M. Voloshin, and V. Zakharov, Nucl. Phys. **B173**, 189 (1980).
37. B. Holdom, Phys. Rev. D **24**, 1441 (1981).
38. B. Holdom, Phys. Lett. **B150**, 301 (1985); K. Yamawaki, M. Bando, and K. Matumoto, Phys. Rev. Lett. **56**, 1335 (1986); T. Appelquist, D. Karabali, and L. Wijewardhana, Phys. Rev. Lett. **57**, 957 (1986); T. Appelquist and L. Wijewardhana, Phys. Rev. D **36**, 568 (1987).
39. E. Eichten and K. Lane, Phys. Lett. **B222**, 274 (1989).
40. W. Marciano, Phys. Rev. D **21**, 2425 (1980).
41. S. Willenbrock, Phys. Lett. **B340**, 236 (1994).
42. D. Schaile, these proceedings.
43. T. Takeuchi, these proceedings.
44. M. Peskin and T. Takeuchi, Phys. Rev. Lett. **65**, 964 (1990); B. Holdom and J. Terning, Phys. Lett. **B247**, 88 (1990); M. Golden and L. Randall, Nucl. Phys. **B361**, 3 (1991).

The Status of the Electroweak Sector of the Standard Model

Tatsu Takeuchi[1]

Fermi National Accelerator Laboratory
P. O. Box 500, Batavia, IL 60510

I will discuss the current status of the determination of $\alpha(m_Z)$ and $\alpha_s(m_Z)$.

INTRODUCTION

I have been asked by the organizers of this conference to present a talk on the status of the electroweak sector of the Standard Model (SM). But frankly, if by the word 'status' one means 'how well theory and experiment agree with each other', there is not much that I can tell you at the moment but that the SM works extremely well. Though the experiments at LEP/SLD and elsewhere have pushed the experimental accuracy of many observables to a mere fraction of a percent, no significant deviation from the SM is yet to be detected. (I will discuss whether the 2.4σ deviation in R_b seen at LEP is 'significant' or not later in this talk.) A detailed comparison between the SM and current precision electroweak measurements can be found in the talk by Dr. Dorothee Schaile in this proceedings (1).

I will therefore focus on the status of the determination of the values of $\alpha(m_Z)$ and $\alpha_s(m_Z)$ instead. These quantities are used as inputs to calculate the SM predictions for the electroweak observables at LEP/SLD and any uncertainty in them limits the accuracy of the theoretical predictions even when the top and Higgs masses are fixed to given values. Consequently, any conclusion we may reach about the 'status' of the electroweak sector of the SM is actually contingent upon how well known these quantities are.

Currently, LEP uses the following values for $\alpha(m_Z)$ and $\alpha_s(m_Z)$:

$$\alpha(m_Z) = 1/(128.87 \pm 0.12),$$
$$\alpha_s(m_Z) = 0.123 \pm 0.006. \tag{1}$$

The value of $\alpha(m_Z)$ is that determined from the $\sigma(e^+e^- \to hadrons)$ data from various experiments in Ref. (7), and the value of $\alpha_s(m_Z)$ is that determined from the hadronic event shapes, jet rates, and energy–energy correlation at LEP in Ref. (2). They are therefore independent of the Z line shape and asymmetry measurements at LEP and SLD.

---

[1] Work supported by the U.S. Department of Energy under contract No. DE–AC02–76CH03000.

Note that the uncertainty in $\alpha(m_Z)$ arises from the incalculability of the hadronic contribution to the photon vacuum polarization which forces us to rely on the $\sigma(e^+e^- \to hadrons)$ data, and the uncertainty in $\alpha_s(m_Z)$ arises mainly from our limited understanding of parton hadronization and our ignorance of higher order QCD corrections. They are both manifestations of our limited ability to handle QCD.

As mentioned above, these uncertainties in $\alpha(m_Z)$ and $\alpha_s(m_Z)$ propagate into the SM predictions that are derived from them. For instance, for the SM with $m_t = 180$GeV, $m_H = 300$GeV, and Eq. 1 as input, the program ZFITTER 4.9 (3) gives the following predictions:

$$\begin{aligned} \sin^2\theta_{\text{eff}}^{\text{lept}} &= 0.2319 \pm 0.0003 \pm 0.0000, \\ R_\ell &= 20.766 \pm 0.006 \pm 0.040, \end{aligned} \qquad (2)$$

where $R_\ell \equiv \Gamma(Z \to hadrons)/\Gamma(Z \to \ell^+\ell^-)$. The first error is that due to the uncertainty in $\alpha(m_Z)$ and the second error is that due to the uncertainty in $\alpha_s(m_Z)$. The error in $\sin^2\theta_{\text{eff}}^{\text{lept}}$ due to the uncertainty in $\alpha_s(m_Z)$ is negligible because QCD corrections only enter at $O(\alpha\alpha_s)$.

On the other hand, the latest measurements of these quantities at LEP/SLD are given by (1)

$$\begin{aligned} \sin^2\theta_{\text{eff}}^{\text{lept}} &= 0.2315 \pm 0.0004 \\ R_\ell &= 20.800 \pm 0.035. \end{aligned} \qquad (3)$$

Comparison with the theoretical predictions Eq. 2 shows not only that the SM works very well, but also just how 'precise' the precision electroweak measurements at LEP/SLD have become. The experimental error is already comparable to or even smaller than the theoretical error [2]. With the experimental precision improving incrementally every year, clearly an improvement on the theoretical error is called for if one wishes to test the SM at an even higher level of accuracy.

In the following sections, I will review the current status of the determination of $\alpha(m_Z)$ and $\alpha_s(m_Z)$ and discuss how the values quoted in Eq. 1 can be expected to improve or change.

THE VALUE OF $\alpha(m_Z)$

Current status

Quite recently, several authors have independently made attempts to reevaluate the value of $\alpha(m_Z)$ through a careful reanalysis of existing $\sigma(e^+e^- \to hadrons)$ data (8–11). In comparison to the previous determination in Ref. (7), the hope was that the uncertainty in $\alpha(m_Z)$ could be reduced since some new data on $\sigma(e^+e^- \to hadrons)$ have been released (12–14), and the

---

[2] Ref. (4) argues that the experimental error on R_ℓ may be greatly underestimated. I will assume that that is not the case in the following.

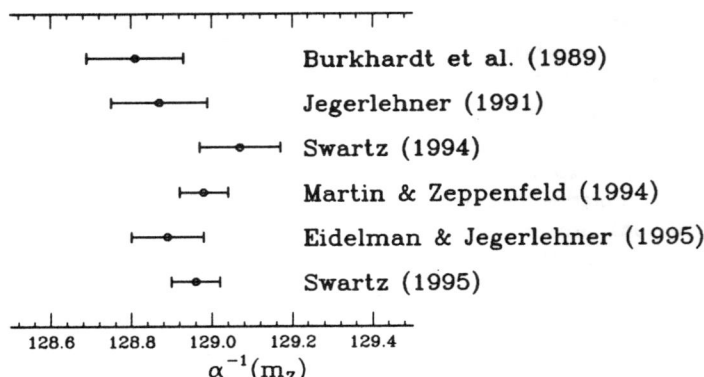

FIG. 1. Evaluations of $\alpha(m_Z)$ by different authors. All values have been rescaled to $m_Z = 91.1887$ GeV and the top quark contribution have been removed.

$O(\alpha_s^3)$ QCD correction to the cross section (15) together with a better determination of $\alpha_s(m_Z)$ (16) was now available. The resulting new values of $\alpha(m_Z)$ are shown in Fig. 1 together with a couple of older values.

As you can see from Fig. 1, instead of finding a significant decrease in the uncertainty of $\alpha(m_Z)$, different authors found central values of $\alpha(m_Z)$ which differed from each other by as much as 2σ. Please note that all these values were obtained by applying different analyses on essentially the same set of data so the statistical errors are almost completely correlated. Therefore, large differences in the central values must be manifestations of systematic errors which have been underestimated.

At the time of this conference, the two latest evaluations by Eidelman and Jegelehner (10) and Swartz (11) were not yet available so the disagreement among authors was rather pronounced. With the two latest numbers, the value of $\alpha(m_Z)$ shows signs of converging to a common value.

In the following subsections, I will discuss how the analyses of different authors differ from each other so that we may get an idea of where the disagreements shown in Fig. 1 are coming from.

The Definition of $\alpha(m_Z)$

The quantity whose value is usually quoted as that of the "effective QED coupling constant at the Z mass scale" is defined as

$$\alpha^{-1}(m_Z) = \alpha^{-1}[1 - \Delta\alpha(m_Z)] \qquad (4)$$

with

$$\Delta\alpha(s) = 4\pi\alpha \text{Re}\left[\Pi'_{QQ}(s) - \Pi'_{QQ}(0)\right], \qquad (5)$$

where $\Pi'_{QQ}(s)$ is the photon vacuum polarization function with only the *light fermion* contributions included. It has been customary to exclude the top and W contributions from $\Delta\alpha(m_Z)$ since the top mass was unknown until quite recently (17) (though care is needed when comparing results since recent authors include it) and the W contribution is also excluded to keep the definition of $\alpha(m_Z)$ gauge independent. These contributions are also numerically small compared to the light fermion contribution since the W^+W^- and $t\bar{t}$ thresholds are above the Z mass so that they do not contribute logarithms to the running of $\alpha(s)$ between $s=0$ and $s=m_Z^2$.

Contribution of the light fermions

The contribution of the leptons to $\Delta\alpha$ can be calculated accurately in perturbation theory and one finds

$$\Delta\alpha_{leptons}(m_Z^2) = \sum_{\ell=e,\mu,\tau} \frac{\alpha}{3\pi}\left[-\frac{8}{3}+\beta_\ell^2 - \frac{1}{2}\beta_\ell(3-\beta_\ell^2)\ln\left(\frac{1-\beta_\ell}{1+\beta_\ell}\right)\right]$$

$$= \sum_{\ell=e,\mu,\tau} \frac{\alpha}{3\pi}\left[\ln\left(\frac{m_Z^2}{m_\ell^2}\right) - \frac{5}{3} + O\left(\frac{m_\ell^2}{m_Z^2}\right)\right]$$

$$= 0.03142, \qquad (6)$$

where $\beta_\ell = \sqrt{1-4m_\ell^2/m_Z^2}$.

On the other hand, the contribution of the five light quarks (u,d,s,c,b) to $\Delta\alpha$ cannot be calculated perturbatively. Instead, unitarity and the analyticity of $\Pi'_{QQ}(s)$ is used to write

$$\Delta\alpha_{hadrons}^{(5)}(s) = \frac{\alpha s}{3\pi}\text{P}\int_{4m_\pi^2}^\infty ds' \frac{R(s')}{s'(s-s')}, \qquad (7)$$

where [3]

$$R(s) \equiv \frac{\sigma(e^+e^- \to \gamma^* \to hadrons)}{\sigma(e^+e^- \to \gamma^* \to \mu^+\mu^-)} = -12\pi\text{Im}\Pi'_{QQ}(s) \qquad (8)$$

and the functional form of $R(s)$ is extracted from experiment. It is this reliance on the experimental values of $R(s)$, which are always accompanied by experimental errors, that we end up with a relatively large error on $\alpha(m_Z)$ even though the find structure constant α is known to extreme accuracy.

The major problem associated with using the experimental values of $R(s)$ is that the data points are only available for discrete, scattered values of s while one needs $R(s)$ for all values of s to calculate $\Delta\alpha_{hadrons}^{(5)}(s)$. Two methods have been used in the literature to deal with this problem. The first is to connect the data points directly with straight lines and perform trapezoidal integration (5,7,10), and the second is to guess the functional form of $R(s)$ and fit it to the data (5,8,9,11).

---

[3] Note that the cross section in the denominator of the definition of $R(s)$ is the cross section of electron pairs annihilating into *massless* muon pairs.

Both methods have their pros and cons. Trapezoidal integration is free of human prejudice about the functional form of $R(s)$ but it does not take into account the experimental errors properly: sparsely distributed precise data points may not get the appropriate weight relative to the densely spaced data points with larger errors. Connecting two data points that are far apart with a straight line will also introduce errors.

On the other hand, fitting a guessed functional form to the data has the advantage that experimental errors are easier to take into account. (Though care is needed in treating normalization errors (18).) However, the result will depend on the choice of the fit function and its parameterization, and systematic errors and biases will be introduced that are difficult to estimate.

The two methods are often combined (*e.g.* fitting Breit–Wigner forms to the narrow resonances and using trapezoidal integration for the continuum) and are supplemented by the use of perturbative QCD for the high energy tail of $R(s)$.

The analysis of Burkhardt et al.

At the time when LEP started running in 1989, the most accurate determination of $\Delta\alpha^{(5)}_{hadrons}(m_Z^2)$ was that given by Burkhardt et al. in Ref. (5):

$$\Delta\alpha^{(5)}_{hadrons}(m_Z^2) = 0.0286 \pm 0.0009. \tag{9}$$

(Actually, Ref. (5) reports the value of $\Delta\alpha^{(5)}_{hadrons}(s)$ at $\sqrt{s} = 92\text{GeV}$. Rescaling to $\sqrt{s} = m_Z = 91.1887\text{GeV}$ gives the above value (6).) This value was calculated using the following three methods of integration and all three were found to agree with each other:

1. Trapezoidal integration for the continuum and the ρ.
   Breit–Wigner forms for the narrow resonances.

2. Trapezoidal integration for the continuum after a partial smoothing out of the $R(s)$ data. (Details are not given in Ref. (5).)
   Breit–Wigner forms for the narrow resonances.

3. Breit–Wigner forms for the ρ and narrow resonances.
   Linear interpolation in $\sqrt{s}$ for every few points in the continuum.

In all three cases, perturbative QCD (at $O(\alpha_s^2)$) was used for $R(s)$ above 40GeV. Eq. 9 corresponds to

$$\alpha^{-1}(m_Z) = \alpha^{-1}\left[1 - \Delta\alpha_{leptons}(m_Z^2) - \Delta\alpha^{(5)}_{hadrons}(m_Z^2)\right]$$
$$= 128.81 \pm 0.12 \tag{10}$$

and results in an error of about 0.1% in $\alpha(m_Z)$. The region below the bottom threshold contributed about 1/3 of $\Delta\alpha^{(5)}_{hadrons}(s)$, and 80% of the error.

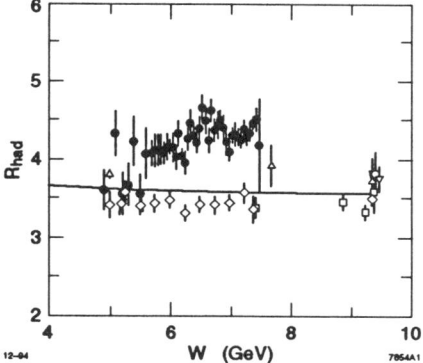

FIG. 2. Comparison of MARK I (black circles) and Crystal Ball (while diamonds) data. The solid line shows the perturbative QCD result.

The analysis of Jegerlehner

The result of Ref. (5) was subsequently updated in 1991 by Jegerlehner (7), who was one of the original authors, in which the data from the MARK I collaboration (21) in the energy region 5-7 GeV were replaced by the more accurate data from the Crystal Ball collaboration (22). Fig. 2 shows the data of both collaborations in the energy range in question, together with the perturbative QCD result. The method of integration was the same as method No. 1 of Ref. (5) and the result was

$$\Delta\alpha^{(5)}_{hadrons}(m_Z^2) = 0.0282 \pm 0.0009, \tag{11}$$

with no change in the size of the error. This corresponds to

$$\alpha^{-1}(m_Z) = 128.87 \pm 0.12 \tag{12}$$

which has been the standard value for the past few years.

The analysis of Swartz

The recent analysis by Swartz in Ref. (8) fits a smooth function described by polynomials in $W = \sqrt{s}$ to the continuum part of $R(s)$. Resonances were described with breit–Wigner forms as usual, and perturbative QCD (at $O(\alpha_s^3)$) was used above 15 GeV.

The major difference from method No. 3 of Ref. (5), or any other analysis that had used fitting functions, was that the correlations between data points within the same experiment due to the overall normalization error had been taken into account. The effect of this was that the fit produced a curve which was somewhat lower than when the correlations were neglected. See Fig. 3 for a comparison.

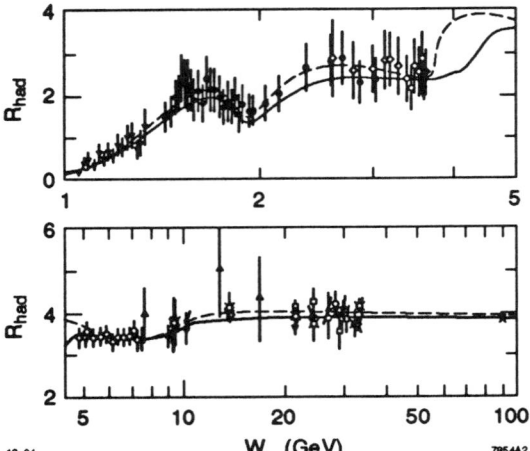

FIG. 3. Fit to the continuum part of $R(s)$ with (solid line) and without (broken line) correlations from normalization errors. The difference between the two in the 3.6–5 GeV region is due to a different treatment of the higher Ψ resonances.

With a lower curve for $R(s)$, the value of $\Delta\alpha^{(5)}_{hadrons}(m_Z^2)$ turned out to be smaller, and $\alpha^{-1}(m_Z)$ larger than previous analyses. Subtracting out the top quark contribution from the value quoted in Ref. (8), we obtain

$$\Delta\alpha^{(5)}_{hadrons}(m_Z^2) = 0.02672 \pm 0.00075, \qquad (13)$$

and

$$\alpha^{-1}(m_Z) = 129.07 \pm 0.10, \qquad (14)$$

which differs from Eq. 12 by $0.20 \approx 2\sigma$.

However, this analysis has been criticized (10) on the grounds that including normalization errors in the correlation matrix will introduce a bias when the normalization error and the number of data points are large (18). Swartz has subsequently updated his analysis to correct for this problem and also to include some new data which was missing from his first analysis (11). The new fit to the continuum part of $R(s)$ is shown in Fig. 4, and results in

$$\Delta\alpha^{(5)}_{hadrons}(m_Z^2) = 0.02752 \pm 0.00046, \qquad (15)$$

and

$$\alpha^{-1}(m_Z) = 128.96 \pm 0.06. \qquad (16)$$

The difference with Eq. 12 has been reduced by 1/2, and the uncertainty has also decreased due to the inclusion of new precise data.

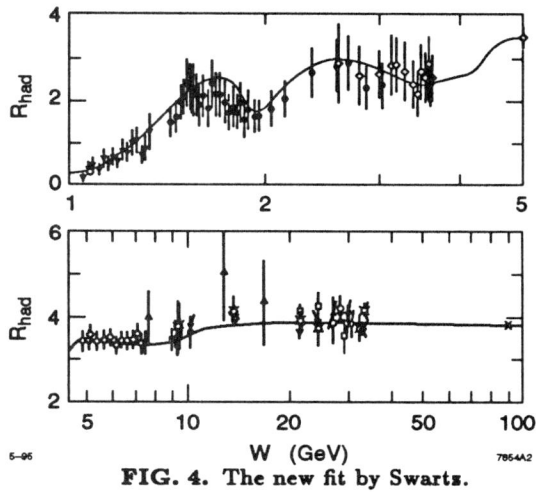
FIG. 4. The new fit by Swartz.

The analysis of Martin and Zeppenfeld

The analysis of Martin and Zeppenfeld in Ref. (9) distinguishes itself from all the other analyses in its extensive use of perturbative QCD. In table 1 I list the energy ranges in which different authors have applied perturbative QCD. (Though I also list the values of $\alpha_s(m_Z)$ that were used, at the current level of accuracy, the difference is insignificant. Changing the value of $\alpha_s(m_Z)$ has little effect on the resulting value of $\alpha(m_Z)$.)

In the two energy regions 3–3.9 GeV and 6.5–∞ GeV, Martin and Zeppenfeld express $R(s)$ as the perturbative QCD value plus the J/Ψ, Ψ, and Υ resonances. The DASP (19), PLUTO (20), MARK I (21), and Crystal Ball (22) data are all rescaled to fit the perturbative QCD result in these energy ranges, and the rescaled data is used to resolve a couple of resonances in the energy region between 3.9 and 6.5 GeV. See Fig. 5.

Due to this heavy reliance on perturbative QCD, and consequently the relatively light reliance on experimental data, the uncertainty in $\Delta\alpha^{(5)}_{hadrons}(m_Z^2)$ is reduced. The value quoted in Ref. (9) is

$$\Delta\alpha^{(5)}_{hadrons}(m_Z^2) = 0.02739 \pm 0.00042, \quad (17)$$

which corresponds to

$$\alpha^{-1}(m_Z) = 128.98 \pm 0.06. \quad (18)$$

This disagrees with both Eq. 12 and 14 by about 0.10, but agrees with Eq. 16.

TABLE 1. Comparison of the reliance on perturbative QCD in the evaluation of $\alpha(m_Z)$ between different authors.

Author	Ref.	Energy Range (GeV)	Order in α_s	$\alpha_s(m_Z)$
Burkhardt et al.	(5,6)	$40 - \infty$	2	0.12 ± 0.02
Jegerlehner	(7)	$40 - \infty$	2	0.117 ± 0.010
Swartz	(8,11)	$15 - \infty$	3	0.125 ± 0.005
Martin & Zeppenfeld	(9)	$3 - 3.9, 6.5 - \infty$	3	0.118 ± 0.007
Eidelman & Jegerlehner	(10)	$40 - \infty$	3	0.126 ± 0.005

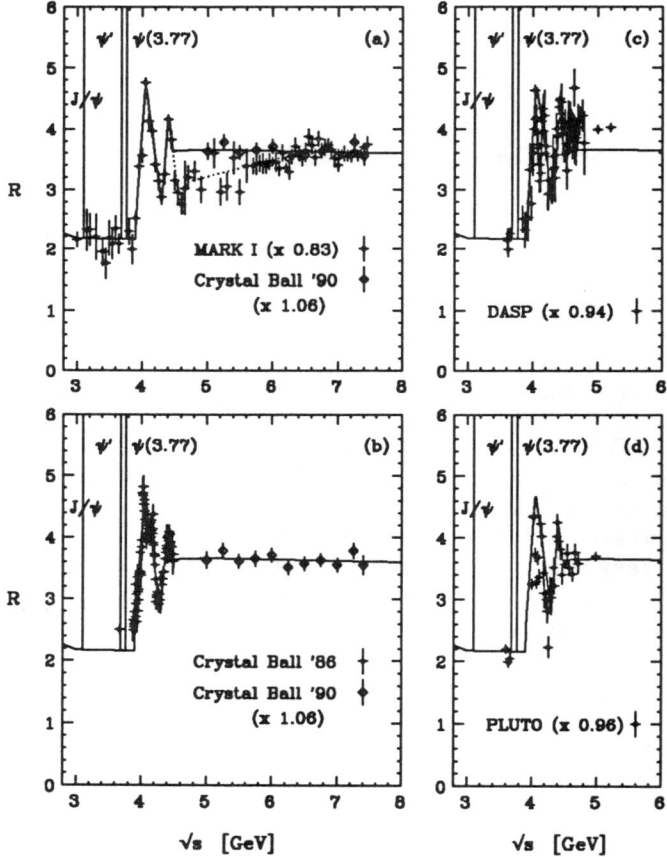

FIG. 5. The rescaling of data in Ref. (9).

The analysis of Eidelman and Jegerlehner

The last analysis I will discuss is by Eidelman and Jegerlehner in Ref. (10). This work is another update of Ref. (7), and again trapezoidal integration is used.

However, in order to take experimental errors into account properly, the data on $R(s)$ was first processed in the following way: First, for all values of s with data points, a value of $R(s)$ and its error was assigned to all the experiments in the region by linear interpolation between the closest data points belonging to that experiment. Then, the average of all the assigned values of $R(s)$ was taken weighted with the experimental errors. The resulting weighted average was used in the integration.

This analysis also used a more comprehensive set of data than any of the other analyses.

After subtracting out the top quark contribution from the value reported in Ref. (10), we obtain

$$\Delta\alpha^{(5)}_{hadrons}(m_Z^2) = 0.02804 \pm 0.00065, \qquad (19)$$

and

$$\alpha^{-1}(m_Z) = 128.89 \pm 0.09, \qquad (20)$$

which agrees very well with the previous estimate (7).

Discussion

While the large disagreement in $\alpha^{-1}(m_Z)$ of about 0.20 between Refs. (7) and (8) was disturbing, the new analyses by both authors (10,11) have decreased the disagreement to less than 0.10. With the estimate of Ref. (9) falling in the same ballpark, around 0.10 would be a good estimate of the actual uncertainty in $\alpha^{-1}(m_Z)$. The smaller errors quoted in Refs. (9) and (11) are most probably underestimates due to the stronger assumptions they make on the functional form of $R(s)$.

A conservative conclusion would be that the current experimental data on $R(s)$ does not allow for a substantially better determination of $\alpha(m_Z)$ than that already given in Ref. (7). Any further improvement on the uncertainty in $\alpha(m_Z)$ requires better measurements of $R(s)$ in the low energy region below 10 GeV.

TABLE 2. Current status of the determination of $\alpha_s(m_Z)$.

Method of Measurement	Reference	$\alpha_s(m_Z)$
DIS	(23)	0.113 ± 0.005
Υ decay	(24)	0.108 ± 0.010
Lattice		
charmonium spectrum	(26)	0.108 ± 0.006
Υ spectrum	(27)	0.115 ± 0.002
LEP		
Z lineshape data	(1)	0.125 ± 0.004
R_ℓ only	(1)	0.128 ± 0.005
event shapes and jet rates	(2)	0.123 ± 0.006
SLD		
event shapes and jet rates	(28)	0.120 ± 0.008

THE VALUE OF $\alpha_s(m_Z)$

Due to the limited number of pages I have been allocated, I will go through this section briefly.

Current status

In the case of $\alpha_s(m_Z)$, there exist more than one way to determine it's value. Ref. (16) gives a comprehensive overview of the many ways to measure $\alpha_s(m_Z)$. In table 2, I list some of the most recent determinations of $\alpha_s(m_Z)$ using various techniques.

Since the experimental error on R_ℓ is smaller than the theoretical error on the prediction based on Eq. 1, we can *assume* the SM and use the value of R_ℓ to determine $\alpha_s(m_Z)$ to a better accuracy than Eq. 1. This gives the value $\alpha_s(m_Z) = 0.128 \pm 0.005$ listed in table 2. A global fit to all the Z line shape data, including R_ℓ, gives $\alpha_s(m_Z) = 0.125 \pm 0.004$, also listed in table 2.

In this approach where one *assumes* the SM and *fits* the value of $\alpha_s(m_Z)$ to the data, any new physics beyond the SM will manifest itself not as a disagreement between the experimental measurement and theoretical prediction of some observable, but as disagreements between different determinations of $\alpha_s(m_Z)$.

Such a disagreement may already have been seen. As you can see from table 2, the 'low-energy' measurements such as Deep Inelastic Scattering (DIS) and Υ decay favor a value of $\alpha_s(m_Z)$ close to 0.11 while the 'high energy' measurements at LEP favor a value around 0.125. While this discrepancy may not seem too significant, considering the fact that unlike the $\alpha(m_Z)$ case these are all independent determinations based on different experiments, it is nevertheless puzzling that the LEP values of $\alpha_s(m_Z)$ are systematically higher than the low energy values.

Shifman argues in Ref. (30) that a value of $\alpha_s(m_Z)$ as high as 0.125 would correspond to a QCD scale of about 500MeV which is disfavored from the success of QCD sum rules and that the value 0.11 corresponding to a QCD scale of about 200MeV is more likely.

An additional support for the lower value of $\alpha_s(m_Z)$ is the recent measurement of the ratio $R_b = \Gamma(Z \to b\bar{b})/\Gamma(Z \to hadrons)$ at LEP which was found to be 2.4σ larger than the prediction for the SM (1). If this is a signal that $\Gamma(Z \to b\bar{b})$ is larger than its SM value, then one should also see a disagreement between theory and experiment in $R_\ell = \Gamma(Z \to hadrons)/\Gamma(Z \to \ell^+\ell^-)$. But since one does not see a noticeable difference when comparing Eqs. 2 and 3, a larger value of $\Gamma(Z \to b\bar{b})$ is not favored by R_ℓ and one is led to conclude that the deviation in R_b is just a statistical fluctuation.

However, if the value of $\alpha_s(m_Z)$ were as small as 0.11, then the story would be different. For $\alpha_s(m_Z) = 0.110$, the SM prediction of R_ℓ will be lowered to 20.679 and the difference from Eq. 3 will be just right to accommodate an excess in $\Gamma(Z \to b\bar{b})$ implied by R_b. In this case, the disagreement between the SM and experiment will be at the 4σ level (29–32).

SUMMARY

Though several authors have attempted to determination $\alpha(m_Z)$ to a better accuracy than Eq. 1, any substantial improvement requires better $R(s)$ data at low energies.

For the $\alpha_s(m_Z)$ case, the disagreement between the low and high energy determination could be a signal for new physics which also causes R_b to deviate from the SM.

ACKNOWLEDGEMENTS

I would like to thank Dr. M. L. Swartz for providing me with the results of his latest analysis, and the postscript files for Figs. 2, 3, and 4. I would also like to thank Dr. D. Zeppenfeld for providing me with the postscript files for Fig. 5. This work is supported by the United States Department of Energy under Contract Number DE-AC02-76CH030000.

REFERENCES

1. D. Schaile, in this proceedings.
2. S. Bethke, in the Proceedings of the *Workshop on Physics and Experiments with Linear e^+e^- Colliders*, Waikoloa, Hawaii, April 16–30, 1993, edited by F. A. Harris, S. L. Olsen, S. Pakvasa, and X. Tata (World Scientific, Singapore, 1993) p.687.
3. D. Bardin et al, CERN-TH.6443/92 (May 1992).
4. M. Consoli and F. Ferroni, *Phys. Lett.* B349, 375 (1995).
5. H. Burkhardt, F. Jegerlehner, G. Penso, and C. Verzegnassi, in 'Polarization at LEP', edited by G. Alexander, G. Altarelli, A. Blondel, G. Coignet,

E. Keil, D. E. Plane and D. Treille, CERN 88–06, Volume 1 (September 1988); Z. Phys. **C43**, 497 (1989).

6. G. Burgers and F. Jegerlehner, in 'Z Physics at LEP 1, edited by G. Altarelli, R. Kleiss, and C. Verzegnassi, CERN 89–08, Volume 1 (September 1989).
7. F. Jegerlehner, Prog. in Particle and Nucl. Phys. **27**, 1 (1991).
8. M. L. Swartz, SLAC–PUB–6710, hep–ph/9411353 (November 1994).
9. A. D. Martin and D. Zeppenfeld, Phys. Lett. **B345**, 558 (1995).
10. S. Eidelman and F. Jegerlehner, PSI–PR–95–1, BUDKERINP 95–5, hep–ph/9502298 (January 1995).
11. M. L. Swartz, private communication.
12. S. I. Dolinsky, et al. (ND), Phys. Rep. **C202**, 99 (1991).
13. A. E. Blinov et al. (MD–1), Z. Phys. **C49**, 239 (1991); BUDKERINP 93–54 (1993).
14. DM2 collaboration: D. Bisello et al., Z. Phys. **C48**, 23 (1990); LAL 90–35 (June 1990); LAL 90–71 (November 1990); A. Antonelli, et al., Z. Phys. **C56**, 15 (1992).
15. S. G. Gorshiny, A. L. Kataev, and S. A. Larin, Phys. Lett. **B259**, 144 (1991), L. R. Surguladze and M. A. Samuel, Phys. Rev. Lett. **66**, 560 (1991); ERRATUM Phys. Rev. Lett. **66**, 2416 (1991).
16. I. Hinchliffe, in the Review of Particle Properties, Phys. Rev. **D50** 1297 (1994); LBL–36374, hep–ph/9501354 (January 1995).
17. CDF collaboration: F. Abe et al., Phys. Rev. **D50**, 2966 (1995); Phys. Rev. Lett. **73**, 225 (1994); FERMILAB–PUB–95/022–E, hep–ex/9503002 (March 1995), D0 collaboration: S. Abachi et al., FERMILAB–PUB–95/028–E, hep–ex/9503003 (March 1995).
18. G. D'Agostini, Nucl. Instr. and Meth. in Phys. Res. **A346**, 306 (1994).
19. DASP collaboration: R. Brandelik et al., Phys. Lett. **B76**, 361 (1978); H. Albrecht et al., Phys. Lett. **B116**, 383 (1982).
20. PLUTO collaboration: J. Burmeister et al., Phys. Lett. **B66**, 395 (1977); Ch. Berger et al., Phys. Lett. **B81**, 410 (1979).
21. Mark I collaboration: J. L. Siegrist et al., Phys. Rev. **D26**, 969 (1982).
22. Crystal Ball collaboration: C. Edwards et al., SLAC–PUB–5160, (January 1990).
23. M. Virchaux and A. Milsztajn, Phys. Lett. **B274**, 221 (1992).
24. M. Kobel, DESY–F31–91–03 (1991).
25. M. B. Voloshin, TPI–MINN–95/1–T, UMN–TH–1326–95, hep–ph/9502224 (February 1995).
26. A. X. El-Khadra, G. Hockney, A. S. Kronfeld, P. B. Mackenzie, Phys. Rev. Lett. **69** 729 (1992).
27. C. T. H. Davies, K. Hornbostel, G. P. Lepage, A. Lidsey, J. Shigemitsu, and J. Sloan, Phys. Lett. **B345**, 42 (1995).
28. SLD Collaboration: K. Abe et al., Phys. Rev. **D51**, 962 (1995).
29. A. Blondel and C. Verzegnassi, Phys. Lett. **B311**, 346 (1993).
30. M. Shifman, Mod. Phys. Lett. **A10**, 605 (1995).
31. B. Holdom, UTPT–95–01, hep–ph/9502273 (February 1995).
32. P. Bamert, C. P. Burgess, and I. Maksymyk, MCGILL–95–18, hep–ph/9505339 (May 1995).

Precision data from LEP

Dorothee Schaile°*

°*CERN, CH1211 Geneva, Switzerland*

This conference report summarizes recent experimental progress on precise electroweak tests at the e^+e^- collider LEP. These data provide primarily a test of the Z propagator and the $Zf\bar{f}$ vertex. However, by imposing certain assumptions on the structure of the underlying theory, they also constrain vector boson self interactions.

We review briefly new results concerning the Z line shape, polarization asymmetries and electroweak measurements with heavy quarks. We discuss the consistency of LEP high precision electroweak measurements within themselves with those from other experiments and with the Standard Model.

INTRODUCTION

Electroweak precision tests at the e^+e^- collider LEP have become a rapidly evolving field, imposing stringent tests on the Standard Model comparable to the $g-2$ experiments as a probe of QED. At present energies LEP primarily explores the parameters of the Standard Model related to the properties of the Z resonance and to Z couplings to fermions. Assuming the Standard Model, the results can be used to derive constraints on the mass of the top quark, m_t, and the Higgs boson, m_H, as well as the strong coupling constant, $\alpha_s(m_Z^2)$. All these parameters are absent from the tree level diagrams of the processes observed at present e^+e^- colliders and only enter via loop corrections. Loops also allow us to constrain the self interaction of vector bosons with these data as discussed in subsequent talks at this conference.

At the time of the conference, the status of electroweak precision tests remained essentially unchanged with respect to that reported at the 27th International Conference on High Energy Physics, Glasgow, Scotland, 20-27 July 1994, which is extensively documented in (1) and (2). In this report we first would like to give a brief account of recent developments in the field since then. We will then discuss the combined harvest of electroweak precision tests which has been updated to correspond to the status of (3).

---

*Heisenberg Fellow

DISCUSSION OF RECENT MEASUREMENTS

Line shape and lepton asymmetries

Recent improvements in this field are based on a high statistics scan in 1993. During this scan more than 18 pb^{-1} were recorded by each experiment at two centre-of-mass energy points roughly 1.8 GeV above and below the Z mass, m_Z, while about 15 pb^{-1} were within 200 MeV of m_Z. Preliminary results are also available from the 1994 running period in which the LEP machine delivered approximately 65 pb^{-1} to each of the four LEP experiments at the Z peak.

The sizable increase in precision of the Z line shape parameters has three main sources: the gain in statistics; the reduction of both the experimental and the theoretical systematic error of the luminosity determination; and last but not least, a precise determination of the LEP energy scale based on the method of resonant depolarization (4). With the final LEP energy calibration calibration for the 1993 scan (5), the LEP energy calibration uncertainty translates in a correlated error among the four LEP experiments on m_Z of ~1.5 MeV and on the total Z width, Γ_Z, of ~1.7 MeV. An additional error, also correlated among the four LEP experiments, arises from an uncertainty of 5 MeV in the determination of the LEP centre-of-mass energy spread, which translates into an error of ~1 MeV on Γ_Z.

The τ polarization

All preliminary results presented at the 1994 summer conferences (1) have been finalized and published (6–8). The OPAL experiment presents preliminary results including the data of the 1994 running period (9).

Electroweak results with heavy flavours

The important experimental aspects which lead to significant improvements in this field are steady refinements of the vertex tagging techniques for b quarks, which in turn rely on several upgrades of the LEP microvertex detectors, and the gain in statistics allowing double tagging techniques. Since the 1994 summer conferences, several new preliminary results have now been published. The dominant changes with respect to the results in (2) are due to the inclusion of new data (10–12). At the time of writing the most up to date collection of the results of the individual experiments is given in (3). Details on the method of their combination can be found in (2,3).

DISCUSSION OF RESULTS

In this section we wish to address the consistency of LEP electroweak precision tests within themselves, with precise electroweak measurements in other fields and with the Standard Model. Then we summarize the resulting constraints on Standard Model parameters. The resulting bounds on anomalous vector boson couplings will be discussed in subsequent reports.

Table 1 displays the most precise electroweak measurements obtained so far. Table 1a summarizes LEP results from the measurement of the Z line shape and lepton forward-backward asymmetries, τ polarization asymmetries, production rates and forward-backward asymmetries of b and c quarks and the charge forward-backward asymmetry measure, $\langle Q_{FB} \rangle$, in hadronic events. These results are complemented by precise measurements of the W boson mass, m_W, in $p\bar{p}$ collisions (16) and of $1 - m_W^2/m_Z^2$ in νN scattering experiments (13–15) (Table 1b) and the determination of the left-right polarization asymmetry in e^+e^- collisions at the SLC (17) (Table 1c).

In order to test the internal consistency of all these measurements it is interesting to project them onto a common scale. All asymmetry measurements depend essentially on the ratio of the vector and axial vector coupling constants of the Z to leptons $g_{V\ell}/g_{A\ell}$[†] and can therefore be parametrized by the effective weak mixing angle, $\sin^2\theta_{\text{eff}}^{\text{lept}}$, defined as:

$$\sin^2\theta_{\text{eff}}^{\text{lept}} \equiv \frac{1}{4}(1 - g_{V\ell}/g_{A\ell}). \quad (1)$$

The resulting values of $\sin^2\theta_{\text{eff}}^{\text{lept}}$ from the asymmetry measurements are shown in Figure 1. The $\chi^2/d.o.f.$ of an overall weighted average is 7.6/6. Note, however, that the value of χ^2 receives a contribution of 5 from the inclusion of the left-right polarization asymmetry, A_{LR}, alone. The value of $\mathcal{A}_e = g_{Ve}g_{Ae}/(g_{Ve}^2 + g_{Ae}^2)$ determined from A_{LR} is based on the same physical phenomena as the determination of $\mathcal{A}_e$ from the τ polarization forward-backward asymmetry, with minimal model assumptions, and to $\mathcal{A}_\tau$ from the integral τ polarization, assuming lepton universality. Both τ polarization results exhibit a very small pull on the LEP average of $\sin^2\theta_{\text{eff}}^{\text{lept}}$.

A comparison of all measurements listed in Table 1 with the Standard Model prediction is presented in Figures 2 and 3. The one standard deviation band of all observables has a significant overlap with the Standard Model prediction for a top mass range of $150 \leq m_t$ [GeV] ≤ 200, with two exceptions: $R_b \equiv \Gamma_{b\bar{b}}/\Gamma_{\text{had}}$ and $R_c \equiv \Gamma_{c\bar{c}}/\Gamma_{\text{had}}$ which both favour lower top masses. As can be seen from the 68% confidence level contour in the R_b-R_c plane, these observables have a non-negligible anticorrelation of -0.4.

The Standard Model calculations are based on most up to date electroweak libraries (18) as verified by the workshop on 'Precision calculations for the Z

[†]Quark forward-backward asymmetries have a reduced sensitivity to corrections particular to the hadronic vertex.

TABLE 1. Summary of measurements included in the combined analysis of Standard Model parameters. Section a) summarises LEP averages, section b) electroweak precision tests from $p\bar{p}$ colliders and νN-scattering, section c) gives the result for $\sin^2\theta_{\text{eff}}^{\text{lept}}$ from the measurement of the left-right polarization asymmetry at SLD. The Standard Model fit results in column 3 and the pulls (difference between fit and measurement in units of the measurement error) in column 4 are derived from the fit including all data (Table 2, column 4) for a fixed value of $m_H = 300$ GeV.

	Measurement	Standard Model Fit	Pull
a) LEP			
line-shape and lepton asymmetries:			
m_Z [GeV]	91.1887 ± 0.0022	91.1885	0.0
Γ_Z [GeV]	2.4971 ± 0.0032	2.4979	-0.3
σ_h^0 [nb]	41.492 ± 0.081	41.441	0.6
R_ℓ	20.800 ± 0.035	20.783	0.5
$A_{\text{FB}}^{0,\ell}$	0.0172 ± 0.0013	0.0157	1.1
+ correlation matrix (3)			
τ polarization:			
$\mathcal{A}_\tau$	0.140 ± 0.008	0.145	-0.6
$\mathcal{A}_e$	0.137 ± 0.009	0.145	-0.9
b and c quark results:			
R_b	0.2204 ± 0.0020	0.2157	2.4
R_c	0.1606 ± 0.0095	0.172	-1.2
$A_{\text{FB}}^{0,b}$	0.1015 ± 0.0036	0.1015	0.0
$A_{\text{FB}}^{0,c}$	0.0760 ± 0.0089	0.0724	0.4
+ correlation matrix (3)			
$q\bar{q}$ charge asymmetry:			
$\sin^2\theta_{\text{eff}}^{\text{lept}}$ ($\langle Q_{\text{FB}}\rangle$)	0.2320 ± 0.0016	0.2318	0.1
b) $p\bar{p}$ and νN			
m_W [GeV] ($p\bar{p}$ (16))	80.26 ± 0.16	80.34	-0.6
$1 - m_W^2/m_Z^2$ (νN (13–15))	0.2253 ± 0.0047	0.2238	0.4
c) SLC			
$\sin^2\theta_{\text{eff}}^{\text{lept}}$ (A_{LR} (17))	0.2294 ± 0.0010	0.2318	-2.4

resonance' (19). As a result of this workshop it has been found that the differences among several electroweak libraries are small compared with present experimental accuracies. The workshop also attempted to estimate the theoretical uncertainties, which are mainly due to missing higher order corrections and the interplay between QCD and electroweak corrections. These theoretical uncertainties were estimated to be small compared with the present experimental accuracy and with the uncertainty due to the effective fine structure constant, $\alpha(m_Z^2)$, which will be discussed below.

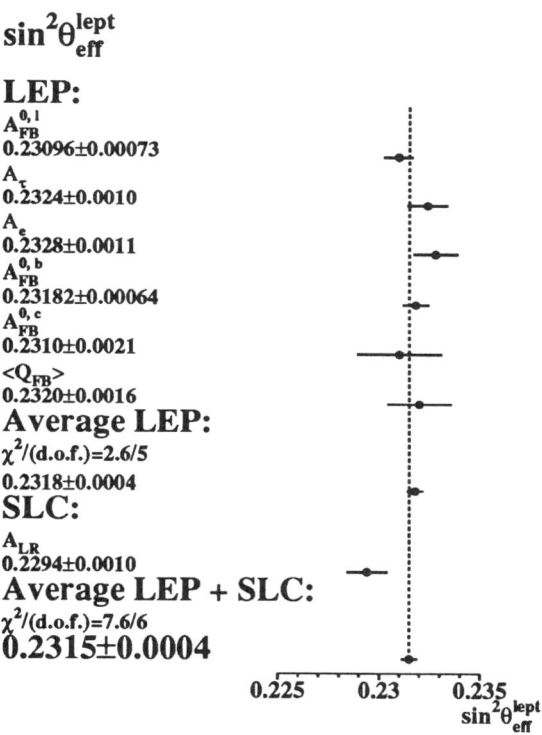

FIG. 1. LEP and SLC measurements of $\sin^2\theta_{\text{eff}}^{\text{lept}}$.

Assuming that the observed fluctuations of the data around the Standard Model prediction with a unique choice of parameters are of statistical nature the measurements X_i can be combined in a global fit which parametrizes them as:

$$X_i = f_i^{SM}(\alpha(m_Z^2), G_F, m_Z, m_t, m_H, \alpha_s(m_Z^2)). \tag{2}$$

In this fit the Fermi constant, G_F, is fixed to the central value given in (20), $\alpha(m_Z^2)$ and m_Z are constrained to $\alpha(m_Z^2) = 1/(128.896 \pm 0.090)$ (21)

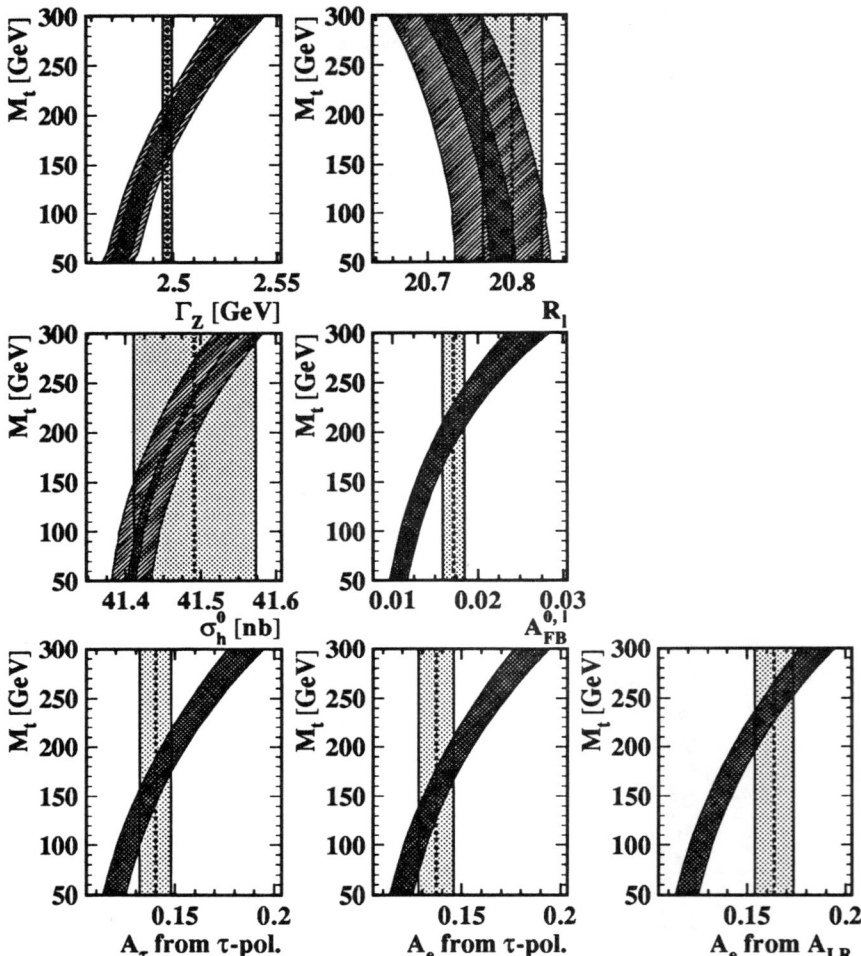

FIG. 2. Comparison of precision measurements with the Standard Model prediction as a function of m_t. The cross-hatched area shows the variation of the Standard Model prediction with m_H spanning the interval $60 < m_H \,(\text{GeV}) < 1000$ and the singly-hatched area corresponds to a variation of $\alpha_s(m_Z^2)$ within the interval $\alpha_s(m_Z^2) = 0.123 \pm 0.006$. The total width of the band corresponds to the linear sum of both uncertainties. The experimental errors on the parameters are indicated as vertical bands.

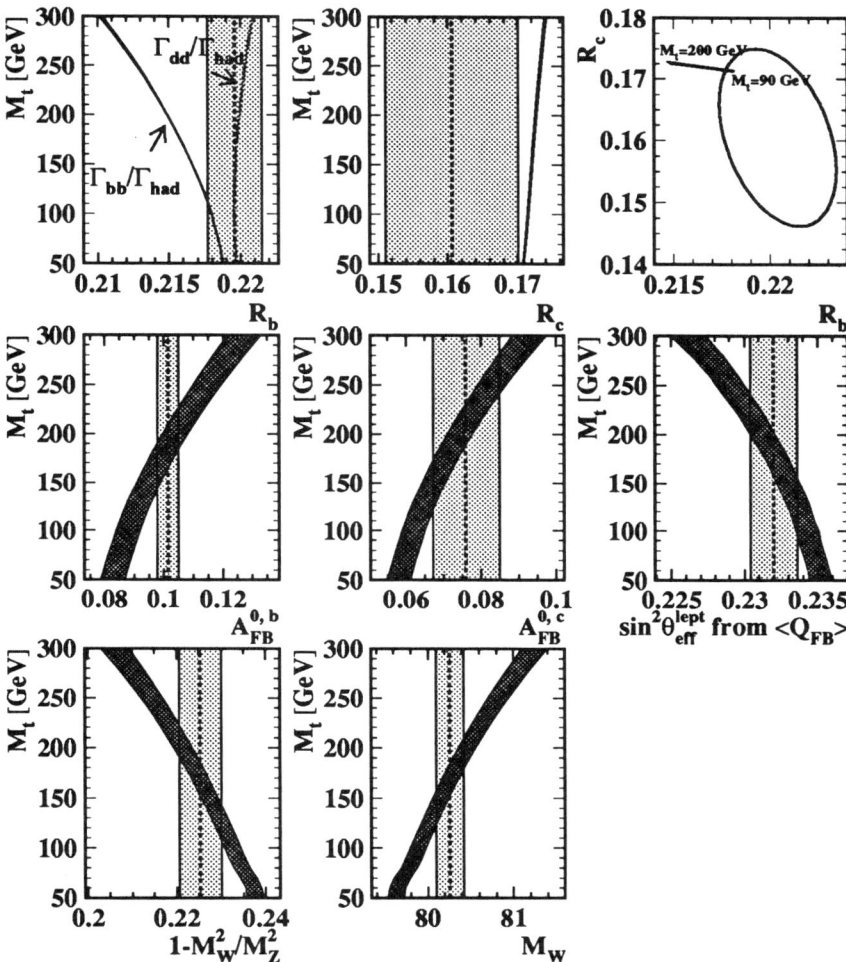

FIG. 3. Comparison of precision measurements with the Standard Model prediction as a function of m_t. The cross-hatched area shows the variation of the Standard Model prediction with m_H spanning the interval $60 < m_H \,(\text{GeV}) < 1000$ and the singly-hatched area corresponds to a variation of $\alpha_s(m_Z^2)$ within the interval $\alpha_s(m_Z^2) = 0.123 \pm 0.006$. The total width of the band corresponds to the linear sum of both uncertainties. The experimental errors on the parameters are indicated as vertical bands. For the comparison of R_b with the Standard Model the value of R_c has been fixed to the Standard Model resulting in $R_b = 0.2196 \pm 0.0019$. Also shown is the 68% confidence level contour in the R_b-R_c plane together with the Standard Model prediction.

and $m_Z = 91.1887 \pm 0.0022$, respectively. The value of m_H is fixed to 300 GeV and the change of parameter results when varying m_H in the interval $60 \leq m_H$ [GeV] ≤ 1000 is quoted as an additional uncertainty.

Table 2 shows the results obtained for the remaining free and unconstrained parameters m_t and $\alpha_s(m_Z^2)$ when fitting the measurements in Table 1 to the Standard Model calculation. We present the results obtained using only LEP data (Table 1a), as well as those obtained by including the measurements of m_W from UA2 (22), CDF (23,24) and DØ (25), and the measurements of the neutrino neutral to charged current ratios from CDHS (13), CHARM (14) and CCFR (15) (Table 1b). Finally we also add the SLC result for the left-right asymmetry, A_{LR} (17) (Table 1c).

TABLE 2. Results of fits to LEP and other precise electroweak data for m_t and $\alpha_s(m_Z^2)$. No external constraint on $\alpha_s(m_Z^2)$ has been imposed. The second column presents the results obtained using LEP data only (Table 1a). In the third column also the combined data from the $p\bar{p}$ collider and νN experiments (Table 1b) are included. The fourth column gives the result when the SLD measurement of the left-right asymmetry (Table 1c) is also added. The central values and the first errors quoted refer to $m_H = 300$ GeV. The second errors correspond to the variation of the central value when varying m_H in the interval $60 \leq m_H$ [GeV] ≤ 1000. The bottom part of the table lists derived results for $\sin^2\theta_{\text{eff}}^{\text{lept}}$, $1 - m_W^2/m_Z^2$ and m_W.

	LEP	LEP + $p\bar{p}$ and νN data	LEP + $p\bar{p}$ and νN data + A_{LR} from SLD
m_t (GeV)	$176 \pm 10 \,^{+17}_{-19}$	$174 \pm 9 \,^{+17}_{-19}$	$179 \pm 9 \,^{+17}_{-19}$
$\alpha_s(m_Z^2)$	$0.125 \pm 0.004 \pm 0.002$	$0.126 \pm 0.004 \pm 0.002$	$0.125 \pm 0.004 \pm 0.002$
χ^2/d.o.f.	8.8/9	9.0/11	15/12
$\sin^2\theta_{\text{eff}}^{\text{lept}}$	$0.2320 \pm 0.0003 \,^{+0.0001}_{-0.0002}$	$0.2320 \pm 0.0003 \,^{+0.0001}_{-0.0002}$	$0.2318 \pm 0.0003 \,^{+0.0001}_{-0.0002}$
$1 - m_W^2/m_Z^2$	$0.2242 \pm 0.0012 \,^{+0.0003}_{-0.0002}$	$0.2245 \pm 0.0011 \,^{+0.0002}_{-0.0002}$	$0.2238 \pm 0.0011 \,^{+0.0003}_{-0.0002}$
m_W (GeV)	$80.32 \pm 0.06 \,^{+0.01}_{-0.01}$	$80.31 \pm 0.06 \,^{+0.01}_{-0.01}$	$80.34 \pm 0.06 \,^{+0.01}_{-0.01}$

The value of m_t resulting from these fits is in excellent agreement with the mass values $m_t = 176 \pm 8$ (stat.)± 10 (syst.) and $m_t = 199^{+19}_{-21}$ (stat.)± 22 (syst.) reported recently from the observation of the top quark at the Tevatron by the CDF (26) and the DØ collaborations (27), respectively. For the sake of comparison with electroweak precision data the weighted average of both measurements, $m_t = 180 \pm 12$ GeV, is used in the following.

Similarly, the value of $\alpha_s(m_Z^2)$ resulting from the fits of Table 2 is in very good agreement with that obtained from event shape measurements at LEP ($\alpha_s(m_Z^2) = 0.123 \pm 0.006$ (28)), which is of similar precision and complementary in all aspects of theoretical and experimental uncertainties. The strong coupling constant can also be determined from the parameter $R_\ell \equiv \Gamma_{\text{had}}/\Gamma_{\ell\ell}$

alone. For $m_Z = 91.1887$ GeV, and imposing $m_t = 180 \pm 12$ GeV as constraint, $\alpha_s = 0.128 \pm 0.005 \pm 0.002$ is obtained, where the second error accounts for the variation of the result when varying m_H in the interval $60 \leq m_H$ [GeV] ≤ 1000. The present theoretical error of $\alpha_s(m_Z^2)$ arising from missing higher order QCD corrections to R_ℓ is estimated in (29) as $\Delta\alpha_s(m_Z^2) = 0.002$. Uncertainties in the electroweak part of the calculation can possibly be reduced below that value. The variation of radiative corrections to R_ℓ due to their dependence on m_H and m_t can be constrained by other precision observables and eventually by direct mass determinations.

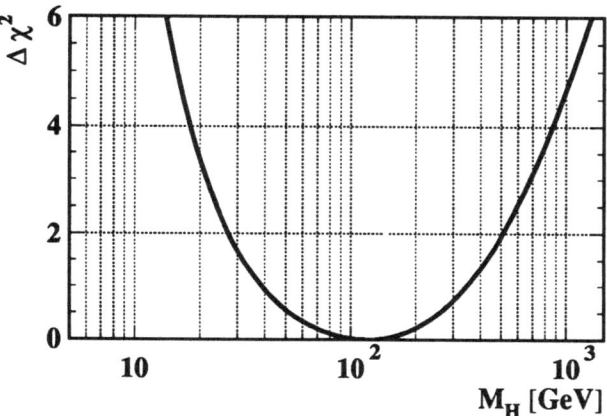

FIG. 4. $\Delta\chi^2 = \chi^2 - \chi^2_{min}$ vs m_H for the data used in the last column of Table 2 and including the Tevatron value of m_t.

The variation in the fitted value of m_t arising from varying m_H in the interval $60 \leq m_H$ [GeV] ≤ 1000 is approximately twice as big as the uncertainty in fitted value of m_t with m_H fixed, as seen in Table 2. It is therefore interesting to constrain m_t by the direct measurement of CDF and DØ and thereby use electroweak precision measurements to constrain m_H[‡]. Figure 4 shows $\Delta\chi^2 \equiv \chi^2 - \chi^2_{min}$ as a function of m_H, when $m_t = 180 \pm 12$ GeV is used as an additional constraint in the fit. The observed $\Delta\chi^2$ curve exhibits a minimum for low values of m_H. However, the entire range of m_H up to 1000 GeV is either already excluded by directed searches or accommodated within an interval in $\Delta\chi^2$ of about four, approximately corresponding to a two-sided 95% probability range. These fits may, however, impose more interesting constraints, provided the Higgs is light[§] and that the experimental

---

[‡]The main m_H dependence of radiative corrections for the measurements listed in Table 1 is given by terms proportional to $\log(m_H)$. The effects of m_H and m_t, however, are correlated for most observables, which underlines the need for a direct measurement of m_t to constrain m_H from electroweak precision tests.

[§]The logarithmic dependence of radiative corrections on m_H translates into

errors can be reduced by a factor of two or more.

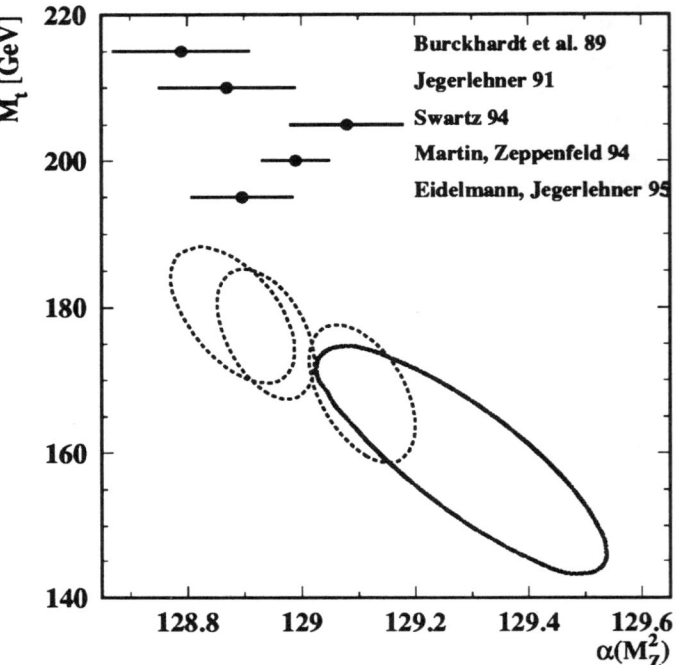

FIG. 5. History of the determination of $\alpha(m_Z^2)$ and the effect on the value of m_t resulting from electroweak precision tests. Shown are one standard deviation contours (39 % confidence level) in the $m_t - \alpha(m_Z^2)^{-1}$ plane ($m_H = 300$ GeV, fixed). For the dashed contours the value of $\alpha(m_Z^2)$ is constrained by the result of (from left to right) Burckhardt et al. (30), Eidelmann and Jegerlehner (21) and Swartz (31), respectively. For the solid contour $\alpha(m_Z^2)$ is treated as free parameter without imposing any constraint.

An important aspect of the interpretation of electroweak precision tests is the value of $\alpha(m_Z^2)$ and its error which arises from the uncertainty of the hadronic contribution to the photon vacuum polarization. Until recently most electroweak calculations were based on $\alpha(m_Z^2) = 1/(128.79 \pm 0.12)$ (30) or on an updated analysis $\alpha(m_Z^2) = 1/(128.87 \pm 0.12)$ (32). Since the 1994 summer conferences, there have been several reevaluations of $\alpha(m_Z^2)$ (21,31,33,34). Figure 5 shows the history of $\alpha(m_Z^2)$ evaluations. Note, that the errors quoted from the different evaluations can be assumed to be almost fully correlated. Also shown are the one standard deviation contours (39% confidence level) in the $m_t - \alpha(m_Z^2)^{-1}$ plane when using these $\alpha(m_Z^2)$ values as constraints in the global fit to all results in Table 1 and the contour leaving $\alpha(m_Z^2)$ unconstrained. Figure 5 illustrates that the uncertainty of $\alpha(m_Z^2)$ may become

---

$\Delta m_H/m_H \approx$ constant for a set of measurements with specified errors.

the dominant limitation of electroweak precision tests. We urgently need a clarification of its central value. For a reduction of the error of $\alpha(m_Z^2)$ more precise measurements of the total hadronic cross section in e^+e^- collisions in the centre-of-mass energy range of $\sim 1 - 10$ GeV are needed.

SUMMARY AND OUTLOOK

At LEP we have now collected a total of more than $14 \cdot 10^6$ hadronic events and the error of several precision observables is still dominated by the statistical component. For the 1995 running period a new scan of the Z resonance is planned with the aim of reducing the error for Γ_Z. The impact of significant experimental progress in the absolute luminosity determination is still masked by the theoretical uncertainty, but improvements are promised.

We wait eagerly for the solution or strengthening of the $\sin^2\theta_{\text{eff}}^{\text{lept}}$ puzzle which arises when comparing asymmetry measurements at LEP with the A_{LR} result from SLC.

To exploit fully the accuracy of precision tests we need a clarification of the current situation for $\alpha(m_Z^2)$ and a reduction of its error.

The precise measurement of m_t at the Tevatron and further improvements are an important input for the interpretation of precise measurements.

The first W-pairs at LEP are expected in late 1996. The new era of precise physics with W-pairs will confirm or disprove our extrapolations for vector boson self interactions from precise measurements at the Z-pole.

Acknowledgements

It is a pleasure to express my thanks and congratulations to the organizers of this conference, who succeeded in creating a stimulating atmosphere for the exchange of knowledge and ideas. I am grateful to my colleagues and friends in the LEP Electroweak Working Group for substantial contributions in combining the material presented. I would like to thank R.W. Jones for a critical reading of the manuscript. This work is supported by a grant from the Deutsche Forschungsgemeinschaft under the Heisenberg program.

REFERENCES

1. D. Schaile, *Precision tests of the electroweak interaction*, CERN-PPE/94-162, 11 October 1994; Proceedings of the XXVII International Conference on High Energy Physics, 20-27 July 1994, Glasgow Scotland UK, ed. by P.J. Bussey and I.G. Knowles, Institute of Physics Publishing, Bristol and Philadelphia. CERN-PPE/94-162, 11 October 1994.
2. The LEP Collaborations ALEPH, DELPHI, L3, OPAL and the LEP Electroweak Working Group, *Combined Preliminary Data on Z Parameters from the LEP Experiments and Constraints on the Standard Model*, CERN-PPE/94-187.

3. G. Wilkinson, 9th Rencontres de Physique de la Vallee d'Aoste - Results and Perspectives in Particle Physics, La Thuile, Italy, 5 - 11 Mar 1995;
   M. Calvi and U. Uwer, 30th Rencontres de Moriond - Electroweak Interactions and Unified Theories, Meribel les Allues, France, 11 - 18 Mar 1995;
   LEP Electroweak Working Group, A combination of Preliminary LEP Electroweak results for the 1995 Winter Conferences, Internal Note, prepared from the Contributions of the LEP Experiments to the 1995 Winter Conferences.
4. LEP Energy Working Group, L. Arnaudon et al., Accurate determination of the LEP beam Energy by resonant depolarization, CERN-SL/94-71 (BI), August 1994.
5. The working group on LEP energy, R. Assmann et al., The Energy Calibration of LEP in the 1993 Scan, CERN-PPE/95-10, 20 January 1995.
6. ALEPH Collaboration, D. Buskulic et al., Improved tau polarisation measurements, CERN-PPE/95-023 (1995).
7. DELPHI Collaboration, P. Abreu et al., Measurements of the τ Polarisation in Z^0 decays, CERN-PPE/95-030 (1995).
8. L3 Collaboration, O. Acciari et al., Phys. Lett. **B341** (1994) 24.
9. OPAL Collaboration, Updated Measurement of the Tau Polarisation Asymmetries, OPAL Internal Physics Note PN172, March 1995.
10. DELPHI Collaboration, Measurement of the Forward-Backward asymmetry of $e^+e^- \to Z \to b\bar{b}$ using prompt leptons and a microvertex tag, DELPHI 94-62 PHYS 383; Measurement of the Forward-Backward Asymmetry of $e^+e^- \to Z \to b\bar{b}$ using prompt leptons, DELPHI 94-107 PHYS 424; New measurement of the forward-backward asymmetry in $Z \to b\bar{b}$ events using a lifetime tag and the jet charge algorithm, DELPHI Note 95-27 PHYS 478.
11. OPAL Collaboration, Measurement of the forward backward asymmetries of $e^+e^- \to Z \to b\bar{b}$ and $e^+e^- \to Z \to c\bar{c}$ from events tagged by a lepton, including 1994 data, OPAL Physics Note PN165, Feb 27 1995;
    OPAL Collaboration, R. Akers et al., Z. Phys. **C60** (1993) 199.
12. OPAL Collaboration, A Measurement of the Forward-Backward asymmetry of $e^+e^- \to b\bar{b}$ applying a jet charge algorithm to lifetime tagged events, OPAL physics note PN168, 28 Feb 1995.
13. CDHS Collaboration, H. Abramowicz et al., Phys. Rev. Lett. **57** (1986) 298;
    CDHS Collaboration, A. Blondel et al., Z.Phys. **C45** (1990) 361.
14. CHARM Collaboration, J.V. Allaby et al., Phys. Lett. **B177** (1986) 446;
    CHARM Collaboration, J.V. Allaby et al., Z. Phys.**C36** (1987) 611.
15. CCFR Collaboration, C.G. Arroyo et al., Phys. Rev. Lett. **72** (1994) 3452.
16. The average for m_W has been performed taking into account the measurements (22–25). The method of the combination follows the description in (35) treating the uncertainty due to parton distribution functions as the only common error. Measurements are averaged weighted by their uncorrelated uncertainties and then the largest parton distribution function uncertainty quoted, $\Delta m_W = 0.085$ GeV (22), is included after averaging.
17. SLD Collaboration, K. Abe et al., Phys. Rev. Lett **73** (1994) 25.
18. Electroweak libraries:
    ZFITTER: D. Bardin et al., Z. Phys. **C44** (1989) 493; Comp. Phys. Comm. **59** (1990) 303; Nucl. Phys. **B351**(1991) 1; Phys. Lett. **B255** (1991) 290 and CERN-TH 6443/92 (May 1992);
    BHM: G. Burgers, W. Hollik and M. Martinez; M. Consoli, W. Hollik and

F. Jegerlehner, *Proceedings of the Workshop on Z physics at LEP I*, CERN Report 89-08 Vol.I, 7; G. Burgers, F. Jegerlehner, B. Kniehl and J. Kühn, the same proceedings, CERN Report 89-08 Vol.I, 55.
These computer codes have recently been upgraded by including the results of (19) and references therein.

19. *Reports of the working group on precision calculations for the Z resonance*, eds. D. Bardin, W. Hollik and G. Passarino, CERN Yellow Report 95-03, Geneva, 31 March 1995.
20. Particle Data Group, Phys. Rev. **D45** (1992) 2.
21. *Hadronic contributions to $(g-2)$ of the leptons and to the effective fine structure constant $\alpha(m_Z^2)$*, S. Eidelmann and F. Jegerlehner, PSI-PR-95-1, BUDKERINP 95-5, January 1995.
22. UA2 Collaboration, J. Alitti et al., Phys. Lett. **B276** (1992) 354.
23. CDF Collaboration, F. Abe et al., Phys. Rev. Lett. **65** (1990) 2243 and Phys. Rev. **D43** (1991) 2070.
24. CDF Collaboration, F. Abe et al., *Measurement of the W Boson Mass*, FERMILAB-PUB-95/033-E and FERMILAB-PUB-95/035-1995.
25. DØ Collaboration, preliminary result presented by C.K. Jung, *W Mass Measurements from DØ and CDF Experiments at the Tevatron*, talk given at the 27th ICHEP, Glasgow, Scotland, 20-27 July 1994.
26. CDF Collaboration, F. Abe et al., *Observation of Top Quark Production in $p\bar{p}$ collisions*, FERMILAB-PUB-95/022-E.
27. DØ Collaboration, S. Abachi et al., *Observation of the Top Quark*, FERMILAB-PUB-95/028-E.
28. S. Bethke, Proceedings of the Linear Collider Workshop in Waikoloa/Hawaii, April 1993;
S. Catani, Proc. of the EPS Conference on High Energy Physics, Marseille, July 22-28, 1993, Editions Frontieres, Ed. J. Carr-M. Perrottet, page 771;
S. Banerjee, Proc. of the EPS Conference on High Energy Physics, Marseille, July 22-28, 1993, Editions Frontieres, Ed. J. Carr-M. Perrottet, page 299.
29. T. Hebbeker, M. Martinez, G. Passarino and G. Quast, Phys. Lett. **B331** (1994) 165.
30. H. Burkhardt, F. Jegerlehner, G. Penso and C. Verzegnassi, Z. Phys. **C 43** (1989) 497.
31. M. L. Swartz, *Reevaluation of the Hadronic Contribution to $\alpha(M_Z^2)$*, SLAC-PUB-6710, November 1994.
32. F. Jegerlehner in *Testing the Standard Model*, eds. M. Cvetic, P. Langacker, World Scientific, Singapore, 1991, p. 476; Prog. Part. Nucl. Phys. **27** (1991) 32.
33. A.D. Martin and D. Zeppenfeld, *A determination of the QED coupling at the Z pole*, MAD/PH/855, DTP/94/110, November 1994.
34. H. Burkhardt, B. Pietrzyk, talk at the 30th Rencontres de Moriond, 11-18 Mar 1995.
35. M. Demarteau et al., *Combining W Mass Measurements*, CDF/PHYS/CDF/PUBLIC/2552 and DØ NOTE 2115.

Probing Trilinear Gauge Boson Couplings at Colliders

Dieter Zeppenfeld

Department of Physics, University of Wisconsin, Madison, WI 53706

A direct measurement of the trilinear $WW\gamma$ and WWZ couplings is possible in the pair production of electroweak bosons at e^+e^- and hadron colliders. This talk addresses some of the theoretical issues: the parameterization of "anomalous couplings" in terms of form factors and effective Lagrangians, the complementary information which can be obtained in e^+e^- vs. hadron collider experiments, and a novel way to implement finite W-width effects in a gauge invariant manner.

---

1. INTRODUCTION

Over the past twenty years a wealth of high precision electroweak data has beautifully confirmed the SM predictions for the couplings of fermions to the electroweak gauge bosons. Measurements of the $f\bar{f}V$ couplings at LEP and the SLC generally agree with the SM at the 0.1–1% level (1) and universality of the lepton couplings has been tested at a similar level. This agreement provides strong evidence that the gauge theory description of electroweak interactions is indeed correct. In spite of these successes the most direct consequence of the underlying $SU(2)$ gauge symmetry, the nonabelian couplings of photons, Z's and W's, remain to be tested with meaningful precision.

Pair production of electroweak bosons (W^+W^- production at e^+e^- colliders, $W\gamma$, WZ and W^+W^- production at hadron colliders) are the prime processes to directly measure the WWV, $V = \gamma, Z$ couplings. With high enough precision one may hope to be sensitive to new physics in the bosonic sector. However, one likely will need a lepton collider in the TeV range to reach the required sensitivity (2,3). For machines such as the Tevatron or LEP II the foremost task will be to confirm the SM predictions for the WWV couplings and to quantify this agreement. For both purposes, discovery of new physics and SM tests, one can introduce a WWV vertex with generalized coupling parameters and then experimentally constrain their deviations from the SM predictions. This is analogous to the introduction of axial and vector couplings g_A and g_V for the $f\bar{f}V$ vertex. In Section 2 I will discuss parameterizations of the WWV vertices, both in terms of form-factors and effective Lagrangians. Ways to extract these couplings from data on weak boson pair production will be discussed in Section 3, with special emphasis on the complementarity of hadron and e^+e^- colliders. Many of these questions

© 1996 American Institute of Physics

are considered in greater detail in other contributions to these Proceedings so the discussion here will be limited to some of the more fundamental questions.

Experimentally one only observes the decay products of W's and Z's and finite width effects must be taken into account, in particular when working close to threshold, such as at the Tevatron or at LEP II. Implementing finite width effects while maintaining gauge invariance becomes a nontrivial task when nonabelian couplings are present. These questions will be discussed in Section 4 for the example of $W\gamma$ production in $p\bar{p}$ collisions. It is shown how inclusion of fermion triangle graphs together with resummation of vacuum polarization contributions (which are the basis for the W Breit Wigner propagator) lead to a gauge invariant result. At the same time this example of SM radiative corrections will serve to illustrate some of the points made in the previous Sections.

2. ANOMALOUS COUPLINGS AND FORM-FACTORS

Because of the rapid decay of W's and Z's, weak boson pair production is seen experimentally as the process

$$f_1\bar{f}_2 \to V_1 V_2 \to f_3\bar{f}_4\, f_5\bar{f}_6 \,, \qquad (1)$$

with both final state fermion-antifermion pairs in a $J = 1$ angular momentum state. We are thus interested in deviations $\Delta\mathcal{M}$ from SM six-fermion amplitudes $\mathcal{M}_{6f}$ in particular partial waves. Apart from anomalies in the three gauge boson couplings such deviations may also arise from new physics in the gauge boson–fermion interactions or from non-standard behaviour of the gauge boson propagators. The latter two, however, are already tested, at the 1% level or slightly better, in four-fermion processes like $e^+e^- \to f\bar{f}$ and we thus assume SM behaviour for both. [1] We are left with deviations $\Delta\mathcal{M}$ which occur due to the Three Gauge Vertex (TGV) and therefore appear in the overall $J = 1$ partial wave. Denoting the decay currents by e.g.

$$J^{(34)}_{V_1\alpha}(q) = \bar{f}_3 \gamma^{\alpha'} (g_V^{f_3 f_4 V_1} + g_A^{f_3 f_4 V_1}\gamma_5) f_4 \, D^{V_1}_{\alpha'\alpha}(q) \,, \qquad (2)$$

where the gauge boson propagator D^V has been included in the definition of the current, we may write the deviation $\Delta\mathcal{M}$ as

$$\Delta\mathcal{M} = \sum J^{(12)}_{V\mu}(P)\, J^{(34)}_{V_1\alpha}(q)\, J^{(56)}_{V_2\beta}(\bar{q})\, g_{VV_1V_2}\, \Delta\Gamma^{\mu\alpha\beta}_{VV_1V_2}(P,q,\bar{q}) \,. \qquad (3)$$

Here $g_{WW\gamma} = -e$, $g_{WWZ} = -e\tan\theta_W$ and the sum indicates that for neutral currents we need to add photon and Z exchange.

By convention we include the SM tree level vertex in the definition of the vertex function $\Gamma^{\mu\alpha\beta}_{VV_1V_2}(P,q,\bar{q})$. The momentum assignment for the vertex

---

[1] This implies that once three boson couplings are tested beyond 10^{-2} accuracy the assumption of SM behaviour of propagators and fermion vertices should be revisited.

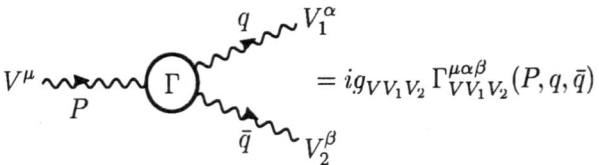

FIG. 1. Feynman rule for the general $V \to V_1 V_2$ vertex.

function is depicted in Fig. 1. In the limit of massless external fermions the currents $J_{V\mu}^{(ij)}$ are conserved, i.e. terms like $P^\mu J_{V\mu}^{(12)}(P)$ can be neglected. As a result the most general tensor structure of the vertex function can be written in terms of seven form factors $f_i(P^2, q^2, \bar{q}^2)$ (4,5)

$$\Gamma^{\mu\alpha\beta}(P,q,\bar{q}) = f_1 (q-\bar{q})^\mu g^{\alpha\beta} - \frac{f_2}{m_W^2}(q-\bar{q})^\mu P^\alpha P^\beta + f_3(P^\alpha g^{\mu\beta} - P^\beta g^{\mu\alpha})$$
$$+ if_4(P^\alpha g^{\mu\beta} + P^\beta g^{\mu\alpha}) + if_5 \varepsilon^{\mu\alpha\beta\rho}(q-\bar{q})_\rho$$
$$- f_6 \varepsilon^{\mu\alpha\beta\rho} P_\rho - \frac{f_7}{m_W^2}(q-\bar{q})^\mu \varepsilon^{\alpha\beta\rho\sigma} P_\rho (q-\bar{q})_\sigma \ . \quad (4)$$

This decomposition is completely general and applicable at all energies. Discrete symmetries of the underlying dynamics imply constraints among them. Parity conservation leads to $f_5 = f_6 = f_7 = 0$. Charge conjugation invariance relates the form factors f_1, f_2, f_3, and f_4.

While the decomposition into form factors is general, convenient parameterizations of their low-energy behavior are provided by the effective Lagrangian approach (6). Many different forms have been used in the literature. Here it suffices to use the phenomenological Lagrangian of Ref. (5) as an example. Keeping C and P conserving terms only, the WWV vertex ($V = Z, \gamma$) is given in terms of three parameters, g_1^V, κ_V and λ_V,

$$\mathcal{L}_{eff}^{WWV} = i g_{WWV} \left(g_1^V (W_{\mu\nu}^+ W^{-\,\mu} - W^{+\,\mu} W_{\mu\nu}^-) V^\nu + \kappa_V W_\mu^+ W_\nu^- V^{\mu\nu} \right.$$
$$\left. + \frac{\lambda_V}{m_W^2} W_\mu^{+\,\nu} W_\nu^{-\,\rho} V_\rho^{\,\mu} \right) , \quad (5)$$

where e.g. $V^{\mu\nu} = \partial^\mu V^\nu - \partial^\nu V^\mu$ is the γ or Z field strength tensor. Within the SM, the couplings are given by $g_1^Z = g_1^\gamma = \kappa_Z = \kappa_\gamma = 1$, and $\lambda_Z = \lambda_\gamma = 0$. The effective Lagrangian of Eq. 5 provides us with the lowest order terms in an expansion of the form factors f_i in powers of the Lorentz invariants P^2, q^2 and $\bar{q}^2$. For (on- or off-shell) $W^+ W^-$ production they are given by

$$f_1^V(P^2, q^2, \bar{q}^2) \approx g_1^V + \lambda_V \frac{P^2}{2m_W^2} , \quad (6)$$
$$f_2^V(P^2, q^2, \bar{q}^2) \approx \lambda_V , \quad (7)$$

$$f_3^V(P^2, q^2, \bar{q}^2) \approx g_1^V + \kappa_V + \lambda_V \frac{q^2 + \bar{q}^2}{2m_W^2} \,, \tag{8}$$

$$f_4^V(P^2, q^2, \bar{q}^2) \approx -i\lambda_V \frac{q^2 - \bar{q}^2}{2m_W^2} \,. \tag{9}$$

The notation developed up to here is getting cumbersome when comparing crossing related processes. The tensor decomposition of Eq. 4 treats incoming and outgoing vector bosons differently. As a result the form factors $f_i(P^2, q^2, \bar{q}^2)$ are process dependent: they mix under crossing. It is more convenient to define form factors $g_1^V(P^2, q^2, \bar{q}^2)$, $\kappa_V(P^2, q^2, \bar{q}^2)$ and $\lambda_V(P^2, q^2, \bar{q}^2)$ such that the relations of Eqs. 6–8 become exact for $V(P) \to W^-(q)W^+(\bar{q})$. This approach has the advantage that the Feynman rules derived from the effective Lagrangian of Eq. 5 can be used directly to calculate the full form factor dependence for any process involving WWV vertices. At the same time the relations between form factors for crossing related processes become manifest. The only disadvantage is that the coupling constants appearing in the effective Lagrangian might be confused with the full form factors, while in reality they just represent the low energy limits of these form factors.

The functional behaviour of the form factors depends on the details of the underlying new physics. Effective Lagrangian techniques (6) are of little help here because the low energy expansion which leads to the effective Lagrangians exactly breaks down where the form factor effects become important. So in practice one will have to make ad hoc assumptions. One possibility is to assume a behaviour similar to nucleon form factors, with constraints derived from unitarity considerations (7). Such constraints do become important at hadron colliders. More generally, they must be included when one searches for very large enhancements of vector boson pair production cross sections.

3. VECTOR BOSON PAIR PRODUCTION

Deviations of the TGV's from their SM, tree level form are most directly observed in vector boson pair production. Candidate processes are $W\gamma$, WZ and W^+W^- production at hadron colliders (namely the Tevatron and, eventually, the LHC) and $e^+e^- \to W^+W^-$ at LEP II or a NLC. Since experimental strategies have been discussed at great depth by other speakers at this symposium (8), I will concentrate here on some of the more basic effects of anomalous TGV's on vector boson pair production.

3.1 W^+W^- Production in e^+e^- Collisions

To lowest order, the production of W pairs in e^+e^- collisions proceeds via the Feynman graphs of Fig. 2. It is instructive to consider the individual contributions of s-channel photon and Z exchange and of t-channel neutrino exchange to the various helicity amplitudes (5),

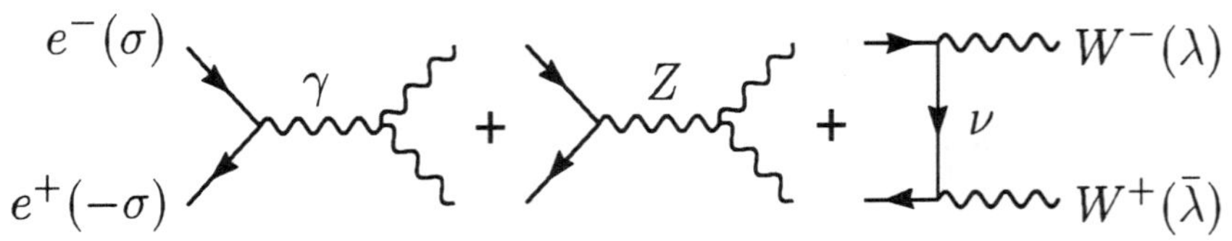

FIG. 2. Feynman graphs for the process $e^+e^- \to W^+W^-$.

$$\mathcal{M}(\sigma, \lambda, \bar{\lambda}) = \mathcal{M} = \mathcal{M}_\gamma + \mathcal{M}_Z + \mathcal{M}_\nu . \tag{10}$$

Here the e^- and e^+ helicities are given by $\sigma/2$ and $-\sigma/2$, and λ and $\bar{\lambda}$ denote the W^- and W^+ helicities. Following Ref. (5) let us define reduced amplitudes $\tilde{\mathcal{M}}$ by splitting off the leading angular dependence in terms of the d-functions d^{J_0} where $J_0 = 1, 2$ denotes the lowest angular momentum contributing to a given helicity combination,

$$\mathcal{M}(\sigma, \lambda, \bar{\lambda}; \theta) = \sqrt{2}e^2 \tilde{\mathcal{M}}_{\sigma,\lambda,\bar{\lambda}}(\theta) \, d^{J_0}_{\sigma, \lambda-\bar{\lambda}}(\theta) . \tag{11}$$

s-channel photon and Z exchange is only possible for $|\lambda - \bar{\lambda}| = 0, 1$. The corresponding reduced amplitudes can be written as

$$\begin{aligned}
\tilde{\mathcal{M}}_\gamma &= -\beta A^\gamma_{\lambda\bar{\lambda}} , \\
\tilde{\mathcal{M}}_Z &= +\beta A^Z_{\lambda\bar{\lambda}} \left[1 - \delta_{\sigma,-1} \frac{1}{2\sin^2\theta_W} \right] \frac{s}{s - m_Z^2} , \\
\tilde{\mathcal{M}}_\nu &= +\delta_{\sigma,-1} \frac{1}{2\beta \sin^2\theta_W} \left[B_{\lambda\bar{\lambda}} - \frac{1}{1 + \beta^2 - 2\beta\cos\theta} C_{\lambda\bar{\lambda}} \right] .
\end{aligned} \tag{12}$$

Here s denotes the e^+e^- center of mass energy and $\beta = \sqrt{1 - 4m_W^2/s}$ is the $W^\pm$ velocity. The subamplitudes A^V, B and C are given in Table 1.

One of the most striking features of the SM are the gauge theory cancellations between γ, Z and neutrino exchange graphs. Within the SM $g_1 = \kappa = 1$, $\lambda = 0$ (or $f_1 = 1$, $f_2 = 0$, $f_3 = 2$) for both the photon and the Z-exchange graphs. As a result $A^\gamma_{\lambda\bar{\lambda}} = A^Z_{\lambda\bar{\lambda}}$ and the βA^V terms in Eq. 12 cancel, except for the difference between photon and Z propagators. Similarly, the $B_{\lambda\bar{\lambda}}$ term in $\tilde{\mathcal{M}}_\nu$ and the $\delta_{\sigma,-1}$ term in $\tilde{\mathcal{M}}_Z$ cancel in the high energy limit for all helicity combinations. While the contributions from individual Feynman graphs grow with energy for longitudinally polarized W's, this unacceptable high energy behavior is avoided in the full amplitude due to the cancellations which can be traced to the gauge theory relations between fermion–gauge boson vertices and the TGV's.

At asymptotically large energies any deviations of $f_3, \ldots f_6$ from their SM values would lead to a growth of at least some of the helicity amplitudes $\tilde{\mathcal{M}}_{0\pm}$

TABLE 1. Subamplitudes for $J_0 = 1$ helicity combinations of the process $e^-e^+ \to W^-W^+$, as defined in Eq. 12. β denotes the W velocity and $\gamma = \sqrt{s}/2m_W$. The relations between the form factors f_i and g_1, κ, and λ are given in Eqs. 6–8, with $q^2 = \bar{q}^2 = m_W^2$.

$\lambda\bar\lambda$	$A^V_{\lambda\bar\lambda}$	$B_{\lambda\bar\lambda}$	$C_{\lambda\bar\lambda}$
++	$g_1^V + 2\gamma^2\lambda_V + \frac{i}{\beta}f_6^V + 4i\gamma^2\beta f_7^V$	1	$1/\gamma^2$
--	$g_1^V + 2\gamma^2\lambda_V - \frac{i}{\beta}f_6^V - 4i\gamma^2\beta f_7^V$	1	$1/\gamma^2$
+0	$\gamma(f_3^V - if_4^V + \beta f_5^V + \frac{i}{\beta}f_6^V)$	2γ	$2(1+\beta)/\gamma$
0-	$\gamma(f_3^V + if_4^V + \beta f_5^V - \frac{i}{\beta}f_6^V)$	2γ	$2(1+\beta)/\gamma$
0+	$\gamma(f_3^V + if_4^V - \beta f_5^V + \frac{i}{\beta}f_6^V)$	2γ	$2(1-\beta)/\gamma$
-0	$\gamma(f_3^V - if_4^V - \beta f_5^V - \frac{i}{\beta}f_6^V)$	2γ	$2(1-\beta)/\gamma$
00	$g_1^V + 2\gamma^2\kappa_V$	$2\gamma^2$	$2/\gamma^2$

or $\tilde{\mathcal{M}}_{\pm 0}$ with energy and hence violate partial wave unitarity. Similarly, non-standard values of f_7, λ or κ in the $s \to \infty$ limit would lead to an unacceptable growth in some of the remaining three helicity amplitudes. Thus, in this limit, partial wave unitarity excludes anomalous TGV's (9); any deviation from the SM must be described by an energy-dependent form factor which approaches its gauge theory value as $s \to \infty$.

Table 1 shows that only seven W^-W^+ helicity combinations contribute to the $J = 1$ channel and the form factors f_i enter in as many different combinations. This explains why exactly seven form factors or coupling constants are needed to parameterize the most general WWV vertex. Since we have both WWZ and $WW\gamma$ couplings at our disposal, the most general $J = 1$ amplitudes $\mathcal{M}_L = \mathcal{M}(\sigma = -1, \lambda, \bar\lambda)$ and $\mathcal{M}_R = \mathcal{M}(\sigma = +1, \lambda, \bar\lambda)$ for both left- and right-handed incoming electrons can be parameterized. Turning the argument around one concludes that all 14 helicity amplitudes need to be measured independently for a complete determination of all the form factors $f_i^\gamma(s)$ and $f_i^Z(s)$, at any value of the center of mass energy, $\sqrt{s}$.

Formidable as this goal may be it can be approached to a remarkable degree by performing a partial wave analysis, in particular of the semileptonic process $e^-e^+ \to W^-W^+ \to \ell^\pm \nu q\bar{q}'$. The charge of the lepton allows to identify the two W charges and hence the production angle θ. From Eq. 11 one finds that the $J = 1$ amplitudes lead to the angular distribution

$$\frac{d\sigma}{d\cos\theta} \sim \frac{\sin^2\theta}{2} \left(|\tilde{\mathcal{M}}_{\sigma,++}|^2 + |\tilde{\mathcal{M}}_{\sigma,--}|^2 + |\tilde{\mathcal{M}}_{\sigma,00}|^2\right)$$
$$+ \frac{(1+\sigma\cos\theta)^2}{4}\left(|\tilde{\mathcal{M}}_{\sigma,+0}|^2 + |\tilde{\mathcal{M}}_{\sigma,0-}|^2\right)$$
$$+ \frac{(1-\sigma\cos\theta)^2}{4}\left(|\tilde{\mathcal{M}}_{\sigma,0+}|^2 + |\tilde{\mathcal{M}}_{\sigma,-0}|^2\right). \quad (13)$$

Hence the amplitudes with different values of $|\lambda - \bar\lambda|$ can be separated, even though in practice one must take into account the additional θ-dependence of

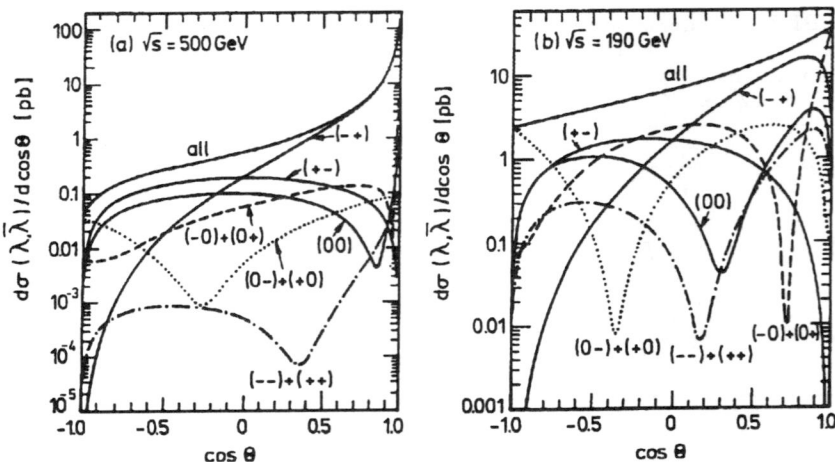

FIG. 3. Angular distributions $d\sigma/d\cos\theta$ for fixed W^-W^+ helicities $(\lambda\bar\lambda)$ in e^-e^+ collisions at a) $\sqrt{s} = 500$ GeV and b) $\sqrt{s} = 190$ GeV. From Ref. (5).

the known neutrino exchange graphs, as is evident from Fig. 3.

Due to the $V - A$ structure of the W-fermion vertices the decay angular distributions of the W's are excellent polarization analyzers. Consider for example the polar angle θ_- of the charged lepton ℓ^- in the W^- rest frame with respect to the W^- direction in the lab frame. Its distribution is proportional to $(1 - \lambda\cos\theta_-)^2$ for transversely polarized W^- and proportional to $\sin^2\theta_-$ for longitudinally polarized W^-. Combined with the information contained in the production angle distribution, the individual helicity amplitudes can be isolated, at least when polarized electron beams are available.

In practice, statistical errors will limit the accuracy with which such an analysis can be carried out. The best sensitivity to anomalous contributions is achieved when interference with large SM amplitudes can be exploited. Unfortunately, the dominant SM amplitudes are the $J_0 = 2$ amplitudes $\mathcal{M}_{+-}$ and $\mathcal{M}_{-+}$ which are purely due to t-channel neutrino exchange (see Fig. 3). At asymptotically large energies the only surviving $J = 1$ helicity amplitude is $\mathcal{M}_{00}$ and even its contribution is small numerically. Polar angle distributions alone yield relatively low sensitivity to anomalous TGV's.

The way out is to measure azimuthal angle distributions and azimuthal angle correlations which exploit the interference of the various $J = 1$ amplitudes with the t-channel neutrino exchange graph. In order to independently measure the various form-factors it is necessary to measure the full five-fold angular distributions

$$\frac{d^5\sigma}{d\cos\theta\, d\cos\theta_+\, d\phi_+\, d\cos\theta_-\, d\phi_-}, \qquad (14)$$

or more precisely the projection of this five-fold angular distribution on the triple $J = 1$ partial wave.

The sensitivity of such an analysis has been investigated for both LEP II and NLC energies (10,11). One finds, for example, that $\Delta\kappa$ should be measurable with an accuracy of $\approx 5 \cdot 10^{-4}$ at a 1.5 TeV NLC (11). Does this mean that electroweak radiative corrections will be probed by measuring W^+W^- production at a NLC? In spite of the small value of $\Delta\kappa$ this is not necessarily the case. According to Table 1, an anomalous value of $\Delta\kappa$ has its largest effect on $W_L W_L$ production. With $\gamma = \sqrt{s}/2m_W \approx 10$ and taking the gauge theory cancellations into account, the effect of $\Delta\kappa = 5 \cdot 10^{-4}$ on $\mathcal{M}_{00}$ is proportional to

$$\tilde{\mathcal{M}}_{00} \sim \frac{m_Z^2}{2m_W^2} + 2\gamma^2 \Delta\kappa = 0.65 + 2 \cdot 10^2 \cdot 5 \cdot 10^{-4} = 0.65 \cdot 1.15 , \quad (15)$$

i.e. the anomalous TGV corresponds to a change of the $W_L W_L$ amplitude of 15%, which probably is more than should be expected from electroweak radiative corrections.

3.2 Weak Boson Pair Production at Hadron Colliders

There are substantial differences in the study of TGV's at e^+e^- vs. hadron colliders. At LEP II or a NLC a detailed study of individual helicity amplitudes is possible and hence the individual form factors can be separated at any center of mass energy $\sqrt{s}$. In the clean environment of these machines errors are largely dominated by statistics and weak boson pair production cross sections can hence be measured with errors in the few percent range, i.e. the search for anomalous TGV's corresponds to the search of $\mathcal{O}(10^{-2})$ deviations of the production cross sections from the SM predictions.

Hadron colliders like the Tevatron or the LHC allow to study all pair production processes: W^+W^-, $W^\pm\gamma$, and $W^\pm Z$ production. Via the last two processes one can thus independently measure $WW\gamma$ and WWZ vertices. At the same time larger center of mass energies are available at the hadron machines compared to their e^+e^- contemporaries and hence the form factors are explored at higher energy scales. In turn this implies large enhancement factors (γ or γ^2) for the anomalous contributions. The signals we are searching for appear in the $J = 1$ partial wave *i.e.* for large production angles of the final state electroweak bosons and they are enhanced at large c.m. energies. Both features move observable effects to large transverse momenta of the produced vector bosons or their decay products. This effect is demonstrated in Fig. 4 where expected transverse momentum distributions in $p\bar{p} \to W^+\gamma$ and $p\bar{p} \to W^+Z$ production at the Tevatron are shown for several choices of anomalous couplings and dipole form factors $\Delta f_i(s) = \Delta f_i^0/(1 + s/1\text{TeV}^2)^2$.

Due to the more difficult background situation and also because of insufficient knowledge of QCD radiative corrections (12), structure function effects

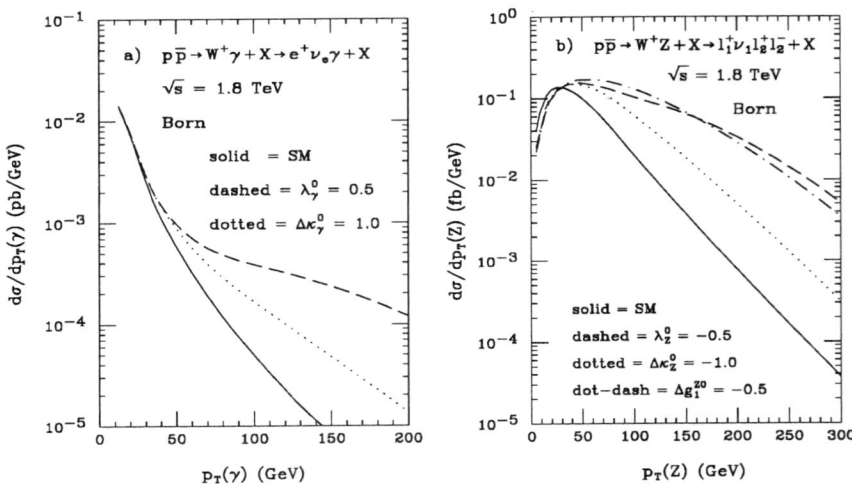

FIG. 4. Transverse momentum distribution of a) the photon in $W^+\gamma$ production and b) the Z in W^+Z production in $p\bar{p}$ collisions at the Tevatron for various sets of anomalous coupling parameters. From Ref. (2).

etc., a comparison of measured and theoretically predicted cross sections at the $\mathcal{O}(10^{-2})$ level is not feasible at hadron colliders. Rather their sensitivity to anomalous TGV's derives from fairly large deviations from the SM in at least some regions of phase space. A typical example is shown in Fig. 4 where for most choices of anomalous couplings $d\sigma/dp_T(V)$ is increased by one order of magnitude or more in parts of the accessible transverse momentum range.

The actual shape of the p_T distributions depends crucially on the energy dependence of the form factors. In weakly interacting models of new physics (like supersymmetry (13)) one should expect that virtual effects of heavy particles (of mass M) will never lead to changes of cross sections by such large factors. At low energy anomalous couplings are expected to scale like (14)

$$\Delta f_i(0) \sim \frac{g^2}{16\pi^2} \frac{m_W^2}{M^2}, \qquad (16)$$

while, at energies above the heavy particle mass M, form factor damping will set in and qualitatively a behaviour like

$$\Delta f_i(s \gg M^2) \sim \frac{g^2}{16\pi^2} \frac{m_W^2}{s} \qquad (17)$$

must be expected. Turning to the effect on the weak boson pair production cross section one finds that even when including a $\gamma^2 = s/4m_W^2$ enhancement factor in the amplitude, the amplitude changes only by a term of order

$$\Delta\mathcal{M} \sim \frac{s}{m_W^2}\Delta f_i(s) \sim \frac{g^2}{16\pi^2}\frac{s}{M^2} \quad \text{for } s \ll M^2 ,$$

$$\frac{g^2}{16\pi^2} \quad \text{for } s \gg M^2 , \tag{18}$$

and the change in the amplitude is at most of the order of the naive perturbative expectation, $g^2/16\pi^2$. Thus, for weakly coupled new physics, no large enhancement of weak boson pair production cross sections occurs. The dramatic increase of WZ or $W\gamma$ production rates shown in Fig. 4 needs some strong interaction dynamics in the weak boson sector, and even then it is not guaranteed to occur (14).

One thus finds that in their search for anomalous TGV's e^+e^- and hadron colliders are complementary. LEP or a 500 GeV NLC will probe W^+W^- pair production cross sections quite precisely, but at relatively low center of mass energies. This limits the enhancement factors for anomalous TGV's but, with sufficient statistics, even relatively weakly coupled new physics may be accessible. At the hadron colliders much higher energy ranges can be probed, but these experiments are only sensitive to strongly coupled new physics.

4. FINITE WIDTH EFFECTS AND GAUGE INVARIANCE

Some of the features of anomalous couplings, namely form factors and the necessity to consider the full S-matrix elements can nicely be illustrated by some very non-anomalous physics, namely fermion loop corrections within the SM. At the same time I would like to address the problem of how to implement finite width effects while maintaining gauge invariance when dealing with processes involving TGV's (15). The discussion will closely follow Ref. (16).

Let us consider $W\gamma$ production at hadron colliders as an example. Denoting the photon polarization vector by $\varepsilon^{*\mu}$ we can write the amplitude as

$$\mathcal{M} = \varepsilon_\mu^* \mathcal{M}^\mu = \varepsilon_\mu^* \mathcal{M}_q^\mu + \varepsilon_\mu^* \frac{1}{\hat{s} - m_W^2}\mathcal{M}_W^\mu , \tag{19}$$

where $\mathcal{M}_q$ denotes t- and u-channel quark exchange graphs and $\mathcal{M}_W$ stands for the s-channel W exchange graph which involves the TGV. Electromagnetic gauge invariance is guaranteed by the relation

$$k_\mu \mathcal{M}^\mu = k_\mu \mathcal{M}_q^\mu + k_\mu \frac{1}{\hat{s} - m_W^2}\mathcal{M}_W^\mu = 0 . \tag{20}$$

Replacing the W propagator factor by a Breit-Wigner form, $1/(\hat{s} - m_W^2 + im_W\Gamma_W)$, disturbs the gauge cancellations between the individual Feynman graphs and thus leads to an amplitude which is not electromagnetically gauge invariant. In addition, a constant imaginary part in the inverse propagator is ad hoc: it results from fermion loop contributions to the W vacuum polarization and the imaginary part should vanish for space-like momentum transfers.

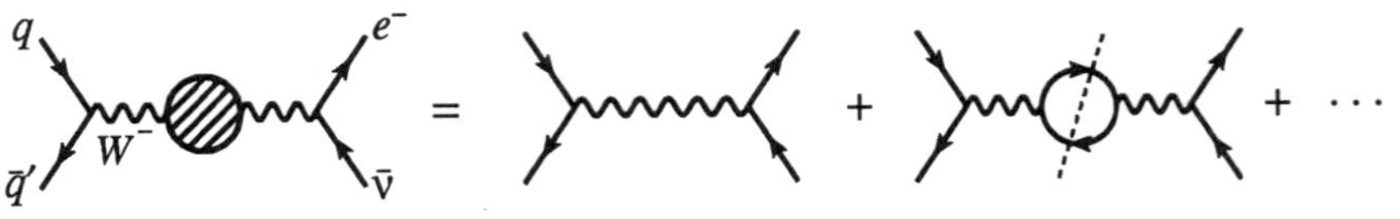

FIG. 5. Feynman graphs for the process $q\bar{q}' \to \ell^-\bar{\nu}$ at lowest order. The resummation of the imaginary part of the W vacuum polarization leads to the Breit-Wigner type W propagator of Eq. 22 which is represented by the shaded blob.

The general structure is best understood by first considering the lower order process $q\bar{q}' \to W^- \to \ell^-\bar{\nu}$ without photon emission (see Fig. 5). Finite width effects are included by resumming the imaginary parts of the fermion loops. Neglecting fermion masses, the transverse part of the W vacuum polarization receives an imaginary contribution

$$Im\,\Pi_W^T(q^2) = \sum_f \frac{g^2}{48\pi}q^2 = q^2\frac{\Gamma_W}{m_W}, \tag{21}$$

while the imaginary part of the longitudinal piece vanishes. In the unitary gauge and for $q^2 > 0$ the W propagator is thus given by

$$D_W^{\mu\nu}(q) = \frac{-i}{q^2 - m_W^2 + iIm\,\Pi_W^T(q^2)}\left(g^{\mu\nu} - \frac{q^\mu q^\nu}{q^2}\right) + \frac{i}{m_W^2 - iIm\,\Pi_W^L(q^2)}\frac{q^\mu q^\nu}{q^2}$$

$$= \frac{-i}{q^2 - m_W^2 + iq^2\gamma_W}\left(g^{\mu\nu} - \frac{q^\mu q^\nu}{m_W^2}(1+i\gamma_W)\right), \tag{22}$$

where the abbreviation $\gamma_W = \Gamma_W/m_W$ has been used. Note that the W propagator has received a q^2 dependent effective width which actually would vanish in the space-like region.

Now consider the same process, but including photon emission. A gauge invariant expression is obtained by attaching the final state photon in all possible ways to all charged particle propagators in the Feynman graphs of Fig. 5. This includes radiation off the two incoming quark lines, radiation off the final state charged lepton, and radiation off the W propagators. In addition, the photon must be attached to the charged fermions inside the W vacuum polarization loops, leading to the fermion triangle graphs of Fig. 6. For a consistent treatment we only need to include the imaginary part of the triangle graphs which is obtained by cutting the triangle graphs into on-shell intermediate states in all possible ways, as shown in the figure.

For the momentum flow of Fig. 6 the lowest order vertex is given by the familiar expression

$$-ie\Gamma_0^{\alpha\beta\mu} = -ie\left((q_1+q_2)^\mu g^{\alpha\beta} - (q_1+k)^\beta g^{\mu\alpha} + (k-q_2)^\alpha g^{\mu\beta}\right). \tag{23}$$

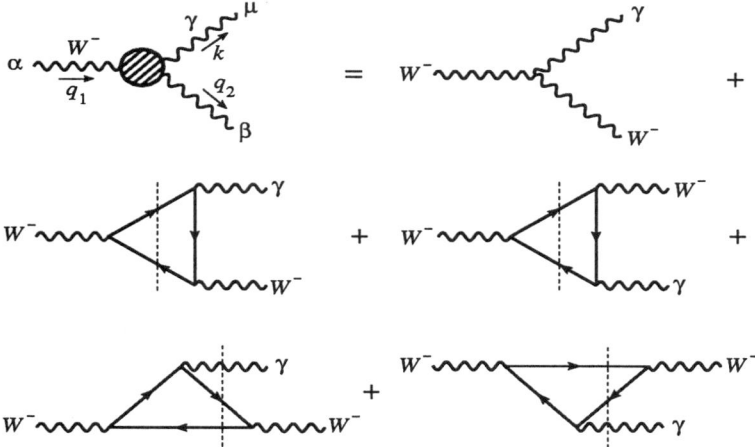

FIG. 6. Effective $WW\gamma$ vertex as needed in the tree level calculation of $W\gamma$ production. In addition to the lowest order vertex the imaginary parts of the fermion triangles must be included (see Eq. 24).

Neglecting the masses of the fermions in the triangle graphs and dropping terms proportional to k^μ (which will be contracted with the photon polarization vector $\varepsilon^{*\mu}$ and hence vanish in the amplitude) the contributions from the four triangle graphs reduce to an extremely simple form. Each fermion doublet f, irrespective of its hypercharge, adds $i(g^2/48\pi)\Gamma_0$ to the lowest order $WW\gamma$ vertex Γ_0. After summing over all fermion species, the lowest order vertex is thus replaced by

$$\Gamma^{\alpha\beta\mu} = \Gamma_0^{\alpha\beta\mu}\left(1 + \sum_f \frac{ig^2}{48\pi}\right) = \Gamma_0^{\alpha\beta\mu}\left(1 + i\frac{\Gamma_W}{m_W}\right) = \Gamma_0^{\alpha\beta\mu}(1 + i\gamma_W) \; . \quad (24)$$

By construction, the resulting amplitude for the process $q\bar{q}' \to \ell^-\bar{\nu}\gamma$ is gauge invariant. Indeed, gauge invariance of the full amplitude can be traced to the electromagnetic Ward identity (17)

$$k_\mu \Gamma_{\alpha\beta}{}^\mu = (iD_W)^{-1}_{\alpha\beta}(q_1) - (iD_W)^{-1}_{\alpha\beta}(q_2) \; . \quad (25)$$

Since

$$k_\mu \Gamma^{\alpha\beta\mu} = \left((q_1^2 g^{\alpha\beta} - q_1^\alpha q_1^\beta) - (q_2^2 g^{\alpha\beta} - q_2^\alpha q_2^\beta)\right)(1 + i\gamma_W) \; , \quad (26)$$

and

$$(iD_W)^{-1}_{\alpha\beta}(q) = (q^2 - m_W^2 + iq^2\gamma_W)\left(g_{\alpha\beta} - \frac{q_\alpha q_\beta}{q^2}\right) - m_W^2 \frac{q_\alpha q_\beta}{q^2} \; , \quad (27)$$

this Ward identity is satisfied by our W propagator and $WW\gamma$ vertex.

The modification of the lowest order $WW\gamma$ vertex in Eq. 24 looks like the introduction of anomalous couplings $g_1^\gamma = \kappa_\gamma = 1 + i\gamma_W$ and one may thus worry that the full amplitude will violate unitarity at large center of mass energies $\sqrt{\hat{s}}$. While indeed the vertex is modified, this modification is compensated by the effective $\hat{s}$-dependent width in the propagator. As compared to the expressions with a lowest order propagator, $1/(\hat{s} - m_W^2)$, which of course has good high energy behaviour, the overall effect is multiplication of the s-channel W-exchange amplitude $\mathcal{M}_W$ by a factor

$$G(\hat{s}) = \frac{\hat{s} - m_W^2}{\hat{s}(1 + i\gamma_W) - m_W^2}(1 + i\gamma_W) = 1 - \frac{i\Gamma_W m_W}{\hat{s} - m_W^2 + im_W \Gamma_W \frac{\hat{s}}{m_W^2}}. \quad (28)$$

Obviously, $G(\hat{s}) \to 1$ as $\hat{s} \to \infty$ and the high energy behaviour of our finite width amplitude is identical to the one of the naive tree level result for $W\gamma$ production. In fact, the contributions from the triangle graphs are crucial to compensate the bad high energy behaviour introduced by the q^2-dependent width in Eq. 22.

This interplay of propagator and vertex corrections illustrates the remarks made in Section 2. The leading one-loop contributions, namely the imaginary parts of $WW\gamma$ vertex and inverse W propagator, lead to a change of the S-matrix element for $W\gamma$ production which can be parameterized in terms of the generalized vertex function $\Gamma_{\gamma WW}^{\mu\alpha\beta}(k, q_1, q_2)$. The nonvanishing form factors in its tensor decomposition are given by

$$g_1^\gamma(q_1^2) = \kappa_\gamma(q_1^2) = G(q_1^2) = 1 - \frac{i\Gamma_W m_W}{q_1^2 - m_W^2 + i\frac{\Gamma_W}{m_W}q_1^2}, \quad (29)$$

and the form factor scale is set by the masses of the particles involved, here the W boson mass.

5. CONCLUSIONS

The direct measurement of the nonabelian $WW\gamma$ and WWZ vertices at present and future hadron and e^+e^- colliders constitutes an important test of the basic structure of electroweak interactions. There are strong theoretical arguments that experiments will yield exactly the results predicted by the SM, even though no rigorous proof of this assertion exists. Observation of anomalous couplings at either the Tevatron or at LEP II would therefore have grave consequences for our understanding of electroweak physics.

Irrespective of how likely an observation of anomalous couplings might be, e^+e- and hadron colliders measure very different aspects of WWV vertex functions. With sufficient statistics e^+e^- experiments are able to probe small deviations from SM cross sections and are hence sensitive to weakly interacting new physics. However, they will mainly probe just one process, W pair production, in a limited energy range. Hadron colliders, on the other hand, can investigate all electroweak boson pair production processes, albeit with lower accuracy. They look for relatively large enhancements of cross sections at high center of mass energies and are thus only sensitive to new strong interaction dynamics in the bosonic sector. Hadron and e^+e^- machines are indeed complementary means to directly study the nonabelian aspects of electroweak interactions.

REFERENCES

1. D. Schaile, these proceedings.
2. H. Aihara et al., report of the DPF study subgroup on *Anomalous Gauge Boson Interactions*, report MAD/PH/871 (1995) (hep-ph/9503425) and references therein.
3. P. Hernandez, these proceedings.
4. K. Gaemers and G. Gounaris, Z. Phys. **C1**, 259 (1979).
5. K. Hagiwara, K. Hikasa, R.D. Peccei, and D. Zeppenfeld, Nucl. Phys. **B282**, 253 (1987).
6. J. Wudka, these proceedings.
7. M. Suzuki, Phys. Lett. **B153**, 289 (1985); C. Bilchak, M. Kuroda and D. Schildknecht, Nucl. Phys. **B299**, 7 (1988); U. Baur and D. Zeppenfeld, Phys. Lett. **201B**, 383 (1988); U. Baur and E. Berger, Phys. Rev. **D47**, 4889 (1993).
8. See the contributions by H. Aihara, T. Barklow, J. Busenitz, T. Fuess, S. Godfrey, T. Han, H. Johari, G. Landsberg, D. Neuberger, A. Schöning, G. Valencia, R. Wagner, R. Walczak, C. Wendt, J. Womersley, and L. Zhang in these proceedings.
9. J. M. Cornwall, D. N. Levin, and G. Tiktopoulos, Phys. Rev. Lett. **30**, 1268 (1973), Phys. Rev. **D10**, 1145 (1974); C. H. Llewellyn Smith, Phys. Lett. **46B**, 233 (1973); S. D. Joglekar, Ann. Phys. **83** (1974) 427.
10. See e.g. M. Bilenkii, J. L. Kneur, F. M. Renard and D. Schildknecht, Nucl. Phys. **B409**, 22 (1993), *ibid.* **B419**, 240 (1994); R. L. Sekulin, Phys. Lett. **B338**, 369 (1994); M. Diehl and O. Nachtmann, Z. Phys. **C62**, 397 (1994); J. Busenitz; S. Godfrey, these proceedings.
11. T. Barklow, these proceedings.
12. J. Ohnemus, these proceedings.
13. A. Lahanas, these proceedings.
14. C. Arzt, M. B. Einhorn, and J. Wudka, Phys. Rev. **D49**, 1370 (1994); Nucl. Phys. **B433**, 41 (1995); M. B. Einhorn and J. Wudka, preprints NSF-ITP-92-01 (1992) and UM-TH-92-25 (1992); J. Wudka, Int. J. Mod. Phys. **A9**, 2301 (1994).
15. See e.g. A. Aeppli, F. Cuypers and G. J. van Oldenborgh, Phys. Lett. **B314**, 413 (1993).
16. U. Baur and D. Zeppenfeld, report MAD/PH/878 (1995), (hep-ph/9503344).
17. See e.g. G. Lopez Castro, J.L.M. Lucio, and J. Pestieau, Mod. Phys. Lett. **A6**, 3679 (1991); M. Nowakowski and A. Pilaftsis, Z. Phys. **C60**, 121 (1993) and references therein.

QCD Corrections to Diboson Production

J. Ohnemus

Department of Physics
University of California
Davis, CA 95616

The QCD radiative corrections to hadronic diboson production are reviewed. The radiative corrections for $W^\pm\gamma$, $Z\gamma$, ZZ, W^+W^-, and $W^\pm Z$ are discussed. Similarities and differences in the behavior of the order α_s cross sections for these processes are emphasized.

INTRODUCTION

The production of weak boson pairs is an important topic to study at hadron colliders because these processes can be used to test the standard model (SM) as well as probe beyond it (1). Diboson production is important for the following reasons.

- The $W^\pm\gamma$, $W^\pm Z$, and W^+W^- processes can be used to test the trilinear $WW\gamma$ and WWZ couplings. These couplings are completely fixed by the $SU(2) \otimes U(1)$ gauge structure of the SM, thus measurements of these couplings provide stringent tests of the SM. Remarkable progress has recently been made in measuring these couplings at the Fermilab Tevatron collider (2).

- The electroweak symmetry breaking (EWSB) mechanism can be probed by studying weak boson pair production. The EWSB mechanism is unknown, but it is believed that either there exists a scalar particle with mass $m \lesssim 1$ TeV or else the longitudinal components of the W and Z bosons become strongly interacting for parton center-of-mass energies larger that about 1 TeV (3). For example, the observation of resonance production of ZZ, W^+W^-, or $\gamma\gamma$ would be a signal for the standard model Higgs boson, whereas enhanced production of longitudinally polarized W and Z pairs would be evidence for a strongly interacting EWSB scenario.

- Diboson production is a potential background to new physics. New heavy particles, such as H^0, $H^\pm$, ρ_{TC}, η_{TC}, W', Z', $\tilde{q}$, and $\tilde{g}$ can decay into weak boson pairs.

In order to test and probe the SM with hadronic diboson production, it is necessary to have precise calculations of SM diboson production, which means the cross sections must be calculated to next-to-leading-order (NLO). The NLO cross section is, in general, less sensitive to the choices of the arbitrary factorization and renormalization scales.

The results described here are based on complete $\mathcal{O}(\alpha_s)$ calculations of the processes $p\overset{(-)}{p} \to V_1 V_2 + X$ where $V_i = W, Z, \gamma$ (4). The calculations also include the leptonic decays of the W and Z bosons (5,6). This is an important feature to include since the W and Z bosons are observed experimentally via their leptonic decay products. It is therefore important to include the experimental cuts on the decay leptons when comparing a theoretical calculation to the experimental data.

The calculations have been done using a combination of analytic and Monte Carlo integration techniques. Among the advantages of this formalism are:

- It is easy to impose cuts in the calculation.

- It is possible to calculate any number of observables simultaneously by simply histogramming the quantity of interest.

- It is possible to calculate not only the NLO inclusive cross section, but also the 0-jet and 1-jet exclusive cross sections.

Details of the formalism can be found in the original references (4–6).

THE $Z\gamma$ AND $W\gamma$ PROCESSES

The first processes to be considered are the $Z\gamma$ and $W\gamma$ processes. The total LO and NLO cross sections for these processes are plotted as functions of the center of mass energy in Fig. 1. The difference between the NLO and LO curves is the $\mathcal{O}(\alpha_s)$ correction. In the $Z\gamma$ process, the $\mathcal{O}(\alpha_s)$ corrections range from 10% to 30% over the domain of $\sqrt{s}$. This is what one naively expects since α_s is of order 0.10. In the $W\gamma$ process, on the other hand, the corrections range from 20% at small $\sqrt{s}$ to a surprising 300% at large $\sqrt{s}$.

In order to understand the large $\mathcal{O}(\alpha_s)$ corrections in the $W\gamma$ process, it is instructive to compare the behavior of the $2 \to 2$ and $2 \to 3$ processes for $Z\gamma$ and $W\gamma$ production. Figure 2(a) compares the $2 \to 2$ cross sections. Normally, hadronic W production is about twice as large as hadronic Z production because the W-to-quark coupling is about twice as big as the Z-to-quark coupling. However, for the $W\gamma$ and $Z\gamma$ processes, exactly the opposite behavior is seen; the $W\gamma$ cross section is only half as big as the $Z\gamma$ cross section. The $W\gamma$ cross section is smaller because it is suppressed by a radiation amplitude zero (RAZ) (7). Delicate cancellations in the $W^\pm\gamma$ amplitude cause it to vanish at $\cos\theta^* = \pm\frac{1}{3}$ where θ^* is the parton center-of-mass scattering angle.

The $2 \to 3$ cross sections for $W\gamma$ and $Z\gamma$ are compared in Fig. 2(b). Here a jet is defined as a final state quark or gluon with transverse momentum $p_T > 50$ GeV and pseudorapidity $|\eta| < 3$. The cross sections have been decomposed into contributions from qg and $q\bar{q}$ initial states (qg also includes $\bar{q}g$). The $qg \to W\gamma + 1$ jet cross section is about twice as big as the $qg \to$

$Z\gamma + 1$ jet cross section, as naively expected. (The $qg \to W\gamma q$ subprocess does not have a RAZ.) The $q\bar{q} \to W\gamma + 1$ jet and $q\bar{q} \to Z\gamma + 1$ jet cross sections, on the other hand, are nearly equal, indicating that the former is still suppressed relative to the later. (The $q\bar{q} \to W\gamma g$ subprocess has a RAZ in the limit $E_g \to 0$.)

In summary, the $2 \to 2$ $W\gamma$ cross section is suppressed relative to the $2 \to 2$ $Z\gamma$ cross section by a RAZ, while the $2 \to 3$ $W\gamma$ cross section is larger than the $2 \to 3$ $Z\gamma$ cross section due to the larger W-to-quark coupling. The net result of these two behaviors is that the $\mathcal{O}(\alpha_s)$ corrections are much larger for $W\gamma$ production than for $Z\gamma$ production.

Figure 3 again shows the total $Z\gamma$ and $W\gamma$ cross sections versus $\sqrt{s}$, but now the NLO cross sections have been decomposed into the Born cross sections and $\mathcal{O}(\alpha_s)$ corrections from $q\bar{q}$ and qg initial states. This decomposition shows that the $\mathcal{O}(\alpha_s)$ $q\bar{q}$ corrections tend to be proportional to the Born cross section, whereas the $\mathcal{O}(\alpha_s)$ qg corrections increase rapidly with $\sqrt{s}$. The $\mathcal{O}(\alpha_s)$ qg corrections increase with $\sqrt{s}$ because the gluon density increases with $\sqrt{s}$.

Figure 4 shows the $p_T(\gamma)$ spectra for $Z\gamma$ and $W^+\gamma$ production at the Large Hadron Collider (LHC) center of mass energy ($\sqrt{s} = 14$ TeV). The figure shows that the NLO corrections increase with $p_T(\gamma)$. This behavior is common to all the diboson processes; the NLO corrections increase with the p_T of the boson.

The rapidity distribution of the photon in the diboson rest frame is shown in Fig. 5 for the Tevatron center of mass energy ($\sqrt{s} = 1.8$ TeV). For the $Z\gamma$ process, the distribution exhibits the usual bell-shaped rapidity distribution, however, for the $W\gamma$ process, the distribution has a pronounced dip in the central rapidity region. This dip is due to the RAZ in the $W\gamma$ process. At the Tevatron energy, the NLO corrections slightly fill the dip, but do not obscure it. Figure 6 shows the photon rapidity distribution at the LHC energy. The NLO corrections are now very large in the $W\gamma$ process and they completely fill the dip in the central rapidity region. It may still be possible, however, to observe the dip in the $W\gamma + 1$ jet exclusive cross section (6).

Figure 7 compares the $p_T(\gamma)$ spectra for the $Z\gamma$ and $W\gamma$ processes at the Tevatron energy. This comparison shows that at high $p_T(\gamma)$, the $W\gamma$ distribution falls more rapidly than the $Z\gamma$ distribution. This behavior is also due to the RAZ in the $W\gamma$ process.

THE ZZ, W^+W^-, AND WZ PROCESSES

Attention now turns to the ZZ, W^+W^-, and WZ processes. The transverse momentum distributions for these processes are shown in Fig. 8. The figure shows that the NLO corrections increase with the p_T of the weak boson and are quite large at high values of p_T. Also note that the NLO corrections increase in the order ZZ, W^+W^-, WZ. This behavior will be discussed later.

Figure 9 again shows the p_T spectra of the weak bosons, but now the 0-jet and 1-jet exclusive components of the NLO inclusive cross section are also shown. (The 0-jet and 1-jet exclusive cross sections sum to the NLO inclusive cross section.) This decomposition shows that the bulk of the large corrections at high p_T are due to events containing a hard jet in the final state. The jet definition used here is $p_T(jet) > 50$ GeV and $|\eta(jet)| < 3$.

The large enhancements to the cross section at high p_T can be traced to collinear splittings in diagrams such as $qg \to Zq$ followed by $q \to qW$; the Z and the quark are produced with high p_T and the quark subsequently radiates a nearly collinear W. In the collinear limit, the $qg \to WZq$ subprocess can be approximated by (8)

$$d\sigma(qg \to WZq) \approx d\sigma(qg \to Zq) \frac{g^2}{16\pi^2} \log^2\left(\frac{p_T^2(Z)}{M_W^2}\right). \quad (1)$$

Figure 10 compares this collinear approximation to the full NLO calculation and shows that the approximation describes well the shape of the p_T distribution at high p_T.

The scale dependance of the total WZ cross section is illustrated in Fig. 11. A common scale Q has been used for both the renormalization scale μ and the factorization scale M. The Born and NLO inclusive cross sections are shown along with the 0-jet and 1-jet components of the NLO inclusive cross section. The 1-jet cross section is a LO quantity and thus has considerable scale dependence. The 0-jet cross section, on the other hand, is a NLO quantity and exhibits little scale dependence. The decomposition shows that the scale dependance of the NLO inclusive cross section is dominated by the scale dependence of the 1-jet component.

Figure 12 compares the p_T spectra of the weak bosons for the ZZ, W^+W^-, and WZ processes. The ZZ and W^+W^- distributions have the same shape at high p_T and are parallel to one another, whereas the WZ distribution falls more rapidly. A similar behavior was observed earlier in Fig. 7 where the $Z\gamma$ and $W\gamma$ processes were compared. In the present case, the WZ p_T spectrum falls faster than the ZZ and W^+W^- spectra because of an approximate amplitude zero (9) in the WZ process.

Approximate Amplitude Zero

The $q_1\bar{q}_2 \to WZ$ subprocess is very similar to the $q_1\bar{q}_2 \to W\gamma$ subprocess, in fact, they are described by the same set of Feynman diagrams, with Z and γ interchanged. Recall that the RAZ in the $W\gamma$ process gave rise to a large $\mathcal{O}(\alpha_s)$ correction. A difference between the two processes is that whereas the $W^\pm\gamma$ process has an exact amplitude zero at $\cos\theta^* = \pm\frac{1}{3}$, the $W^\pm Z$ process has only an approximate amplitude zero at $\cos\theta^* = \pm 0.1$. Basically, what happens in the WZ case is that the dominant helicity amplitudes have an exact zero, while the other helicity amplitudes remain finite but small. The approximate amplitude zero in the WZ process causes the NLO corrections to be larger than they were in either the ZZ or W^+W^- processes. The approximate amplitude zero suppresses the WZ Born cross section and thus makes the NLO corrections appear large. A more in depth discussion of approximate amplitude zeros can be found in the talk by T. Han (10).

SUMMARY

The QCD radiative corrections to weak boson pair production at hadron colliders has been reviewed. The $\mathcal{O}(\alpha_s)$ cross sections for the diboson combinations $Z\gamma$, $W\gamma$, ZZ, W^+W^-, and WZ have been discussed and compared. Some general features of the $\mathcal{O}(\alpha_s)$ cross sections are summarized here.

- The NLO corrections increase with the center-of-mass energy. This is due to the opening of the $qg \to V_1V_2q$ subprocess at $\mathcal{O}(\alpha_s)$ in conjunction with the gluon density which increases with the center-of-mass energy.

- The NLO corrections are largest at high $p_T(V)$. This is due to collinear splittings in the $qg \to V_1V_2q$ subprocesses which give rise to an enhancement factor $\log^2(p_T^2(V_1)/M_2^2)$.

- The bulk of the large corrections at high $p_T(V)$ come from events which contain a hard jet in the final state.

- p_T distributions are most affected by the NLO corrections. These distributions tend to be enhanced at large values of p_T.

- Invariant mass and angular distributions under go relatively little change in shape at NLO, instead, these distributions tend to be scaled up uniformly.

- The NLO corrections to $W\gamma$ production are large due to a radiation amplitude zero.

- The NLO corrections to WZ production are large due to an approximate amplitude zero.

- The NLO corrections are modest at the Tevatron center of mass energy but are significant at the LHC energy.

REFERENCES

1. R.W. Brown and K.O. Mikaelian, Phys. Rev. D **19**, 922 (1979); R.W. Brown, K.O. Mikaelian, and D. Sahdev, *ibid.* **20**, 1164 (1979).
2. See talks by H. Aihara, T. Feuss, C. Wendt, H. Johari, L. Zhang, G. Landsberg, B. Wagner, and D. Neuberger in these proceedings.
3. D. Dicus and V. Mathur, Phys. Rev. D **7**, 3111 (1973); M. Veltman, Acta Phys. Pol. **B8**, 475 (1977); B.W. Lee, C. Quigg, and H. Thacker, Phys. Rev. D **16**, 1519 (1977); J. van der Bij and M. Veltman, Nucl. Phys. **B231**, 205 (1984); M. S. Chanowitz and M. K. Gaillard, Nucl. Phys. **B216**, 379 (1985).

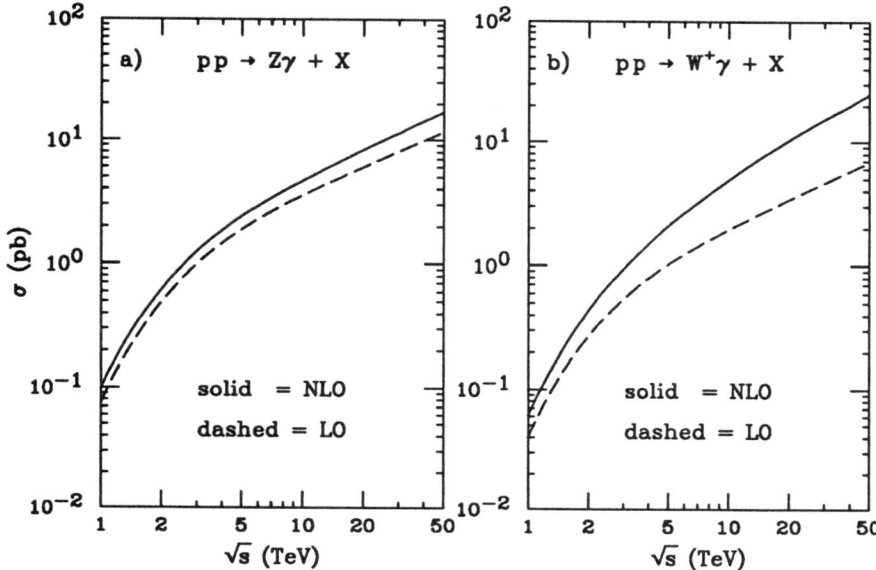

FIG. 1. Total cross section as a function of the center-of-mass energy for (a) $pp \to Z\gamma + X$ and (b) $pp \to W^+\gamma + X$. The LO and NLO cross sections are shown.

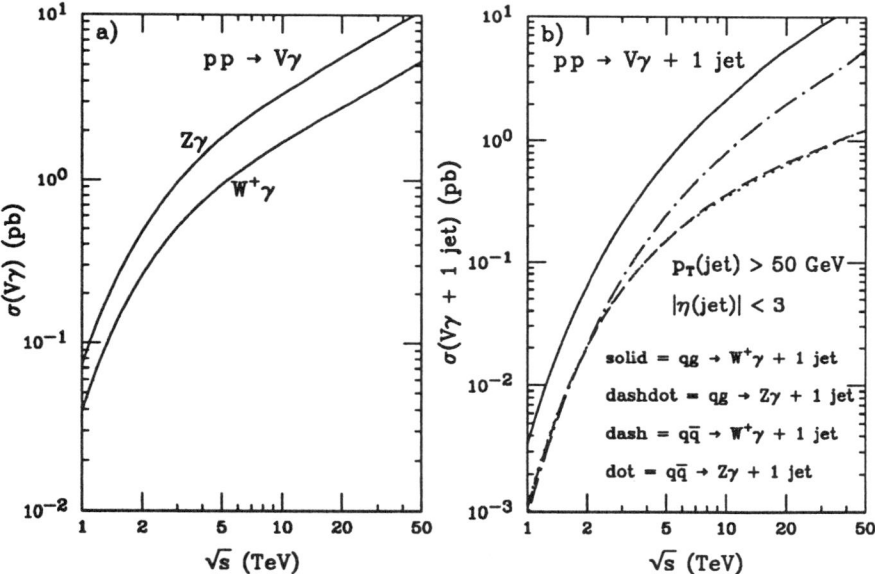

FIG. 2. (a) The $2 \to 2$ Born cross sections for $pp \to Z\gamma$ and $pp \to W^+\gamma$. (b) The $2 \to 3$ cross sections for $Z\gamma$ and $W^+\gamma$ production. The cross sections have been decomposed into contributions from $q\bar{q}$ and qg initial states.

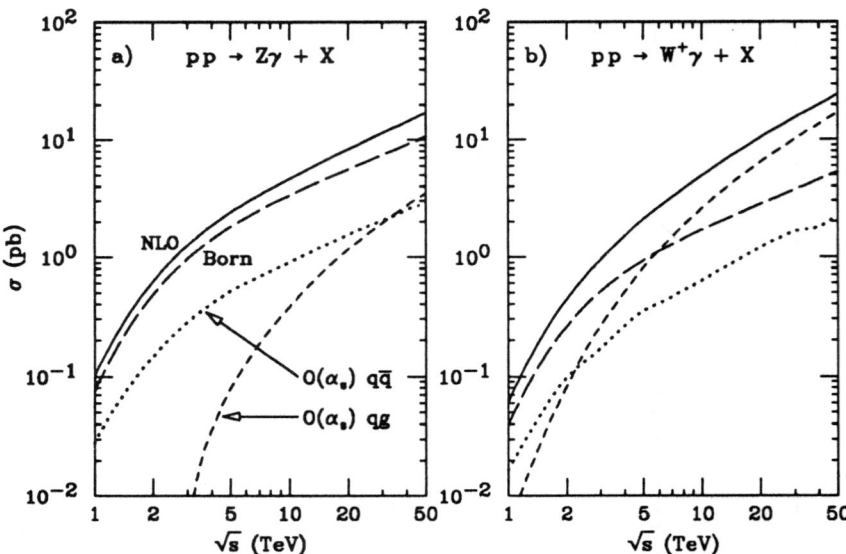

FIG. 3. Same as Fig. 1, but now the NLO cross section has been decomposed into the Born cross section and the order α_s corrections from $q\bar{q}$ and qg initial states.

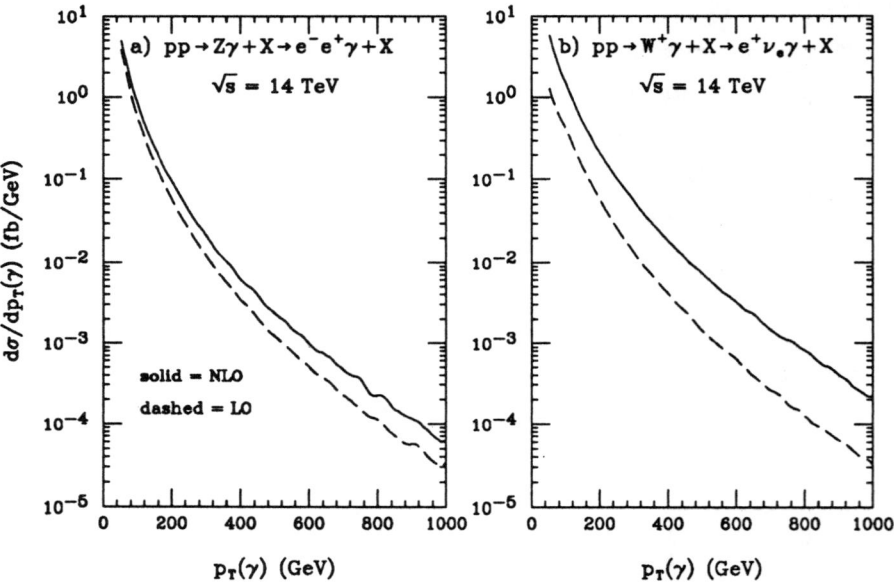

FIG. 4. Photon transverse momentum distributions at the LHC energy for (a) $pp \to Z\gamma + X \to e^-e^+\gamma + X$ and (b) $pp \to W^+\gamma + X \to e^+\nu_e\gamma + X$

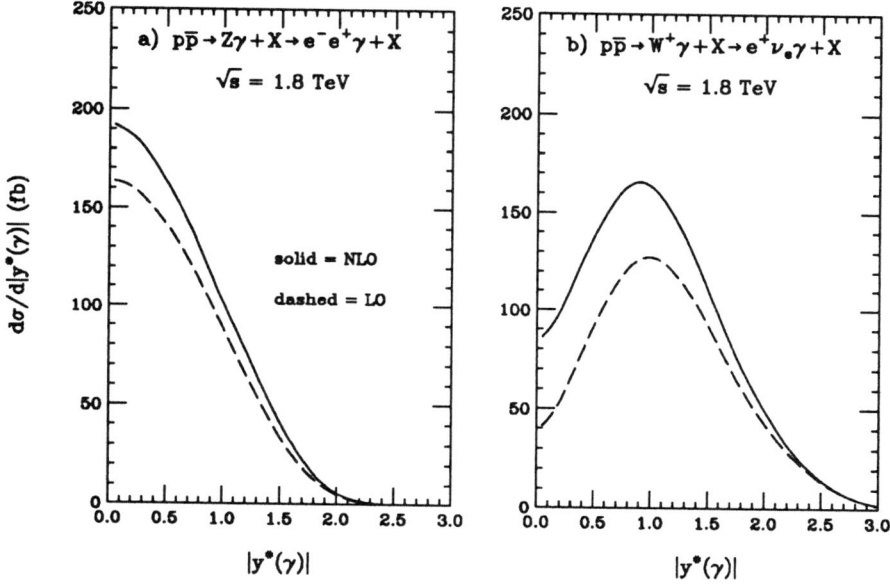

FIG. 5. Photon rapidity distributions in the diboson rest frame at the Tevatron energy for (a) $Z\gamma$ production and (b) $W^+\gamma$ production.

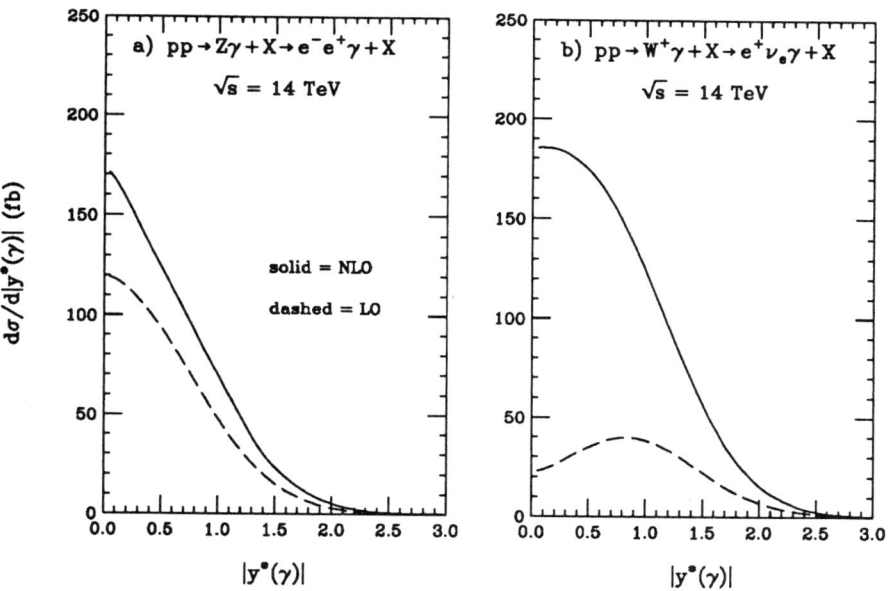

FIG. 6. Same as Fig. 5, but for the LHC energy.

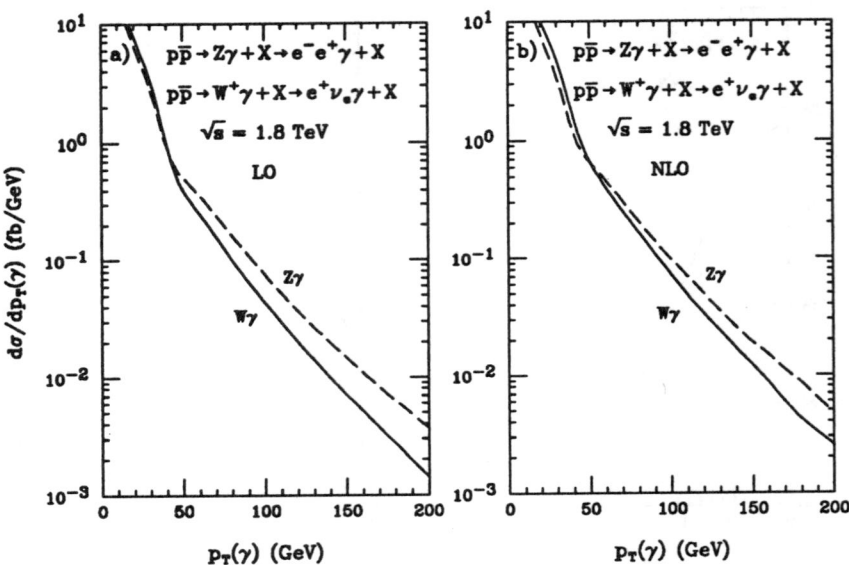

FIG. 7. Photon transverse momentum distributions for $Z\gamma$ and $W\gamma$ production at the Tevatron energy. Parts (a) and (b) are the LO and NLO cross sections, respectively.

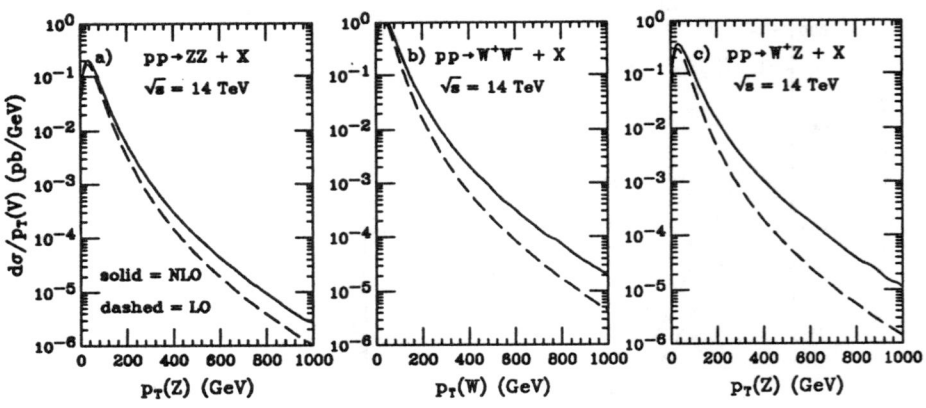

FIG. 8. Weak boson transverse momentum distributions for (a) ZZ, (b) W^+W^-, and (c) W^+Z production at the LHC energy.

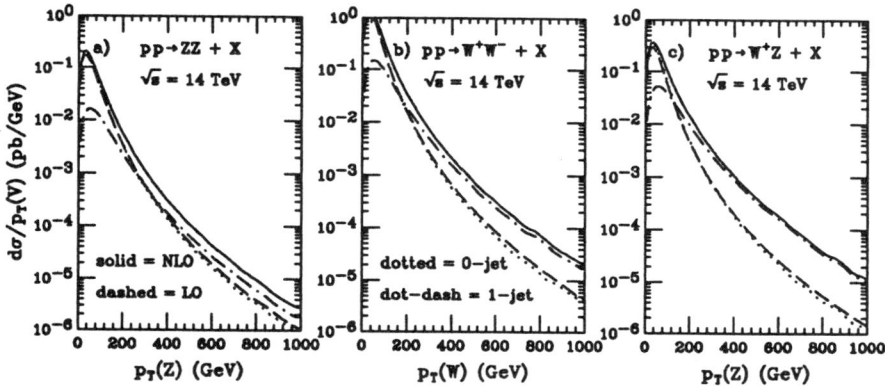

FIG. 9. Same as Fig. 8. but now the 0-jet and 1-jet exclusive components of the NLO inclusive cross section are also shown.

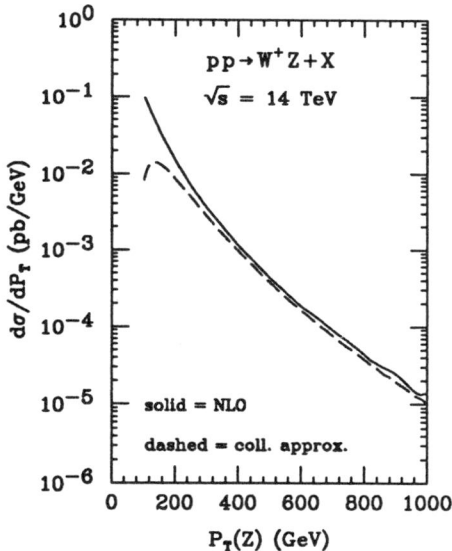

FIG. 10. The $p_T(Z)$ distribution for $pp \to W^+Z + X$ at the LHC energy. The full NLO cross section is compared to the cross section obtained from the collinear approximation given in Eq. (1).

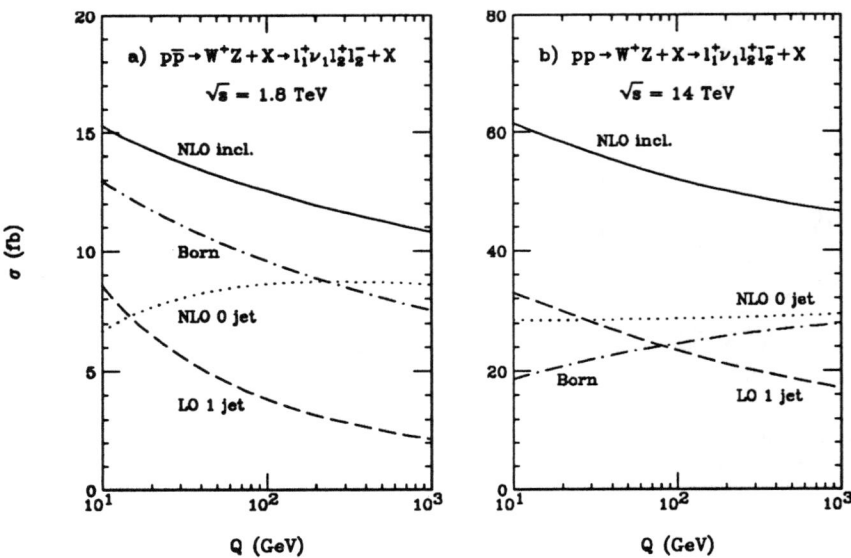

FIG. 11. Total cross section for W^+Z production as a function of the scale Q for (a) the Tevatron energy and (b) the LHC energy. The Born, NLO inclusive, 0-jet exclusive, and 1-jet exclusive cross sections are shown.

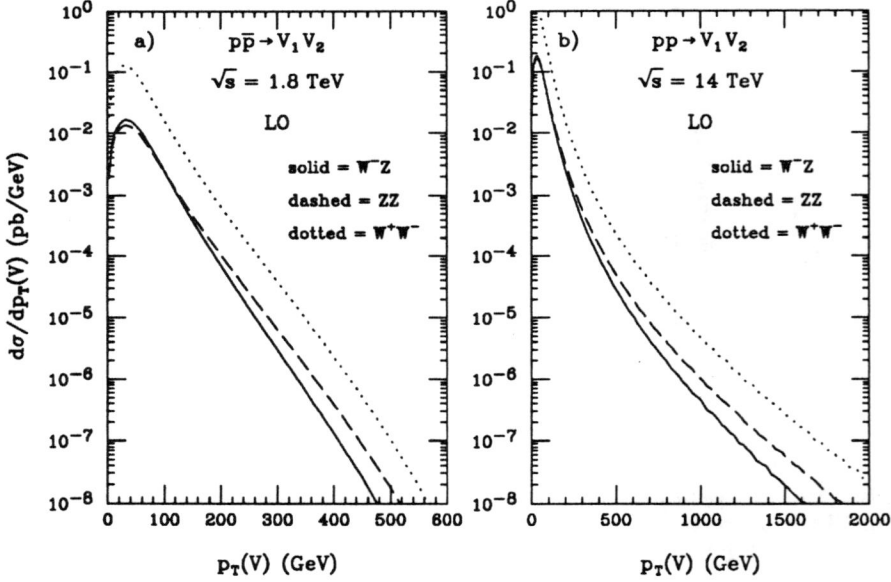

FIG. 12. The weak boson transverse momentum distributions at LO for ZZ, W^+W^-, and W^-Z production. Parts (a) and (b) are for the Tevatron and LHC energies, respectively.

4. J. Ohnemus and J.F. Owens, Phys. Rev. D **43**, 3626 (1991); J. Ohnemus, *ibid.* **44**, 1403 (1991); **44**, 3477 (1991); **47**, 940 (1993).
5. J. Ohnemus, Phys. Rev. D **50**, 1931 (1994); *ibid.* **51**, 1068 (1995).
6. U. Baur, T. Han, and J. Ohnemus, Phys. Rev. D **48**, 5140 (1993).
7. K.O. Mikaelian, M.A. Samuel, and D. Sahdev, Phys. Rev. Lett. **43**, 746 (1979); R.W. Brown, K.O. Mikaelian, and D. Sahdev Phys. Rev. D **20**, 1164 (1979); D. Zhu, Phys. Rev. D **22**, 2266 (1980); T.R. Grose and K.O. Mikaelian, Phys. Rev. D **23**, 123 (1981); C.J. Goebel, F. Halzen, and J.P. Leveille, Phys. Rev. D **23**, 2682 (1981); S.J. Brodsky and R.W. Brown, Phys. Rev. Lett. **49**, 966 (1982); M.A. Samuel, Phys. Rev. D **27**, 2724 (1983); R.W. Brown, K.L. Kowalski, and S.J. Brodsky, Phys. Rev. D **28**, 624 (1983); R.W. Brown and K.L. Kowalski, Phys. Rev. D **29**, 2100 (1984).
8. S. Frixione, P. Nason, and G. Ridolfi, Nucl. Phys. B **383**, 3 (1992).
9. U. Baur, T. Han, and J. Ohnemus, Phys. Rev. Lett. **72**, 3941 (1994).
10. T. Han, in these proceedings.

$W\gamma$ and $Z\gamma$ Production at Tevatron[1,2]

H. Aihara

Lawrence Berkeley Laboratory

We present results from CDF and DØ on $W\gamma$ and $Z\gamma$ productions in $p\bar{p}$ collisions at $\sqrt{s} = 1.8$ TeV. The goal of the analyses is to test the non-abelian self-couplings of the W, Z and photon, one of the most direct consequences of the $SU(2)_L \otimes U(1)_Y$ gauge symmetry. We present direct measurements of $WW\gamma$ couplings and limits on $ZZ\gamma$ and $Z\gamma\gamma$ couplings, based on $p\bar{p} \to \ell\nu\gamma + X$ and $p\bar{p} \to \ell\ell\gamma + X$ events, respectively, observed during the 1992–1993 run of the Fermilab Tevatron Collider.

INTRODUCTION

Direct measurement of the $WW\gamma$ gauge boson couplings is possible through study of $W\gamma$ production in $p\bar{p}$ collisions at $\sqrt{s} = 1.8$ TeV. The most general effective Lagrangian (1), invariant under $U(1)_{EM}$, for the $WW\gamma$ interaction contains four coupling parameters, CP-conserving κ and λ, and CP-violating $\tilde{\kappa}$ and $\tilde{\lambda}$. The CP–conserving parameters are related to the magnetic dipole (μ_W) and electric quadrupole (Q^e_W) moments of the W boson, while the CP–violating parameters are related to the electric dipole (d_W) and the magnetic quadrupole (Q^m_W) moments: $\mu_W = (e/2m_W)(1 + \kappa + \lambda)$, $Q^e_W = (-e/m_W^2)(\kappa - \lambda)$, $d_W = (e/2m_W)(\tilde{\kappa} + \tilde{\lambda})$, $Q^m_W = (-e/m_W^2)(\tilde{\kappa} - \tilde{\lambda})$ (2). In the Standard Model (SM) the $WW\gamma$ couplings at the tree level are uniquely determined by the $SU(2)_L \otimes U(1)_Y$ gauge symmetry: $\kappa = 1$ ($\Delta\kappa \equiv \kappa - 1 = 0$), $\lambda = 0$, $\tilde{\kappa} = 0$, $\tilde{\lambda} = 0$. The direct and precise measurement of the $WW\gamma$ couplings is of interest since the existence of anomalous couplings, i.e. measured values different from the SM predictions, would indicate the presence of physics beyond the SM. A $WW\gamma$ interaction Lagrangian with constant, anomalous couplings violates unitarity at high energies, and, therefore, the coupling parameters must be modified to include form factors (e.g. $\Delta\kappa(\hat{s}) = \Delta\kappa/(1 + \hat{s}/\Lambda_W^2)^n$, where $\hat{s}$ is the square of the invariant mass of the W and the photon, Λ_W is the form factor scale, and $n = 2$ for a dipole form factor) (3).

---

[1] Invited talk given at the International Symposium on Vector Boson Self-Interactions, UCLA, February 1-3, 1995.

[2] This work was supported by the Director, Office of Energy Research, Office of High Energy and Nuclear Physics, Division of High Energy Physics of the U.S. Department of Energy under Contract DE-AC03-76SF00098.

The study of the $Z\gamma$ production in $p\bar{p}$ collision is also an important test of the SM description of gauge-boson self-interactions. Since the photon does not couple directly to the Z in the SM, this study is sensitive to anomalous couplings beyond the SM. The most general $ZZ\gamma$ ($Z\gamma\gamma$) vertex function is characterized by a set of four coupling parameters $h_{1-4}^{Z(\gamma)}$ (1). All these coupling parameters vanish at tree level within the framework of the SM. The couplings h_3^V and h_4^V conserve CP, while h_1^V and h_2^V are CP-violating. Similarly to the $WW\gamma$ anomalous couplings, the $ZZ\gamma(Z\gamma\gamma)$ couplings must be regulated by generalized dipole form factors: $(h_i^V(\hat{s}) = h_{i0}^V/(1 + \hat{s}/\Lambda_Z^2)^n$, where h_{i0}^V represents the low energy ($\hat{s} = 0$) limit for the couplings, and $n = 3$ for $h_{1,3}^V$ and $n = 4$ for $h_{2,4}^V$. Here the values for n were chosen so that the unitarity is preserved and that all terms in the matrix element proportional to h_{i0}^V have the same asymptotic energy behavior. At the Tevatron, the $W\gamma$ production is insensitive to the form factor effects for $\Lambda_W >$ a few 100 GeV, whereas the form factor effects cannot be ignored for $Z\gamma$ production due to the higher power of $\hat{s}$ dependence in the $ZZ\gamma(Z\gamma\gamma)$ vertex function.

We present studies of the $WW\gamma$ and $ZZ\gamma(Z\gamma\gamma)$ couplings based on $p\bar{p} \to \ell\nu\gamma + X$ and $p\bar{p} \to \ell\ell\gamma$ ($\ell = e, \mu$) events observed with the CDF (4) and DØ detector (5) during the 1992–1993 run of the Fermilab Tevatron Collider, corresponding to integrated luminosities of ~ 20 pb^{-1} for CDF and ~ 14 pb^{-1} for DØ. The $\ell\nu\gamma$ events contain the $W\gamma$ production process, $p\bar{p} \to W\gamma + X$ followed by $W \to \ell\nu$, and the radiative $W \to \ell\nu\gamma$ decay where the photon originates from bremsstrahlung of the charged lepton. Anomalous coupling parameters enhance the $W\gamma$ production with a large $\hat{s}$, and thereby result in an excess of events with high transverse energy, E_T, photons, well separated from the charged lepton. The $\ell\ell\gamma$ events contain the radiative $Z \to \ell\ell\gamma$ decay, the direct $Z\gamma$ production where the photon is radiated from one of the annihilating quarks, and the possible $Z\gamma$ events due to the anomalous Z-γ couplings. The presence of the Z-γ couplings will also be signaled by an excess of Z production with high E_T photons.

PHOTON DETECTION AT CDF AND DØ

Since the good detection of the photon is the key to the $W\gamma$ and $Z\gamma$ measurements, we briefly review how photons are detected by the CDF and DØ detectors. A photon is identified as a calorimeter energy cluster satisfying the following condition. A calorimeter cluster must (i) have a high electromagnetic energy fraction; (ii) be isolated; (iii) have shower shape consistent with a single photon; and (iv) have no tracks pointing to it. Table 1 summarizes the actual conditions required by CDF and DØ.

To test shower shape of the cluster CDF uses the central electromagnetic strip chambers (6) (CES) placed after ~ 6.3 radiation lengths in the central electromagnetic calorimeter. The CES determines shower position and transverse development of an electromagnetic shower at shower maximum by mea-

TABLE 1. Summary of photon detection at CDF and DØ

	CDF	DØ
detection region	$|\eta| < 1.1$ $(1.1 < |\eta| < 2.4$ [a]$)$	$|\eta| < 1.1$ $1.5 < |\eta| < 2.5$
minimum E_T^γ	7 GeV	10 GeV
EM fraction	HAD/EM $< 0.055 + 0.00045 \times E(GeV)$	$EM/Total > 0.9$
Isolation	$(E_T(0.4) - E_T^\gamma)/E_T^\gamma < 0.15$ [b] $p_T(0.4) < 2$ GeV/c	$(E(0.4) - EM(0.2))/EM(0.2) < 0.10$ [c]
Shower shape	transverse	longitudinal/transverse
No track	No matching tracks	No matching tracks

[a] Analysis in progress.
[b] $E_T(0.4)$ is the E_T in a cone of $\Delta R = \sqrt{(\Delta\eta)^2 + (\Delta\phi)^2} = 0.4$ around the photon candidate. $p_T(0.4)$ is the sum of p_T of the charged tracks within the same cone.
[c] $E(0.4)$ is the total energy inside a cone of radius $\Delta R = 0.4$, and $EM(0.2)$ is the EM energy inside a cone of 0.2.

surement of the charge deposition on orthogonal, fine-grained (1.5 cm spacing) strips and wires. DØ tests both longitudinal and transverse shower shapes including correlations between energy deposits in the fine-grained calorimeter cells (7). The DØ electromagnetic calorimeter module has 4 longitudinal layers. Each of layers 1, 2 and 4 is segmented transversely to $\Delta\eta \times \Delta\phi = 0.1 \times 0.1$, while the third layer, which typically contains 65% of the EM energy, has segmentation of $\Delta\eta \times \Delta\phi = 0.05 \times 0.05$. ($\eta$ is the pseudorapidity defined as $\eta = -\ln(\tan(\theta/2))$, θ being the polar angle with repect to the beam axis. ϕ is the azimuthal angle.)

Both CDF and DØ found that the detection efficiency for photons depends on E_T^γ due to the isolation requirement. DØ found its cluster shape requirement also results in the E_T dependence. The overall photon detection efficiency was obtained by combining this E_T-dependent efficiency with the probabilities of losing a photon due to e^+e^- pair conversions and due to an overlap with a random track in the event. Table 2 summarizes the photon detection efficiencies at CDF and DØ.

TABLE 2. Summary of photon detection efficiency.

	CDF	DØ	
	$|\eta| < 1.1$	$|\eta| < 1.1$	$1.5 < |\eta| < 2.5$
$E_T^\gamma > 25$ GeV	0.804 ± 0.023	0.74 ± 0.07	0.58 ± 0.05
$= 10$		0.43 ± 0.04	0.38 ± 0.03
$= 7$	0.731 ± 0.021		

TABLE 3. Summary of $W\gamma$ event selection.

	CDF		DØ	
	$e\nu\gamma$	$\mu\nu\gamma$	$e\nu\gamma$	$\mu\nu\gamma$
Geometry	$|\eta_e| < 1.1$	$|\eta_\mu| < 0.6$	$|\eta_e| < 1.1$ $1.5 < |\eta_e| < 2.5$	$|\eta_\mu| < 1.7$
	$|\eta_\gamma| < 1.1$		$|\eta_\gamma| < 1.1, 1.5 < |\eta_\gamma| < 2.5$	
Kinematics (in GeV)	$E_T^e > 20$ $\not{E}_T > 20$ $E_T^\gamma > 7$	$p_T^\mu > 20$ $\not{E}_T > 20$	$E_T^e > 25$ $\not{E}_T > 25$ $E_T^\gamma > 10$	$p_T^\mu > 15$ $\not{E}_T > 15$
	$\Delta R_{\ell\gamma} > 0.7$		$\Delta R_{\ell\gamma} > 0.7$	
$\int L dt$ pb^{-1}	19.6 ± 0.7	18.6 ± 0.7	13.8 ± 0.7	13.7 ± 0.7

$W\gamma$ ANALYSIS

The $W\gamma$ candidates were obtained by searching for events containing an isolated lepton (e or μ) with high E_T, large missing transverse energy, $\not{E}_T$, and an isolated photon. Table 3 summarizes geometrical and kinematic selection as well as integrated luminosity used in each channel. Both CDF and DØ required that the separation between a photon and a lepton be $\Delta R_{\ell\gamma} > 0.7$. This requirement suppresses the contribution of the radiative W decay process. The CDF observed 18 $W(e\nu)\gamma$ candidates and 7 $W(\mu\nu)\gamma$ candidates (8), while the DØ observed 11 $W(e\nu)\gamma$ candidates and 12 $W(\mu\nu)\gamma$ candidates (9).

The background estimate, summarized in Table 4, includes contributions from: W+jets, where a jet is misidentified as a photon; $Z\gamma$, where the Z decays to $\ell^+\ell^-$, and one of the leptons is undetected or is mismeasured by the detector and contributes to $\not{E}_T$; $W\gamma$ with $W \to \tau\nu$ followed by $\tau \to \ell\nu\bar{\nu}$. The W+jets background was estimated using the probability, $\mathcal{P}(j \to ``\gamma")$, for a jet to be misidentified as a photon determined as a function of E_T of the jet by measuring the fraction of jets in a sample of multijet events that

TABLE 4. Summary of $W\gamma$ data and backgrounds.

	CDF		DØ	
	$e\nu\gamma$	$\mu\nu\gamma$	$e\nu\gamma$	$\mu\nu\gamma$
Source:				
W+jets	4.6 ± 1.8	1.9 ± 0.6	1.7 ± 0.9	1.3 ± 0.7
$Z\gamma$	0.43 ± 0.02	1.14 ± 0.06	0.11 ± 0.02	2.7 ± 0.8
$W(\tau\nu)\gamma$	0.29 ± 0.02	0.15 ± 0.01	0.17 ± 0.02	0.4 ± 0.1
Total background	5.3 ± 1.8	3.2 ± 0.6	2.0 ± 0.9	4.4 ± 1.1
Data	18	7	11	12
Signal	12.7 ± 4.6	3.8 ± 2.7	$9.0^{+4.2}_{-3.1} \pm 0.9$	$7.6^{+4.4}_{-3.2} \pm 1.1$

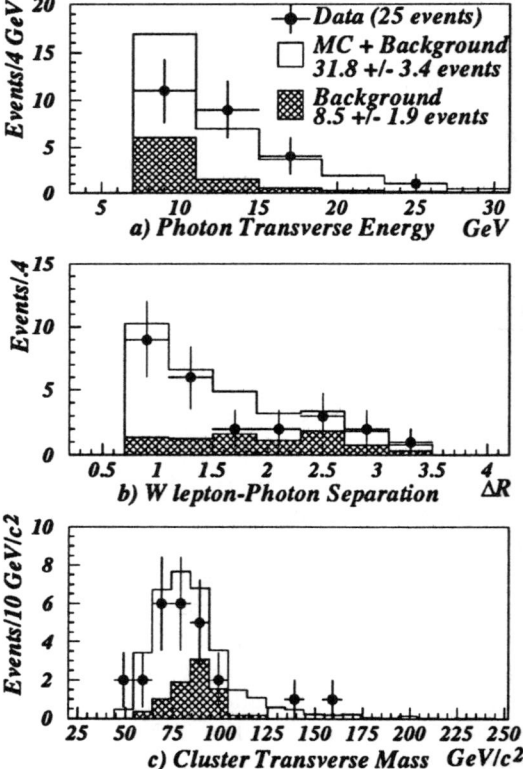

FIG. 1. CDF Distribution of (a) E_T^γ, (b) $\mathcal{R}_{\ell\gamma}$ and (c) $M_T(\gamma\ell;\nu)$ for the $W(e\nu)\gamma + W(\mu\nu)\gamma$ combined sample. The points are data. The shaded areas represent the estimated background, and the solid histograms are the expected signal from the Standard Model plus the estimated background.

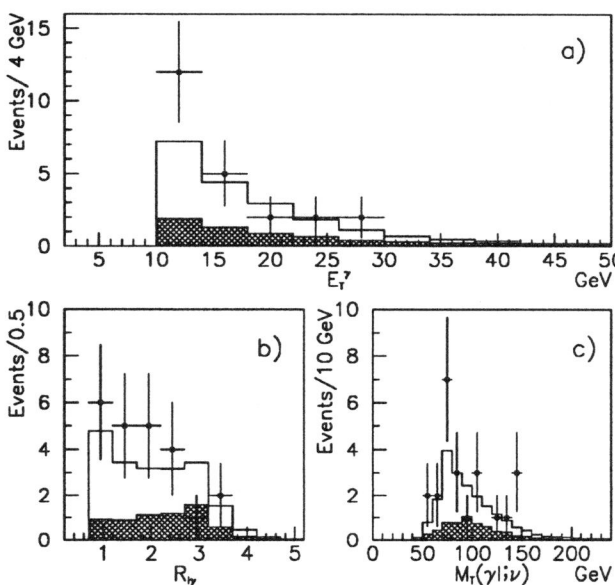

FIG. 2. DØ Distribution of (a) E_T^γ, (b) $\mathcal{R}_{\ell\gamma}$ and (c) $M_T(\gamma\ell;\nu)$ for the $W(e\nu)\gamma + W(\mu\nu)\gamma$ combined sample. The points are data. The shaded areas represent the estimated background, and the solid histograms are the expected signal from the Standard Model plus the estimated background.

TABLE 5. Comparison of data and the SM prediction for $W\gamma$.

	CDF		DØ	
	$e\nu\gamma$	$\mu\nu\gamma$	$e\nu\gamma$	$\mu\nu\gamma$
Signal	12.7 ± 4.6	3.8 ± 2.7	$9.0^{+4.2}_{-3.1} \pm 0.9$	$7.6^{+4.4}_{-3.2} \pm 1.1$
SM prediction	15.4 ± 0.7	7.9 ± 0.4	6.9 ± 1.0	6.7 ± 1.2
$\sigma_{W\gamma}$ ($E_T^\gamma > 7\text{GeV}, \Delta R_{\ell\gamma} > 0.7$) pb	141.7 ± 53	83 ± 59		
$\sigma_{W\gamma}$ ($E_T^\gamma > 10\text{GeV}, \Delta R_{\ell\gamma} > 0.7$) pb			147^{+73}_{-56}	127^{+78}_{-61}
$e + \mu$ combined	122 ± 42 pb		138^{+55}_{-43} pb	
SM prediction	172 ± 26 pb		112 ± 10 pb	

pass our photon identification requirements. For the photon criteria used by CDF, $\mathcal{P}(j \to \text{``}\gamma\text{''}) \sim 8 \times 10^{-4}$ at $E_T^j = 9$ GeV, decreasing exponentially to $\mathcal{P}(j \to \text{``}\gamma\text{''}) \sim 1 \times 10^{-4}$ at $E_T^j = 25$ GeV. For the photon criteria used by DØ, $\mathcal{P}(j \to \text{``}\gamma\text{''}) \sim 4 \times 10^{-4}$ ($6 \times 10^{-4}$) in the central (endcap) calorimeter, and varies only slowly with $E_T^j$. The total number of $W$+jets background events was calculated by applying $\mathcal{P}(j \to \text{``}\gamma\text{''})$ to the observed E_T spectrum of jets in the inclusive $W(\ell\nu)$ sample. The backgrounds due to $Z\gamma$ and $W \to \tau\nu$ were estimated from Monte Carlo simulations.

The kinematic and geometrical acceptance was calculated as a function of coupling parameters, $\Delta\kappa$ and λ, using the Monte Carlo program of Baur and Zeppenfeld, in which the $W\gamma$ production and radiative decay processes are generated to leading order, and higher order QCD effects are approximated by a K-factor. Both CDF and DØ used the MRSD_' structure functions and simulated the p_T distribution of the $W\gamma$ system using the observed p_T spectrum of the W in the inclusive $W(\ell\nu)$ sample. The generated events underwent a detector simulation. Table 5 shows the comparison between the observed signal and the SM prediction. CDF obtained the $W\gamma$ cross section for photons with $E_T^\gamma > 7$ GeV and $\Delta R_{\ell\gamma} > 0.7$ from a combined $e + \mu$ sample: $\sigma(W\gamma) = 122 \pm 42$ pb, while the SM predicts 172 ± 26 pb. DØ obtained $\sigma(W\gamma) = 138^{+55}_{-43}$ pb for photons with $E_T^\gamma > 10$ GeV and $\Delta R_{\ell\gamma} > 0.7$, and the SM predicts 112 ± 10 pb. Here we used BR$(W \to \ell\nu) = 0.108$. The observed cross section agrees with the SM prediction within errors.

Figures 1 and 2 show that data and the SM prediction plus the background in the distributions of E_T^γ, $\Delta R_{\ell\gamma}$, and the cluster transverse mass defined by $M_T(\gamma\ell;\nu) = (((m_{\gamma\ell}^2 + |\mathbf{E}_T^\gamma + \mathbf{E}_T^\ell|^2)^{\frac{1}{2}} + \not\!\!E_T)^2 - |\mathbf{E}_T^\gamma + \mathbf{E}_T^\ell + \not\!\!\mathbf{E}_T|^2)^{\frac{1}{2}}$. Of 25 events CDF observed, 16 events having $M_T(\gamma\ell;\nu) \leq M_W$ are primarily the radiative W decay events plus background. Similarly, of 23 events DØ observed, 11 events are primarily the radiative W decay events plus background. The absence of an excess of high E_T photons rules out deviations from the SM couplings.

To set limits on the anomalous coupling parameters, a binned maximum likelihood fit was performed on the E_T^γ spectrum for each of the $W(e\nu)\gamma$ and $W(\mu\nu)\gamma$ samples, by calculating the probability for the sum of the Monte Carlo prediction and the background to fluctuate to the observed number of events. The uncertainties in background estimate, efficiencies, acceptance and integrated luminosity were convoluted in the likelihood function with Gaussian distributions. A dipole form factor with a form factor scale $\Lambda_W = 1.5$ TeV was used in the Monte Carlo event generation. The limit contours for the CP-conserving anomalous coupling parameters $\Delta\kappa$ and λ are shown in Fig. 3, assuming that the CP-violating anomalous coupling parameters $\tilde{\kappa}$ and $\tilde{\lambda}$ are zero. For comparison, previous limits obtained by UA2 and CDF from the 1988-89 data are included. Current limits on CP-conserving anomalous $WW\gamma$ couplings are:

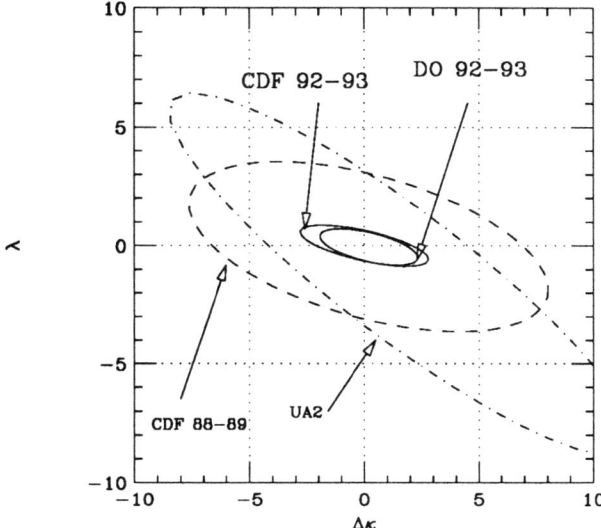

FIG. 3. Limits on CP–conserving anomalous coupling parameters $\Delta\kappa$ and λ. The ellipses represent the 95% exclusion contours. Present limits from CDF and DØ are shown together with previous limits.

CDF $\quad -2.3 < \Delta\kappa < 2.3\ (\lambda = 0),\quad -0.7 < \lambda < 0.7\ (\Delta\kappa = 0),$

DØ $\quad -1.6 < \Delta\kappa < 1.8\ (\lambda = 0),\quad -0.6 < \lambda < 0.6\ (\Delta\kappa = 0),$

at the 95% confidence level. Limits on CP–violating coupling parameters were within $3-6\%$ of those obtained for $\Delta\kappa$ and λ. It was found that the limits are insensitive to the form factor for $\Lambda_W > 200$ GeV and are well within the constraints imposed by the S-matrix unitarity (10) with $\Lambda_W = 1.5$ TeV. DØ also performed a two dimensional fit including $\Delta R_{\ell\gamma}$, and found that the results are within 3% of those obtained from a fit to the E_T^γ spectrum only.

$Z\gamma$ ANALYSIS

The $Z\gamma$ candidates were obtained by searching for events containing two isolated, high E_T, leptons, and an isolated photon. Table 6 summarizes geometrical and kinematic selection as well as integrated luminosity used in each channel. The CDF observed 4 $ee\gamma$ candidates and 4 $\mu\mu\gamma$ candidates (11), while the DØ observed 4 $ee\gamma$ candidates and 2 $\mu\mu\gamma$ candidates (12). The background estimate, summarized in Table 7, includes contributions from: Z+jets, where a jet is misidentified as a photon; $Z\gamma$ with $Z \to \tau\tau$. Because

TABLE 6. Summary of $Z\gamma$ event selection.

	CDF		DØ	
	$ee\gamma$	$\mu\mu\gamma$	$ee\gamma$	$\mu\mu\gamma$
Geometry	$\|\eta_{e1}\| < 1.1$ $1.1 < \|\eta_{e2}\| < 4.2$	$\|\eta_{\mu1}\| < 0.6$ $\|\eta_{\mu2}\| < 1.2$	$\|\eta_{e1,2}\| < 1.1$ $1.5 < \|\eta_{e1,2}\| < 2.5$	$\|\eta_{\mu1,2}\| < 1.0$
	$\|\eta_\gamma\| < 1.1$		$\|\eta_\gamma\| < 1.1, 1.5 < \|\eta_\gamma\| < 2.5$	
Kinematics (in GeV)	$E_T^{e1} > 20$ $E_T^{e2} > 20, 15, 10$ $E_T^\gamma > 7$	$p_T^{\mu1,2} > 20$	$E_T^{e1,2} > 25$ $E_T^\gamma > 10$	$p_T^{\mu1} > 15$ $p_T^{\mu2} > 8$
	$\Delta R_{\ell\gamma} > 0.7$		$\Delta R_{\ell\gamma} > 0.7$	
$\int Ldt$ pb^{-1}	19.7 ± 0.7	18.6 ± 0.7	13.9 ± 1.7	13.3 ± 1.6

we require three isolated objects in the final state, the background in the $Z\gamma$ candidates is small. The background-subtracted signal agrees well with the SM prediction calculated using the Monte Carlo program of Baur and Berger. CDF derived the $Z\gamma$ cross section times $Z \to \ell\ell$ branching ratio for photons with $\Delta R_{\ell\gamma} > 0.7$ and $E_T^\gamma > 7$ GeV from a combined $e + \mu$ sample: $\sigma(Z\gamma) \cdot Br(Z \to \ell\ell) = 5.1 \pm 1.9(stat) \pm 0.3(syst)$ pb, in good agreement with the SM prediction of $5.2 \pm 0.6(stat \oplus syst)$ pb. Figure 4 and 5 show the data and the SM prediction plus the background in the distributions of E_T^γ and $\ell^+\ell^-\gamma$ invariant mass for CDF, and E_T^γ for DØ, respectively. No significant deviation from the SM prediction was observed.

Similarly to the $W\gamma$ analysis, limits on anomalous $Z\gamma$ couplings were obtained by a fit to the E_T^γ spectrum. Figure 6 shows the current CDF and DØ 95% limit contours for anomalous $ZZ\gamma$ couplings together with the limits from L3 (13) experiment and the constraints from S-matrix unitarity for $\Lambda_Z = 500$ GeV. The pair of h_{30}^Z and h_{40}^Z is CP-conserving, while that of h_{10}^Z and h_{20}^Z is CP-violating. Limits on CP-conserving $ZZ\gamma$ couplings are:

CDF $\quad -3.0 < h_{30}^Z < 2.9$ $(h_{40}^Z = 0)$, $\quad -0.7 < h_{40}^Z < 0.7$ $(h_{30}^Z = 0)$,

DØ $\quad -1.9 < h_{30}^Z < 1.8$ $(h_{40}^Z = 0)$, $\quad -0.5 < h_{40}^Z < 0.5$ $(h_{30}^Z = 0)$,

at the 95% confidence level. Limits on $Z\gamma\gamma$ couplings are the same to within 0.1. The sensitivity of limits to the form factor scale, Λ_Z, was studied. Both CDF and DØ data reach the limit set by unitarity for $\Lambda_Z \sim 500$ GeV, which can be interpreted as the sensitivity limit from the current data.

TABLE 7. Summary of $Z\gamma$ data, backgrounds and the SM predictions.

	CDF		DØ	
Source:	$ee\gamma$	$\mu\mu\gamma$	$ee\gamma$	$\mu\mu\gamma$
Z+jets	0.4 ± 0.2	0.1 ± 0.1	0.43 ± 0.06	0.02 ± 0.01
$Z(\tau\tau)\gamma$	negligible	negligible	negligible	0.03 ± 0.01
Total background	0.4 ± 0.2	0.1 ± 0.1	0.43 ± 0.06	0.05 ± 0.01
Data	4	4	4	2
Signal	3.6 ± 2.0	3.9 ± 2.0	$3.6^{+3.2}_{-1.9}$	$1.95^{+2.6}_{-1.3}$
SM prediction	4.3 ± 0.2	2.8 ± 0.1	3.2 ± 0.5	2.5 ± 0.5

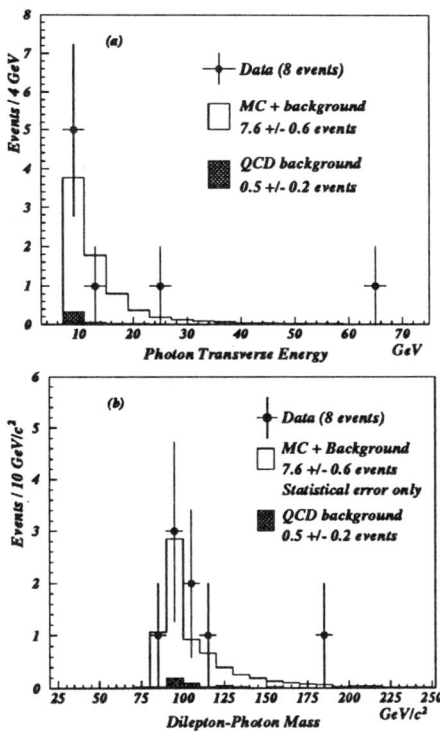

FIG. 4. CDF Distributions of (a) E_T^γ and (b) the $\ell^+\ell^-\gamma$ invariant mass. for electron and muon channels combined. The points are data. The shaded areas represent the estimated background, and the solid histograms are the expected signal from the Standard Model plus the estimated background.

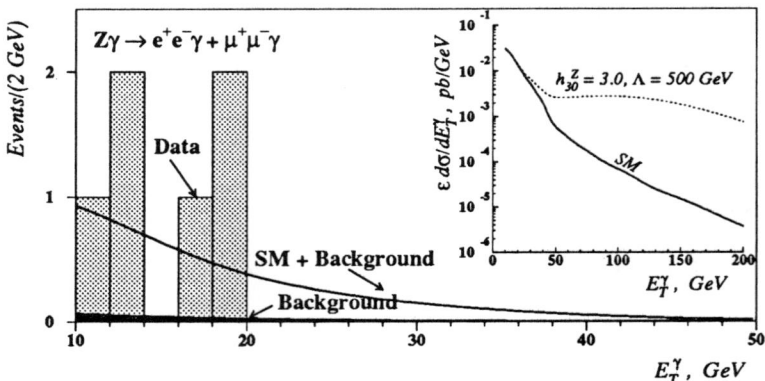

FIG. 5. DØ Distributions of (a) E_T^γ of $ee\gamma$ and $\mu\mu\gamma$ events. The histogram corresponds to the data, the shaded area represents the estimated background, and the solid line shows the sum of the SM prediction and the background. The insert shows $d\sigma/dE_T^\gamma$ folded with efficiencies for the SM and anomalous ($h_{30}^Z = 3.0$) couplings.

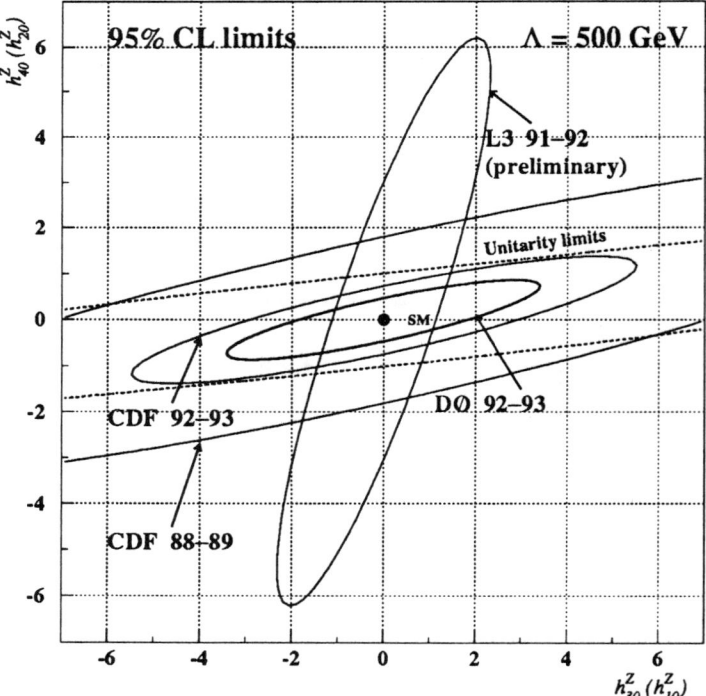

FIG. 6. Comparison of the limits on the CP-conserving (CP-violating) anomalous $ZZ\gamma$ coupling parameters h_{30}^Z and h_{40}^Z (h_{10}^Z and h_{20}^Z). The solid ellipses represent 95% CL exclusion contours for DØ, CDF and L3 experiments. The dashed curve shows limits from unitarity for the form factor scale of $\Lambda_Z = 500$ GeV.

CONCLUSION

In conclusion, CDF and DØ has studied $W\gamma$ and $Z\gamma$ productions at $\sqrt{s} = 1.8$ TeV in electron and muon channels. The observed photon E_T spectra agree well with the standard model predictions, yielding limits on anomalous $WW\gamma$, $ZZ\gamma$ and $Z\gamma\gamma$ couplings.

It is a pleasure to thank the members of the organizing committee, U. Baur, S. Errede and T. Müller, and the conference staff for running the conference so smoothly. I am indebted to my colleagues on DØ and the members of CDF electroweak physics group for their help in preparing the talk. This work was supported by the Director, Office of Energy Research, Office of High Energy and Nuclear Physics, Division of High Energy Physics of the U.S. Department of Energy under Contract DE-AC03-76SF00098.

REFERENCES

1. K. Hagiwara, R.D. Peccei, D. Zeppenfeld and K. Hikasa, Nucl. Phys. **B282**, 253 (1987).
2. K. Kim and Y-S. Tsai, Phys. Rev. D **7**, 3710 (1973).
3. U. Baur and E.L. Berger, Phys. Rev. D **41**, 1476 (1990).
4. CDF Collaboration, F. Abe et al., Nucl. Instrum. Methods **A271**, 387 (1988).
5. DØ Collaboration, S. Abachi et al., Nucl. Instrum. Methods **A338**, 185 (1994).
6. CDF Collaboration, F. Abe et al., Fermilab-PUB-94-244-E. To appear in Phys. Rev. D.
7. DØ Collaboration, M. Narain, "Proceedings of the American Physical Society Division of Particles and Fileds Meeting," Fermilab (1992), eds. R. Raja and J. Yoh, Vol.2, 1678.
8. CDF Collaboration, F. Abe et al., Phys. Rev. Lett. **74**, 1936 (1995).
9. DØ Collaboration, S. Abachi et al., Fermilab-PUB-95-101-E, Submitted to Phys. Rev. Lett.
10. U. Baur and D. Zeppenfeld, Phys. Lett. **B201**, 383 (1988).
11. CDF Collaboration, F. Abe et al., Phys. Rev. Lett. **74**, 1941 (1995).
12. G. Landsberg, these proceedings; DØ Collaboration, S. Abachi et al., Fermilab-PUB-95-042-E. Submitted to Phys. Rev. Lett.
13. P. Mattig, these proceedings; O. Adrianni et. al, Phys. Lett. **B345**, 609 (1995).

WW and WZ Production at the Tevatron

Theresa A. Fuess

Argonne National Laboratory
Argonne, Illinois 60439

Direct limits are set on WWZ and $WW\gamma$ three-boson couplings in a search for WW and WZ production in $p\bar{p}$ collisions at $\sqrt{s} = 1.8$ TeV using the DØ and CDF detectors at the Fermilab Tevatron.

INTRODUCTION

Among the most characteristic and fundamental signatures of non-Abelian symmetry of SU(2) × U(1) gauge theory of the Standard Model electroweak interactions are the interactions of W, Z, and γ bosons with each other. The interaction between the W and γ was previously studied in the process $p\bar{p} \to W\gamma$ (1). Here we report on bounds on the WWZ and $WW\gamma$ couplings obtained from the production of WW and WZ in $p\bar{p}$ interactions at $\sqrt{s} = 1.8$ TeV (2)

In the standard model, the dominant contribution to diboson production in $p\bar{p}$ collisions at $\sqrt{s} = 1.8$ TeV comes from two types of Feynman diagrams (figure 1). There are substantial cancellations between the t- or u-channel diagrams, which involve only the couplings of the bosons to fermions, and the s-channel diagrams which contain the three-boson coupling. These cancellations result in standard model cross sections of 9.5 pb and 2.5 pb for WW and WZ production respectively. To the extent that the fermionic couplings of the W, Z, and γ have been well tested, we may regard diboson production as primarily a test of the three-boson couplings.

The most general $WW\gamma$ and WWZ couplings consistent with Lorentz invariance have been formulated and may be parameterized in terms of fourteen

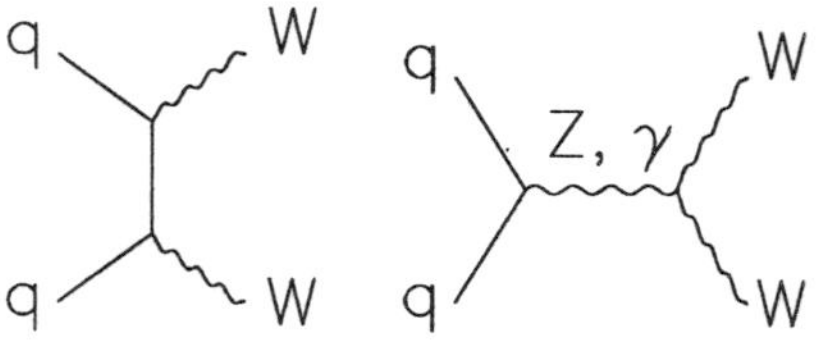

FIG. 1. Feynman diagrams for WW production. In the standard model there are substantial cancellations between these two types of diagrams.

independent couplings(or form factors), seven for the $WW\gamma$ vertex and seven for the WWZ vertex (3). They are g_1^V, g_4^V, g_5^V, λ^V, κ^V, $\tilde{\lambda}^V$, and $\tilde{\kappa}^V$ where V is either γ (for $WW\gamma$) or Z (for WWZ). The standard model SU(2) × U(1) electroweak theory corresponds to the choice $g_1^\gamma = g_1^Z = 1$ and $\kappa^\gamma = \kappa^Z = 1$ with all other couplings set to zero. The terms $\Delta\kappa \equiv \kappa - 1$ and $\Delta g_1 \equiv g_1 - 1$ are also used.

If any of these couplings differ substantially from the standard model values then the cross section increases. The enhancement is greatest at high boson P_T where the strongest cancellations occur in the standard model. Any couplings, differing from the standard model values, that are independent of $\sqrt{\hat{s}}$ cause the diboson production cross section to violate unitarity at some large $\sqrt{\hat{s}}$. To avoid this the anomalous parts of the couplings are made functions of $\sqrt{\hat{s}}$ and a form factor scale Λ_{FF} in such a way that they approach their standard model values when $\sqrt{\hat{s}}$ is bigger than Λ_{FF}.

$$\xi(\hat{s}) = \xi_{SM} + \frac{\xi(0) - \xi_{SM}}{(1 + \hat{s}/\Lambda_{FF}^2)^2} \tag{1}$$

where ξ stands for any of the couplings, ξ_{SM} is its value in the standard model, and Λ_{FF} represents the energy scale of unknown phenomena. The sensitivity of the measurement of the couplings can depend on the value of Λ_{FF} used. However, if Λ_{FF} is big enough, there is little effect at lower energies where the measurement is made.

The WW and WZ production at the Tevatron was studied in two channels, decay to leptons plus jets and decay to leptons only. The decay of WW, WZ to leptons plus jets gives better sensitivity to anomalous three-boson couplings than the purely leptonic channels because the leptonic branching fractions of the W and Z are small and because the acceptance of the detector for jets is larger than for leptons. Background from the QCD processes $p\bar{p} \to W + $ jets and $p\bar{p} \to Z + $ jets is greatly reduced by requiring a large boson P_T, while retaining good sensitivity to anomalous three-boson couplings (3). This measurement was made by CDF. The purely leptonic decay mode of WW does not have the overwhelming QCD W plus jets background and therefore allows observation of the predicted standard model signal with a direct measurement of the production cross section. This measurement was performed by both DØ and CDF. In both measurements the leptons include electrons and muons.

THE COLLIDER DETECTOR AT FERMILAB

The Collider Detector at Fermilab (CDF) has been described in detail elsewhere (4). Here we give a brief description of the components relevant to this analysis. The location of the event vertex is measured along the beam direction with a time projection chamber (VTX). The momenta of charged particles are measured in the central tracking chamber (CTC), which is surrounded by a 1.4 T superconducting solenoidal magnet. Outside the CTC, the calorimeter is organized in electromagnetic (EM) and hadronic (HAD) compartments with projective towers covering the pseudorapidity range $|\eta| \leq 3.6$.

Outside the central calorimeter, the region $|\eta| \leq 1.0$ is instrumented with drift chambers for muon identification.

Each electron is identified by an isolated cluster in either the central EM calorimeter ($|\eta| \leq 1.1$) which matches a track in the CTC or the endplug EM calorimeter ($1.1 \leq |\eta| \leq 2.4$) with associated hits in the VTX. Each muon is identified by an isolated track in the CTC with minimum ionizing energy in the calorimeter. Events with one or more muons must have at least one muon with matching hits in the muon chambers. The presence of neutrinos is inferred from missing transverse energy ($\displaystyle{\not}E_T$), which is measured by the magnitude of the vector sum of the calorimeter tower energies perpendicular to the beam axis. Jet energy is measured by clustering the EM and HAD calorimeter energy within a cone $\Delta R < 0.4$, where $\Delta R = \sqrt{\Delta\phi^2 + \Delta\eta^2}$, and ϕ is the azimuthal angle (5).

THE DØ DETECTOR

The DØ detector (6) consists of three major components: the calorimeter, tracking, and muon systems. A hermetic, compensating, uranium-liquid argon sampling calorimeter with fine transverse and longitudinal segmentation in projective towers measures energy out to $|\eta| \sim 4.0$, where η is the pseudo-rapidity. The energy resolution for electrons and photons is $15\%/\sqrt{E(\text{GeV})}$. The resolution for the transverse component of missing energy, $\displaystyle{\not}E_T^{cal}$, is 1.1 GeV + $0.02(\sum E_T)$, where $\sum E_T$ is the scalar sum of transverse energy, E_T, in GeV, deposited in the calorimeter. The central and forward drift chambers are used to identify charged tracks for $|\eta| \leq 3.2$. There is no central magnetic field. Muons are identified and their momentum measured with three layers of proportional drift tubes, one inside and two outside of the magnetized iron toroids, providing coverage for $|\eta| \leq 3.3$. The muon momentum resolution, determined from $J/\psi \to \mu\mu$ and $Z \to \mu\mu$ events, is $\sigma(1/p) = 0.18(p-2)/p^2 \oplus 0.008$ (p in GeV/c). The p_T of identified muons is used to correct $\displaystyle{\not}E_T^{cal}$ to form the missing transverse energy, $\displaystyle{\not}E_T$.

Muons are required to be isolated, to have energy deposition in the calorimeter corresponding to at least that of a minimum ionizing particle, and to have $|\eta| \leq 1.7$. For the $\mu\mu$ channel, cosmic rays are rejected by requiring that the muons have timing consistent with the beam crossing. Electrons are identified through the longitudinal and transverse shape of isolated energy clusters in the calorimeter and by the detection of a matching track in the drift chambers. Electrons are required to be within a fiducial region of $|\eta| \leq 2.5$. A criterion on ionization (dE/dx), measured in the drift chambers, is imposed to reduce backgrounds from photon conversions and hadronic showers with large electromagnetic content.

CDF: $WW, WZ \to l\nu jj, lljj$

The data for this analysis were recorded with the Collider Detector at Fermilab during the 1992-93 Fermilab Tevatron collider run, corresponding to an integrated luminosity of $19.6\,\mathrm{pb}^{-1}$. We search for WW and WZ event candidates consistent with the decay of one boson to leptons and the other to hadrons. Background QCD processes are calculated at Born level (7), including simulation of the CDF detector and jet fragmentation using an adaptation of the HERWIG program (8,9). The boson P_T requirement for WW and WZ event selection is chosen so that less than one background event is expected in the final sample. With this choice it is unnecessary to perform a background subtraction and any theoretical uncertainty in the background calculation is avoided.

A leptonic W decay is identified by an isolated electron or muon with $P_T > 20\,\mathrm{GeV/c}$ and $\not{E}_T > 20\,\mathrm{GeV}$ forming a transverse mass $M_T > 40\,\mathrm{GeV/c^2}$. A leptonic Z decay is identified by an electron or muon pair of opposite charge forming an invariant mass $70 < M < 110\,\mathrm{GeV/c^2}$. In events with a leptonic W or Z decay, a candidate hadronic W or Z decay is defined by the two jets (leading jets) in the event with the highest jet transverse energies (E_T). Each jet must have $E_T > 30\,\mathrm{GeV}$ and the invariant mass of the jet pair must be in the range $60 < M_{JJ} < 110\,\mathrm{GeV/c^2}$. The P_T of the two-jet system, interpreted as a hadronic W or Z decay, is required to satisfy $P_T > 130\,\mathrm{GeV/c}$ for leptonic W events or $P_T > 100\,\mathrm{GeV/c}$ for leptonic Z events.

The two-jet mass spectrum is shown in Figure 2a for events with a leptonic W decay and with both leading jets satisfying $E_T > 30\,\mathrm{GeV}$. The sum of the predicted Standard Model WW and WZ signals plus QCD background is also shown, where the background is normalized to the observed number of W events with two jets minus the predicted signal. Figure 2b shows the two-jet P_T distribution in the subset of events which satisfy the two-jet mass criterion. The two-jet P_T requirement is indicated by the arrow. One event passes this cut. For events with a leptonic Z decay there are no events which satisfy all selection criteria.

The limits on the couplings follow from a Monte Carlo calculation of expected event yields for various values of the couplings. The Monte Carlo event generator (3,10) calculates to leading order the processes $p\bar{p} \to W^+W^-$ and $p\bar{p} \to WZ$ with subsequent decay of a W to $e\nu$, $\mu\nu$, or jj and a Z to ee, $\mu\mu$, or jj. Higher order QCD corrections to the cross section are accounted for by a "K-factor" of $K = 1 + \frac{8}{9}\pi\alpha_s$ (3). MTB2 structure functions are used (11). Initial and final state QCD radiation effects and jet fragmentation are modelled with an adaptation of HERWIG (8,9). The event generator is combined with a detector simulation which includes trigger efficiencies, lepton identification efficiencies, and jet response modeling. A fast parametrization of the full detector simulation was also employed. The trigger and lepton identification efficiencies are determined from the data and amount to 78% for electrons and 79% for muons. The modeling of the jet response and resolution are tuned to agree with studies of collider and test beam data (12).

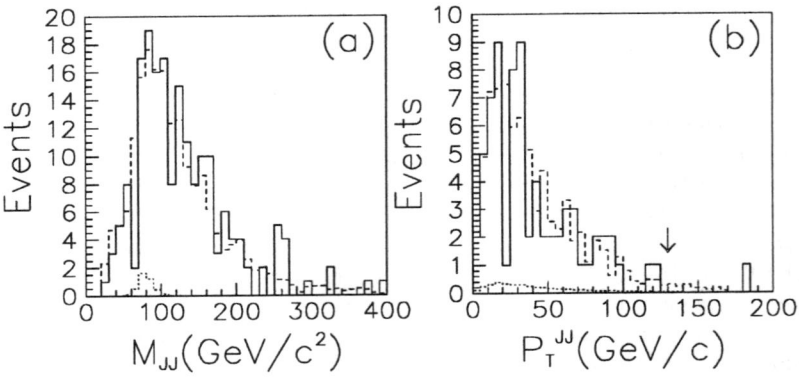

FIG. 2. CDF selection of $WW/WZ \to l\nu jj$ candidates. All event selection cuts except the two-jet mass and two-jet P_T cuts were used to select the events in (a). The subset of events from (a) passing the two-jet mass cut is shown in (b). One event remains after making all cuts. The solid line shows the data, the dots show the predicted standard model diboson signal, and the dashes show the predicted signal plus background shape.

The two-jet mass resolution is expected to be 9 GeV/c^2 for diboson events that would pass our candidate selection criteria. The efficiency of the two-jet mass cut is 88% for events passing all other cuts.

The systematic uncertainties on the yield are the uncertainties in the structure functions (6%), jet E_T scale and resolution (16%), luminosity (4%), lepton identification efficiency (1%), and trigger efficiency (1%). The Monte Carlo acceptance modeling has 3% statistical uncertainty, and a 5% systematic uncertainty allows for differences between fast and full detector simulations. In addition a 14% uncertainty is assigned for the effects of higher order QCD corrections (8,13,14). These uncertainties are combined in quadrature.

The acceptance is a strong function of the couplings, because of the boson P_T cut in combination with a varying boson P_T distribution. For standard model couplings, 0.13 $WW/WZ \to l\nu jj$ events and 0.02 $WZ \to lljj$ events are expected to pass the selection criteria, where l is either an electron or a muon. The observation of one event in the $l\nu jj$ channel and zero events in the $lljj$ channel is therefore not indicative of a departure from standard model couplings, even without consideration of the QCD background.

The predicted yield of high P_T boson pairs is a quadratic function of the anomalous couplings. The lack of an excess of events therefore results in bounds on the couplings which take the form of ellipses in the plane of any two couplings. Since the one event passing all selection criteria could be either signal or background, we calculate the confidence limits from the probability of observing one or less signal events. We do not perform a background subtraction and therefore obtain conservative limits. The probability distribution used is the convolution of a Poisson distribution with a Gaussian, where the Gaussian smears the mean of the Poisson distribution around the expected yield within the systematic uncertainty.

In Figure 3 we present bounds on four pairs of couplings. Except as noted in the figure caption, for each case all the other couplings are fixed at the standard model values. Each pair is constrained to the interior of an ellipse, which is a two dimensional section through an ellipsoidal allowed region in the fourteen dimensional space of three boson couplings. Because the bosons are required to have high P_T our search is most sensitive to the couplings at energies near $\sqrt{\hat{s}} = 500\,\text{GeV}$. The limit contours, however, correspond to the value of the couplings at $\sqrt{\hat{s}} = 0$ and therefore depend on the choice of Λ_{FF} according to equation (1). The bounds are shown for $\Lambda_{FF} = 1000$ GeV and $\Lambda_{FF} = 1500$ GeV. The unitarity bounds, which depend strongly on Λ_{FF}, are also shown (15,16). For values of Λ_{FF} larger than about 1600 GeV the bounds from unitarity are stronger than the bounds from the search.

Figure 3a shows limits in the plane λ^γ vs. λ^Z. The limits are stronger for λ^Z, illustrating the fact that the search is in general more sensitive to the WWZ couplings. It is therefore complementary to studies of the process $p\bar{p} \to W\gamma$ (1).

The limits of Figure 3b focus on the WWZ vertex, assuming that the $WW\gamma$ couplings take their standard model values. Bounds are shown for the couplings g_1^Z and κ^Z, which are the only WWZ couplings predicted to be nonzero in the standard model. The fact that the point $g_1^Z = \kappa^Z = 0$ lies outside the allowed region can be interpreted as direct evidence for a non-zero WWZ coupling, and for the resulting destructive interference between s-channel and t- or u-channel diagrams which takes place in the standard model. Specifically, the search is directly sensitive to the WWZ coupling in the region $\sqrt{\hat{s}} = 500\,\text{GeV}$. If the WWZ coupling were zero in this region, the s-channel diagram containing the WWZ vertex would not contribute to the amplitude, and the other diagrams by themselves would predict the observation of 15 ± 3 events, where the uncertainty is systematic. Independent of the choice of Λ_{FF}, this possibility is excluded at greater than 99% CL.

Figures 3c and 3d show limits on the couplings κ and λ, assuming specific relations between the WWZ and $WW\gamma$ couplings. In Figure 3c, the WWZ couplings are assumed to equal the $WW\gamma$ couplings. The resulting 95% CL limits on κ and λ separately, assuming that only one departs from its standard model value, are $-0.11 < \kappa < 2.27$ and $-0.81 < \lambda < 0.84$ for the choice $\Lambda_{FF} = 1000\,\text{GeV}$. With the assumption of matching WWZ and $WW\gamma$ couplings, limits also result for the W boson electric quadrupole moment $Q_e^W = \frac{-e}{M_W^2}(\kappa - \lambda)$ and magnetic dipole moment $\mu^W = \frac{e}{2M_W}(1 + \kappa + \lambda)$. In the standard model, these moments take the values $Q_e^W = \frac{-e}{M_W^2}$ and $\mu^W = \frac{e}{M_W}$. The point $Q_e^W = \mu^W = 0$ is outside the allowed region. Assuming only one of the moments departs from its standard model value, the limits at 95% CL are $-2.42 < Q_e^W/(e/M_W^2) < 0.35$ and $0.37 < \mu^W/(e/M_W) < 1.70$ for $\Lambda_{FF} = 1000\,\text{GeV}$.

For Figure 3d, the relation assumed between the WWZ and $WW\gamma$ cou-

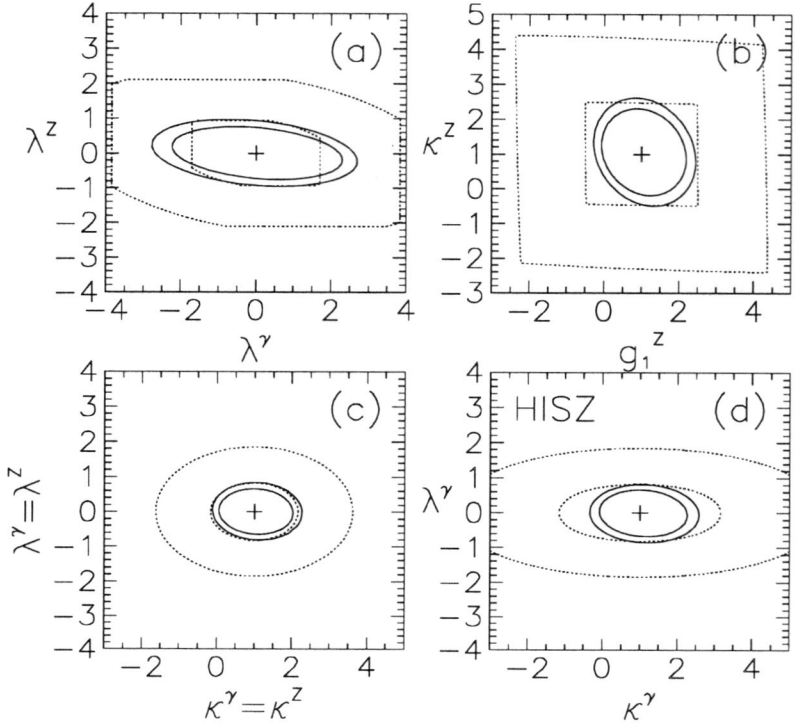

FIG. 3. CDF allowed regions for pairs of anomalous couplings from the analysis of WW, $WZ \to l\nu jj$, $lljj$ events. All couplings, other than those listed for each contour, are held at their standard model values. The solid lines are the 95% CL limits and the dotted lines are the unitarity limits; each is shown for $\Lambda_{FF} = 1000$ GeV (outer) and 1500 GeV (inner). The + signs indicate the Standard Model values of the couplings. (a) λ^γ and λ^Z; (b) g_1^Z and κ^Z; (c) κ and λ assuming the WWZ and $WW\gamma$ couplings are the same; (d) κ^γ, κ^Z, λ^γ, λ^Z and g_1^Z in the HISZ prescription (see text), with independent variables κ^γ and λ^γ.

plings is given by the HISZ equations (17), which specify λ^Z, κ^Z, and g_1^Z in terms of the independent variables κ^γ and λ^γ. This prescription preserves SU(2) × U(1) gauge invariance and is well motivated in an effective Lagrangian approach. The corresponding subspace of anomalous couplings is not well constrained by previous indirect measurements (17). The individual 95% CL bounds on λ^γ and κ^γ are $-0.35 < \kappa^\gamma < 2.57$ and $-0.85 < \lambda^\gamma < 0.81$ for $\Lambda_{FF} = 1000$ GeV, if only one of the two is varied from its standard model value.

DØ : $WW \to l\nu l\nu$

The data for the DØ analysis were recorded during the 1992-93 collider run and correspond to an integrated luminosity of approximately 14 pb^{-1}. The DØ event samples come from triggers with dilepton signatures. The $e\mu$ sample is selected from events passing the trigger requirement of an electromagnetic cluster with $E_T \geq 7$ GeV and a muon with $p_T \geq 5$ GeV/c. The ee candidates are required to have two isolated electromagnetic clusters, each with $E_T \geq 10$ GeV. The $\mu\mu$ candidates are selected from events where at least one muon is identified with $p_T \geq 5$ GeV/c at the trigger level.

In the offline selection for the $e\mu$ channel, a muon with $p_T \geq 15$ GeV/c and an electron with $E_T \geq 20$ GeV are required. Both $\not{E}_T$ and $\not{E}_T^{cal}$ are required to be ≥ 20 GeV. In order to suppress $Z \to \tau\bar{\tau}$ and $b\bar{b}$ backgrounds, it is required that $20° \leq \Delta\phi(p_T^\mu, \not{E}_T) \leq 160°$ if $\not{E}_T \leq 50$ GeV, where $\Delta\phi(p_T^\mu, \not{E}_T)$ is the angle in the transverse plane between the muon and $\not{E}_T$. One event survives these selection cuts in a data sample corresponding to an integrated luminosity of 13.5 ± 1.6 pb^{-1}.

For the ee channel, two electrons are required, each with $E_T \geq 20$ GeV. The $\not{E}_T$ is required to be ≥ 20 GeV. The Z boson background is reduced by removing events where the dielectron invariant mass is between 77 and 105 GeV/c^2. It is required that $20° \leq \Delta\phi(p_T^e, \not{E}_T) \leq 160°$ for the lower energy electron if $\not{E}_T \leq 50$ GeV. This selection suppresses $Z \to ee$ as well as $\tau\tau$. The integrated luminosity in this channel is 13.9 ± 1.7 pb^{-1}. One event survives these selection requirements.

For the $\mu\mu$ channel, two muons are required, one with $p_T \geq 20$ GeV/c and another with $p_T \geq 15$ GeV/c. In order to remove Z boson events, it is required that the $\not{E}_T$ projected on the dimuon bisector in the transverse plane be greater than 30 GeV. This selection requirement is less sensitive to the momentum resolution of the muons than is a dimuon invariant mass cut. It is required that $\Delta\phi(p_T^\mu, \not{E}_T) \leq 170°$ for the higher p_T muon. No events survive these selection requirements in a data sample corresponding to an integrated luminosity of 11.8 ± 1.4 pb^{-1}.

Finally, in order to suppress background from $t\bar{t}$ production, the vector sum of the E_T from hadrons, $\vec{E}_T^{had}$, defined as $-(\vec{E}_T^{l1} + \vec{E}_T^{l2} + \vec{\not{E}}_T)$ is required to be less than 40 GeV in magnitude for all channels. Figure 4 shows a Monte Carlo simulation of E_T^{had} for ~ 20 fb^{-1} of SM WW and $t\bar{t}$ events. For WW events, non-zero values of E_T^{had} are due to gluon radiation and detector resolution.

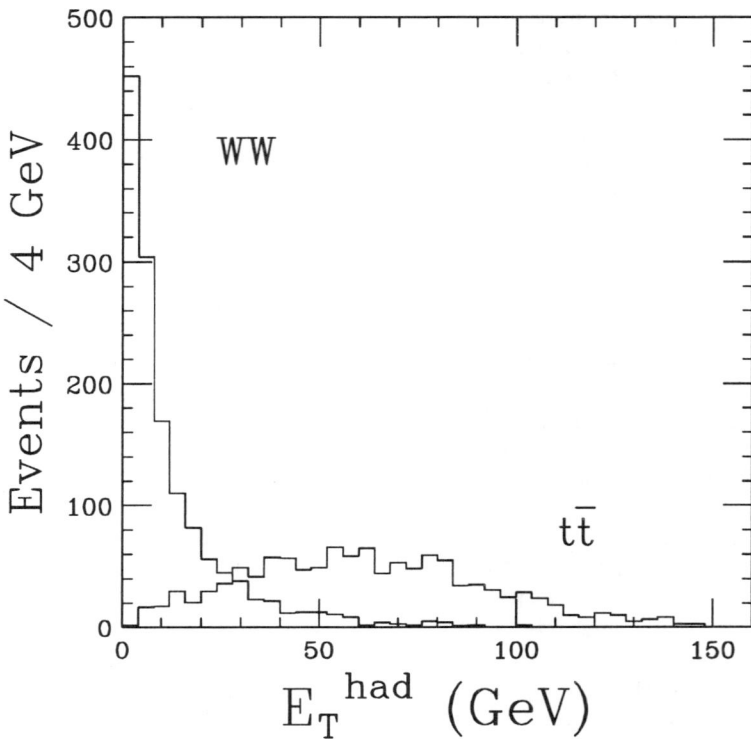

FIG. 4. E_T^{had} for Monte Carlo WW and $t\bar{t}$ events with $M_{\text{top}} = 160$ GeV/c^2 ($\int L dt \sim 20$ fb^{-1}). Events with $E_T^{\text{had}} \geq 40$ GeV were rejected.

For $t\bar{t}$ events, the most significant contribution is the b-quark jets from the t-quark decays. This selection reduces the background from $t\bar{t}$ production by a factor of four for a t-quark mass of 160 GeV/c^2 and is slightly more effective for a more massive t-quark. The efficiency of this selection criterion for SM W boson pair production events is $0.95^{+0.01}_{-0.04}$ and decreases slightly with increasing W boson pair invariant mass. The surviving ee candidate passes this selection requirement but the $e\mu$ candidate (18) is rejected.

The detection efficiency for SM W boson pair production events is determined using the **PYTHIA** (19) event generator followed by a detailed **GEANT** (20) simulation of the DØ detector. Muon trigger and electron identification efficiencies are derived from the data. The overall detection efficiency for SM $WW \to e\mu$ is 0.092 ± 0.010. For the ee channel the efficiency is 0.094 ± 0.008. For the $\mu\mu$ channel it is 0.033 ± 0.003. For the three channels combined, the expected number of events for SM W boson pair production, based on a cross section of 9.5 pb (13), is 0.46 ± 0.08. The Monte Carlo program of Ref. (3)

Background	$e\mu$	ee	$\mu\mu$
$Z \to ee$ or $\mu\mu$	—	0.02 ± 0.01	0.066 ± 0.026
$Z \to \tau\tau$	0.11 ± 0.05	$< 10^{-3}$	$< 10^{-3}$
Drell-Yan dileptons	—	$< 10^{-3}$	$< 10^{-3}$
$W\gamma$	0.04 ± 0.03	0.02 ± 0.01	—
QCD	0.07 ± 0.07	0.15 ± 0.08	$< 10^{-3}$
$t\bar{t}$	0.04 ± 0.02	0.03 ± 0.01	0.009 ± 0.003
Total	0.26 ± 0.10	0.22 ± 0.08	0.075 ± 0.026

TABLE 1. DØ summary of backgrounds to $WW \to ee$, $WW \to e\mu$ and $WW \to \mu\mu$ events. The units are expected number of background events in the data sample. The uncertainties include both statistical and systematic contributions.

followed by a fast detector simulation (21) is used to estimate the detection efficiency for W boson pair production as a function of the coupling parameters λ and κ. The backgrounds due to Z boson, Drell-Yan dilepton, $W\gamma$, and $t\bar{t}$ events are estimated using the **PYTHIA** and **ISAJET** (22) Monte Carlo event generators followed by the **GEANT** detector simulation. The backgrounds from $b\bar{b}$, $c\bar{c}$, multi-jet, and $W+$ jet events, where a jet is mis-identified as an electron, are estimated using the data. The $t\bar{t}$ cross section estimates are from calculations of Laenen *et al.* (23). The $t\bar{t}$ background is averaged for $M_{\text{top}} = 160$, 170, and 180 GeV/c^2. The background estimates are summarized in Table 1.

The 95% confidence level upper limit on the W boson pair production cross section is estimated based on one signal event including a subtraction of the expected background of 0.56 ± 0.13 events. The branching ratio $W \to l\bar{\nu} = 0.108 \pm 0.004$ (24) is assumed. Poisson-distributed numbers of events are convoluted with Gaussian uncertainties on the detection efficiencies, background and luminosity. For SM W boson pair production, the upper limit for the cross section is 91 pb at the 95% confidence level. From the observed limit, as a function of λ and κ, and the theoretical prediction of the W boson pair production cross section, the 95% confidence level limits on the coupling parameters shown in Figure 5 (solid line) are obtained. Also shown in Figure 5 (dotted line) is the contour of the unitarity constraint on the coupling limits for the form factor scale $\Lambda = 900$ GeV. This value of Λ is chosen so that the observed coupling limits lie within this ellipse. The limits on the CP-conserving anomalous coupling parameters are $-2.6 < \Delta\kappa < 2.8$ ($\lambda = 0$) and $-2.2 < \lambda < 2.2$ ($\Delta\kappa = 0$).

CDF: $WW \to l\nu l\nu$

The data for the CDF analysis of WW in the purely leptonic mode (25) were taken during the 1992-93 and 1994 Tevatron collider runs and corresponds to an integrated luminosity of 45 pb^{-1}. The electron and muon selection are similar to that used in the CDF top search (26). Events were required to have

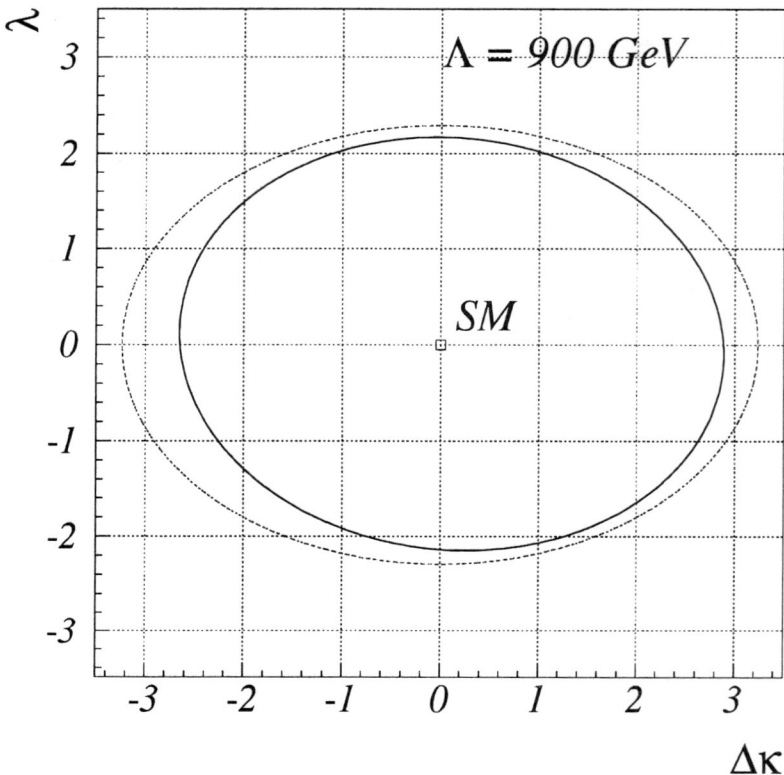

FIG. 5. DØ 95% CL limits on the CP-conserving anomalous couplings λ and $\Delta\kappa$, assuming that $\lambda_\gamma = \lambda_Z$ and $\kappa_\gamma = \kappa_Z$. The dotted contour is the unitarity limit for the form factor scale $\Lambda = 900$ GeV which was used to set the coupling limits.

Background in 19.3 pb^{-1}	$e\mu, ee, \mu\mu$
$Z \to ee, \mu\mu, \tau\tau$	0.03
$t\bar{t}$	0.06
$b\bar{b}$	0.07
Fake leptons	0.22
Total	0.38

TABLE 2. CDF summary of backgrounds to $WW \to ee, e\mu$, and $\mu\mu$ events. The units are expected number of background events in 19.3 pb^{-1}.

$\displaystyle{\not\!\!E}_T > 25 Gev$ and $\Delta\Phi > 20\deg$, where $\Delta\Phi$ is the angle between the $\displaystyle{\not\!\!E}_T$ vector and the momentum of the charged lepton. Events are rejected if there is any jet with $E_T > 10 \text{GeV}$ or if there are two oppositely charged leptons with an invariant mass in the Z mass window, 75-105 GeV/c^2. Two events pass all cuts compared to the Standard Model prediction of 1.3 WW events and 0.38 background events. The predicted background is outlined in Table 2

The resulting WW production cross section is found to be $7.9\,^{+11.4}_{-7.9} \pm 2.2$ pb. The 95% confidence level upper limit on this cross section is 39.5 pb. The 95% confidence level limits on the couplings are $-1.8 < \Delta\kappa < 1.9$ ($\lambda = 0$) and $-1.4 < \Delta\lambda < 1.4$ ($\Delta\kappa = 0$) assuming $\Delta\kappa^Z = \Delta\kappa^\gamma$ and $\lambda^Z = \lambda^\gamma$ and that the acceptance is independent of $\Delta\kappa$ and λ.

CONCLUSIONS

In conclusion, a search for WW and WZ in $p\bar{p}$ collisions at $\sqrt{s} = 1.8$ TeV is made. The resulting limits on the trilinear couplings and on WW production are summarized in Table 3.

ACKNOWLEDGEMENTS

We thank U. Baur, T. Han and D. Zeppenfeld for Monte Carlo programs and for many stimulating discussions. We thank the Fermilab staff and the technical staffs of the participating institutions for their vital contributions. This work was supported by the U.S. Department of Energy and National Science Foundation; the Italian Istituto Nazionale di Fisica Nucleare; the Ministry of Education, Science and Culture of Japan; the Natural Sciences and Engineering Research Council of Canada; the National Science Council of the Republic of China; the A. P. Sloan Foundation; and the Alexander von Humboldt-Stiftung; the Commissariat a L'Energie Atomique in France; the Ministry for Atomic Energy and the Ministry of Science and Technology Policy in Russia; CNPq in Brazil; the Departments of Atomic Energy and Science and Education in India; Colciencias in Colombia; CONACyT in Mexico; and the Ministry of Education, Research Foundation and KOSEF in Korea.

Experiment process luminosity	Λ_{ff} (GEV)	assumptions	95% CL limits
DØ $WW \to l\nu l\nu$ 14 pb^{-1}	900	$WWZ = WW\gamma$	$\Delta\kappa \in (-2.6, 2.8)$ $\lambda \in (-2.2, 2.2)$ $\sigma(p\bar{p} \to WW) < 91$pb
CDF $WW \to l\nu l\nu$ 45 pb^{-1}	1000	$WWZ = WW\gamma$	$\Delta\kappa \in (-1.8, 1.9)$ $\lambda \in (-1.4, 1.4)$ $\sigma(p\bar{p} \to WW) < 40$pb
CDF $WW, WZ \to l\nu jj, lljj$ 19.6 pb^{-1}	1000	$WWZ = WW\gamma$	$\Delta\kappa \in (-1.1, 1.3)$ $\lambda \in (-0.8, 0.8)$ $Q_e^W = \mu^W = 0$ is ruled out.
		$\kappa^\gamma = g_1^\gamma = 1$	$\kappa^Z = g_1^Z = 0$ is ruled out.
		HISZ	$\Delta\kappa^\gamma \in (-1.4, 1.5)$ $\lambda^\gamma \in (-0.8, 0.8)$ $\Delta\kappa^Z \in (-0.5, 0.5)$ $\lambda^Z \in (-0.8, 0.8)$ $\Delta g_1^Z \in (-0.9, 1.0)$

TABLE 3. Summary of limits set on $WW\gamma$ and WWZ trilinear couplings and on WW production at the Tevatron. All couplings, other than those for which limits are show and those under HISZ, are held at their standard model values.

REFERENCES

1. J. Alitti *et al.*, Phys. Lett. B **277**, 194 (1992); F. Abe *et al.*, Phys. Rev. Lett. **74**, 1936 (1995); S. Abachi *et al.*, FERMILAB-PUB-95/0XX-E (1995), submitted to Phys. Rev. Lett.; H. Iahara, these proceedings.
2. F. Abe *et al.*, FERMILAB-PUB-95/036-E (1995), submitted to Phys. Rev. Lett. Mar 7, 1995; S. Abachi *et al.*, FERMILAB-PUB-95/044-E (1995), submitted to Phys. Rev. Lett. Mar 10, 1995.
3. K. Hagiwara *et al.*, Phys. Rev. D **41**, 2113 (1990).
4. F. Abe *et al.*, Nucl. Instrum. Methods A **271**, 387 (1988).
5. F. Abe *et al.*, Phys. Rev. D **45**, 1448 (1992).
6. S. Abachi *et al.*, Nucl. Instrum. Methods A **338**, 185 (1994).
7. F. A. Berends *et al.*, Nucl. Phys. **B357**, 32 (1991).
8. G. Marchesini *et al.*, Comp. Phys. Comm. **67**, 465 (1992).
9. J. Benlloch, *The Fermilab Meeting*, Proceedings of the Division of Particles and Fields of the APS, Batavia, IL, edited by C. H. Albright *et al.* (World Scientific, Singapore, 1993), p. 1091.
10. D. Zeppenfeld, private communication.
11. J. Morfin and W. K. Tung, Z. Phys. C **52**, 13 (1991).
12. F. Abe et al., Phys. Rev. Lett. **68**, 1104 (1992).
13. J. Ohnemus, Phys. Rev. D **44**, 1403 (1991); J. Ohnemus, Phys. Rev. D **44**, 3477 (1991); U. Baur *et al.*, FSU-HEP-941010 (October 1994).
14. T. Han, private communication; V. Barger *et al.*, Phys. Rev. D **41**, 2782 (1990).
15. U. Baur and D. Zeppenfeld, Phys. Lett. B **201**, 383 (1988).
16. U. Baur, private communication.
17. K. Hagiwara *et al.*, Phys. Rev. D **48**, 2182 (1993); and references therein.
18. DØ Collaboration, S. Abachi *et al.*, Phys. Rev. Lett. **72**, 2138 (1994). The kinematic properties of this $t\bar{t}$ candidate are discussed in detail '
19. T. Sjöstrand, "PYTHIA 5.6 and Jetset 7.3 Physics and Manual," CERN-TH.6488/92, 1992, (unpublished).
20. F. Carminati *et al.*, " GEANT Users Guide," CERN Program Library, December 1991 (unpublished).
21. H. Johari, Ph. D. thesis, Northeastern University, 1995 (unpublished).
22. F. Paige and S. Protopopescu, BNL Report BNL38034, 1986 (unpublished), release V6.49.
23. E. Laenen, J. Smith, and W. L. van Neerven, Phys. Lett. B **321**, 254 (1994). For the $t\bar{t}$ background, the central value estimate of the cross section is used.
24. Particle Data Group, L. Montanet *et al.*, Phys. Rev. D **50**, 1173 (1994). The weighted average of the $W \to e\nu$ and $W \to \mu\nu$ branching fraction data is used.
25. See L. Zhang in these proceedings for more details.
26. F. Abe *et al.*, Phys. Rev. D **50**, 2966 (1994).

Standard Model Higher Order Corrections to the $WW\gamma/WWZ$ Vertex

Joannis Papavassiliou

New York University, New York 10003, USA

> Using the S–matrix pinch technique we obtain to one loop order gauge independent γW^-W^+ and ZW^-W^+ vertices in the context of the standard model, with all incoming momenta off–shell. We show that the vertices so constructed satisfy simple QED–like Ward identities. These gauge invariant vertices give rise to expressions for the magnetic dipole and electric quadrupole form factors of the W gauge boson, which, unlike previous treatments, satisfy the crucial properties of infrared finiteness and perturbative unitarity.

INTRODUCTION

A new and largely unexplored frontier on which the ongoing search for new physics will soon focus is the study of the structure of the three-boson couplings (1). A general parametrization of the trilinear gauge boson vertex for two on–shell Ws and one off–shell $V = \gamma$, Z is (2)

$$\Gamma^V_{\mu\alpha\beta} = -ig_V \left[f\left[2g_{\alpha\beta}\Delta_\mu + 4(g_{\alpha\mu}Q_\beta - g_{\beta\mu}Q_\alpha) \right] \right.$$
$$+ 2\Delta\kappa_V \left(g_{\alpha\mu}Q_\beta - g_{\beta\mu}Q_\alpha \right)$$
$$\left. + 4\frac{\Delta Q_V}{M_W^2}(\Delta_\mu Q_\alpha Q_\beta - \tfrac{1}{2}Q^2 g_{\alpha\beta}\Delta_\mu) \right], \qquad (1)$$

with $g_\gamma = gs$, $g_Z = gc$, where g is the $SU(2)$ gauge coupling, $s \equiv sin\theta_W$ and $c \equiv cos\theta_W$. In the above formula terms which are odd under the individual discrete symmetries of C, P, or T have been omitted. The four-momenta Q and Δ are related to the incoming momenta q, p_1 and p_2 of the gauge bosons V, W^- and W^+ respectively, by $q = 2Q$, $p_1 = \Delta - Q$ and $p_2 = -\Delta - Q$. The form factors $\Delta\kappa_V$ and ΔQ_V, also defined as

$$\Delta\kappa_V = \kappa_V + \lambda_V - 1, \qquad (2)$$

and

$$\Delta Q_V = -2\lambda_V, \qquad (3)$$

are compatible with C, P, and T invariance, and are related to the magnetic dipole moment μ_W and the electric quadrupole moment Q_W, by the following expressions (3), (4), (5), (6):

© 1996 American Institute of Physics

$$\mu_W = \frac{e}{2M_W}(2 + \Delta\kappa_\gamma),\qquad(4)$$

and

$$Q_W = -\frac{e}{M_W^2}(1 + \Delta\kappa_\gamma + \Delta Q_\gamma).\qquad(5)$$

In the context of the standard model, their canonical, tree level values, are $f = 1$ and $\Delta\kappa_V = \Delta Q_V = 0$. To determine the radiative corrections to these quantities one must cast the resulting one-loop expressions in the following form:

$$\Gamma_{\mu\alpha\beta}^V = -ig_V[a_1^V g_{\alpha\beta}\Delta_\mu + a_2^V(g_{\alpha\mu}Q_\beta - g_{\beta\mu}Q_\alpha) + a_3^V \Delta_\mu Q_\alpha Q_\beta],\qquad(6)$$

where a_1^V, a_2^V, and a_3^V are complicated functions of the momentum transfer Q^2, and the masses of the particles appearing in the loops. It then follows that $\Delta\kappa_V$ and ΔQ_V are given by the following expressions:

$$\Delta\kappa_V = \frac{1}{2}(a_2^V - 2a_1^V - Q^2 a_3^V)\qquad(7)$$

and

$$\Delta Q_V = \frac{M_W^2}{4}a_3^V.\qquad(8)$$

Calculating the one-loop expressions for $\Delta\kappa_V$ and ΔQ_V is a non-trivial task, both from the technical and the conceptual point of view. If one calculates just the Feynman diagrams contributing to the γW^+W^- vertex and then extracts from them the contributions to $\Delta\kappa_\gamma$ and ΔQ_γ, one arrives at expressions that are plagued with several pathologies, gauge-dependence being one of them. Indeed, even if the two W are considered to be on shell, since the incoming photon is not, there is no *a priori* reason why a gauge-independent answer should emerge. In the context of the renormalizable R_ξ gauges the final answer depends on the choice of the gauge fixing parameter ξ, which enters into the one-loop calculations through the gauge-boson propagators (W, Z, γ, and unphysical scalar particles). In addition, as shown by an explicit calculation performed in the Feynman gauge ($\xi = 1$), the answer for $\Delta\kappa_\gamma$ is *infrared divergent* and violates perturbative unitarity, e.g. it grows monotonically for $Q^2 \to \infty$ (7). Clearly, regardless of the measurability of quantities like $\Delta\kappa_\gamma$ and ΔQ_γ, from the theoretical point of view one should at least be able to satisfy such crucial requirements as gauge-independence and infrared finiteness, when calculating the model's prediction for them. Indeed, all the above pathologies may be circumvented if one adopts the pinch technique (PT), first invented by Cornwall (8). The application of this method gives rise to new expressions, $\hat{\Delta}\kappa_\gamma$ and $\hat{\Delta}Q_\gamma$, which are gauge fixing parameter (ξ) independent, ultraviolet and infrared finite, and well behaved for large momentum transfers Q^2 (9).

I. THE PINCH TECHNIQUE

The simplest example that demonstrates how the PT works is the gluon two point function (10). Consider the S-matrix element T for the elastic scattering such as $q_1\bar{q}_2 \to q_1\bar{q}_2$, where q_1, q_2 are two on-shell test quarks with masses m_1 and m_2. To any order in perturbation theory T is independent of the gauge fixing parameter ξ. On the other hand, as an explicit calculation shows, the conventionally defined proper self-energy depends on ξ. At the one loop level this dependence is canceled by contributions from other graphs, which, at first glance, do not seem to be propagator-like. That this cancellation must occur and can be employed to define a gauge-independent self-energy, is evident from the decomposition:

$$T(s,t,m_1,m_2) = T_0(t,\xi) + \sum_{i=1}^{2} T_i(t,m_i,\xi) + T_3(s,t,m_1,m_2,\xi), \qquad (9)$$

where the function $T_0(t,\xi)$ depends kinematically only on the Mandelstam variable $t = -(\hat{p}_1 - p_1)^2 = -q^2$, and not on $s = (p_1 + p_2)^2$ or on the external masses. Typically, self-energy, vertex, and box diagrams contribute to T_0, T_1, T_2, and T_3, respectively. Such contributions are ξ dependent, in general. However, as the sum $T(s,t,m_1,m_2)$ is gauge-independent, it is easy to show that Eq(9) can be recast in the form

$$T(s,t,m_1,m_2) = \hat{T}_0(t) + \hat{T}_1(t,m_1) + \hat{T}_2(t,m_2) + \hat{T}_3(s,t,m_1,m_2), \qquad (10)$$

where the $\hat{T}_i$ ($i = 0, 1, 2, 3$) are *individually* ξ-independent. The propagator-like parts of vertex and box graphs which enforce the gauge independence of $\hat{T}_0(t)$, are called pinch parts. They emerge every time a gluon propagator or an elementary three-gluon vertex contributes a longitudinal k_μ to the original graph's numerator. The action of such a term is to trigger an elementary Ward identity of the form

$$\slashed{k} = (\slashed{p} + \slashed{k} - m) - (\slashed{p} - m) \qquad (11)$$

when it gets contracted with a γ matrix. The first term removes (pinches out) the internal fermion propagator, whereas the second vanishes on shell. From the gauge-independent functions $\hat{T}_i$ ($i = 0, 1, 2, 3$) one may now extract a gauge-independent effective gluon (G) self-energy $\hat{\Pi}_{\mu\nu}(q)$, gauge-independent $Gq_i\bar{q}_i$ vertices $\hat{\Gamma}_\mu^{(i)}$, and a gauge-independent box $\hat{B}$, in the following way:

$$\begin{aligned}
\hat{T}_0 &= g^2 \bar{u}_1 \gamma^\mu u_1 [(\tfrac{1}{q^2})\hat{\Pi}_{\mu\nu}(q)(\tfrac{1}{q^2})] \bar{u}_2 \gamma^\nu u_2, \\
\hat{T}_1 &= g^2 \bar{u}_1 \hat{\Gamma}_\nu^{(1)} u_1 (\tfrac{1}{q^2}) \bar{u}_2 \gamma^\nu u_2, \\
\hat{T}_2 &= g^2 \bar{u}_1 \gamma^\mu u_1 (\tfrac{1}{q^2}) \bar{u}_2 \hat{\Gamma}_\nu^{(2)} u_2, \\
\hat{T}_3 &= \hat{B},
\end{aligned} \qquad (12)$$

where u_i are the external spinors, and g is the gauge coupling. Since all hatted quantities in the above formula are gauge-independent, their explicit form may be calculated using any value of the gauge-fixing parameter ξ, as long as one properly identifies and allots all relevant pinch contributions. The choice $\xi = 1$ simplifies the calculations significantly, since it eliminates the longitudinal part of the gluon propagator. Therefore, for $\xi = 1$ the pinch contributions originate only from momenta carried by the elementary three-gluon vertex The one-loop expression for $\hat{\Pi}_{\mu\nu}(q)$ is given by (10) :

$$\hat{\Pi}_{\mu\nu}(q) = \Pi^{(\xi=1)}_{\mu\nu}(q) + t_{\mu\nu}\Pi^P(q) , \tag{13}$$

where

$$t_{\mu\nu} = (g_{\mu\nu}q^2 - q_\mu q_\nu) \tag{14}$$

and

$$\Pi^P(q) = -2ic_a g^2 \int_n \frac{1}{k^2(k+q)^2} , \tag{15}$$

where $\int_n \equiv \int \frac{d^n k}{(2\pi)^n}$ is the dimensionally regularized loop integral, and c_a is the quadratic Casimir operator for the adjoint representation [for $SU(N)$, $c_a = N$]. After integration and renormalization we find

$$\Pi^P(q) = -2c_a(\frac{g^2}{16\pi^2})\ln(\frac{-q^2}{\mu^2})] . \tag{16}$$

Adding this to the Feynman-gauge proper self-energy

$$\Pi^{(\xi=1)}_{\mu\nu}(q) = -[\frac{5}{3}c_a(\frac{g^2}{16\pi^2})\ln(\frac{-q^2}{\mu^2})]t_{\mu\nu} , \tag{17}$$

we find for $\hat{\Pi}_{\mu\nu}(q)$

$$\hat{\Pi}_{\mu\nu}(q) = -bg^2\ln(\frac{-q^2}{\mu^2})t_{\mu\nu} , \tag{18}$$

where $b = \frac{11c_a}{48\pi^2}$ is the coefficient of $-g^3$ in the usual β function.

This procedure can be extended to an arbitrary n-point function; of particular physical interest are the gauge-independent three and four point functions $\hat{\Gamma}_{\mu\nu\alpha}$ (11) and $\hat{\Gamma}_{\mu\nu\alpha\beta}$ (12), which at one-loop satisfy the following *tree-level* Ward identities:

$$q_1^\mu \hat{\Gamma}_{\mu\nu\alpha}(q_1, q_2, q_3) = t_{\nu\alpha}(q_2)\hat{d}^{-1}(q_2) - t_{\nu\alpha}(q_3)\hat{d}^{-1}(q_3)$$
$$q_1^\mu \hat{\Gamma}^{abcd}_{\mu\nu\alpha\beta} = f_{abp}\hat{\Gamma}^{cdp}_{\nu\alpha\beta}(q_1+q_2, q_3, q_4) + c.p. , \tag{19}$$

where $\hat{d} = [q^2 - \hat{\Pi}(q)]^{-1}$ and f^{abc} the structure constants of the gauge group.

Finally, the generalization of the PT to the case of non-conserved external currents is technically more involved (13). The main reasons are the following:

(a) The charged W couples to fermions with different, non-vanishing masses $m_i, m_j \neq 0$, and consequently the elementary Ward identity of Eq.(11) gets modified to :

$$k_\mu \gamma^\mu P_L \equiv \not{k} P_L = S_i^{-1}(p+k)P_L - P_R S_j^{-1}(p) + m_i P_L - m_j P_R \qquad (20)$$

where

$$P_{R,L} = \frac{1 \pm \gamma_5}{2} \qquad (21)$$

are the chirality projection operators. The first two terms of Eq(20) will pinch and vanish on shell, respectively, as they did before. But in addition, a term proportional to $m_i P_L - m_j P_R$ is left over. In a general R_ξ gauge such terms give rise to extra propagator and vertex-like contributions, not present in the massless case.

(b) Additional graphs involving the "unphysical" would-be Goldstone bosons χ and ϕ, and physical Higgs H, which do not couple to massless fermions, must now be included. Such graphs give rise to new pinch contributions, even in the Feynman gauge, due to the momenta carried by interaction vertices such as $\gamma \phi^+ \phi^-$, $Z\phi^+\phi^-$, $W^+\phi^-\chi$, $HW^+\phi^-$, e.g. vertices with one vector gauge boson and two scalar bosons.

II. THE CURRENT ALGEBRA FORMULATION OF THE PINCH TECHNIQUE

We now present an alternative formulation of the PT introduced in the context of the standard model (14). In this approach the interaction of gauge bosons with external fermions is expressed in terms of current correlation functions (15), i.e. matrix elements of Fourier transforms of time-ordered products of current operators. This is particularly economical because these amplitudes automatically include several closely related Feynman diagrams. When one of the current operators is contracted with the appropriate four-momentum, a Ward identity is triggered. The pinch part is then identified with the contributions involving the equal-time commutators in the Ward identities, and therefore involve amplitudes in which the number of current operators has been decreased by one or more. A basic ingredient in this formulation are the following equal-time commutators;

$$\begin{aligned}
\delta(x_0 - y_0)[J_W^0(x), J_Z^\mu(y)] &= c^2 J_W^\mu(x)\delta^4(x-y) \ , \\
\delta(x_0 - y_0)[J_W^0(x), J_W^{\mu\dagger}(y)] &= -J_3^\mu(x)\delta^4(x-y) \ , \\
\delta(x_0 - y_0)[J_W^0(x), J_\gamma^\mu(y)] &= J_W^\mu(x)\delta^4(x-y) \ , \\
\delta(x_0 - y_0)[J_V^0(x), J_{V'}^\mu(y)] &= 0 \ ,
\end{aligned} \qquad (22)$$

where $J_3^\mu \equiv 2(J_Z^\mu + s^2 J_\gamma^\mu)$ and $V, V' \in \{\gamma, Z\}$. To demonstrate the method with an example, consider the vertex Γ_μ, where now the gauge particles in the loop are W instead of gluons and the incoming and outgoing fermions are massless. It can be written as follows (with $\xi = 1$):

$$\Gamma_\mu = \int \frac{d^4k}{2\pi^4} \Gamma_{\mu\alpha\beta}(q, k, -k-q) \int d^4x e^{ikx} < f|T^*[J_W^{\alpha\dagger}(x) J_W^\beta(0)]|i> \ . \quad (23)$$

When an appropriate momentum, say k_α, from the vertex is pushed into the integral over dx, it gets transformed into a covariant derivative $\frac{d}{dx_\alpha}$ acting on the time ordered product $< f|T^*[J_W^{\alpha\dagger}(x) J_W^\beta(0)]|i>$. After using current conservation and differentiating the θ-function terms, implicit in the definition of the T^* product, we end up with the left-hand side of the second of Eq(22). So, the contribution of each such term is proportional to the matrix element of a single current operator, namely $< f|J_3^\mu|i >$; this is precisely the pinch part. Calling Γ_μ^P the total pinch contribution from the Γ_μ of Eq(23), we find that

$$\Gamma_\mu^P = -g^3 c I_{WW}(Q^2) < f|J_3^\mu|i > , \quad (24)$$

where

$$I_{ij}(q) = i \int_n \frac{1}{(k^2 - M_i^2)[(k+q)^2 - M_j^2]} \ . \quad (25)$$

Obviously, the integral in Eq(25) is the generalization of the QCD expression Eq(15) to the case of massive gauge bosons.

III. GAUGE–INVARIANT GAUGE BOSON VERTICES AND THEIR WARD IDENTITIES

We consider the S-matrix element for the process

$$e^- + \nu + e^- \to e^- + e^- + \bar{\nu} \ . \quad (26)$$

and isolate the part $T(q, p_1, p_2)$ of the S–matrix which depends only on the momentum transfers q, p_1, and p_2. The tree-level vector-boson propagator $\Delta_{\mu\nu}^i(q)$ in the R_ξ gauges is given by

$$\Delta_i^{\mu\nu}(q, \xi_i) = \frac{1}{q^2 - M_i^2}[g^{\mu\nu} - (1-\xi_i)\frac{q^\mu q^\nu}{q^2 - \xi_i M_i^2}] \ , \quad (27)$$

with $i = W, Z, \gamma$, and $M_\gamma = 0$. Its inverse $\Delta_i^{-1}(q, \xi_i)^{\mu\nu}$ is given by

$$\Delta_i^{-1}(q, \xi)^{\mu\nu} = (q^2 - M_i^2)g^{\mu\nu} - q^\mu q^\nu + \frac{1}{\xi_i}q^\mu q^\nu \ . \quad (28)$$

The propagators $\Delta_s(q,\xi_i)$ of the unphysical (would–be) Goldstone bosons are given by

$$\Delta_s(q,\xi_i) = \frac{-1}{q^2 - \xi_i M_i^2} , \qquad (29)$$

with $(s,i) = (\phi, W)$ or (χ, Z) and explicitly depend on ξ_i. On the other hand, the propagators of the fermions (quarks and leptons), as well as the propagator of the physical Higgs particle are ξ_i-independent at tree-level.

Since the final result (with pinch contributions included) is gauge-independent, we choose to work in the Feynman gauge ($\xi_i = 1$); this particular gauge simplifies the calculations because it removes all longitudinal parts from the tree-level gauge boson propagators. So, pinch contributions can only originate from appropriate momenta furnished by the tree–level gauge boson vertices. Applying the pinch technique algorithm we isolate all vertex-like parts contained in the box diagrams and allot them to the usual vertex graphs. The final expressions for one loop gauge-independent trilinear gauge boson vertices are :

$$\frac{1}{g^3 s}\widehat{\Gamma}^{\gamma W^- W^+}_{\mu\alpha\beta} = \Gamma^{\gamma W^- W^+}_{\mu\alpha\beta}|_{\xi_i=1} + q^2 B_{\mu\alpha\beta} + U_W^{-1}(p_1)^\rho_\alpha B^+_{\mu\rho\beta} + U_W^{-1}(p_2)^\rho_\beta B^-_{\mu\alpha\rho}$$
$$-2\Omega\Gamma_{\mu\alpha\beta} + p_{2\beta}g_{\mu\alpha}\,\mathcal{M}^- + p_{1\alpha}g_{\mu\beta}\,\mathcal{M}^+ , \qquad (30)$$

$$\frac{1}{g^3 c}\widehat{\Gamma}^{Z W^- W^+}_{\mu\alpha\beta} = \Gamma^{Z W^- W^+}_{\mu\alpha\beta}|_{\xi_i=1} + U_Z^{-1}(q)^\rho_\mu B_{\rho\alpha\beta} + U_W^{-1}(p_1)^\rho_\alpha B^+_{\mu\rho\beta}$$
$$+ U_W^{-1}(p_2)^\rho_\beta B^-_{\mu\alpha\rho} - 2\Omega\Gamma_{\mu\alpha\beta} + q_\mu g_{\alpha\beta}\,M_Z^2\,\mathcal{M}$$
$$+ p_{2\beta}g_{\mu\alpha}\,M_W^2\,\mathcal{M}^- + p_{1\alpha}g_{\mu\beta}\,M_W^2\,\mathcal{M}^+ , \qquad (31)$$

where

$$\Omega = I_{WW}(q) + s^2 I_{W\gamma}(p_1) + c^2 I_{WZ}(p_1) + s^2 I_{W\gamma}(p_2) + c^2 I_{WZ}(p_2) , \qquad (32)$$

and

$$\mathcal{M}^-(q,p_1,p_2) = \frac{s^2}{c^2} J_{WW\gamma} + \frac{1-2s^2}{2c^2} J_{WWZ} + \frac{1}{2} J_{WWH} + \frac{1}{2c^2} J_{ZHW} , \qquad (33)$$

with

$$J_{ABC} = \int_n \frac{1}{[(k+p_1)^2 - M_A^2][(k-p_2)^2 - M_B^2][k^2 - M_C^2]} , \qquad (34)$$

and the property

$$\mathcal{M}^+(q,p_1,p_2) = -\mathcal{M}^-(q,p_2,p_1) . \qquad (35)$$

The gauge-independent vertices satisfy the following simple Ward identities (WI), relating them to the W self energy and χWW vertex constructed also via the PT :

$$q^\mu \widehat{\Gamma}_{\mu\alpha\beta}^{ZW^-W^+} + iM_Z \widehat{\Gamma}_{\alpha\beta}^{\chi W^-W^+} = gc \left[\widehat{\Pi}_{\alpha\beta}^W(1) - \widehat{\Pi}_{\alpha\beta}^W(2) \right] , \quad (36)$$

$$q^\mu \widehat{\Gamma}_{\mu\alpha\beta}^{\gamma W^-W^+} = gs \left[\widehat{\Pi}_{\alpha\beta}^W(1) - \widehat{\Pi}_{\alpha\beta}^W(2) \right] . \quad (37)$$

These WI are the one-loop generalizations of the respective *tree level* WI; their validity is crucial for the gauge independence of the S-matrix. It is important to emphasize that they make no reference to ghost terms, unlike the corresponding Slavnov-Taylor identities satisfied by the conventional, gauge-dependent vertices.

For the case of *on-shell* Ws one sets $p_1^2 = p_2^2 = M_W^2$ and neglects all terms proportional to $p_{1\alpha}$ and $p_{2\beta}$, as well as the left over pinch terms of the W legs. Then the γWW vertex reduces to the form

$$\frac{1}{g^3 s} \widehat{\Gamma}_{\mu\alpha\beta}^{\gamma W^-W^+} = \Gamma_{\mu\alpha\beta}^{\gamma W^-W^+}|_{\xi=1} + q^2 \, B_{\mu\alpha\beta}(q, p_1, p_2) - 2\Gamma_{\mu\alpha\beta} I_{WW}(q) . \quad (38)$$

This is of course the same answer one obtains by applying the PT *directly* to the S-matrix of $e^+e^- \to W^+W^-$. Thus for the form factors $\Delta\kappa_\gamma$, ΔQ_γ the only function we need is $B_{\mu\alpha\beta}$, given below

$$g^2 B_{\mu\alpha\beta} = \sum_{V=\gamma Z} g_V^2 \int_n \frac{i R_{\alpha\beta\mu}}{[(k+p_1)^2 - M_W^2][(k-p_2)^2 - M_W^2][k^2 - M_V^2]} , \quad (39)$$

with

$$R_{\alpha\beta\mu} = g_{\alpha\beta} \left(k - \frac{3}{2}(p_1 - p_2)\right)_\mu - g_{\alpha\mu} (3k + 2q)_\beta - g_{\beta\mu} (3k - 2q)_\alpha . \quad (40)$$

IV. MAGNETIC DIPOLE AND ELECTRIC QUADRUPOLE FORM FACTORS FOR THE W

Having constructed the gauge-independent γWW vertex we proceed to extract its contributions to the magnetic dipole and electric quadrupole form factors of the W. We use carets to denote the gauge independent one-loop contributions. Clearly,

$$\widehat{\Delta}\kappa_\gamma = \Delta\kappa_\gamma^{(\xi=1)} + \Delta\kappa_\gamma^P , \quad (41)$$

and

$$\widehat{\Delta}Q_\gamma = \Delta Q_\gamma^{(\xi=1)} + \Delta Q_\gamma^P , \quad (42)$$

where $\Delta Q_\gamma^{(\xi=1)}$ and $\Delta Q_\gamma^{(\xi=1)}$ are the contributions of the usual vertex diagrams in the Feynman gauge (7), whereas ΔQ_γ^P and ΔQ_γ^P the analogous contributions from the pinch parts. The task of actually calculating $\hat{\Delta}\kappa_\gamma$ and $\hat{\Delta} Q_\gamma$ is greatly facilitated by the fact that the quantities $\Delta\kappa_\gamma^{(\xi=1)}$ and $\Delta Q_\gamma^{(\xi=1)}$ have already been calculated in (7). It must be emphasized however that the expression for $\Delta\kappa_\gamma^{(\xi=1)}$ (but not $\Delta Q_\gamma^{(\xi=1)}$) is infrared divergent for $Q^2 \neq 0$ due to the presence of the following double integral over the Feynman parameters (t,a), given in Eq.(26) of (7):

$$R = -(\frac{\alpha_\gamma}{\pi})\frac{Q^2}{M_W^2}\int_0^1 da \int_0^1 \frac{dtt}{t^2 - t^2(1-a)a(\frac{4Q^2}{M_W^2})}$$

$$= -(\frac{\alpha}{2\pi})\frac{Q^2}{M_W^2}\int_0^1 \frac{da}{1-(1-a)a(\frac{4Q^2}{M_W^2})})\int_0^1 \frac{dt}{t} \quad (43)$$

By performing the momentum integration in $B_{\mu\alpha\beta}$, we find for $p_1^2 = p_2^2 = M_W^2$

$$B_{\mu\alpha\beta} = -\frac{Q^2}{8\pi^2 M_W^2}\sum_{V=\gamma,Z}g_V^2 \int_0^1 da \int_0^1 (2tdt)\frac{F_{\mu\alpha\beta}}{L_V^2}, \quad (44)$$

where

$$F_{\mu\alpha\beta} = 2(\frac{3}{2} + at)g_{\alpha\beta}\Delta_\mu + 2(3at + 2)[g_{\alpha\mu}Q_\beta - g_{\beta\mu}Q_\alpha], \quad (45)$$

and

$$L_V^2 = t^2 - t^2 a(1-a)(\frac{4Q^2}{M_W^2}) + (1-t)\frac{M_V^2}{M_W^2}, \quad (46)$$

from which immediately follows that

$$a_1^P(Q^2) = -\frac{1}{2}\frac{Q^2}{M_W^2}\sum_V \frac{\alpha_V}{\pi}\int_0^1 da \int_0^1 (2tdt)\frac{2(\frac{3}{2} + at)}{L_V^2} \quad (47)$$

and

$$a_2^P(Q^2) = -\frac{1}{2}\frac{Q^2}{M_W^2}\sum_V \frac{\alpha_V}{\pi}\int_0^1 da \int_0^1 (2tdt)\frac{2(2 + 3at)}{L_V^2}, \quad (48)$$

and since there is no term proportional to $\Delta_\mu Q_\alpha Q_\beta$,

$$a_3^P(Q^2) = 0. \quad (49)$$

Therefore,

$$\Delta\kappa_\gamma^P = -\frac{1}{2}\frac{Q^2}{M_W^2}\sum_V \frac{\alpha_V}{\pi}\int_0^1 da \int_0^1 (2tdt)\frac{(at-1)}{L_V^2}, \quad (50)$$

and
$$\Delta Q_\gamma^P = 0 \ . \tag{51}$$

It is important to notice that even though $\Delta Q_\gamma^P = 0$ both μ_W and Q_W will assume values different than those predicted in the $\xi = 1$ gauge. Indeed, even though the value of λ_γ does not change, the value of κ_γ changes, and this change affects both μ_W and Q_W through Eq(4) and Eq(5). In the expression given in Eq(50) the first term (for V=Z) is infrared finite (since $M_Z \neq 0$), whereas the second term (for $V = \gamma$) is infrared divergent, since $M_\gamma = 0$. Calling this second term Θ we have

$$\Theta = -\frac{1}{2}(\frac{\alpha_\gamma}{\pi})\frac{Q^2}{M_W^2}\int_0^1 da \int_0^1 dt \frac{2t(at-1)}{t^2[1 - a(1-a)\frac{4Q^2}{M_W^2}]} , \tag{52}$$

which can be rewritten as

$$\Theta = -R - (\frac{\alpha_\gamma}{\pi})\frac{Q^2}{M_W^2}\int_0^1 da \frac{a}{1 - a(1-a)\frac{4Q^2}{M_W^2}} , \tag{53}$$

where R is the infrared divergent integral defined in Eq(43). On the other hand, the second term in Eq(53) is infrared finite. Clearly, including the first term of Eq(53) in the value of $\hat{\Delta}\kappa_\gamma$ exactly cancels the infrared divergent contribution of Eq(43), thus giving rise to an infrared finite expression for $\hat{\Delta}\kappa_\gamma$. So, after the infrared divergent part of Eq(52) is cancelled, $\Delta\kappa_\gamma^P$ is given by the following expression:

$$\Delta\kappa_\gamma^P = \Theta_\gamma + \Theta_Z , \tag{54}$$

with Θ_γ the second term in Eq(53), and Θ_Z the second term in Eq(50), namely

$$\Theta_\gamma = -(\frac{\alpha_\gamma}{\pi})\frac{Q^2}{M_W^2}\int_0^1 da \frac{a}{1 - a(1-a)\frac{4Q^2}{M_W^2}} , \tag{55}$$

and

$$\Theta_Z = -\frac{Q^2}{M_W^2}(\frac{\alpha_Z}{\pi})\int_0^1 da \int_0^1 dt \frac{t(at-1)}{L_Z^2} , \tag{56}$$

and from Eq(41)

$$\hat{\Delta}\kappa_\gamma = [\Delta\kappa_\gamma^{(\xi=1)}]_{if} + \Theta_\gamma + \Theta_Z , \tag{57}$$

where the subscript (if) in the first term of the R.H.S. indicates that the contribution from the $\xi = 1$ gauge is now genuinely infrared finite. Finally, the magnetic dipole moment μ_W and electric quadrupole moment Q_W are given by

$$\mu_W = \frac{e}{2M_W}(2 + \hat{\Delta}\kappa_\gamma) \tag{58}$$

and

$$Q_W = -\frac{e}{M_W^2}(1 + \hat{\Delta}\kappa_\gamma + 2\hat{\Delta}Q_\gamma) . \tag{59}$$

Both $\Delta Q_\gamma^{(\xi=1)}$ and $\Delta\kappa_\gamma^{(\xi=1)}$ have been computed numerically in (7). We now proceed to compute the integrals in Eq(55) and Eq(56), which determine $\Delta\kappa_\gamma^P$. It is elementary to evaluate Θ_γ. Setting $\Theta_\gamma = -(\frac{\alpha_\gamma}{\pi})\hat{\Theta}_\gamma$ we have:

$$\begin{aligned}\hat{\Theta}_\gamma &= \tfrac{2}{\Delta}[arctg(\tfrac{1}{\Delta}) - arctg(\tfrac{-1}{\Delta})], & Q^2 < M_W^2 \\ &= -4, & Q^2 = M_W^2 \\ &= \tfrac{2}{\Delta}\ln[\tfrac{|\Delta-1|}{\Delta+1}], & Q^2 > M_W^2\end{aligned} \tag{60}$$

for space-like Q^2, where $\Delta = \sqrt{|\frac{M_W^2}{Q^2} - 1|}$, and

$$\hat{\Theta}_\gamma = \frac{2}{\Delta}\ln[\frac{|\Delta - 1|}{\Delta + 1}] \tag{61}$$

for time-like Q^2, where $\Delta = \sqrt{\frac{M_W^2}{|Q^2|} + 1}$.

The double integral Θ_Z can in principle be expressed in a closed form in terms of Spence functions [see for example (16)], but this is of limited usefulness for our present calculation. Instead, we evaluated this integral numerically. We used the same values for the constants appearing in our calculations as in (7), namely $\alpha_\gamma = \frac{1}{128}$, $M_W = 80.6 GeV$, $M_Z = 91.1 GeV$ and $s = 0.23$.

The result of the computation is very interesting. $\Delta\kappa_\gamma^P$, which originates from pinching box diagrams, furnishes exactly the contributions needed to restore the unitarity of the final answer. Indeed, as the authors of (7) emphasized, $\Delta\kappa_\gamma^{(\xi=1)}$ is by itself not a gauge invariant object in the limit $Q^2 \to \infty$, where the local $SU(2) \times U(1)$ symmetry is restored. For large values of Q^2, $\Delta\kappa_\gamma^P$ is nearly equal in magnitude and opposite in sign to $\Delta\kappa_\gamma^P$. Therefore, when according to Eq(41) and Eq(42) both contributions are added, $\hat{\Delta}\kappa_\gamma \to 0$ as $Q^2 \to \pm\infty$. Clearly, the inclusion of the pinch parts from the box graphs is *crucial* for restoring the good asymptotic behavior of the W form factors.

V. CONCLUSIONS

We presented a study of the structure of trilinear gauge boson vertices in the context of the standard model. Using the S-matrix pinch technique gauge-independent γWW and ZWW vertices were constructed to one-loop order, with all three incoming momenta off-shell. These vertices satisfy naive QED-like Ward identities, which relate them to the gauge independent W self-energy, which were also obtained via the pinch technique. The tree-level Ward identities are to be contrasted with the complicated Ward identities satisfied by the conventionally defined gauge-dependent vertices; in particular, no ghost terms need be included. Finally, when the appropriate Lorentz structures are extracted, these vertices give rise to gauge-independent, infrared finite, and asymptotically well-behaved magnetic dipole and electric quadrupole form factors for the W, which can, at least in principle, be promoted to physical observables. It would be interesting to determine how these quantities could be directly extracted from future e^+e^- experiments.

VI. ACKNOWLEDGMENT

I thank J. M. Cornwall, K. Hagiwara, A. Lahanas, K. Philippides, R. Peccei, and D. Zeppenfeld for useful discussions, and E. Karagiannis for his warm hospitality during my visit in Los Angeles. This work was supported by the National Science Foundation under Grant No.PHY-9017585.

REFERENCES

1. G. Belanger, F. Boudjema, D. London, Phys. Rev. Lett. **65** 2943 (1990).
2. W. A. Bardeen, R. Gastmans, and B. Lautrup, Nucl. Phys. **B 46** 319 (1972).
3. K. J. F. Gaemers and G. J. Gounaris, Z. Phys. C1, 259 (1979).
4. K. Hagiwara et al, Nucl. Phys. **B 282**, 253 (1987).
5. U. Baur and D. Zeppenfeld, Nucl. Phys. **B 308**, 127 (1988)
6. U. Baur and D. Zeppenfeld, Nucl. Phys. **B 325**, 253 (1989)
7. E. N. Argyres et al, Nucl. Phys. **B 391**, 23 (1993)
8. J. M. Cornwall, in *Proceedings of the 1981 French-American Seminar on Theoretical Aspects of Quantum Chromodynamics*, Marseille, France, 1981, edited by J. W. Dash (Centre de Physique Théorique, Marseille, 1982).
9. J. Papavassiliou and K. Philippides, Phys. Rev. **D 48** 4255 (1993)
10. J. M. Cornwall, Phys. Rev. **D 26** 1453 (1982)
11. J. M. Cornwall and J. Papavassiliou, Phys. Rev. **D 40** 3474 (1989)
12. J. Papavassiliou, Phys. Rev. **D 47** 4728 (1993)
13. J. Papavassiliou, Phys. Rev. **D 50** 5958 (1994)
14. G. Degrassi and A. Sirlin, Phys. Rev. **D 46** 3104 (1992)
15. A. Sirlin, Rev. Mod. Phys. **50** 573 (1978).
16. G. t'Hooft and M. Veltman, Nucl. Phys. **B 153**,365 (1979).

Static Quantities of the W bosons in the MSSM

A.B.Lahanas*

University of ATHENS, Physics Department
Nuclear and Particle Physics Section
Athens 157 71 Greece

I discuss the static quantities of the W boson, magnetic dipole and electric quadrupole moments, in the context of the minimal supersymmetric standard model, in which supersymmetry is broken by soft terms A_o, m_o, $M_{1/2}$. Following a renormalization group analysis it is found that the supesymmetric values of Δk_γ and ΔQ_γ can be largely different, in some cases, from the standard model predictions but of the same order of magnitude for values of $A_0, m_0, M_{1/2} \leq \mathcal{O}(1 TeV)$. Therefore possible supersymmetric structure can be probed provided the accuracy of measurements for Δk_γ, ΔQ_γ reaches $10^{-2} - 10^{-3}$ and hence hard to be detected at LEP2. In cases where $M_{1/2} \ll A_0, m_0$, the charginos and neutralinos may give substantial contributions saturating the LEP2 sensitivity limits. This occurs when their masses $m_{\tilde{C}}, m_{\tilde{Z}}$ turn out to be both light satisfying $m_{\tilde{C}} + m_{\tilde{Z}} \simeq M_W$. However these extreme cases are perturbatively untrustworthy and besides unnatural for they occupy a small region in the parameter space.

INTRODUCTION

Supersymmetry is a reasonable extension of the SM, theoretically motivated but without any direct experimental confirmation for its existence as yet. The recently revived interest in supersymmetric theories derives from the fact that high precision measurements of the SM parameters at LEP e^+e^- CERN collider shows that SU(3),SU(2),U(1) gauge couplings merge at a single point at energies $\sim 10^{16} GeV$ if supersymmetry is adopted with an effective SUSY breaking scale M_S (1)

$$M_Z < M_S < 1 TeV$$

In order to produce SUSY particles with such large masses at observable

---

[1] My thanks are due to the organizers for inviting and giving me the opportunity to participate in the Symposium, and also to the UCLA Physics Department for the hospitality extended to me during my stay in Los Angeles. Work supported in part by EEC contract SCI-CT92-0792.

rates high energies and luminocities are required and the question is if there are signals for SUSY below the supersymmetric particle thresholds.

The three gauge boson vertex will be probed in future experiments with high accuracy and it is perhaps a good place to look for supersymmetric signatures. In particular the static quantities of the W-boson are affected by the radiative effects which are due to supersymmetric particles and deviations from the Standard Model predictions are expected.

Are these deviations detectable ? How they depend on the effective SUSY scale ?

To answer this within the context of the $MSSM$ requires a systematic analysis in which alllimitations imposed by the RG and the radiative symmetry breaking are duly taken into account.

THE MSSM

The MSSM is the minimal extension of the SM in that it is based on the gauge group $SU(3) \times SU(2) \times U(1)$ and has the minimal physical content. It involves two Higgs multiplets $\hat{H}_{1,2}$, and the minimum number of chiral quark and lepton multiplets to accomodate the matter fermions $(\hat{Q}, \hat{U}^c, \hat{D}^c, \hat{L}, \hat{E}^c)$. Its Lagrangian is

$$\mathcal{L} = \mathcal{L}_{SUSY} + \mathcal{L}_{soft}.$$

$\mathcal{L}_{SUSY}$ is its supersymmetric part derived from a superpotential $\mathcal{W}$ bearing the form

$$\mathcal{W} = h_U \hat{Q} \hat{H}_2 \hat{U}^c + h_D \hat{Q} \hat{H}_1 \hat{D}^c + h_E \hat{L} \hat{H}_1 \hat{E}^c + \mu \hat{H}_1 \hat{H}_2$$

and $\mathcal{L}_{soft}$ is its supersymmetry breaking part given by

$$-\mathcal{L}_{soft} = \sum_i m_i^2 |\Phi_i|^2$$
$$+ (h_U A_U Q H_2 U^c + h_D A_D Q H_1 D^c + h_E A_L L H_1 E^c + h.c.)$$
$$+ (\mu B H_1 H_2 + h.c.) + \frac{1}{2} \sum_a M_a \bar{\lambda}_a \lambda_a.$$

The sum extends over all scalar fields involved and all family indices have been suppressed (2).

All soft scalar masses m_i, gaugino masses M_a, and trilinear scalar couplings $A_{U,D,L}$ are assumed equal at the unification scale, that is we adopt universal boundary conditions as suggested by grand unification and absence of FCNC.

$$m_i = m_0 \, , \, A_{U,D,L} = A_0 \, , \, M_a = M_{1/2} (at \, M_{GUT})$$

This choice parametrizes our ignorance concerning the origin of the supersymmetry breaking terms in the most economical way but it is in no way

TABLE 1. A typical mass spectrum of the MSSM for the inputs shown below

$m_t = 170$, $\tan\beta = 2.1$	$A_0, m_0, M_{1/2}$: 500, 500, 75
Particle	Physical mass (case $\mu > 0$)
Top : M_t	174.6
Higgses $H_\pm$ H_o A h_o	786.4 784.1 782.3 88.4
Squarks $\tilde{u}_L, \tilde{c}_L$ $\tilde{u}_R, \tilde{c}_R$ $\tilde{d}_L, \tilde{s}_L$ $\tilde{d}_R, \tilde{s}_R$ $\tilde{t}_1, \tilde{t}_2$ $\tilde{b}_1, \tilde{b}_2$	524.3 523.0 527.8 523.9 442.2 , 144.0 523.9 , 387.7
Sleptons : $\tilde{e}_L, \tilde{\nu}_L, \tilde{e}_R$	504.0 , 502.0 , 500.4
Gluinos : $\tilde{g}$	189.0
Neutralinos : $\tilde{Z}_{1,2,3,4}$	27.0 , 52.4 , 503.7 , 489.1
Charginos : $\tilde{C}_{1,2}$	50.7 , 501.5

mandatory. The sparticle mass spectrum is completely known once all soft SUSY breaking and mixing parameters at the unification scale M_{GUT} are given as well as the top Yukawa coupling. The number of parameters is reduced to five if we make use of the fact that $M_z = 91.18 GeV$. A convenient choice is to take as independent parameters :

$$m_0 \quad , \quad M_{1/2} \quad , \quad A_0 \quad , \quad m_t(M_z) \quad , \quad \tan\beta(M_z) = \frac{<H_2>}{<H_1>}$$

Then by running the RGE's of all couplings and masses involved the full set of parameters down at energies $\sim M_z$ is known and predictions can be made. There are some subtleties in this approach which are associated with the breaking of the electroweak symmetry ,which takes place via radiative corrections, the appearance of particle thresholds etc. which affect the low energy predictions for the sparticle mass spectrum but these in no way affect the static quantities of the W boson at the one loop order.

A typical mass spectrum is shown in table 1 where the one loop corrections to the Higgs particles, due to the heavy top and stop sector , have been taken into account. As is well known these yield large radiative corrections especially to the lightest of the neutral Higgses involved.

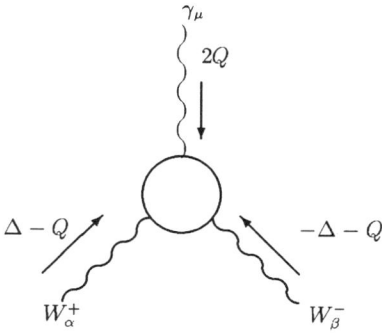

FIG. 1. Kinematics of the WWV vertex

STATIC QUANTITIES OF THE W -BOSON IN THE MSSM

The $WW\gamma$ vertex

The most general form of the WWV vertex ($V = \gamma, Z$), with the two W's on shell and neglecting the scalar components of the boson V, is (3)

$$\Gamma^V_{\mu\alpha\beta} = -ig_V \{ f_V [2g_{\alpha\beta}\Delta_\mu + 4(g_{\alpha\mu}Q_\beta - g_{\beta\mu}Q_\alpha)] +$$

$$2\Delta k_V (g_{\alpha\mu}Q_\beta - g_{\beta\mu}Q_\alpha) + 4\frac{\Delta Q_V}{M_W^2}\Delta_\mu(Q_\alpha Q_\beta - \frac{Q^2}{2}g_{\alpha\beta})\} + ...$$

$$(g_\gamma = e\,,\; g_Z = e\cot\theta_W)$$

(ellipsis are C and CP odd terms)
The labelling of the momenta and Lorentz indices is as shown in figure 1.
$\Delta k_V(Q^2), \Delta Q_V(Q^2)$ are functions of Q^2. The static quantities of the W boson magnetic dipole μ_W and electric quadrupole Q_W moments are related

TABLE 2. Particle contributing to the static quantities of the W boson

SM	MSSM
Gauge bosons	Gauge bosons
Matter fermions	Matter fermions
1 physical Higgs	5 physical Higgses
	$\tilde{q}, \tilde{l}$
	$\tilde{Z}, \tilde{C}$

to these by [2],

$$\mu_W = \frac{e}{2M_W}(2 + \Delta\kappa_\gamma(0)), \quad Q_W = -\frac{e}{M_W^2}(1 + \Delta\kappa_\gamma(0) + \Delta Q_\gamma(0))$$

$\Delta k_V(Q^2), \Delta Q_V(Q^2)$ receive contributions from radiative corrections due to the SM itself as well as from possible existence of new physics which opens at some scale $\Lambda > G_F^{-\frac{1}{2}}$. In table 2 we display the various sectors contributions to these quantities in the SM and $MSSM$.

SM calculations

Within the SM $\Delta k_\gamma(0), \Delta Q_\gamma(0)$ were first calculated long time ago by Bardeen,Gastmans and Lautrup, (4). The effect of the heavy fermion family (t,b) was subsequently discussed by Couture and Ng (5). The form factors $\Delta k_V(Q^2)$, $\Delta Q_V(Q^2)$ have been also calculated (6) and their Q^2 dependence has been studied in detail. In that work it was found that as Q^2 grows $\Delta k_V(Q^2)$ increases , violating unitarity , and has singular infrared (IF) behaviour . This reflects the fact that away from $Q^2 = 0$ the results are not gauge independent. Actually the calculations in that reference were performed in the 't Hooft-Feynman ($\xi = 1$) gauge. In order to get gauge independent results additional contributions stemming from box diagrams have to be added as was noted by Papavassiliou and Phillipides (7).

SUSY calculations

$\Delta k_\gamma(0), \Delta Q_\gamma(0)$ have been also calculated in supersymmetric versions of the SM. Bilchak, Gastmans and Van Proyen (8), studied $\Delta k_\gamma(0), \Delta Q_\gamma(0)$ in a particular supersymmetric model in which electroweak symmetry is broken through a singlet which gets nonvanishing v.e.v. SUSY however remains unbroken in this model. Aliev (9), dealt with the MSSM in which SUSY is

---

[2] In other schemes in which parametrization in terms of k_γ, λ_γ is prefered: $\mu_W = \frac{e}{2M_W}(1 + k_\gamma + \lambda_\gamma)$, $Q_W = -\frac{e}{M_W^2}(k_\gamma - \lambda_\gamma)$

broken by the appearance of soft terms $A_0, B_0, m_0, M_{1/2}$. However no renormalization group analysis is presented in that paper ; results are only given in a particular case which is actually the supersymmetric limit of the MSSM ,that is no soft SUSY breaking terms and absence of Higgsino mixing parameter. It also seems that the contributions of the sensitive Neutralino-Chargino sector presented in that reference are incorrectly given. Couture,Ng,Hewett and Rizzo (10), did a more systematic analysis ; however the constraints imposed by the Renormalization Group study of the MSSM , especially those from the radiative breaking of the EW symmetry, have not been considered. Also mixings of the various sparticles occurring after electroweak breaking takes place have been ignored. In a more recent paper (11), we systematically analyzed the static quantities μ_W, Q_W or equivalently $\Delta k_\gamma(0), \Delta Q_\gamma(0)$ in the context of the MSSM as functions of the soft SUSY breaking parameters $A_0, B_0, m_0, M_{1/2}$ and the top quark mass . We followed a Renormalization Group (RG) analysis and took into account all constraints imposed by the radiative breaking scenario. The contributions of the various sectors involved are as follows:

Gauge Bosons

In units of $g^2/16\pi^2 \simeq 2.6 \times 10^{-3}$ and for $Q^2 = 0$ the gauge boson contributions to $\Delta k_\gamma(0), \Delta Q_\gamma(0)$ are (4),

$$\gamma : \quad \Delta k_\gamma = \frac{20}{3}\sin^2\theta_W \;, \quad \Delta Q_\gamma = \frac{4}{9}\sin^2\theta_W$$

$$Z : \quad \Delta k_\gamma = \frac{20}{3R} - \frac{5}{6} + \frac{1}{2}\int_0^1 dt \frac{t^4 + 10t^3 - 36t^2 + 32t - 16}{t^2 + R(1-t)}$$

$$\Delta Q_\gamma = (\frac{8}{3R} + \frac{1}{3})\int_0^1 dt \frac{t^3(1-t)}{t^2 + R(1-t)}$$

$$(R = (M_Z/M_W)^2)$$

These result to

$$\Delta k_\gamma(0) = 1.18 \;, \quad \Delta Q_\gamma(0) = .235$$

,in units of $g^2/16\pi^2$. For nonvanishing Q^2 the Pinch Parts of the box graphs should be included in order to get gauge independent results as already discussed.

Matter Fermions

Matter fermions are the same in both SM and MSSM and such contributions have been calculated. However we think there is a sign error in the original

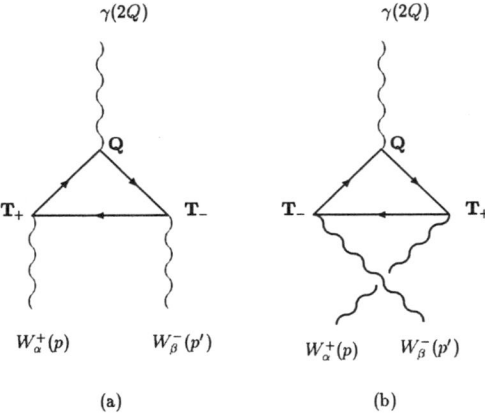

FIG. 2. Triangle fermion graphs contributing to the magnetic dipole and electric quadrupole moments. Q, T_+, T_- denote electric charge and isospin raising and lowering operators respectively.

paper of Bardeen et al which has been propagated in all following references (11). This has been also noted independently of us by Culatti (12).

There are two triangle fermion graphs contributing which are crossed of each other as shown in figure 2. One may think that since 'Up' and 'Down' quarks carry opposite electric charges the triangle graphs in which an 'Up' quark couples to the photon and the same graph in which the 'Down' plays that role give opposite contributions to $\Delta k_\gamma(0), \Delta Q_\gamma(0)$. If for the sake of the argument assume that all fermions are massless ,which is actually the case for the first two families, this would mean that the total fermionic contribution is proportional to

$$Trace \ \{Q\}$$

which is well known to vanish (anomaly cancellation condition). This is stated in almost all previous references and for this reason the contributions of the first two generations of fermions are not considered. Ignoring group factors the triangle graphs shown in figure 2, follow from each other under the inter-

changes
$$\alpha \rightleftharpoons \beta \quad , \quad Q_\mu \rightleftharpoons Q_\mu, \Delta_\mu \rightleftharpoons -\Delta_\mu$$

With the relavant group factors taken into account we get for the two graphs
$$Graph(2a) = Tr(T_-T_+Q)\, V_{\mu\alpha\beta}(Q,\Delta)$$
$$Graph(2b) = Tr(T_+T_-Q)\, V_{\mu\beta\alpha}(Q,-\Delta)$$

where the tensor $V_{\mu\alpha\beta}(Q,\Delta)$ which also includes the anomaly term is given by,
$$V_{\mu\alpha\beta}(Q,\Delta) = \alpha_0 \epsilon_{\alpha\beta\mu\lambda}\Delta^\lambda$$
$$+ \beta_1 g_{\alpha\beta}\Delta_\mu + \beta_2(g_{\alpha\mu}Q_\beta - g_{\beta\mu}Q_\alpha) + \beta_3\Delta_\mu Q_\alpha Q_\beta + ...$$

It is seen that anomaly term preserves its sign under the interchange of indices and momenta given above unlike the rest of the terms whose sign is flipped. This results to a total contribution,
$$Trace(\,Q\,\{T_-,T_+\})\epsilon^{\alpha\beta\mu\lambda}\Delta_\lambda + Trace(\,Q\,[T_-,T_+])(\beta_1 g_{\alpha\beta}\Delta_\mu + ...)$$

Thus the fermion contributions to the dipole/quadrupole moments are weighted by
$$Trace\{\,Q\,T^{(3)}\,\}$$
and the anomaly by
$$Trace\{\,Q\,\} = 0$$

Thus 'Up' and 'Down' quarks yield same sign contributions despite the fact that they carry opposite electric charges. This we think had been overlooked in previous works. As a result the first two generations, which we assume to have vanishing masses, yield nonzero contributions to the dipole and quadrupole moments contrary to what has been previusly claimed. The fermionic contributions to $\Delta k_\gamma, \Delta Q_\gamma$ of an $SU(2)$ doublet $\binom{f}{f'}_L$ are thus given by,

$$\Delta k_\gamma = \frac{C_g}{2}Q_{f'}\int_0^1 dt\, \frac{t^4 + (r_f - r_{f'} - 1)t^3 + (2r_{f'} - r_f)t^2}{t^2 + (r_{f'} - r_f - 1)t + r_f} - [f \rightleftharpoons f']$$

$$\Delta Q_\gamma = \frac{2C_g}{3}Q_{f'}\int_0^1 dt\, \frac{t^3(1-t)}{t^2 + (r_{f'} - r_f - 1)t + r_f} - [f \rightleftharpoons f']$$

$(r_{f,f'} \equiv (m_{f,f'}/M_W)^2)$

The first two families yield,
$$\Delta k_\gamma(0) = -1.334 \quad , \quad \Delta Q_\gamma(0) \simeq 1.776$$

always in units of $g^2/16\pi^2$. Actually this is the largest contributions of all sectors to $\Delta Q_\gamma(0)$. The third family contributes
$$\Delta k_\gamma(0) \approx -.62 \quad , \quad \Delta Q_\gamma(0) = .145$$
for $m_t = 170 GeV$ and $m_b \approx 5 GeV$.

Higgs Bosons

The Higgs sector of the MSSM is not like that of the SM. One needs two Higgs doublets H_1, H_2 whose mass eigenstates $A(neutral\ CP\ odd)$, $h_0, H_0, (neutrals\ CP\ even)$, and the charged Higgses $H_\pm$ have the following masses,

$$A \quad : m_A^2 = m_1^2 + m_2^2$$

$$H_0, h_0 \quad : m_{H,h}^2 = \tfrac{1}{2}\{(m_A^2 + M_Z^2)^2 \pm \sqrt{(m_A^2 + M_Z^2)^2 - 4M_Z^2 m_A^2 \cos^2(2\beta)}\}$$

$$H_\pm \quad : m_{H_\pm}^2 = m_A^2 + M_W^2$$

h_0, H_0 are mixings of the fields $\xi_{1,2} \equiv Re H_{1,2}^0$ and this mixing is specified by the angle θ

$$\xi_2 = (\cos\theta) h_0 + (\sin\theta) H_0, \quad \xi_1 = (\sin\theta) h_0 - (\cos\theta) H_0$$

The field h_0 is predominantly ξ_1 if $\sin\theta$ is close to unity and in that case it is the SM Higgs boson. The contributions of a two Higgs model has been first discussed by Couture et al. (13); however no dependence on the the mixing angle θ appears in their results.

At the tree level the neutral h_0, the lightest of H_0, h_0, is lighter than the Z gauge boson. However radiative effects due to the heavy top/stop system are substantial resulting to corrections that can push its mass to values exceeding M_Z ($m_h \approx 60 - 130\ GeV$ for small $\tan\beta$). This neutral yields the largest contributions of all Higgses involved. $H_\pm, A, H_0$ have masses of the order of $\mathcal{O}(\mathcal{M}_{SUSY})$, lying therefore in the TeV range. Their contributions to the dipole/quadrupole moments are much smaller.

The Higgs contributions to the moments under discussion are given by,

$$A : \quad \Delta k_\gamma = D_2(R_A, R_+) \quad , \quad \Delta Q_\gamma = Q(R_A, R_+)$$
$$h_0 : \quad \Delta k_\gamma = \sin^2\theta\, D_1(R_h) + \cos^2\theta\, D_2(R_h, R_+)$$
$$\Delta Q_\gamma = \sin^2\theta\, Q(R_h, 1) + \cos^2\theta\, Q(R_h, R_+)$$
$$H_0 : \quad As\ in\ h_0\ with\ R_h \to R_H \quad and \quad \sin^2\theta \rightleftharpoons \cos^2\theta$$

$$R_a \equiv (m_a/M_W)^2 \quad a = h_0, H_0, A, H_\pm$$

while the corresponding Standard Model Higgs contribution is,

$$\Delta k_\gamma = D_1(\delta) \quad , \quad \Delta Q_\gamma = Q(\delta, 1) \quad (\delta = (m_{Higgs}/M_W)^2)$$

In the equations above the functions $D_{1,2}, Q$ are defined as,

▷ $$D_1(r) \equiv \frac{1}{2}\int_0^1 dt \frac{2t^4 + (-2-r)t^3 + (4+r)t^2}{t^2 + r(1-t)}$$

$$D_2(r,R) \equiv \frac{1}{2}\int_0^1 dt \frac{2t^4 + (-3-r+R)t^3 + (1+r-R)t^2}{t^2 + (-1-r+R)t + r}$$

$$Q(r,R) \equiv \frac{1}{3}\int_0^1 dt \frac{t^3(1-t)}{t^2 + (-1-r+R)t + r}$$

Scanning the parameter space we found that the Higgs contributions to the dipole and quadrupole moments receive values

$$\Delta\kappa_\gamma \simeq 1.-.5 \quad , \quad \Delta Q_\gamma \simeq \mathcal{O}(10^{-2})$$

Squarks-Sleptons

This sector gives $\mathcal{O}(10^{-2})$ contributions to both $\Delta\kappa_\gamma, \Delta Q_\gamma$ even in cases where due to large mixings one of the stops turns out to be light. For a sfermion $SU(2)$ doublet $\binom{\tilde{f}}{\tilde{f}'}_L$ these contributions read as follows,

$$\Delta\kappa_\gamma = -C_g Q_{f'} \sum_{i,j=1}^2 (K_{i1}^{\tilde{f}} K_{j1}^{\tilde{f}'})^2 \int_0^1 dt \frac{t^2(t-1)(2t-1+R_{\tilde{f}'_j}-R_{\tilde{f}_i})}{t^2 + (R_{\tilde{f}'_j}-R_{\tilde{f}_i}-1)t + R_{\tilde{f}_i}} - [f \rightleftharpoons f']$$

$$\Delta Q_\gamma = -\frac{2C_g Q_{f'}}{3} \sum_{i,j=1}^2 (K_{i1}^{\tilde{f}} K_{j1}^{\tilde{f}'})^2 \int_0^1 dt \frac{t^3(1-t)}{t^2 + (R_{\tilde{f}'_j}-R_{\tilde{f}_i}-1)t + R_{\tilde{f}_i}} - [f \rightleftharpoons f']$$

$R_{\tilde{f}_i,\tilde{f}'_i} \equiv (m_{\tilde{f}_i,\tilde{f}'_i}/M_W)^2$; $m_{\tilde{f}_i,\tilde{f}'_i}$ *are sfermion masses.*

The matrices $\mathbf{K}^{\tilde{f},\tilde{f}'}$ shown in the expressions above diagonalize the sfermion mass matrices. The calculation is complicated only by the presence of $\tilde{f}_L - \tilde{f}_R$ mixings due to the electroweak symmetry breaking effects. In the absence of SUSY breaking their contributions to the quadrupole moment cancels against that of fermions as they should.

Neutralinos-Charginos

This is perhaps the most difficult sector to deal with due to substantial mixings originating from the EW symmetry breaking effects. In the chargino sector the charged $SU(2)$ gauge fermions $\tilde{W}_\pm$ mix with the charged Higgs fermions $\tilde{H}_1^-, \tilde{H}_2^+$ through the mass matrix

$$\mathcal{M}_C = \begin{pmatrix} M_2 & -g_2 v_2 \\ -g_2 v_1 & \mu \end{pmatrix}$$

which is diagonalized by two unitary matrices U, V,

$$U\mathcal{M}_C V^\dagger = diag\{m_1, m_2\}$$

The mass eigenstates $\tilde{C}_1^+, \tilde{C}_2^+$ (Charginos) are Dirac fermions with masses m_1, m_2:

$$m_{1,2}^2 = \frac{1}{2}[M_2^2 + \mu^2 + 2M_W^2 \pm$$

$$\sqrt{((M_2 - \mu)^2 + 2M_W^2(1 + \sin 2\beta))((M_2 + \mu)^2 + 2M_W^2(1 - \sin 2\beta))}]$$

In the Neutralino sector, the gauginos $\tilde{W}_3$, $\tilde{B}$ and the neutral Higgsinos $\tilde{H}_1^0, \tilde{H}_2^0$ get mixed with a mass matrix,

$$\mathcal{M_N} = \begin{pmatrix} M_1 & 0 & g'v_1/\sqrt{2} & -g'v_2/\sqrt{2} \\ 0 & M_2 & -gv_1/\sqrt{2} & gv_2/\sqrt{2} \\ g'v_1/\sqrt{2} & -gv_1/\sqrt{2} & 0 & -\mu \\ -g'v_2/\sqrt{2} & gv_2/\sqrt{2} & -\mu & 0 \end{pmatrix}$$

which is diagonalized by an orthogonal matrix $\mathcal{O}$:

$$\mathcal{O M_N O}^T = diagonal$$

The eigenstates of the mass matrix $\mathcal{M_N}$ are four Majorana fermions $\tilde{Z}_\alpha$, $\alpha = 1, 2, 3, 4$. Their Weak and Electromagnetic currents of these states are,

$$J_+^\mu = \sum_{\alpha,i} \bar{\tilde{Z}}_\alpha \gamma^\mu (R C_{\alpha i}^R + L C_{\alpha i}^L) \tilde{C}_i^+ \quad , \quad J_{em}^\mu = \sum_i \bar{\tilde{C}}_i^+ \gamma^\mu \tilde{C}_i^+$$

where the left and right handed couplings $C_{\alpha i}^{R,L}$ are given in terms of U, V, O matrices by,

$$C_{\alpha i}^R = -\frac{1}{\sqrt{2}} O_{3\alpha} U_{i2}^* - O_{2\alpha} U_{i1}^* \quad , \quad C_{\alpha i}^L = +\frac{1}{\sqrt{2}} O_{4\alpha} V_{i2}^* - O_{2\alpha} V_{i1}^*$$

There is only one triangle graph contributing in this case since the neutralini are Majoranna fermions yielding,

$$\Delta k_\gamma = -\sum_{i,\alpha} F_{\alpha i} \int_0^1 dt \frac{t^4 + (R_\alpha - R_i - 1)t^3 + (2R_i - R_\alpha)t^2}{t^2 + (R_i - R_\alpha - 1)t + R_\alpha}$$

$$+ \sum_{i,\alpha} sign(m_i m_\alpha) \, G_{\alpha i} \sqrt{R_\alpha R_i} \int_0^1 dt \frac{4t^2 - 2t}{t^2 + (R_i - R_\alpha - 1)t + R_\alpha}$$

$$\Delta Q_\gamma = -\frac{4}{3} \sum_{i,\alpha} F_{\alpha i} \int_0^1 dt \frac{t^3(1-t)}{t^2 + (R_i - R_\alpha - 1)t + R_\alpha},$$

$(R_{\alpha,i} \equiv (m_{\alpha,i}/M_W)^2)$

TABLE 3. Order of magnitude contributions to Δk_γ, ΔQ_γ

	Δk_γ	ΔQ_γ
Gauge bosons	$+\mathcal{O}(1)$	$+\mathcal{O}(10^{-1})$
Matter fermions	$-\mathcal{O}(1)$	$+\mathcal{O}(1)$
Higgses	$+\mathcal{O}(1)$	$+\mathcal{O}(10^{-2})$
$\tilde{q}, \tilde{l}$	$\pm\mathcal{O}(10^{-2})$	$\pm\mathcal{O}(10^{-2})$
$\tilde{Z}, \tilde{C}$	$\pm\mathcal{O}(10^{-2}-?)$	$\pm\mathcal{O}(10^{-2}-?)$

The prefactors in these formulae are :

$$F_{\alpha i} = |C^R_{\alpha i}|^2 + |C^L_{\alpha i}|^2 \quad , \quad G_{\alpha i} = (C^L_{\alpha i} C^{R*}_{\alpha i} + (h.c))$$

The neutralino-chargino sector can accomodate light mass eigenstates and in such a case the contributions to the dipole and quadrupole moments are not in general suppressed. Actually we can have a sizeable effect from this sector when there are light neutralino-chargino states ($< M_W$) and this can happen provided the soft mass $M_{1/2}$ is smaller than A_0, m_0. In that case

$$\Delta k_\gamma(0) = \mathcal{O}(1) \quad , \quad \Delta Q_\gamma(0) = \mathcal{O}(1)$$

When $M_{1/2} \approx A_0, m_0 \sim \mathcal{O}(TeV)$ both are $\mathcal{O}(10^{-2})$ or even smaller.

NUMERICAL ANALYSIS

We scanned the parameter space $A_0, m_0, M_{1/2}$ in the range $\mathcal{O}(100 GeV)$ to $1 TeV$. The space is divided into three regions:

- $A_0 \simeq m_0 \simeq M_{1/2}$ (comparable)
- $A_0 \simeq m_0 << M_{1/2}$ (Gaugino dominant)
- $A_0 \simeq m_0 >> M_{1/2}$ (Light gluinos)

For the top mass m_t we considered values in the range ,

$$130 GeV < m_t < 190 GeV$$

We found that the MSSM predictions for the dipole and quadrupole moments differ, in general, from those of the SM but they of the same order of magnitude. Therefore supersymmetric structure can not be possibly probed at LEP2. The order of magnitude of the contributions of the various sectors , in units of $g^2/(16\pi^2)$, are as shown in table 3. Running the numerical routines we found regions of the parameter space allowing for light chargino and neutralino masses satisfying

$$m_{\tilde{C}} + m_{\tilde{Z}} \approx M_W$$

In such cases the values of Δk_γ, ΔQ_γ are substantially enhanced. In fact in those cases the integrations over the Feynman parameter are of the form

$$\int_0^1 f(t)dt/[(t-\alpha)^2 + \epsilon^2]$$

with $0 < \alpha < 1$ and ϵ small (in situations like that we are actually close to an anomalous threshold). However even for such relatively large contributions of this sector we can not have values approaching the sensitivity limits of LEP2. Only in a very limited region of the parameter space and when accidentally the sum $m_{\tilde{C}} + m_{\tilde{Z}}$ turns out to be almost equal to W - boson mass, the chargino and neutralino contributions can be very large saturating the sensitivity limits of LEP2. We disregard such large contributions since they are not perturbatively trusted. Even if it were not for that reason these cases are unnatural occupying a very small portion of the available parameter space which is further reduced if the lower experimental bound $m_{\tilde{C}} > 45 GeV$ on the chargino mass is observed which does not allow for arbitrarilly small values of $M_{1/2}$). Therefore
Although neutralinos and charginos may, in some cases, yield large contributions approaching the sensitivity limits of LEP2 we do not think that these cases are natural.

CONCLUSIONS

The main results of our analysis are :

- The MSSM predictions for the Dipole and Quadrupole moments differ, in general, from those of the SM but they are of the same order of magnitude ($\mathcal{O}(10^{-3})$) in the entire parameter space $A_o, m_o, M_{1/2}$. Experiments should reach this level of accuracy for such differences to be observed. Hence deviations from the Standard Model predictions due to SUSY are unlikely to be observed at LEP2.

- The Neutralino and Chargino sector is the principal source of deviations from the SM predictions when this sector involves light states ($< M_W$). This occurs when $M_{1/2}$ is light and for positive values of $\mu > 0$.

- The Sector of Neutralinos and Charginos may yield contributions to the Dipole and Quadrupole moments whose magnitudes saturates the sensitivity limits of LEP2. This happens when $m_{\tilde{C}} + m_{\tilde{Z}} \simeq M_W$. We consider these cases unnatural and perturbatively untrustworthy.

- To be of relevance for future collider experiments the analysis should be extended to include values $s \equiv 4Q^2 > 4M_W^2$. The results of such an analysis will appear in a future publication (14).

TABLE 4. MSSM contributions to $\Delta k_\gamma, \Delta Q_\gamma$ for the inputs shown below. For comparison the SM predictions are shown for $m_{Higgs} = 50, 100/, and/, 300/, GeV$

		$m_t = 160$ $A_0, m_0, M_{1/2}$ =	$\tan\beta = 2$ 300, 300, 80		
		Δk_γ		ΔQ_γ	
	$\mu > 0$		$\mu < 0$	$\mu > 0$	$\mu < 0$
q, l		-1.973		1.922	
W, γ, Z		1.179		0.235	
$h_0, H_{\pm,0}, A$		.946		.028	
$\tilde{q}, \tilde{l}$	.009		-.035	.027	.025
$\tilde{Z}, \tilde{C}$	.697		.026	-.592	-.170
Total	.859		.143	1.621	2.041
SM (m_{Higgs} 50,100,300 GeV)		$\Delta k_\gamma^{SM} =$ $\Delta Q_\gamma^{SM} =$	.188 2.186	-.106, 2.174	-.449 2.161

REFERENCES

1. J. Ellis, S. Kelley and D.V. Nanopoulos, Phys. Lett. B **260**, 131 (1991) ;
   U. Amaldi, W. de Boer and M. Fürstenau, Phys. Lett. B **260**, 447 (1991).
2. For a review see H.P. Nilles, Phys. Rep. C **110**, 1 (1984);
   A.B. Lahanas and D.V. Nanopoulos, Phys. Rep. C **145**, 1 (1987).
3. K. F. Gaemers and G. J. Gounaris, Z. Phys. C **1** 259 (1979) .
   K. Hagiwara, R.D. Peccei, D. Zeppenfeld and K. Hikasa, Nucl. Phys. B **282**, 253 (1987).
   U. Baur and D. Zeppenfeld, Nucl. Phys. B **308**, 127 (1988).
4. W.A. Bardeen, R. Gastmans and B. Lautrup, Nucl. Phys. B **46**, 319 (1972).
5. G. Couture and J.N.Ng, Z. Phys. C**35**, 65 (1987).
6. E.N. Argyres, G. Katsilieris, A.B. Lahanas, C.G. Papadopoulos and
   V.C. Spanos, Nucl. Phys. B**391**, 23 (1993).
7. J. Papavassiliou and K. Philippides, Phys. Rev. D **48**, 4225 (1993).
8. C.L. Bilchak, R. Gastmans and A. Van Proeyen, Nucl. Phys. B **273**, 46 (1986).
9. T.M. Aliev, Phys. Lett. B **155** 364 (1985).
10. G. Couture, J. N. Ng, J. L. Hewett and T. G. Rizzo, Phys. Rev. D **38**, 860 (1988).
11. A. B. Lahanas and V. C. Spanos, Phys. Lett. B **334** 378 (1994).
12. A. Culatti, Padova Univ. preprint, DFPD/94/TH/25.
13. G. Couture, J. N. Ng, J. L. Hewett and T. G. Rizzo, Phys. Rev. D **36**, 1859 (1987).
14. A. B. Lahanas and V. C. Spanos (in preparation).

Rare b Decays and Anomalous Couplings

Stephen M. Playfer

Dept. of Physics, Syracuse University, Syracuse NY13244-1130

The current experimental status of the rare process $b \to s\gamma$ is reviewed. The Standard Model prediction for the decay rate is compared to the data and used to obtain bounds on anomalous couplings and other non Standard Model effects. Prospects for measuring the related processes $b \to d\gamma$ and $b \to s\ell^+\ell^-$ are discussed.

I. INTRODUCTION

The dominant decays of the b quark are charged current couplings via a W^- to a c or a u quark. Observation of these decays has led to measurements of the Cabibbo-Kobayashi-Maskawa (CKM) matrix elements $|V_{cb}|$ and $|V_{ub}|$ (1). The recent observation of the rare process $b \to s\gamma$ (2,3) is the first direct evidence for a flavor-changing neutral current decay described in the Standard Model by a one-loop diagram where a W^- is emitted and reabsorbed. Such diagrams are commonly known as "penguin" diagrams. The γ can be radiated from the W or the t quark in the loop, or from the initial or final state quarks. The top quark loop dominates the decay amplitude, which is proportional to the CKM matrix elements V_{tb} and V_{ts}, and to a kinematic factor which is a function of $(m_t/m_W)^2$.

The new experimental results have inspired a large number of papers investigating the sensitivity of $b \to s\gamma$ to extensions of the Standard Model (4). Useful bounds on the parameters of such models are found in some cases. The constraints that can be placed on anomalous W and t quark couplings will be discussed in detail.

II. OBSERVATION OF RADIATIVE PENGUIN DECAYS

The inclusive process $b \to s\gamma$ leads to many exclusive final states where the s quark hadronizes with the spectator quark. Angular momentum conservation forbids the decay $B \to K\gamma$, but it is expected that $K^*(892)\gamma$ will be a significant fraction of the inclusive rate. The remaining inclusive rate comes from higher mass K^* resonances and non-resonant $K(n\pi)$ final states. There are large variations among the theoretical predictions for the fraction of $b \to s\gamma$ that hadronizes as $B \to K^*\gamma$.

A. Observation of $B \to K^*\gamma$

The first successful search for $b \to s\gamma$ by the CLEO collaboration looked for the exclusive $K^*\gamma$ final state. This is much easier than trying to measure the inclusive branching ratio for $b \to s\gamma$, because the final state is completely kinematically constrained, and the analysis is similar to that used for reconstructing hadronic B meson final states at the $\Upsilon(4S)$ (5). Neutral clusters in a CsI calorimeter are selected between 2.1 and 2.9 GeV which have a shower shape consistent with a single γ, and which cannot be combined with another γ to form a π^0. The $K^*(892)$ candidates are searched for in three channels: $K^{*0} \to K^+\pi^-$, $K^{*-} \to K^-\pi^0$ and $K^{*-} \to K^0\pi^-$. If the energy sum of the K^* and the γ is within 75 MeV of the known energy of the beam, E_{beam}, then the beam constrained invariant mass

$$m_b = \sqrt{E_{beam}^2 - \left(\overrightarrow{P_{K^*}} + \overrightarrow{P_\gamma}\right)^2} \qquad (1)$$

is plotted for each candidate event and an excess is looked for at the known B meson mass. The difference in shape between jetlike continuum events and spherical $B\bar{B}$ events is exploited by making cuts on several event shape variables (2,6) to suppress the continuum background.

FIG. 1. Beam constrained mass distribution for $B \to K^*\gamma$, dark shaded $K^{*0} \to K^+\pi^-$, light shaded $K^{*-} \to K^-\pi^0$, unshaded $K^{*-} \to K^0\pi^-$

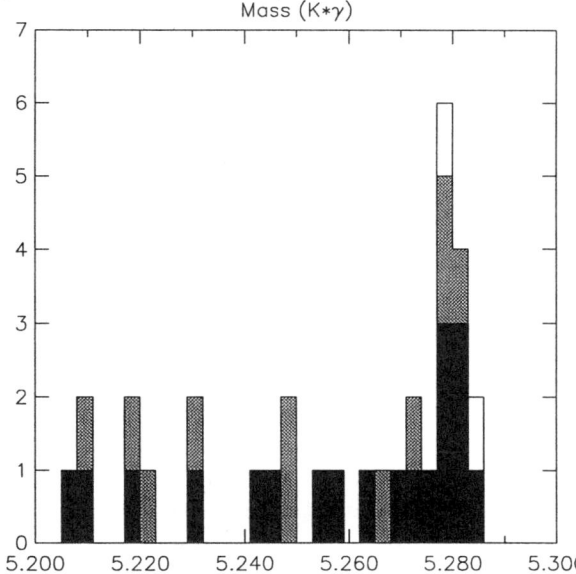

In $1.4fb^{-1}$ of $\Upsilon(4S)$ data there are eight $K^{*0}\gamma$ and five $K^{*-}\gamma$ candidates within 6 MeV of M_B. The continuum background level is one event in each of $K^{*0} \to K^+\pi^-$ and $K^{*-} \to K^-\pi^0$, and zero in $K^{*-} \to K^0\pi^-$ where there are two candidates. This is a clear signal for the decay $B \to K^*\gamma$ (Figure 1). The yields of $B^0 \to K^{*0}\gamma$ and $B^- \to K^{*-}\gamma$ are consistent. If the decay rates and the relative fractions of B^- and B^0 produced at the $\Upsilon(4S)$ are assumed to be equal, the average branching ratio is $(4.5 \pm 1.5 \pm 0.9) \times 10^{-5}$.

B. Measurement of Inclusive $b \to s\gamma$

Recently, the CLEO collaboration has also made the first measurement of the inclusive $b \to s\gamma$ branching ratio. The signature for the inclusive process is a photon with energy between 2.2 and 2.7 GeV. This region contains 75-90% of the signal according to calculations that include the smearing due to the Fermi motion of the quarks in the B meson, and the motion of a B meson produced at the $\Upsilon(4S)$.

There are large backgrounds to the inclusive signal from continuum jets ($e^+e^- \to q\bar{q}$) and initial state radiation (ISR). These backgrounds are suppressed by two methods: a shape variable analysis using a neural network, and a B reconstruction analysis. After these cuts have been made, the remaining continuum background is subtracted using scaled off-resonance data. There are also small backgrounds from other B decays, mostly resulting from π^0 and η decays. As a first approximation these are taken from a Monte Carlo simulation, but then a correction is made for any differences that are observed between the π^0 and η spectra measured in data, and those predicted by the Monte Carlo. This takes into account any omissions in the Monte Carlo (e.g. $b \to sg$).

The neural network analysis uses a set of eight variables defining the event shape, including measurements of event sphericity and of the energies contained in cones parallel and anti-parallel to the photon. Since none of the eight variables has clear discriminating power compared to the others, they are combined into a joint variable, r, which tends to +1 for signal, and -1 for continuum background. A neural net is used for this purpose.

The B reconstruction analysis combines the high energy photon with a candidate X_s system, where X_s contains either a $K_s \to \pi^+\pi^-$ or a charged track consistent with a Kaon, and an additional 1-4 π's, of which one may be a π^0. If this combination of particles is consistent with an exclusive $B \to X_s\gamma$ decay the event is accepted. There are reconstruction ambiguities and cross-feed between decay modes, but these are not important if the method is used only to suppress continuum background, and have not been considered in the CLEO analysis. If they were to be corrected for it would be possible to obtain additional information about the exclusive decay modes that contribute. Figure 2 shows the apparent X_s mass distribution, with a fit indicating the presence of a large component from $K^*(892)$.

The two methods for suppressing continuum are complementary. The neural net has high efficiency (32%) but modest background suppression, whereas the B reconstruction method has low efficiency (9%), but suppresses

FIG. 2. Apparent X_S mass distribution from the B reconstruction analysis. The solid curve is a fit to the expected distribution from a spectator model. The dashed curve shows the non-K*(892) component of the fit.

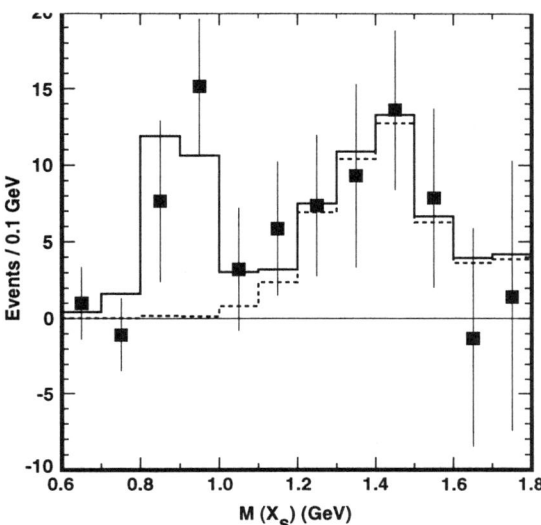

the background by an additional factor of 14. Figure 3 shows the photon energy spectra from the two analyses. The measured branching ratios are $\mathcal{B}(b \to s\gamma) = (1.88 \pm 0.74) \times 10^{-4}$ from the event-shape analysis and $\mathcal{B}(b \to s\gamma) = (2.75 \pm 0.67) \times 10^{-4}$ from the B reconstruction analysis. Since the two analyses are found to be only slightly correlated an average result of $\mathcal{B}(b \to s\gamma) = (2.32 \pm 0.57 \pm 0.35) \times 10^{-4}$ is quoted, where the first error is statistical, and the second systematic. Details of the contributions to the systematic error can be found in (7).

III. THEORY OF RADIATIVE PENGUIN DECAYS

A. Standard Model Prediction for $b \to s\gamma$

The partial decay width for $b \to s\gamma$ is described by:

$$\Gamma(b \to s\gamma) = \frac{\alpha G_F^2 m_b^5}{128\pi^4}|V_{ts}^* V_{tb} C_7^{eff}(\mu)|^2 \qquad (2)$$

The dependence on m_b^5 is removed by normalizing to the decay rate for $b \to c\ell\nu$:

$$\frac{\Gamma(b \to s\gamma)}{\Gamma(b \to c\ell\nu)} = \frac{|V_{ts}^* V_{tb}|^2}{|V_{cb}|^2}\frac{\alpha}{6\pi g(m_c/m_b)}|C_7^{eff}(\mu)|^2 \qquad (3)$$

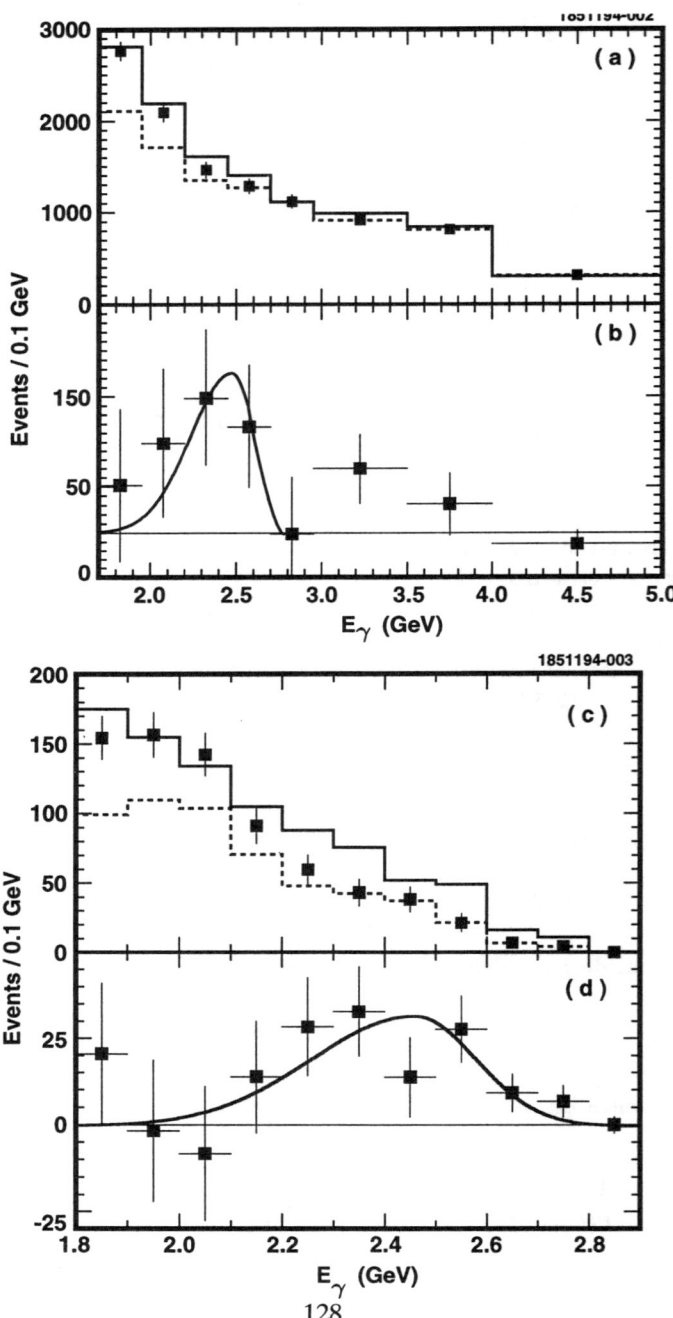

FIG. 3. Photon energy spectra from the neural net analysis, (a) & (b), and from the B reconstruction analysis, (c) & (d). In a) & (c) the on resonance data are the solid lines, the scaled off resonance data are the dashed lines, and the sum of backgrounds from off resonance data and $b \to c$ Monte Carlo are shown as the square points with error bars. In (b) & (d) the backgrounds have been subtracted to show the net signal for $b \to s\gamma$. The solid lines are fits of the expected signal shape.

where the factor $g(m_c/m_b)$ corrects for phase space. In these expressions $C_7^{eff}(\mu)$ is an effective coefficient of the electromagnetic loop operator:

$$\mathcal{O}_7 = \frac{e}{8\pi^2} m_b \bar{s}_\alpha \sigma^{\mu\nu}(1+\gamma_5)b^\alpha F_{\mu\nu} \tag{4}$$

The value of C_7 can be calculated perturbatively at the mass scale $\mu = M_W$. The explicit expression for $C_7(M_W)$ as a function of (m_t^2/M_W^2) can be found in (8). The evolution from M_W down to a mass scale $\mu = m_b$ introduces large QCD corrections. These are calculated using an operator product expansion based on an effective Hamiltonian:

$$\mathcal{H}_{eff}(b \to s\gamma) = -2\sqrt{2} G_F V_{ts}^* V_{tb} \sum_{i=1}^{8} C_i(\mu)\mathcal{O}_i(\mu) \tag{5}$$

Renormalization of the coefficients, C_i, and operator mixing, lead to a value of $C_7^{eff}(\mu)$ significantly larger than $C_7(M_W)$ (8). This increases the predicted rate for $b \to s\gamma$ by a factor of 2-3.

Evidently the prediction for the rate is very sensitive to the QCD corrections. The leading log calculation is uncertain to about 25%, primarily because it is not clear at which renormalization scale, μ, the effective coefficient, $C_7^{eff}(\mu)$, resulting from the operator product expansion, should be evaluated. Values between $\mu = \frac{1}{2} m_b$ and $\mu = 2m_b$ have been suggested. A next-to-leading order calculation requires the evaluation of additional two-loop diagrams, as well as some three-loop diagrams. It is hoped that these calculations can be done, since they are expected to reduce the uncertainty in the Standard Model prediction to about 10%.

B. Comparison between Experiment and Theory

The leading log prediction for $\mathcal{B}(b \to s\gamma)$ is $(2.8 \pm 0.8) \times 10^{-4}$ (8). If the next-to-leading order terms that have been calculated are included they tend to reduce the prediction to about 1.9×10^{-4} (9). Both these predictions are in excellent agreement with the experimental result of $(2.3 \pm 0.6 \pm 0.4) \times 10^{-4}$. Since the theoretical uncertainties are dominated by the choice of the renormalization scale, μ, it is difficult to obtain useful constraints on other Standard Model parameters such as m_t and V_{ts}. The CDF measurement of $m_t = (176 \pm 8 \pm 10)$ GeV (10) is well within the range required for consistency with $b \to s\gamma$. Ali et al. (11) have set bounds on V_{ts}:

$$0.62 < \frac{|V_{ts}|}{|V_{cb}|} < 1.10 \tag{6}$$

but this ratio is expected to be one if the CKM matrix is unitary.

The fraction of the inclusive $b \to s\gamma$ rate hadronizing as $B \to K^*\gamma$ depends on the $B \to K^*$ form factor. This has been calculated by many authors

TABLE 1. Predictions for the ratio of $B \to K^*\gamma$ to $b \to s\gamma$.

Author(s)	Reference	Method	$B \to K^*\gamma$ Fraction
Altomari	(12)	Spectator Quark Model	4.5%
Deshpande & Trampetic	(13)	Relativistic Quark Model	6 - 14%
Aliev et al	(14)	QCD Sum Rules	39%
Ali & Greub	(15)	Spectator Quark Model	(13±3)%
O'Donnell & Tung	(16)	Heavy Quark Symmetry	10%
Ball	(17)	QCD Sum Rules	(20±6)%
Atwood & Soni	(18)	Bound State Resonances	1.6 - 2.5%
Bernard, Hsieh & Soni	(19)	Lattice QCD	(6.0±1.2±3.4)%
UKQCD collaboration	(20)	Lattice QCD	15 - 35%

using either QCD sum rules, Lattice QCD, or Heavy Quark Effective Theory (HQET). Table 1 summarizes these predictions for the ratio of $B \to K^*\gamma$ to $b \to s\gamma$. It can be seen that the predictions range from a few percent to 40%, and that recent predictions do not seem to be converging. The data suggest a value of $(21 \pm 7)\%$ for this ratio, which is not accurate enough to limit the range of acceptable form factor models. It has been suggested by Isgur (21) that the discrepancies between the models could be resolved by using the measured $D \to K^*$ form factors as a basis for calculating all heavy to light quark form factors. Until accurate predictions are available for the $B \to K^*$ form factor, the exclusive measurement is not very useful for constraining the Standard Model or new physics.

C. Extensions of the Standard Model

The measurement of $b \to s\gamma$ has inspired a large number of theoretical investigations of extensions of the Standard Model that could lead to significant changes in the predicted rate for $b \to s\gamma$. These studies use the upper and lower limits (95% C.L.):

$$1.0 \times 10^{-4} < \mathcal{B}(b \to s\gamma) < 4.2 \times 10^{-4} \qquad (7)$$

to constrain the allowed parameter space of the Standard Model extension being considered. Among the most widely discussed models are Higgs doublets, Supersymmetry, anomalous $WW\gamma$ couplings, and anomalous top quark couplings. We give a brief summary of these cases below. For investigations into other phenomena such as leptoquarks, a fourth generation and left-right symmetric models the reader is referred to the review article by Hewett (4).

In two-Higgs doublet models there is a charged Higgs that can be inserted into the loop instead of the W boson. There are two models for the couplings of the Higgs doublets to the quarks, depending on how the fermion masses are generated. In both cases the free parameters are the charged Higgs mass, M_H, and the ratio of the doublet vacuum expectation values, Tanβ. With Model I couplings, the $b \to s\gamma$ rate is enhanced at low Tanβ, suppressed for values of Tanβ between 0.5 and 1.0, and is rather insensitive to large Tanβ. Model II couplings always enhance the $b \to s\gamma$ rate. In this case the experimental upper limit requires M_H to be at least 240 GeV even for large values of Tanβ (22).

Supersymmetry introduces many additional particles that can appear inside the loop. In the limit of exact supersymmetry these additional contributions cancel the Standard Model contribution, and $b \to s\gamma$ does not occur at all. In supersymmetric models there are charged Higgs bosons with Model II type couplings that enhance the rate for $b \to s\gamma$. Contributions in which down type squarks and either neutralinos or gluinos are inserted into the loop are usually found to be negligible. However, there are significant contributions when an up type squark and a chargino are inserted in the loop. There are several recent analyses of the size and sign of the chargino contributions relative to the Standard Model and charged Higgs contributions. It appears that there are some regions of the parameter space where the supersymmetric model predicts a rate comparable to or below the Standard Model, even for small values of M_H. This requires a small stop quark mass, a large value of Tanβ, and a higgsino mass parameter $\mu < 0$ (23).

D. Constraints on Anomalous Couplings

The existence of anomalous couplings at the $WW\gamma$ vertex can be constrained by tree-level processes such as $e^+e^- \to W^+W^-$ and $p\bar{p} \to W\gamma$, and by loop diagrams in processes such as $b \to s\gamma$ (24). The anomalous couplings are described by two parameters, λ and $\Delta\kappa$, which are zero in the Standard Model, but can acquire non-zero values in some extensions of the Standard Model. They modify the value of $C_7(M_W)$, and hence the predicted rate for $b \to s\gamma$:

$$C_7(M_W) = C_7(M_W)^{SM} + A_1 \Delta\kappa + A_2 \lambda \tag{8}$$

The coefficients A_1 and A_2 are functions of (m_t^2/M_W^2). The larger value of A_1 leads to three times more sensitivity to $\Delta\kappa$ compared to λ. Figure 4 shows the bounds that can be set on λ and $\Delta\kappa$ from existing data. The ellipse is the limit obtained from the D0 experiment at the Tevatron (25). A similar ellipse is obtained by the CDF experiment (26). The limits from $b \to s\gamma$ are complementary to these ellipses. Large positive and negative values of $\Delta\kappa$ and λ are excluded by the upper limit on $b \to s\gamma$. In the region around $\Delta\kappa = -1$ there is a complicated interference between the three terms in equation (8). This leads to the exclusion of a narrow band by the lower limit on $b \to s\gamma$.

Anomalous top quark couplings have also been considered (27). The first possibility is that there are anomalous $tt\gamma$ couplings in analogy to the $WW\gamma$

FIG. 4. Limits on anomalous $WW\gamma$ couplings. The shaded regions are allowed by the $b \to s\gamma$ measurement. The region between the shaded regions is excluded by the lower limit, the outer unshaded regions by the upper limit. The ellipse shows the limits obtained by the D0 and CDF experiments.

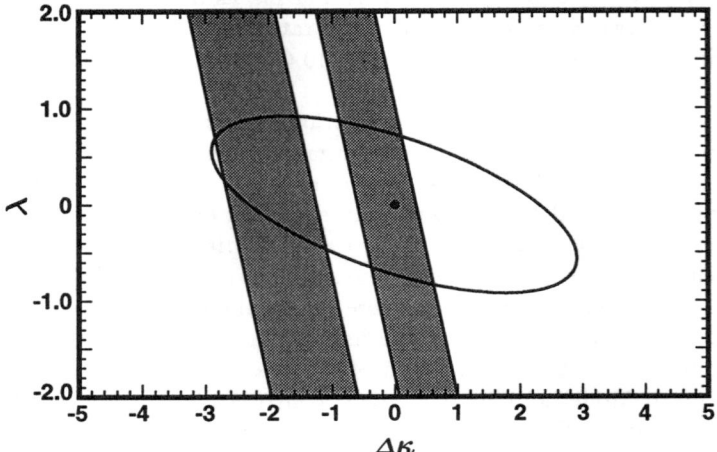

case considered above. Once again this would modify C_7 through two additional parameters. There is also the possibility of anomalous gluon couplings to the top quark that would modify C_8, but the constraints on these couplings from $b \to s\gamma$ are found to be rather weak. Finally there is the interesting point that $b \to s\gamma$ probes the V-A structure of the tbW and tsW couplings (28).

IV. SEARCHES FOR OTHER RADIATIVE PENGUIN DECAYS

A. The Decay $B \to \rho\gamma$

It was suggested by Ali (29) that the ratio of CKM elements $|V_{td}|/|V_{ts}|$ could be extracted from a measurement of:

$$\frac{\mathcal{B}(B^- \to \rho^-\gamma)}{\mathcal{B}(B^- \to K^{*-}\gamma)} = \frac{\mathcal{B}(B^0 \to \rho^0\gamma) + (B^0 \to \omega\gamma)}{\mathcal{B}(B^0 \to K^{*0}\gamma)} = \frac{|V_{td}|^2}{|V_{ts}|^2}\xi\Omega \qquad (9)$$

where Ω corrects for phase space, and ξ corrects for SU(3) symmetry breaking. More recent theoretical work (30) suggests that there are differences between long-distance contributions to $b \to s\gamma$ and $b \to d\gamma$, and also significant contributions from the light quark loops to $b \to d\gamma$, which make this extraction more complicated, and perhaps even impossible. However, it is still interesting to search for these decays experimentally.

CLEO has made a preliminary search for $B^- \to \rho^-\gamma$ (31), $B^0 \to \rho^0\gamma$ and $B^0 \to \omega\gamma$. A data sample of $2.0 fb^{-1}$ at the $\Upsilon(4S)$ results in upper limits between 1.0 and 2.5×10^{-5} for the three modes. This corresponds to a limit on the ratio in equation (9) of 0.34 at 90% confidence level. The search is beginning to be background limited. In $\omega\gamma$ the background is primarily from the continuum, whereas in $\rho^-\gamma$ and particularly $\rho^0\gamma$ there is significant feeddown from misidentified $K^*\gamma$ events. Future detectors with better particle identification (32) will be able to suppress this feeddown, but the continuum background will remain a problem.

B. Searches for $b \to s\ell^+\ell^-$

The process $b \to s\ell^+\ell^-$ occurs through a loop diagram with a virtual γ or Z boson, or through a box diagram containing two W bosons. In addition the hadronic decays $B \to \psi(')K^{(*)}$ contribute to the related exclusive decays $B \to K^{(*)}\ell^+\ell^-$ through the secondary decays $\psi(') \to \ell^+\ell^-$. A full understanding of $b \to s\ell^+\ell^-$ has to include both the short distance contributions from the loop and box diagrams, and the long distance contributions from the ψ decays, and the interference between them (11,33).

At low dilepton masses the dominant contribution from the virtual γ can be directly related to $b \to s\gamma$. There are sharp peaks from the ψ contributions at m_ψ and $m_{\psi'}$ which can be directly related to the measurements of the exclusive hadronic decays. At high dilepton masses the Z and box contributions are important, as are possible additional contributions from other heavy mass particles. The interference between the various diagrams can be studied by measuring the shape of the dilepton mass spectrum, and by measuring the lepton-pair asymmetry.

The high dilepton mass range has been studied at hadron colliders where there is a good signature for dimuon pairs. The first search for events with dimuon masses between 3.9 and 4.4 GeV was performed by the UA1 experiment. They found upper limits of 5.0×10^{-5} for the inclusive process $b \to s\mu^+\mu^-$, and 2.3×10^{-5} for the exclusive channel $B^0 \to K^{*0}\mu^+\mu^-$. Both these limits should be interpreted as referring only to the short distance contributions from the loop and box diagrams, since there is an extrapolation to the remainder of the phase space under the assumption that the long distance contributions are negligible above $m_{\psi'}$. Recently the CDF collaboration has presented preliminary results from the Tevatron collider. Their search over the dimuon mass ranges 3.2-3.5 and 3.8-4.4 GeV gives upper limits of 3.5×10^{-5} and 5.3×10^{-5} for the exclusive channels $B^0 \to K^{*0}\mu^+\mu^-$ and $B^- \to K^-\mu^+\mu^-$ respectively.

In contrast to the hadron collider experiments CLEO has searched for all dilepton masses except for the ranges 2.9-3.2 and 3.5-3.8 GeV where the $\psi(')$ contributions dominate (34). The analysis uses standard methods to reconstruct exclusive B meson decays from a candidate K or K^* meson and a pair of identified leptons. Some typical plots of the beam-constrained mass distributions are shown in Figure 7. In an $\Upsilon(4S)$ data sample of 2.0 fb^{-1} the background is less than one event in the signal region for each of the exclu-

FIG. 5. Beam constrained mass distributions for $B \to K^{(*)} \ell^+ \ell^-$: (a) $K^+ e^+ e^-$ (b) $K^+ \mu^+ \mu^-$ (c) $K^{*0} e^+ e^-$ (d) $K^{*0} \mu^+ \mu^-$

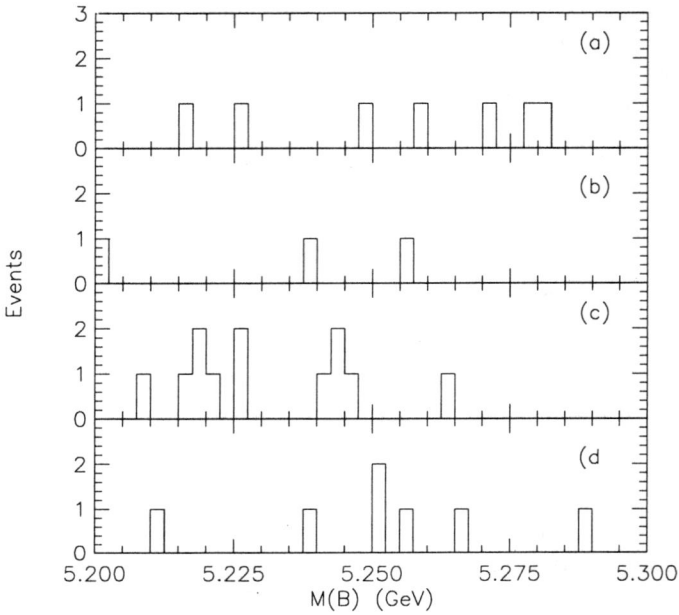

sive channels. The residual background is half from the continuum and half from $B\bar{B}$ events where both B mesons decay semileptonically. Table 2 summarizes the preliminary upper limits from CLEO for the exclusive channels $B \to K^{(*)} \mu^+ \mu^-$ and $B \to K^{(*)} e^+ e^-$. The rate for the decays to electron pairs is predicted to be larger than that to muon pairs due to the contribution from low mass pairs below the dimuon mass threshold. In some cases the limits from CLEO are close to the theoretical expectations. In the fu-

TABLE 2. Results of $b \to s \ell^+ \ell^-$ searches at CLEO.

B Decay Mode	Candidate Events	Detection Efficiency	90% C.L. Upper Limit	Standard Model Prediction (35)
$K^+ e^+ e^-$	2	24.4%	12.0×10^{-6}	0.6×10^{-6}
$K^+ \mu^+ \mu^-$	0	15.1%	9.0×10^{-6}	0.6×10^{-6}
$K^{*0} e^+ e^-$	0	9.8%	16.0×10^{-6}	5.6×10^{-6}
$K^{*0} \mu^+ \mu^-$	0	5.0%	31.0×10^{-6}	2.9×10^{-6}

ture significant increases in statistics at both hadron colliders and at $\Upsilon(4S)$ machinesare expected to lead to the observation of $b \to s \ell^+ \ell^-$, and eventually to measurements of the dilepton mass distribution and the lepton pair asymmetry.

REFERENCES

1. S. Stone, "Semileptonic B Decays", "B Decays" 2nd ed. (World Scientific, 1994).
2. R. Ammar et al, Phys. Rev. Lett. **71**, 674 (1993).
3. M.S. Alam et al, CLEO Preprint CLNS 94/1314, (to appear in Phys. Rev. Lett.).
4. J. Hewett, preprint SLAC-PUB-6521 (1994).
5. T. Browder, K. Honscheid & S. Playfer, "A Review of Hadronic and Rare B Decays", in "B Decays" 2nd ed. (World Scientific, 1994).
6. M. Artuso, "Experimental Facilities for b-quark Physics", in "B Decays" 2nd ed. (World Scientific, 1994).
7. J.A. Ernst, Ph.D. Thesis, Univ. of Rochester (1995).
8. A. J. Buras, M. Misiak, M. Münz & S. Pokorski, Nucl. Phys. **B424**, 374 (1994).
9. M. Ciuchini et al, Phys. Lett. **B334**, 137 (1994).
10. F. Abe et al, Phys. Rev. Lett. **74**, 2626 (1995).
11. A. Ali, G. Giudice & T. Mannel, Preprint CERN-TH.7346 (1994).
12. T. Altomari, Phys. Rev. **D37**, 677 (1988).
13. N. Deshpande & J. Trampetic, Mod. Phys. Lett. **A4**, 2095 (1989).
14. T. M. Aliev et al, Phys. Lett. **B237**, 569 (1990).
15. A. Ali & C. Greub, Phys. Lett. **B259**, 182 (1991).
16. J. O'Donnell & H. Tung, Phys. Rev. **D48**, 2145 (1993).
17. P. Ball, preprint TUM-T31-43/93 (1993).
18. D. Atwood & A. Soni, Z. Phys. **C64**, 241 (1994).
19. C. Bernard, P. Hsieh & A. Soni, Phys. Rev. Lett. **72**, 1402 (1994).
20. D. R. Burford et al, preprint FERMILAB-PUB-95/023-T (1995).
21. N. Isgur, private communication.
22. J. Hewett, Phys. Rev. Lett. **70**, 1045 (1993), V. Barger, M. Berger & R. Phillips, Phys. Rev. Lett. **70**, 1368 (1993).
23. S. Bertolini it et al, Nucl. Phys. **B353**, 591 (1991), R. Barbieri & G. Giudice, Phys. Lett. **B309**, 86 (1993), Y. Okada, Phys. Lett. **B315**, 119 (1993), R. Garisto & J. Ng, Phys. Lett. **B315**, 372 (1993), F. Borzumati, Z. Phys. **C63**, 291 (1994), Y. Okada, preprint KEK-TH-428 (1995).
24. S.P. Chia, Phys. Lett. **B240**, 465 (1990), K.A. Peterson, Phys. Lett. **B282**, 207 (1992), T.G. Rizzo, Phys. Lett. **B315**, 471 (1993), X.G. He & B. Mckellar, Phys. Lett. **B320**, 165 (1994).
25. J. Ellison, Proc. of DPF Meeting, Albuquerque, NM (1994).
26. F. Abe et al, Phys. Rev. Lett. **74**, 1936 (1995).
27. J. Hewett & T.G. Rizzo, Phys. Rev. **D49**, 319 (1994).
28. K. Fujiyama & A. Yamada, Phys. Rev. **D49**, 5890 (1994).
29. A. Ali, V. Braun and H. Simma, Z. Phys. **C63**, 437 (1994).
30. D. Atwood, B. Blok & A. Soni, Preprint SLAC-PUB-6635 (1994), N. Deshpande, X. He & J. Trampetic, Preprint OITS-564-REV (1994).
31. M. Athanas et al, CLEO-CONF 94-2, submission to ICHEP94 conference, Glasgow (1994).
32. CLEO III Detector: Design & Physics Goals, CLEO preprint CLNS 94/1277.
33. A. Buras & M. Münz, Preprint MPI-PhT/94-096 (1994).
34. R. Balest et al, CLEO-CONF 94-4, submission to ICHEP94 conference, Glasgow (1994).
35. A. Ali, C. Greub & T. Mannel, Preprint DESY-93-016 (1993).

Theory of Rare B Decays

Adam F. Falk

Department of Physics and Astronomy
The Johns Hopkins University
3400 North Charles Street
Baltimore, Maryland 21218 U.S.A.

Theoretical aspects of rare B decays are reviewed. The focus is on the relation between short-distance interactions and physical observables. It is argued that there remain significant uncertainties in the theoretical treatment of certain important quantities.

INTRODUCTION

While hadrons containing bottom quarks decay weakly, and hence are quite long-lived, all beautiful things must one day come to an end. For the typical B meson, the end comes after about 1.5 ps. While this provides enough time for the meson to pass through a measurable distance within a detector, given today's silicon technology, their lifetime is still so short that B mesons can be studied experimentally only by the careful examination of their decay products. Hence the study of the bottom quark is essentially the study of its decays.

It is believed that almost all bottom quarks decay weakly into charm quarks, via the W-emission process depicted schematically in Fig. 1. Rare b decays, then, are those which do not include the release of a c quark into the final state. These may include both Cabibbo-suppressed decays, such as those mediated by the transition $b \to uW^-$, and flavor-changing neutral decays, such as penguin-induced transitions. While the dominant decay mode of the b quark is believed to be well-understood, it is hoped that the rare decays may provide a window onto new physics beyond the standard model. Not only may one test the standard model by comparing the small predicted rates for rare channels to experiment, but the very fact that the charmless channels are suppressed makes them ideal places to look for anomalous enhancements coming from new particles and interactions at high energy scales.

The theory of rare b decays has two distinct parts, which are separated from each other conceptually and practically by their dependence on physics at very different energy scales. From the "high-energy" viewpoint, rare b decays are mediated by intermediate particles of large virtuality, and the challenge is to understand the structure of the quark-level transitions which such virtual particles can induce. From the "low-energy" viewpoint, rare b decays are mediated by local and nonrenormalizable point interactions, with coefficients

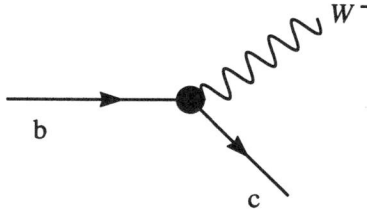

FIG. 1. The ordinary decay of a b quark, via $b \to cW^-$.

which are determined at high energies, but at low energies may be viewed simply as coupling constants of the theory. The relation between the high-energy and low-energy viewpoints is demonstrated schematically in Fig. 2 for two typical transitions.

From the low-energy viewpoint, the theoretical challenge is to relate the strengths of the suppressed nonrenormalizable quark-level couplings to *physical* properties of observable hadrons such as B and Λ_b. The situation is complicated by the long-distance effects of the strong QCD interactions. Typically, the structure of bottom hadrons cannot be computed from first principles, and one must find techniques which minimize one's sensitivity to uncomputable low-energy effects, while allowing one to extract from experiment as much information as possible about high-energy physics. At low energies, what one would like to measure experimentally are the coefficients of the nonrenormalizable operators such as those pictured in Fig. 2. It is the goal of low-energy high-energy physics to make this possible.

In what follows, I shall review the theory of rare b decays both from the high-energy and the low-energy points of view. In contrast to the spirit of the rest of this conference, however, my emphasis will be on the physics at low energies. In focusing on these possibly less-familiar effects, I hope to convince this "high energy" audience of the important limitations which low energy strong interactions place on understanding the physical manifestations of virtual high energy interactions. The good news is that much work is still in progress to minimize these limitations and to maximize the fundamental discovery potential of experimental b physics.

HIGH ENERGY VIEWPOINT

The decays of b quarks, both ordinary and rare, are generated by virtual interactions at some high scale $M \gg m_b$. At lower scales $\mu < M$, these interactions generate nonrenormalizable local operators. From the high energy viewpoint, there two questions which must be answered:

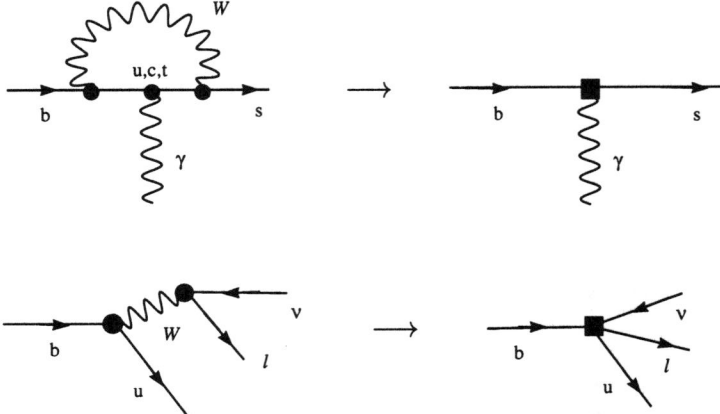

FIG. 2. Rare decays of a b quark, from the high-energy and low-energy viewpoints.

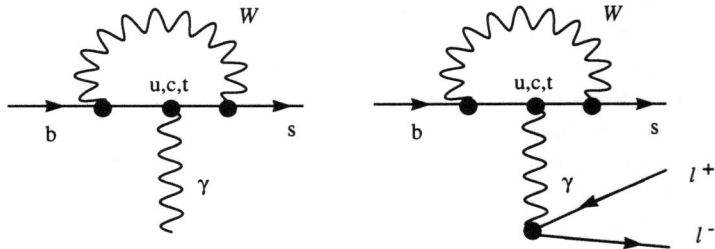

FIG. 3. Penguin-induced decays of the b quark.

1. What operators are generated?
2. With what coefficients?

For example, the penguin diagrams pictured in Fig. 3 generate, among others, the operators

$$\begin{aligned} C_7 \mathcal{O}_7 &= C_7(\mu)\, (\bar{s}\sigma^{\mu\nu} b)_R\, F_{\mu\nu}\,, \\ C_8 \mathcal{O}_8 &= C_8(\mu)\, (\bar{s}\gamma^\mu b)_L\, \bar{\ell}\gamma_\mu \ell\,, \\ C_9 \mathcal{O}_9 &= C_9(\mu)\, (\bar{s}\gamma^\mu b)_L\, \bar{\ell}\gamma_\mu \gamma_5 \ell\,. \end{aligned} \quad (1)$$

Perturbative QCD corrections are included by dressing the graphs in Fig. 3 with gluons. The leading logarithm approximation, which resums all terms of the form $\alpha_s^n(\mu)\ln^n(\mu/M_W)$, suffers from a strong ambiguity in the choice of renormalization scale μ. The resolution of this ambiguity will only properly be resolved by a full next-to-leading order calculation. The present state of the art for the coefficient $C_7(\mu)$ is summarized in Ref. (1). The calcula-

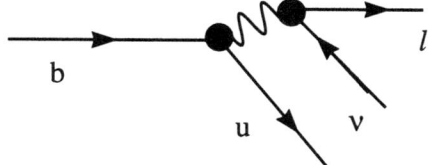

FIG. 4. Charmless semileptonic b quark decay.

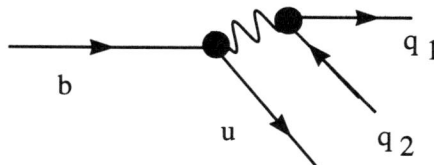

FIG. 5. Charmless nonleptonic b quark decay.

tion is complete to order α_s, and partially complete at next-to-leading order. Varying the renormalization scale μ from $m_b/2$ to m_b, one finds a residual scale-dependence uncertainty of approximately ±15%. Since $\mathcal{O}_7$ is the operator which is primarily responsible for the rare decay $B \to X_s\gamma$, there is a corresponding uncertainty of at least ±30% in the prediction of this decay rate in the standard model. The coefficients $C_8(\mu)$ and $C_9(\mu)$, which are responsible for the decay $B \to X_s \ell^+\ell^-$, are known with similar accuracy.

There are also charmless weak b decays, which are rare because their rates are suppressed compared to the dominant weak decay mode by the factor $|V_{ub}/V_{cb}|^2 \sim 10^{-2}$. Charmless semileptonic decays, shown in Fig. 4, arise from operators of the form

$$A(\mu)\bar{u}\gamma^\mu(1-\gamma_5)b\,\bar{\ell}\gamma_\mu(1-\gamma_5)\nu\,. \tag{2}$$

Since this operator may be written, up to weak and electromagnetic corrections and fermion masses, as a product of conserved currents, the coefficient $A(\mu)$ suffers from no scale ambiguity. It has been computed to order $\alpha_s(m_b)$, and the residual uncertainty is small.

The same is not true of charmless nonleptonic decays, mediated by operators such as shown in Fig. 5. At low energies, these diagrams induce four-quark operators of the form

$$\begin{aligned} C_1\,\mathcal{O}_1 &= C_1(\mu)\,\bar{u}_i\gamma^\mu(1-\gamma_5)b_i\,\bar{q}_{1j}\gamma_\mu(1-\gamma_5)q_{2j}\,, \\ C_2\,\mathcal{O}_2 &= C_2(\mu)\,\bar{u}_i\gamma^\mu(1-\gamma_5)b_j\,\bar{q}_{1j}\gamma_\mu(1-\gamma_5)q_{2i}\,, \end{aligned} \tag{3}$$
$$\tag{4}$$

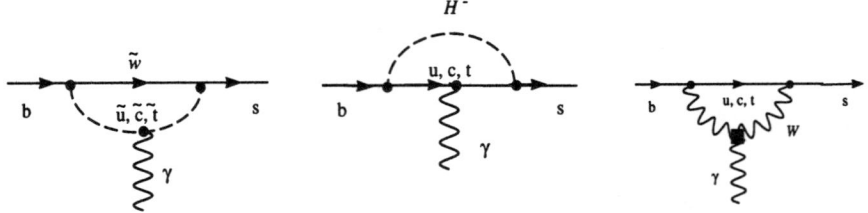

FIG. 6. New physics contributions to the rare process $b \to X_s \gamma$.

where the indices i and j indicate sums over colors. These operators are *not* products of currents, and they receive renormalizations from perturbative QCD which are as large as those received by penguin operators. As summarized in Ref. (2), the coefficients $C_1(\mu)$ and $C_2(\mu)$ have now been computed at next-to-leading order. The residual uncertainty, largely arising from scheme-dependence, is about $\pm 15\%$.

New physics at high energies can also contribute to the coefficients $C_i(\mu)$. For example, as shown in Fig. 6, supersymmetric particles, extra scalars, and anomalous trilinear gauge couplings can all modify $C_7(\mu)$ as compared to the standard model. The size of these new contributions depends, of course, on the particular model involved. However, because of the uncertainties inherent in the perturbative corrections to the standard model, new physics will *only* be observable in rare b decays if causes deviations from the standard model at significantly more than the 15% level. This is the most important lesson to be taken from the high energy point of view.

LOW ENERGY VIEWPOINT

At low energies, we start with an interaction Lagrangian density which is a sum over nonrenormalizable operators with coefficients determined at high energies,

$$\mathcal{L} = \sum C_i(\mu) \mathcal{O}_i(\mu). \tag{5}$$

The matrix elements of the operators $\mathcal{O}_i$ are defined so as to cancel the μ dependence of any physical observable. The operators and their coefficients are renormalized at a low energy scale $\mu \sim m_b$, and nonrenormalizable terms are suppressed by powers of the scale M at which the interactions become nonlocal and new physics comes into play.

The challenge, at low energies, is to use the Lagrangian (5) to make *physical* predictions. One option is to try to predict *exclusive* decay modes, such as

$B \to K^*\gamma$ or $B \to \rho\ell\nu$. However, the theoretical methods available are not entirely satisfactory: the Heavy Quark Effective Theory, so useful for $b \to c$ transitions (3), is of limited applicability here with only light quarks in the final state. Lattice calculations eventually may provide important information on exclusive matrix elements, but that is for the most part still in the future. For now, one is left to rely on phenomenological models, which for all their occasional successes do not provide any controlled approximation to QCD.

Alternatively, one may consider *inclusive* decay modes, such as $B \to X_s\gamma$ or $B \to X_u\ell\nu$. There has been considerable recent progress (4) in the computation of such quantities in a simultaneous expansion in $1/m_b^n$ and $\alpha_s(m_b)^n$. It has recently been understood, as well, that there are important limitations to such calculations. We shall now review this situation in some detail.

The theoretical analysis of inclusive B decays relies on the Operator Product Expansion and perturbative QCD. The partial width Γ for an operator $\mathcal{O}$ to mediate the decay of a B to any final state X with the correct quantum numbers is proportional to the square of the matrix element, summed over the possible final states,

$$\Gamma \sim \sum_X |\langle X|\mathcal{O}|B\rangle|^2. \qquad (6)$$

By the Optical Theorem, Γ may be rewritten as the imaginary part of a forward scattering amplitude,

$$\Gamma \sim \operatorname{Im} \langle B|T\{\mathcal{O},\mathcal{O}^\dagger\}|B\rangle, \qquad (7)$$

which is then expanded simultaneously in powers of $\alpha_s(m_b)$ and $1/m_b$. One obtains expressions for the inclusive partial widths; for example (4,5),

$$\Gamma(B \to X_s\gamma) \propto m_b^5 |C_7(\mu)|^2 \left\{ 1 + \frac{\lambda_1 - 9\lambda_2}{2m_b^2} + K(\mu)\,\alpha_s(m_b) + \ldots \right\},$$
$$\Gamma(B \to X_u\ell\nu) \propto m_b^5 |V_{ub}|^2 \left\{ 1 + \frac{\lambda_1 + 3\lambda_2}{2m_b^2} + K_{\text{s.l.}}\,\alpha_s(m_b) + \ldots \right\}. \qquad (8)$$

Here the nonperturbative parameters λ_1 and λ_2 are defined by hadronic matrix elements (8),

$$\lambda_1 = \langle B|\bar{b}(iD)^2 b|B\rangle/2m_b,$$
$$\lambda_2 = \langle B|\bar{b}(-\tfrac{i}{2}\sigma^{\mu\nu})G_{\mu\nu}b|B\rangle/2m_b. \qquad (9)$$

It is straightforward to find $K_{\text{s.l.}} = \frac{2}{3\pi}(\frac{25}{4} - \pi^2)$ (9), while the perturbative correction $K(\mu)$ is too messy to be illuminating (1).

There are a number of sources of uncertainty in the expressions (8). The operator $\bar{b}(-\tfrac{i}{2}\sigma^{\mu\nu})G_{\mu\nu}b$ violates the Heavy Quark Spin Symmetry and may be measured directly from the B–B^* mass difference, $\lambda_2 \approx 0.12\,\text{GeV}^2$. However, λ_1 can not be measured directly; instead, one must rely on phenomenological

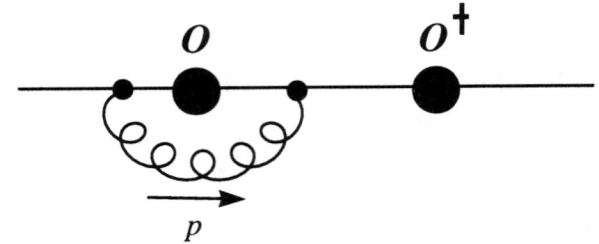

FIG. 7. A typical one-loop radiative correction to $T\{\mathcal{O},\mathcal{O}^\dagger\}$.

FIG. 8. The gluon propagator replaced by the sum of self-energy graphs.

models. While this is unfortunate, if we assume $\lambda_1 \leq 1\,\text{GeV}^2$, then inspection of Eq. (8) shows that the error induced in the partial widths Γ is 10% or less. Higher order nonperturbative corrections, of order $1/m_b^3$, are expected to be at the level of a few percent.

The primary sources of uncertainty in Eq. (8) are the value to take for the bottom mass m_b, and uncomputed higher order radiative corrections. Because of the overall factor of m_b^5, the theoretical partial widths are extremely sensitive to this parameter. For example, allowing m_b to vary over the range $4.5\,\text{GeV} \leq m_b \leq 5.0\,\text{GeV}$ induces an uncertainty in Γ of approximately 50%. While lower values for m_b seem currently to be preferred, the issue is still quite unsettled.[1]

A. Higher order radiative corrections

Higher order radiative corrections to the partial widths have recently been considered by a number of authors. In particular, attention has been paid to a set of corrections which are dominant in the limit of large N_f (large number of quark flavours), and which in the real world still may be particularly large. These come from taking the one loop radiative correction to the time-ordered product (7), an example of which is shown in Fig. 7, and replacing the gluon propagator with a sum of self-energy bubbles. If this replacement, which is illustrated in Fig. 8, is carried out to all orders and then extrapolated to the physical N_f, it amounts to replacing the strong coupling constant $\alpha_s(m_b)$ by its running value $\alpha_s(p^2)$ evaluated at the loop momentum.

The BLM scale-setting prescription (10) requires that one perform the substitution shown in Fig. 8 to leading order; the two-loop contribution to the

---

[1] There is an ongoing controversy over issues as fundamental as the proper *definition* of m_b. I will not review this discussion here.

radiative correction is then expected on general grounds to be parametrically large. Once this part of the two-loop computation has been done, one adjusts the scale μ in the one-loop result to absorb it. A recent application of this criterion to the inclusive rate for $B \to X_u \ell \nu$ indicates that the appropriate scale for this process is $\mu \sim m_b/10$ rather than m_b (11). A more complicated scale-setting procedure which resums all orders in the bubble sum (but which suffers from a certain lack of uniqueness) does not, in general, lead to quite such a low value of μ (12), although the two-loop corrections are, of course, still quite large.

This treatment of higher order radiative corrections leaves us with two questions.
1. Should such a low renormalization scale be taken seriously? If so, then clearly the entire program of computing inclusive rates perturbatively is in trouble. If not, then one still has to do deal with the fact that two-loop corrections are much larger than one might naïvely have thought.
2. If there is a class of diagrams which is unusually large, can perturbation theory be improved in a sensible way? Once such a resummation has been performed, can one show that the remaining uncertainties are likely to be small?

B. The need for endpoint spectra

Final states with charm present an enormous background to rare B decays. For example, the decay $B \to X_c \ell \nu$ obscures $B \to X_u \ell \nu$, and $B \to D\pi^0 \to D\gamma\gamma$ presents a problematic background to $B \to X_s \gamma$. Typically, strict kinematic cuts are used to exclude such process. For example, studies of rare decays accept only leptons and photons with energies in the range $2.2\,\text{GeV} \leq E_\ell, E_\gamma \leq 2.7\,\text{GeV}$, beyond the kinematic endpoint for charm in the final state. Hence it is necessary for theorists to compute not only partial widths Γ, but inclusive lepton *spectra* $d\Gamma/dE$ within 20% or so of the endpoint. A cartoon of a lepton energy spectrum, along with the kinematic cut, is shown in Fig. 9.

The theoretical problem is that the OPE does not converge when restricted to the lepton or photon energy endpoint. What is computable is not the full differential spectrum $d\Gamma/dE$ but rather *moments* of this spectrum, obtained by weighting the differential spectrum by some function and then integrating. Introducing the scaled energy variable $y = 2E_{\ell,\gamma}/m_b$, we thus "smear" with a weighting function with support only in the small region $1 - \delta \leq y \leq 1$. The size δ of the smearing region controls the convergence of the OPE. For $\delta \sim \Lambda_{\text{QCD}}/m_b \sim 10\%$, *all orders* in the $1/m_b$ expansion contribute equally (6), and the leading terms in the expansion (8) is clearly insufficient. Of course, this is only an order of magnitude estimate, and how the OPE converges for the experimentally chosen upper value of δ cannot be determined from such general considerations. At this point, then, a certain amount of faith is

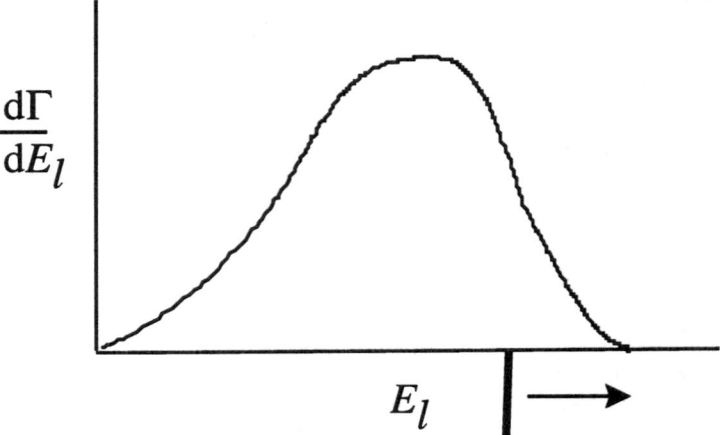

FIG. 9. A cartoon of the lepton energy spectrum in $B \to X_u \ell \nu$, along with the kinematic cut.

required in the interpretation of the smeared theoretical spectra.

An interesting by-product of this analysis is the result that the same infinite sum of terms in the $1/m_b^n$ expansion determines the shape of the endpoint spectrum in $B \to X_s \gamma$ and in $B \to X_u \ell \nu$ (6,7). Whether this relation yields useful predictive power is still to be seen.

C. Sudakov Logarithms

Another source of uncertainty in the shape of the endpoint spectrum comes from Sudakov logarithms (13). For example, the perturbative corrections to the lepton energy spectrum in $B \to X_u \ell \nu$ is extremely singular near the endpoint $y = 1$ (14):

$$\frac{d\Gamma}{dy} = \frac{d\Gamma_0}{dy} \left\{ 1 - \frac{2\alpha_s}{3\pi} \left[\ln^2(1-y) + \frac{31}{6} \ln(1-y) + \ldots \right] + \mathcal{O}(\alpha_s^2) \right\}, \quad (10)$$

where $d\Gamma_0/dy$ is the spectrum at tree level. At order α_s^2, the leading singularity is $\ln^4(1-y)$, and so forth. These Sudakov double logarithms may be resummed into an exponential suppression factor:

$$\frac{d\Gamma}{dy} = \frac{d\Gamma_0}{dy} \exp\left\{ -\frac{2\alpha_s}{3\pi} \ln^2(1-y) \right\} + \ldots . \quad (11)$$

This leading behaviour is actually *stronger* very near $y = 1$ than that given by the nonperturbative power corrections, but it is calculable.

What must be suppressed are the leading *uncalculated* corrections, which is accomplished by smearing over a region δ large enough that they may be

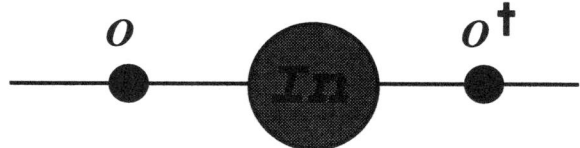

FIG. 10. The one instanton contribution to $T\{\mathcal{O},\mathcal{O}^\dagger\}$.

neglected. In the large m_b limit, this requires that we smear over a region formally much larger than $\delta \sim \Lambda_{\rm QCD}/m_b$, given by the condition (7)

$$\delta > \exp\left\{-\sqrt{\pi/\alpha_s(m_b)}\right\}. \tag{12}$$

In this strict limit, then, all nonperturbative corrections to the endpoint shape would be irrelevant. But for realistic $m_b \approx 4.8\,{\rm GeV}$, and the given experimental smearing region $\delta \approx 0.1 \sim 0.2$, do the uncalculated Sudakov effects *actually* dominate the nonperturbative power corrections? It is difficult to guess, based only on naïve power counting arguments. Explicit calculations of the subleading Sudakov logarithms may help clarify the situation.

For technical reasons, the Sudakov corrections to the smeared photon spectrum in $B \to X_s\gamma$ are under much better theoretical control than in $B \to X_u\ell\nu$ (6). They do not introduce unmanageable uncertainties into the computation of the weighted spectra.

D. Instantons

Finally, there are possibly large contributions to energy endpoint spectra from instantons. These arise because the light quark which is produced in the short-distance interactions can propagate in an instanton background, as pictured in Fig. 10. Chay and Rey computed, in the dilute instanton gas approximation, the one instanton contribution to $d\Gamma/dy$ for $B \to X_u\ell\nu$ and $B \to X_s\gamma$ (15). Their result diverges dramatically at the endpoint, as $y \to 1$. The contribution to $B \to X_s\gamma$ is nonetheless small and under control when one computes weighted spectra, but the same is not true for $B \to X_u\ell\nu$. Instead, one finds that the one instanton contribution is entirely untrustworthy in the experimentally defined window.

The one instanton contribution goes bad in this region presumably because multi-instanton configurations begin to be important. We have used the one instanton calculation as the motivation for a *crude* ansatz for the multi-instanton result in this region (16). This ansatz incorporates, as much as possible, the reliable information from the one-instanton calculation. When we vary this naïve "best guess" ansatz by two orders of magnitude, we find that over most of the ansatz parameter space, the instantons do in fact dominate the weighted endpoint spectra.

One must be careful about interpreting this result. It is potentially interesting only in a *negative* sense. On the one hand, the actual numbers certainly cannot be believed; by no means do we claim to have computed the correct multi-instanton contribution. On the other, we have failed to find any justification for ignoring the instantons in the endpoint region. In light of this equivocal situation, one may well wonder whether one can still trust the relationship between the endpoint spectra for $B \to X_s\gamma$ and $B \to X_u\ell\nu$ proposed in Refs. (6,7). This is a situation badly in need of clarification. Invocations of faith, one way or the other, will not be sufficient; a more sophisticated estimate of instanton contributions is what is required. Such an estimate could show, for example, that our ansatz for the multi-instanton contribution is entirely too crude, and that other techniques can be used to prove that the multi-instanton contribution is necessarily negligible. We certainly hope that this will prove to be the case.

CONCLUSIONS

We may summarize the status of the theory of rare B decays from each of our two viewpoints:

Low Energy Viewpoint:

1. Exclusive decay rates are extremely difficult to compute reliably. One must resort to models and other uncontrolled assumptions, a situation which is most unsatisfactory.

2. Inclusive calculations, by contrast, may be performed in a controlled expansion in powers of α_s and $1/m_b$. However, there remain unresolved uncertainties about
   a. uncomputed higher orders in α_s and the renormalization scale μ;
   b. Sudakov double logarithms near the lepton energy endpoint;
   c. instanton contributions near the lepton energy endpoint.

3. The decay $B \to X_s\gamma$ is in much better shape with respect to Sudakov and instanton corrections than is $B \to X_u\ell\nu$. Hence, while the calculation of $B \to X_s\gamma$ is itself perhaps fairly secure, the proposed relationship between the endpoint spectra in $B \to X_s\gamma$ and $B \to X_u\ell\nu$ may well be threatened by these effects.

High Energy Viewpoint:

1. In view of the significant uncertainties in existing theoretical calculations, only modifications to the Standard Model which affect rare decays at the 50% level or higher are likely to be experimentally detectable. Small modifications, say at the 10% level, are unlikely ever to be seen.

2. There is considerable room for the situation to improve, and much work remains to be done.

ACKNOWLEDGMENTS

It is a pleasure to thank the organizers for a stimulating conference and absolutely lovely weather. This work was supported by the National Science Foundation under Grant No. PHY-9404057 and National Young Investigator Award No. PHY-9457916, and by the Department of Energy under Outstanding Junior Investigator Award No. DE-FG02-094ER40869.

REFERENCES

1. M. Ciuchini, E. Franco, G. Martinelli, L. Reina and L. Silvestrini, Phys. Lett. **B334**, 137 (1994).
2. A. Buras, Nucl. Phys. **B434**, 606 (1995).
3. For reviews and references to the original literature, see N. Isgur and M.B. Wise, "Heavy Quark Symmetry," in *B Decays*, ed. S. Stone, Singapore: World Scientific, 1991, p. 158; B. Grinstein, 1994 TASI Lectures, UCSD Report No. USCD/PTH 94-24; M. Neubert, Phys. Rep. **245**, 359 (1994).
4. J. Chay, H. Georgi and B. Grinstein, Phys. Lett. **B247**, 399 (1990); I.I. Bigi, N.G. Uraltsev and A.I. Vainshtein, Phys. Lett. **B293**, 430 (1992); I.I. Bigi, B. Blok, M. Shifman, N.G. Uraltsev and A.I. Vainshtein, Minnesota Report No. TPI–MINN–92/67–T (1992); A.F. Falk, M. Luke and M.J. Savage, Phys. Rev. **D49**, 3367 (1994).
5. A.V. Manohar and M.B. Wise, Phys. Rev. **D49**, 1310 (1994); I.I. Bigi, M. Shifman, N.G. Uraltsev and A.I. Vainshtein, Phys. Rev. Lett. **71**, 496 (1993); B. Blok, L. Koyrakh, M. Shifman and A.I. Vainshtein, Phys. Rev. **D49**, 3356 (1994); Erratum, Phys. Rev. **D50**, 3572 (1994); T. Mannel, Nucl. Phys. **B413**, 396 (1994).
6. M. Neubert, Phys. Rev. **D49**, 3392 (1994); M. Neubert, Phys. Rev. **D49**, 4623 (1994); T. Mannel and M. Neubert, CERN Report No. CERN-TH-7156-94 (1994); I.I. Bigi, M. Shifman, N.G. Uraltsev and A.I. Vainshtein, Int. J. Mod. Phys. **A9**, 2467 (1994).
7. A.F. Falk, E. Jenkins, A.V. Manohar and M.B. Wise, Phys. Rev. **D49**, 4553 (1994).
8. A.F. Falk and M. Neubert, Phys. Rev. **D47**, 2965 (1993).
9. B. Guberina, R.D. Peccei and R. Rückl, Nucl. Phys. **B171**, 333 (1980).
10. S.J. Brodsky, G.P. Lepage and P.B. Mackenzie, Phys. Rev. **D28**, 228 (1983).
11. M. Luke, M.J. Savage and M.B. Wise, Phys. Lett. **B343**, 329 (1995).
12. M. Neubert, CERN Report Nos. CERN-TH-7487-94 (1994), CERN-TH-7524-94 (1995).
13. V. Sudakov, JETP (Sov. Phys.) **3**, 65 (1956); G. Altarelli, Phys. Rep. **81**, 1 (1982).
14. M Jeżabek and J.H. Kühn, Nucl. Phys. **B320**, 20 (1989).
15. J. Chay and S.J. Rey, Seoul Report Nos. SNUTP-94-08 (1994), SNUTP-94-54 (1994).
16. A.F. Falk and A. Kyatkin, Johns Hopkins Report No. JHU–TIPAC–950005 (1995).

Single Photon and Radiative Events at LEP

Peter Mättig

Universität Bonn, 53115 Bonn, Germany

Experimental studies of exotic photon production at the LEP e^+e^- collider are summarised. Emphasis is given to the potential signals of anomalous boson self interactions: searches for $ZZ\gamma$ couplings in $\nu\bar{\nu}\gamma$ final states, Z^0 resonance decays into three photons and the decay $Z^0 \to \gamma +$ Higgs. A brief account of studies on possible non-standard fermion-photon couplings is also given.

INTRODUCTION

Photon production at LEP allows a broad range of physics, both within and beyond the Standard Model, to be explored. Within the Standard Model photons do not couple directly to the Z^0. Instead, as depicted in Fig.1, photons originate mainly from incoming (1a) or outgoing (1b) fermions. Particularly in hadronic Z^0 decays photons may originate from hadron decays like $\pi^0 \to \gamma\gamma$ (1c). In addition, two photons may be produced by an electron t-channel exchange (1d). With the exception of hadron decays these contributions are theoretically understood to high precision and have been used to study QED at the highest energies, electroweak couplings of all fermions and QCD.

The main characteristics of these photons are their typical bremsstrahlung spectra with a preference for small angles $\sim 1/\alpha_{f,\gamma}$ with respect to the emitting fermion and for low energies $\sim 1/E_\gamma$. On the Z^0, initial state photons are suppressed due to the large drop in cross section below the resonance. More important for most experimental studies are photons emitted from the outgoing fermions. Photons from hadron decays like $\pi^0 \to \gamma\gamma$ in $Z^0 \to$hadrons or $\tau^+\tau^-$ are also emitted dominantly along the fermion direction. They are in general surrounded by other hadrons and can be suppressed by isolation requirements on the photon candidate. Somewhat different is the $e^+e^- \to \gamma\gamma$ process, which, in the absence of higher order photon emission (e.g. 1d), leads to monoenergetic photons of beam energy.

Any deviation from these rather strict expectations, either of the photon yield or of their properties, is an unambiguous sign for physics outside the Standard Model. This, together with their rather clear experimental signature, makes photons at LEP a sensitive probe for non-standard physics.

Radiative events appear in many scenarios of Standard Model extensions, some of which are addressed in (1). In the context of vector boson self inter-

TABLE 1. Data samples used at LEP for $f\bar{f}\gamma$ analyses. Typical values are given for cuts on the minimum photon energy E_γ^{min}, the minimum angle θ_γ with respect to the beam direction, and the minimum angle $\alpha_{f,\gamma}^{min}$ between the outgoing fermion and the photon. The data samples used by the various experiments for published or preliminary analyses are also given.

	$\nu\nu\gamma$	$\mu\mu\gamma$	$\tau\tau\gamma$	$q\bar{q}\gamma$
E_γ^{min} (GeV)	1.5	2	3	5
$\|\cos\theta_\gamma\|^{max}$	0.7	0.98	0.94	0.94
$\alpha_{f,\gamma}^{min}$ (deg.)		5	10	10-15
ALEPH	16 pb^{-1}			41 pb^{-1}
DELPHI		40 pb^{-1}	40 pb^{-1}	80 pb^{-1}
L3	80 pb^{-1}	17 pb^{-1}	17 pb^{-1}	40 pb^{-1}
OPAL	40 pb^{-1}	7 pb^{-1}	7 pb^{-1}	80 pb^{-1}
Physics	ν counting	Z'	Z'	QCD
	ν^*, magn.moment		el., magn. moments	el.weak

emphasis was on low energy photons. To make sure that no particle evades detection in the blind area of the beam pipe, photons at a large polar angle are selected. In the case of $q\bar{q}\gamma$ the main experimental challenge is to suppress the π^0 background. This is achieved by requiring the photon candidate to be fairly energetic and isolated. This implies a large $\alpha_{f,\gamma}$.

SEARCHING FOR ANOMALOUS $ZZ\gamma$ COUPLINGS

In its most general form the $Z^0 \to Z^{0*}\gamma$ production can be expressed (2) by four form factors h_i. All of them violate charge conjucation, two of them (i=1,2) are CP violating, the other two are CP conserving. Neglecting the charge of the fermions, the cross sections are identical for $h_1=h_3$ (and $h_2=h_4$). In the Standard Model $h_i=0$. In its extensions at some typical scale Λ, the h_i are expected to be $\mathcal{O}(m_{Z^0}^4/\Lambda^4)$ for $s << \Lambda^2$.

At LEP the potential signature for an anomalous coupling is an excess of photons in the process $e^+e^-(\to Z^0) \to \gamma f\bar{f}$ where the $f\bar{f}$ are fermion pairs produced with a composition of the standard Z^0 decays (i.e. $\sim$20% $\nu\bar{\nu}$, $\sim$10% l^+l^-, $\sim$70% $q\bar{q}$). Within the cuts of $\alpha_{f,\gamma} > 100$ mrad, $|\cos\theta_\gamma| < 0.95$, and $E_\gamma > 5$ GeV, typically used at LEP, anomalous couplings would, according to the calculation of (3), lead to cross sections (given in pb)

$$\sigma_{anomalous} \sim 1.7|h_3|^2 BR(Z^0 \to f\bar{f}) \quad or \quad \sim 0.25|h_4|^2 BR(Z^0 \to f\bar{f}) \quad (1)$$

which has to be compared to the Standard Model background of

$$\sigma_{ISR} \sim 45 \cdot BR(Z^0 \to f\bar{f}) \quad and \quad \sigma_{FSR} \sim 2400 \cdot e_f^2 \cdot BR(Z^0 \to f\bar{f}) \quad (2)$$

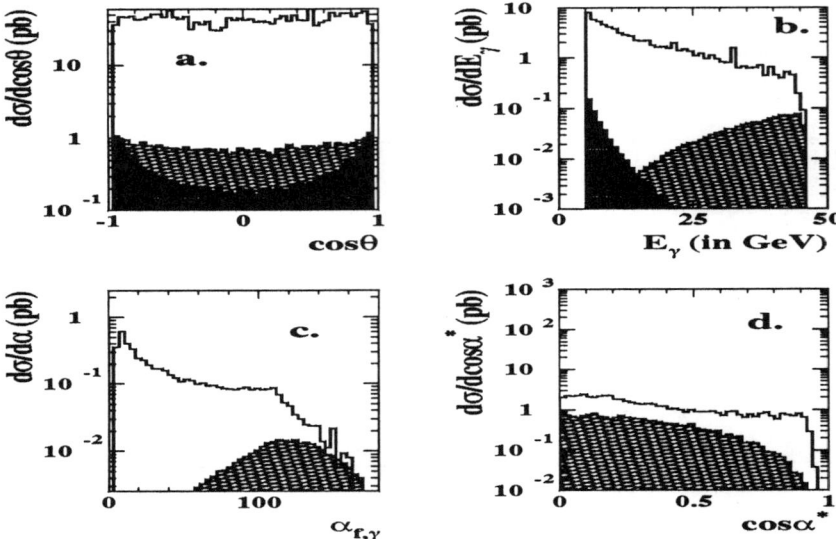

FIG. 2. Properties of various photon sources in the $\mu\mu\gamma$ final state. The open histogram shows the FSR and the dark histogram the ISR contributions. An anomalous coupling of $|h_3|$ =5 would lead to the cross hatched distribution. Distributions are shown after incremental cuts: a. polar angle $\cos\theta$ with wide cuts (see text); b. photon energy E_γ after $|\cos\theta_\gamma| < 0.8$; c. minimum angle $\alpha_{f,\gamma}$ for $E_\gamma >$23 GeV; d. angle α^* in fermionic rest system after cut on $\alpha_{f,\gamma} > 100$ degrees.

here e_f is the electric charge of the outgoing fermion.

The polar angle distributions of initial and final state radiation and for an anomalous coupling $|h_3|$=5 are shown in Fig.2a. A cut on the polar angle like $|\cos\theta| <$0.8 suppresses the contribution from ISR significantly. The resulting distribution of the photon energy is shown in Fig.2b. Applying an additional cut on the photon energy like $2E_\gamma/E_{cm} >$0.5, i.e. $E_\gamma >$23 GeV, eliminates the ISR contribution almost completely, while hardly affecting anomalous contributions. As shown in Fig.2c and d the much more abundant FSR background is much harder to suppress and only at the price of a substantial loss in efficiency for anomalous production. Since it is free of final state photons, the $\nu\bar{\nu}\gamma$ channel becomes the outstanding mode for testing for anomalous $ZZ\gamma$ couplings.

The photon energy spectrum of L3 (4) is shown in Fig.3a together with the Standard Model expectation and an expectation for an anomalous coupling $|h_3|$=1 which leads to high energy photons. Together with the results from ALEPH (5), and OPAL (6) the observed yields for $E_\gamma > 15$ and 23 GeV are listed in Table 2. Also listed is the effective luminosity

$$\mathcal{L}_{eff} = \epsilon_\gamma \sum_i \frac{\sigma(E^i_{cm})}{\sigma(M_{Z^0})} \mathcal{L}^i_{nom}$$

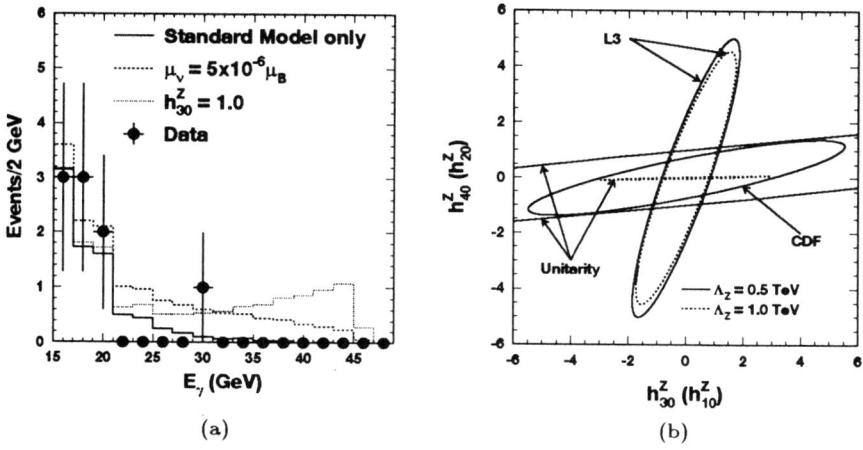

FIG. 3. (a) Photon energy spectrum for $\nu\bar{\nu}\gamma$ events. (b) Allowed region of anomalous couplings $h_3(h_1)$ and $h_2(h_4)$ from both LEP and the Tevatron

TABLE 2. Number of high energy single photons observed (and expected).

Experiment	Ref.	$\mathcal{L}_{eff}$	$E_\gamma > 15$ GeV	$E_\gamma > 23$ GeV
ALEPH	(5)	6.3 pb^{-1}	0 (exp. 0)	0 (exp. 0)
L3	(4)	35.0 pb^{-1}	9 (exp. 8)	1 (exp. 1.2)
OPAL	(6)	16.7 pb^{-1}	5 (exp. 1.4)	0 (exp. 0)
Combined		48.0 pb^{-1}	14 (exp. 9.4)	1 (exp. 1.2)

where $\mathcal{L}^i_{nom}$ is the experimental luminosity collected at the c.m. energy E^i_{cm}. These luminosities are weighted by the ratio of the cross sections at this energy and at the Z^0 peak, ϵ_γ is the experimental photon detection efficiency.

No significant excess of high energy photons is observed. Based on the LEP data sample the production of more than four γZ^{0*} events with an energy of more than 23 GeV can be excluded at 95% C.L.. This can be interpreted as

$$|h^Z_{1,3}| < 0.73 \qquad |h^Z_{2,4}| < 1.96 \qquad (4)$$

Here it was assumed that just one coupling constant contributes. In principle the two either CP conserving or violating contributions could interfer. The allowed region in the (h_3, h_4) and (h_1, h_2) parameter space as obtained by L3 is displayed in Fig.3b.

Also shown are the corresponding limits from $Z\gamma$ final states as observed by CDF (7). In comparing LEP and Tevatron limits a few differences have to be noted. Firstly, the $p\bar{p}$ result does not discriminate between the $\gamma\gamma Z$ and $ZZ\gamma$ couplings. Secondly, at LEP the couplings are determined at a *fixed* c.m. energy where the sensitivity to $h_2(h_4)$ is less than the one to $h_1(h_3)$. At the Tevatron the limits are derived from the energy *dependence* of the production. Since the anomalous contributions grow like $(E_{cm}/M_Z)^3$ and $(E_{cm}/M_Z)^5$ for the two coefficients, one is more sensitive to $h_2(h_4)$. As a result the LEP and Tevatron limits are rather orthogonal. This energy dependence at the Tevatron also requires them to introduce a form-factor behaviour to avoid unitarity violation. CDF uses $h_i = h_i^0/(1 + s/\Lambda^2)^n$, with n=3 for i=1,3, and n=4 for i=2,4 for some assumed values of Λ.

Combining LEP and CDF, limits of $\mathcal{O}(1)$ are obtained for $|h_i|$. This is the first direct constraint on a potential anomalous $(Z,\gamma)Z\gamma$ coupling, which is, however, still several orders of magnitude above the expectation for some new effects.

Based on these numbers one may estimate the final potential of LEP1 [1]. At the end of '95, running at the Z^0 pole will be completed and some 200 pb^{-1} will be collected per experiment. Allowing for an improved efficiency for high energy photons of $\sim$70% and taking into account the distribution of c.m. energies around the Z^0-pole, we expect a $\sim$ tenfold increase in effective luminosity from the combination of all experiments. The sensitivity to anomalous couplings will then improve by $\sim \sqrt{10}$ to $|h_{1,3}|$ <0.2 and $|h_{2,4}|$ <0.6.

THE FERMION - PHOTON COUPLING

Apart from searching for anomalous *boson* self couplings, the photon spectra may also be interpreted in terms of *fermion* - photon couplings. For example, the $\nu\bar{\nu}\gamma$ measurement (4) has been used by L3 to set limits on an anomalous magnetic moment of the τ neutrino of $\mu_{\nu_\tau} < 4.1 \cdot 10^{-6} \mu_B$ with μ_B the Bohr magneton. Similar analyses can be performed for charged leptons.

The photon energy spectrum for τ pairs from a preliminary measurement of DELPHI (8) is shown in Fig.4a to be in agreement with the Standard Model expectation. Using the calculation of (9), anomalous magnetic or electric moments F_2^τ and F_{EDM}^τ of τ's would lead to an essentially energy *independent* distribution. The observed distribution can be translated into

$$F_2^\tau(q^2 = 0) < 0.072 \qquad F_{EDM}^\tau(q^2 = 0) < 4 \cdot 10^{-16} \qquad (5)$$

both at 95% confidence. These measurements complement the tighter bounds obtained from τ pair production at $q^2 > (2m_\tau)^2$ in the e^+e^- continuum.

The $\mu\mu\gamma$ events were used by OPAL (10) and DELPHI (11) to study event properties at c.m. energies several GeV below the pole. No deviation from the Standard Model expectations are observed allowing limits to be set on possible additional Z' gauge bosons.

---

[1] The prospects at LEP2 are addressed by Busenitz at this conference.

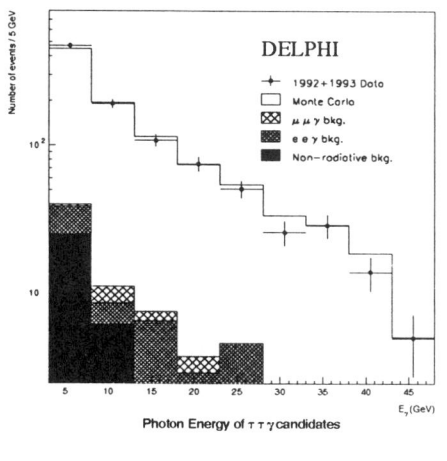
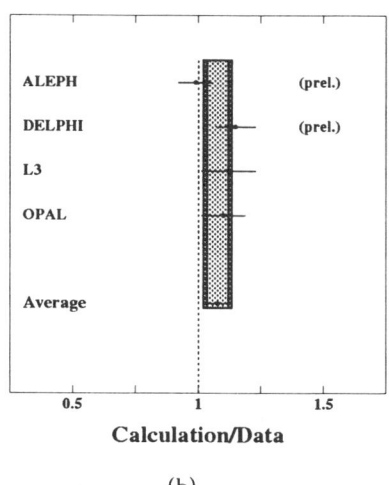

FIG. 4. (a) Photon energy spectrum in $\tau\tau\gamma$ events. (b) Ratio theory/data for isolated photon production in hadronic events. The light grey region indicates the one standard deviation range of the LEP average with only the experimental uncertainties. For the dark grey area theoretical uncertainties are included.

Because of its wide potential for QCD and electroweak quark studies (12) photon bremsstrahlung from quarks has been rather actively scrutinized at LEP (for a review see (13)). All experiments have compared *isolated* photon production with matrix element calculations of $\mathcal{O}(\alpha\alpha_s)$. In Fig.4b the ratio $R_{q\bar{q}\gamma}$ of the calculation (14) over the data (15) are displayed. The selection criteria for the various experiments are slightly different but lead to consistent values of $R_{q\bar{q}\gamma}$ which can be averaged to yield

$$R_{q\bar{q}\gamma} = \frac{\sigma_{theory}}{\sigma_{measurement}} = 1.077 \pm 0.042 \pm 0.04 \qquad (6)$$

indicating agreement with the Standard Model. The two uncertainties are due to the experimental analysis and to the theoretical calculation, particularly the value of α_s, the dependence of the cut - off against the quark - photon singularity and hadronisation (16).

SEARCHING FOR $Z^0 \to \gamma\gamma\gamma$

Whereas the Z^0 decay into two photons is forbidden by Bose - symmetry, it is the Standard Model dynamics that prohibits the decay of the Z^0 into three photons at tree level. Loop contributions give rise to BR($Z^0 \to \gamma\gamma\gamma$) ~ $\mathcal{O}(10^{-10})$, much below the sensitivity of current LEP experiments. Z^0 decays into three photons may be largely enhanced in extensions of the Standard Model, if, for example, the Z^0 is built up from charged constituents of size $1/\Lambda$. Its charge may be 'felt' by the photon depending on its resolution power $1/E_{cm}$. According to (17), within such models

$$\Gamma(Z^0 \to \gamma\gamma\gamma) = \frac{64\alpha^3}{9}(\pi^2 - 9)M_Z \frac{|\phi(0)|^2 N_c N_H <Q^6>}{4\pi M_Z^3} \qquad (7)$$

where $|\phi(0)|$ is the value of the wave function at the origin, N_c and N_H are the number of colours and hypercolours of the constituents and Q their electric charge. This leads to estimates of BR($Z^0 \to \gamma\gamma\gamma$) ~ $2 \cdot 10^{-4} <Q^6>$.

Standard Model background to this resonance decay is the pure QED process (Fig.1d) which decreases like ~ $1/E_{cm}^2$. A direct Z^0 decay, however, should show up as a resonance structure of the cross section. No such enhancement is observed, see, for example, Fig.5 of ref. (18).

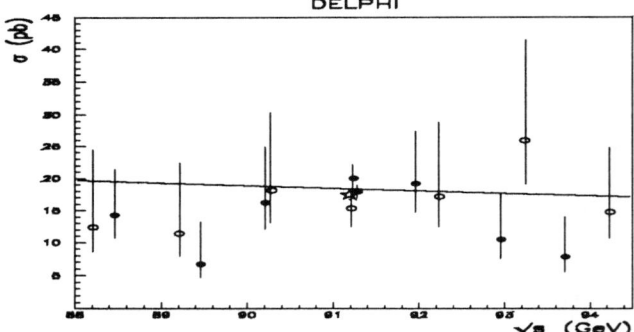

FIG. 5. Cross section for $e^+e^- \to \gamma\gamma(\gamma)$ in the central region of the DELPHI detector. Open and full circles represent data collected in different years, the line is the QED expectation.

Alternatively the direct $Z^0 \to \gamma\gamma\gamma$ decay could be established from the energy spectrum of the least energetic photon. For the Standard Model background its energy should be distributed according to the $1/E_\gamma$ bremsstrahlung spectrum. Assuming for the direct decay a phase space distribution, the energy of the least energetic photon tends towards $M_Z/3$. Such a search has been performed by L3 (19). The most restrictive limits on direct Z^0 decays are, at 95 % confidence,

$$BR(Z^0 \to \gamma\gamma\gamma) \quad < 1.7 \cdot 10^{-5} \; (DELPHI); \qquad < 1 \cdot 10^{-5} \; (L3) \qquad (8)$$

ANOMALOUS HIGGS PRODUCTION

Searches for the Standard Model Higgs boson at LEP have up to now been based on the Bjorken process $Z^0 \to Z^{0*}H^0$ (20) allowing limits of $m_H > 64.5$ GeV to be set (21). An alternative Higgs production mechanism (Fig.1g) is the decay $Z^0 \to \gamma H^0$ which proceeds dominantly via W loops (22). Its branching ratio of $10^{-6} - 10^{-7}$ is too small to be observed at LEP.

It may increase if the couplings among bosons are anomalous. Expressing the new interactions at some scale Λ by an effective Lagrangian, one finds (23)

$$\Gamma(Z^0 \to H^0\gamma) = \frac{\alpha}{96 m_Z}(m_Z^2 - m_H^2)^3 \left| c_{SM} + \frac{\delta_{new}}{\Lambda^2} \right|^2 \qquad (9)$$

where c_{SM} denotes the Standard Model contribution and δ_{new}/Λ^2 parametrises the contribution from new boson interactions. Even taking into account constraints from measurements at low energy, from LEP and from the Tevatron, these non Standard Model contributions may increase the $H^0\gamma$ production by several orders of magnitude while at the same time weaken the standard Higgs limits.

The decay rates of a Standard Model Higgs boson are unambiguously predicted. For the range of m_H of 12-80 GeV, relevant for the current Higgs searches at LEP, 85% of the Higgses should decay into a beauty pair. Extensions of the Standard Model may imply different branching ratios. For example, anomalous boson couplings could enhance the decay branching ratio $H^0 \to \gamma\gamma$ to even become the dominant decay mode. At LEP this would lead to a resonant $\gamma\gamma$ state with associated production of fermion pairs. A couple of years ago, the L3 collaboration claimed an excess of this kind in $(\mu^+\mu^-, e^+e^-)\gamma\gamma$ events with $M_{\gamma\gamma} \sim 60$ GeV (24). With increased statistics and extending the search to other channels this observation was not confirmed by the other LEP experiments (25).

All LEP experiments have performed searches for a narrow resonance in the channel $\gamma f \bar{f}$ with f being either neutrinos, charged leptons or quarks. The mass of the fermionic system recoiling against the photon can either be reconstructed from the photon energy alone using

$$M_{recoil} = E_{cm}\sqrt{1 - 2E_\gamma/E_{cm}}, \qquad (10)$$

or from kinematical fits exploiting the precise knowledge in e^+e^- collisions of the total c.m. energy and momentum ($\sum \vec{p} = 0$). These methods allow a good mass resolution even if the scalar particle decays hadronically yielding $\delta M_{recoil} \sim 1$ GeV for large masses and ~ 2-3 GeV for small masses.

To suppress the principle background from final state bremsstrahlung photons, angular cuts can be imposed. For example, one can use α^*, the angle between the fermion and the photon in the fermionic rest system. Whereas

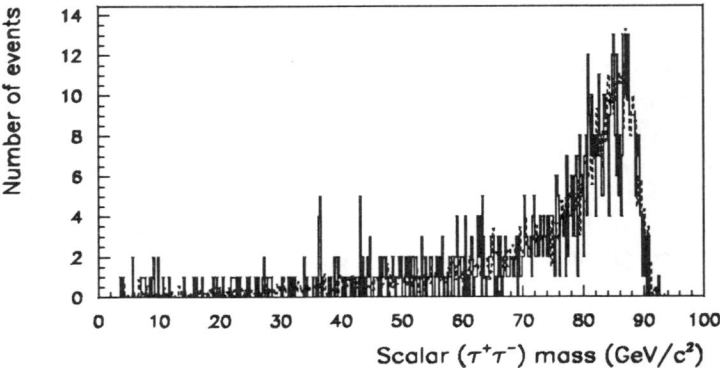

FIG. 6. Invariant mass of the $\tau^+\tau^-$ system produced in association with a photon.

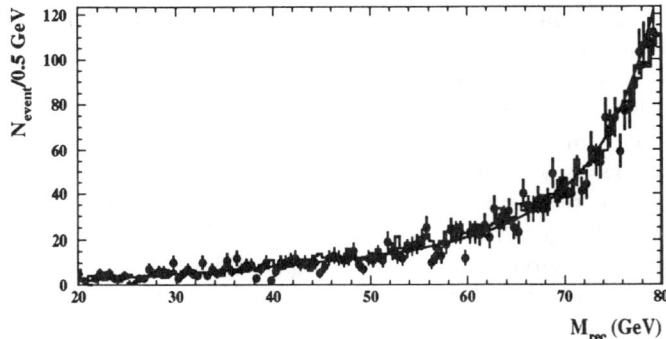

FIG. 7. Invariant mass of the hadronic system produced in association with a photon.

bremsstrahlung photons tend to be aligned with the emitting fermion, a Higgs decay leads to an isotropic distribution.

Preliminary spectra of the τ pair (26) and hadronic recoil masses (27) are shown in Figs.6,7. No resonance structure is observed in either of these distributions. The spectra can be interpreted in terms of limits on the product branching ratios BR($Z^0 \to S^0\gamma$)×BR($S^0 \to f\bar{f}$) at 95% confidence level as being smaller than typically $\sim 5 \cdot 10^{-6}$ for charged leptons (26) (Fig.8) and $\sim 3 \cdot 10^{-5}$ for quarks (27,28).

Given the probable preference of the scalar to decay into beauty quarks, one can improve limits by invoking beauty tagging. Selecting hadronic events with a secondary vertex, beauty events can be enriched with a rather high efficiency. Tagging of $b\bar{b}\gamma$ events has been used by OPAL (27) and DELPHI (28) who retain 60 (20) % of all beauty events while rejecting 90 (95) % of events with final state photon radiation from other quark species. The spectrum from OPAL is shown in Fig.9. No significant resonance structure is

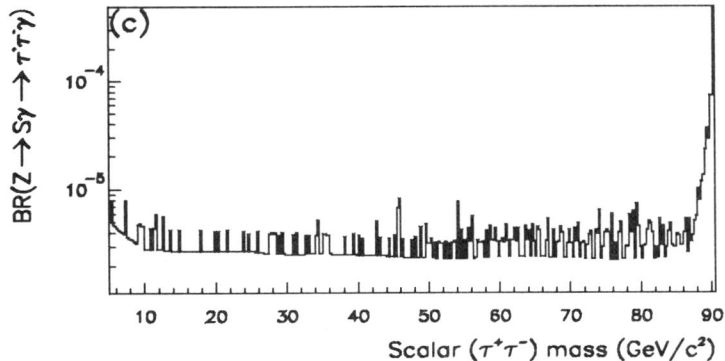

FIG. 8. Lower limits at 95% CL for the branching ratio BR($Z^0 \to S^0\gamma$) BR($S^0 \to \tau^+\tau^-$) as a function of the $\tau^+\tau^-$ mass.

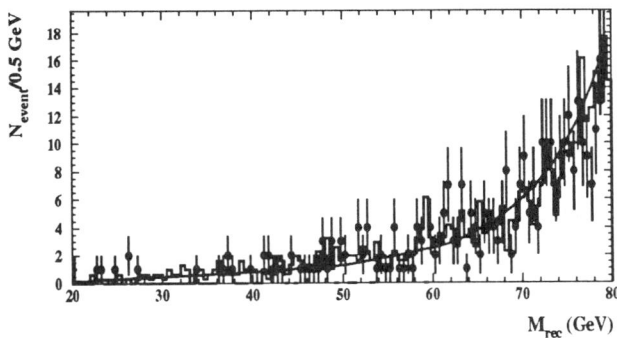

FIG. 9. Invariant mass of the $b\bar{b}$ system produced in association with a photon.

observed. The result can be interpreted as (see Fig.10)

$$BR(Z^0 \to S^0\gamma) \times BR(S^0 \to b\bar{b}) < 2-3 \cdot 10^{-5} \tag{11}$$

which restricts some of the allowed parameter region used in the calculation of (23) or for a composite Higgs (29).

Combining the results of the LEP experiments and assuming a 2-4 fold increase in statistics until the end of LEP1, limits of $5 \cdot 10^{-6}$ for $b\bar{b}$ and some factor five less for leptons are in reach.

CONCLUSIONS

Anomalous photon production is predicted in various extensions of the Standard Model and has been searched for in many different ways at LEP. On the basis of typically 40-50 pb^{-1} no signal has yet been observed.

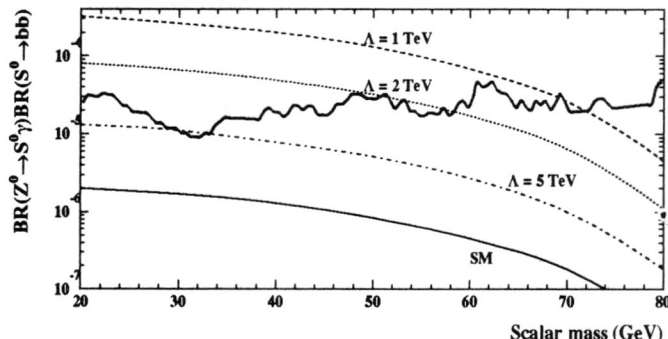

FIG. 10. Lower limits at 95% CL for the branching ratio BR($Z^0 \to S^0\gamma$) BR($S^0 \to b\bar{b}$) as a function of the scalar mass (thick full line). Also given is the Standard Model expectation (thin full line) and those for a composite Higgs assuming various values of the composite scale Λ.

Using the $Z \to \nu\bar{\nu}$ decay, limits are derived for anomalous $ZZ\gamma$ couplings of

$$|h^Z_{1,3}| < 0.73 \qquad |h^Z_{2,4}| < 1.96$$

Photon production in events with charged leptons and quarks also agree with the Standard Model expectation allowing various new measurements particularly for the τ. The respective limits on an anomalous magnetic moment of the ν_τ, of the τ itself and the electric dipole moment of the τ are

$$\mu_{\nu_\tau} < 4.1 \cdot 10^{-6} \mu_B \qquad F^\tau_2(q^2=0) < 0.072 \qquad F^\tau_{EDM}(q^2=0) < 4 \cdot 10^{-16}$$

Negative searches have been performed on resonant $\gamma\gamma\gamma$ production leading to limits of

$$\Gamma(Z^0 \to \gamma\gamma\gamma)/\Gamma(Z^0) < 1 \cdot 10^{-5}$$

Finally, studies on scalar resonance production in conjunction with γ emission have been made. No narrow signal has been observed and limits on the product branching ratio BR($Z^0 \to S^0\gamma$)BR($S^0 \to f\bar{f}$) of $\sim 5 \cdot 10^{-6}$ (charged leptons) and $\sim 2 \cdot 10^{-5}$ (hadrons, beauty pairs) have been set.

ACKNOWLEDGEMENT

I wish to thank Ulrich Baur, Steve Errede and Thomas Müller for organising an excellent and inspiring symposium. I am grateful to many of my LEP colleagues for providing me with the most recent results of their experiment and Ulrich Baur for letting me use his program to calculate effects of anomalous $ZZ\gamma$ couplings. In writing up the talk I profited from several comments by Fawzi Boudjema and Graham Wilson.

REFERENCES

1. Workshop on Photon Radiation from Quarks, Annecy 2-3 December 1991, ed. S.Cartwright, CERN 92-04.
2. K.Hagiwara et al. Nucl.Phys. **B282** 253 (1987).
3. U.Baur and E.Berger, Phys.Rev. **D47** 4889 (1993)
4. L3-Collaboration, M.Acciarri et al., Phys.Lett. **B346** 190 (1995).
5. ALEPH-Collaboration, D.Buskulic et al., Phys.Lett. **B313** 520 (1993).
6. OPAL-Collaboration, R.Akers et al., Z.Phys. **C65** 47 (1995).
7. CDF-Collaboration, F.Abe et al., FERMILAB Pub-94/244-E and Phys.Rev.Lett. **74** 1941 (1995)
8. DELPHI-Collaboration, Internal note DELPHI 94-89 PHYS 406.
9. J.A.Grifols and A.Mendez, Phys.Lett. **B255** 611 (1991).
10. OPAL-Collaboration, P.D.Acton et al., Phys.Lett. **B268** 122 (1991).
11. DELPHI-Collaboration, P.Abreu et al., CERN-PPE 94/121.
12. P.Mättig and W.Zeuner, Z.Phys. **C52** 31 (1991).
13. P.Mättig in Proceedings of the Eighth Lake Louise Winter Institute, Lake Louise, Alberta, Canada, 21-27 February 1993, eds. A.Astbury et al.
14. Here the calculation of G.Kramer and H.Spiesberger in (1) was used. See also E.W.N.Glover and J.Stirling, Phys.Lett. **B295** 128 (1992) and Z.Kunszt and Z.Trocsanyi, Nucl.Phys. **B394** 139 (1993).
15. ALEPH-Collaboration 93-111 PHYSIC 93-092; DELPHI-Collaboration to be published; L3-Collaboration O.Adriani et al., Phys.Lett. **B301** 136 (1993); OPAL-Collaboration P.Acton et al., Z.Phys. **C58** 405 (1993).
16. P.Mättig, H.Spiesberger and W.Zeuner, Z.Phys. **C60** 613 (1995).
17. F.Renard, Phys.Lett. **B116** 264 (1982).
18. DELPHI-Collaboration, P.Abreu et al., Phys.Lett. **B327** 386 (1994).
19. L3-Collaboration, M.Acciarri et. al., Phys.Lett. **B345** 609 (1995).
20. J.D.Bjorken, in Proceedings of the 1976 Summer Institute on Particle Physics, Stanford, ed. M.C.Zipf (SLAC, Stanford CA, 1977) pg.1.
21. F.Richard in Proceedings of the XXVII International Conference on the High Energy Physics 20-27 July 1994 Glasgow, Scotland, UK ed. P.J.Bussey and I.G.Knowles.
22. R.N.Cahn, M.S.Chanowitz and N.Fleishon, Phys.Lett. **B82** 112 (1979).
23. K.Hagiwara, R.Szalapski and D.Zeppenfeld, Phys.Lett. **B318** 155 (1993).
24. L3-Collaboration, O.Adriani et al., Phys.Lett. **B295** 337 (1992).
25. G.W.Wilson in Proceedings of the International Europhysics Conference on High Energy Physics, Marseille, France (July 1993) eds. J.Carr and M.Perrotet.
26. ALEPH-Collaboration, submitted to the 27th International Conference on High Energy Physics, Glasgow, Scotland, 20-27 July 1994.
27. OPAL-Collaboration, Internal note OPAL-PN-140.
28. DELPHI-Collaboration, Internal note DELPHI 94-119 PHYS 436.
29. D.Düsedau and J.Wudka, Phys.Lett. **B180** 290 (1986)

Experimental Signatures of a Parity Violating Anomalous Coupling g_5^Z

G. Valencia

Department of Physics and Astronomy
Iowa State University
Ames, IA 50011

I discuss the experimental signatures of a parity violating but CP conserving interaction in the symmetry breaking sector of the electroweak theory.

INTRODUCTION

The standard model of electroweak interactions has now been tested thoroughly in a number of experiments. The only sector that has not been tested directly is the electro-weak symmetry breaking (or Higgs) sector. It is very important to understand in detail the experimental signatures for the symmetry breaking sector. These vary from the direct search for new particles such as a Higgs boson, to the search for indirect manifestations of the existence of these new particles.

A convenient parameterization of these indirect effects of new particles at energies below threshold for their production is that of anomalous gauge boson couplings, the subject of this meeting. As discussed by Wudka (1), there are several ways in which these anomalous gauge boson couplings may be written in terms of a low energy effective Lagrangian.

I choose to study the case of a strongly interacting symmetry breaking sector in which there is no light Higgs boson, and therefore, use an effective Lagrangian with a non-linearly realized symmetry breaking. My motivation for this choice is simple: if there is a light Higgs boson we will find it directly and not through its contributions to anomalous couplings. I furthermore choose the "Gasser and Leutwyler" (2) construction of the effective Lagrangian because it makes the discussion of global symmetries transparent.

First I briefly review the formalism in order to establish the notation and discuss the possible size of the parity violating anomalous coupling from simple dimensional analysis. I then study the indirect bounds that can be placed on this coupling from its one-loop contribution to rare decays and partial Z widths. Finally I discuss how to isolate the parity violating coupling in future high energy experiments.

FORMALISM

Effective Lagrangian

The starting point is the minimal standard model without a Higgs boson. This model can be written as the usual standard model, but replacing the scalar sector with the effective Lagrangian (3):

$$\mathcal{L}^{(2)} = \frac{v^2}{4} \text{Tr}\left(D^\mu \Sigma^\dagger D_\mu \Sigma \right). \tag{1}$$

The matrix $\Sigma \equiv \exp(i\vec{w} \cdot \vec{\tau}/v)$, contains the would-be Goldstone bosons w_i that give the W and Z their mass via the Higgs mechanism. Their interactions with the $SU(2)_L \times U(1)_Y$ gauge bosons follow from the covariant derivative:

$$D_\mu \Sigma = \partial_\mu \Sigma + \frac{i}{2} g W_\mu^i \tau^i - \frac{i}{2} g' B_\mu \Sigma \tau_3. \tag{2}$$

The details of the physics that break electroweak symmetry determine the next-to-leading order effective Lagrangian. At energies small compared to Λ, it is sufficient to consider those terms that are suppressed by E^2/Λ^2 with respect to Eq. 1. There are three terms in this next to leading order effective Lagrangian that contribute to gauge boson self-energies at tree level (and thus to the LEP observables $\epsilon_{1,2,3}$ of Ref. (4)). For later reference, the one the respects the custodial symmetry is L_{10}.

There are several terms in the next to leading order effective Lagrangian that contribute to three gauge boson couplings at tree level. Only two of them respect the custodial symmetry, for later reference they are L_{9L}, L_{9R}.

Finally, there are also several terms in the next to leading order effective Lagrangian that contribute at tree level to couplings with at least four gauge bosons. Two of these terms respect the custodial symmetry and for later reference they are L_1, L_2.

The next to leading order effective Lagrangian that respects the custodial symmetry is then:

$$\begin{aligned}
\mathcal{L}^{(4)} = \frac{v^2}{\Lambda^2} \Bigg\{ & L_1 \left[\text{Tr}\left(D^\mu \Sigma^\dagger D_\mu \Sigma \right) \right]^2 + L_2 \, \text{Tr}\left(D_\mu \Sigma^\dagger D_\nu \Sigma \right) \text{Tr}\left(D^\mu \Sigma^\dagger D^\nu \Sigma \right) \\
& - ig L_{9L} \, \text{Tr}\left(W^{\mu\nu} D_\mu \Sigma D_\nu \Sigma^\dagger \right) - ig' L_{9R} \, \text{Tr}\left(B^{\mu\nu} D_\mu \Sigma^\dagger D_\nu \Sigma \right) \\
& + gg' L_{10} \, \text{Tr}\left(\Sigma B^{\mu\nu} \Sigma^\dagger W_{\mu\nu} \right) \Bigg\}.
\end{aligned} \tag{3}$$

There are many more terms that break the custodial symmetry, but only one that violates parity while conserving $\mathcal{CP}$. This term gives rise to three and four gauge boson couplings and is the subject of this talk.

The motivation for considering this term is, of course, that we should explore all possibilities for the symmetry breaking sector. In theories where the electroweak symmetry breaking sector conserves parity, like the minimal standard model or most technicolor theories, this term is expected to be very small.

The parity violating and $\mathcal{CP}$ conserving effective Lagrangian at order $1/\Lambda^2$ is

$$\mathcal{L}_{\text{p.v.}}^{(4)} = \frac{v^2}{\Lambda^2} g \hat{\alpha} \epsilon^{\alpha\beta\mu\nu} \text{Tr}\left(\tau_3 \Sigma^\dagger D_\mu \Sigma\right) \text{Tr}\left(W_{\alpha\beta} D_\nu \Sigma \Sigma^\dagger\right) \tag{4}$$

where $W_{\mu\nu}$ is the $SU(2)$ field strength tensor. In terms of $W_\mu \equiv W_\mu^i \tau_i$, it is given by:

$$W_{\mu\nu} = \frac{1}{2}\left(\partial_\mu W_\nu - \partial_\nu W_\mu + \frac{i}{2} g [W_\mu, W_\nu]\right). \tag{5}$$

It is easy to see that this is the only term that violates parity and conserves $\mathcal{CP}$ to order $1/\Lambda^2$.

In unitary gauge, the effects of the Lagrangian Eq. 4, are very simple. There is a three gauge boson interaction:

$$\mathcal{L}^{(3)} = -\frac{\hat{\alpha} g^3 v^2}{\Lambda^2 c_\theta} \epsilon^{\alpha\beta\mu\nu} \left(W_\nu^- \partial_\alpha W_\beta^+ - W_\beta^+ \partial_\alpha W_\nu^-\right) Z_\mu, \tag{6}$$

which generates a $Z(q) \to W^+(p^+) W^-(p^-)$ coupling. In the notation of Ref. (5) we have the correspondence:

$$g_5^Z = \hat{\alpha} \frac{g^2}{c_\theta^2} \frac{v^2}{\Lambda^2} = \frac{4 M_Z^2}{\Lambda^2} \hat{\alpha}. \tag{7}$$

There is also a four gauge boson interaction required by electromagnetic gauge invariance:

$$\mathcal{L}^{(4)} = i \frac{2 \hat{\alpha} g^4 v^2 s_\theta}{\Lambda^2 c_\theta} \epsilon^{\alpha\beta\mu\nu} W_\alpha^- W_\beta^+ Z_\mu A_\nu. \tag{8}$$

This interaction contributes to the processes we discuss and must be considered simultaneously with that of Eq. 6. The Feynman rules for this interaction were written down in Ref. (6).

Natural size of g_5^Z and Unitarity

Within the minimal standard model, the operator Eq. 4 is generated at one-loop by the splitting between top-quark and bottom-quark masses. In the limit $m_t \gg m_W$, and setting $m_b = 0$ one finds (6):

$$\left(\frac{v^2}{\Lambda^2}\hat{\alpha}\right)_{top} = \frac{N_c}{128\pi^2}\left(1 - \frac{8}{3}s_\theta^2\right) \approx 10^{-3} \qquad (9)$$

For comparison, in this same limit one obtains $\frac{v^2}{\Lambda^2}L_1 \sim -3\times 10^{-4}$ and $\frac{v^2}{\Lambda^2}L_2 \sim 6\times 10^{-4}$ (7). We see that in this limit (in which the custodial symmetry is violated "maximally"), the parity violating coupling is of the same size as other anomalous couplings. Of course, this limit is not allowed by the size of $\Delta\rho$. Taking the scale Λ to be a few TeV (2 TeV for definiteness), one expects that in theories where there is no custodial symmetry and $\rho \approx 1$ accidentally, $\hat{\alpha}$ can be of order one. On the other hand, in theories with a custodial symmetry, one expects $\hat{\alpha}$ to be at most as large as $\Delta\rho$.

Our effective Lagrangian formalism breaks down at some scale $\Lambda \leq 3$ TeV, and this manifests itself in amplitudes that grow with energy and violate unitarity at some scale related to Λ. By studying the high energy behavior of longitudinal vector boson scattering one finds that the effective Lagrangian description breaks down between 1 and 2 TeV. For our numerical estimates we will work with energies up to 2 TeV. We thus turn the question around and ask how large can $\hat{\alpha}$ be so that all scattering amplitudes remain below their unitarity bound at energies up to 2 TeV. The answer is $|\hat{\alpha}| < 5$. This means that the bounds that can be placed at high energy experiments on this coupling will only be meaningful if they are better than $|\hat{\alpha}| < 5$. For bounds placed at lower energy machines such as LEP, the bad high energy behavior shows up in the need for counterterms to the one-loop calculations. Since we do not know what those counterterms are, the bounds obtained will be connected to "naturalness" assumptions.

PRESENT BOUNDS

In this section we study the bounds that already exist on g_5^Z. They follow from considering the one-loop effects of the operator Eq. 4 in the coupling of a Z boson to fermions. These observables do not single out the effects of g_5^Z, they are sensitive to most of the anomalous couplings.

Rare K- and B-meson decays

These rare decays receive contributions from the parity violating effective Lagrangian Eq. 4 at the one-loop level. One-loop amplitudes with one vertex from the $\mathcal{O}(1/\Lambda^2)$ effective Lagrangian are $\mathcal{O}(1/\Lambda^4)$. A complete study thus requires the next to next to leading order counterterms, as well as two loop contributions from the leading order effective Lagrangian. It is clear that there are several contributions to these decays that occur at the same order as the one-loop contribution from g_5^Z and that they could cancel: we assume that they do not.

In unitary gauge, the g_5^Z coupling affects this decays by modifying the "Z-penguin" diagram as discussed in Ref. (6,8). The result is dominated by top-quark intermediate states and is finite due to a GIM cancellation. With $x_t = m_t^2/m_W^2$ and defining

$$W(x_t) \equiv \frac{3}{4} x_t \left(\frac{1}{1 - x_t} + \frac{x_t \log x_t}{(1 - x_t)^2} \right) \tag{10}$$

one can write down the result in terms of the notation of Ref. (9), by replacing:

$$Y(x_t) \to \hat{Y}(x_t) = Y(x_t) + g_5^Z c_\theta^2 W(x_t)$$
$$Y(x_t) = \frac{x_t}{8} \left(\frac{x_t - 4}{x_t - 1} + \frac{3x_t}{(x_t - 1)^2} \log x_t \right) \tag{11}$$

One finds for example:

$$\Gamma(B_s \to \mu^+ \mu^-) = \frac{G_F^2}{\pi} \left(\frac{\alpha}{4\pi s_\theta^2} \right)^2 F_B^2 m_\mu^2 m_B |V_{tb} V_{ts}^*|^2 \hat{Y}(x_t)^2. \tag{12}$$

There are similar contributions to the decays $K_L \to \mu^+ \mu^-$ and $K^+ \to \pi^+ \nu \bar{\nu}$. Because $K_L \to \mu^+ \mu^-$ is dominated by long distance physics, it can only be used to place a "theoretical" bound on g_5^Z by requiring the new contribution to be less than the standard model short distance contribution (8). This results in

$$g_5^Z < \mathcal{O}(1). \tag{13}$$

If the rates for the short distance dominated processes $B_s \to \mu^+ \mu^-$ or $K^+ \to \pi^+ \nu \bar{\nu}$ are measured to within factors of two, the same bound Eq. 13 will be obtained. To improve this bound would require a precision measurement of the rate, combined with detailed knowledge of all the standard model parameters (CKM angles, top quark mass, and decay constants) (6).

Partial Z widths at LEP

High precision measurements at the Z pole at LEP combined with polarized forward backward asymmetries at SLC put stringent limits on any new physics beyond the standard model. These measurements are now sufficiently precise to limit the one-loop contribution of anomalous three gauge boson couplings to the Z pole observables.

The bounds arise because at the one-loop level Eq. 4 modifies the $Z f' \bar{f}$ couplings. Because the operator modifies the gauge boson self-couplings, its one-loop effects on the Z couplings to fermions affect both the flavor diagonal and the flavor changing vertices considered in the previous section. It turns out that the flavor diagonal vertices provide better constraints due to the extraordinary precision of the LEP measurements.

The flavor diagonal calculation is different from the flavor changing calculation in that the one-loop effects of g_5^Z are now divergent. Nevertheless, from the effective field theory perspective this is not significant. In both cases, there are other contributions to the physical processes from other non-renormalizable interactions between fermions and gauge bosons. In the previous section we adopted the point of view that those other interactions did not cancel the contributions of g_5^Z to rare decays. In this section we adopt the point of view that the renormalization of such couplings removes any divergence from physical amplitudes. As is usual in effective field theory calculations, we estimate the size of the g_5^Z contribution to the physical amplitudes from the leading non-analytic terms that go like $\log(\mu)$.

In addition to the direct contribution of g_5^Z to the $Zf\bar{f}$ vertex we must consider indirect effects due to renormalization. In particular, the operator of Eq. 3 also modifies the $W^\pm \to \ell^\pm \nu$ coupling, contributing in this way to muon decay and thus introducing a renormalization of G_F. In terms of the input parameters: G_F as measured in muon decay, $\alpha_*(M_Z^2) \approx 1/128.8$ (11) and the physical Z mass, and using a s_θ^2 defined by the relation:

$$s_Z^2 c_Z^2 \equiv \frac{\pi \alpha_*}{\sqrt{2} G_F M_Z^2}, \tag{14}$$

we find:

$$\frac{\delta \Gamma_f^5}{\Gamma_f^{(0)}} = \frac{3\alpha}{2\pi} g_5^Z \log\left(\frac{\mu}{M_W}\right) \left[\frac{2L_f}{L_f^2 + R_f^2} \frac{c_\theta^2}{s_\theta^2} + \left(1 + \frac{2R_f(L_f + R_f)}{L_f^2 + R_f^2}\right) \frac{c_\theta^2}{s_\theta^2 - c_\theta^2}\right]. \tag{15}$$

Where the shifts in the partial decay widths of the Z are defined by:

$$\Gamma(Z \to f\bar{f}) = \Gamma_f^{SM} + \delta\Gamma_f^5 \equiv \Gamma_f^{SM}\left(1 + \frac{\delta\Gamma_f^5}{\Gamma_f^{SM}}\right). \tag{16}$$

To place bounds on g_5^Z (and other couplings) we compare the standard model predictions, Γ_f^{SM}, including the one loop QED and QCD radiative corrections with the most recent results from LEP. We use the theory numbers of Langacker (12).

Our 90% confidence level interval for the allowed values of g_5^Z is shown in Table 1 (10). To place this result in perspective, it is instructive to compare them with results for the couplings that respect the custodial symmetry in Eq. 3 (13):

From Table 1 we see that the best limits are placed on the coupling that contributes to the Z self-energy at tree level, L_{10}. The bounds on the other couplings are obtained by taking only one of them to be non-zero at a time, and they are all comparable. A deviation in the partial Z widths from their standard model value could not be attributed to a single coupling. In order to isolate the effects of g_5^Z we consider in the next two sections other observables that single out the parity violating operator.

TABLE 1. 90% confidence level intervals for g_5^Z from different LEP observables.

Coupling ($\Lambda = 2$ TeV)	90% confidence level interval
$L_{10}^r(M_Z)_{new}$	(-0.46, 0.77)
L_{9L}	(-22, 16)
L_{9R}	(-77, 94)
$L_1 + 5/2 L_2$	(-28, 26)
$\hat{\alpha}$	(-9, 5)
g_5^Z	(-0.07, 0.04)

FUTURE BOUNDS

In this section we discuss the most promising reactions to place bounds on g_5^Z in future colliders. These bounds arise from considering observables that single out the coupling g_5^Z making through its parity violating nature.

Forward-backward asymmetry in $e_L^+ e_R^- \to W^+ W^-$

In this section we study the effect of the parity violating operator Eq. 4 on the process $e^+ e^- \to W^+ W^-$. This process receives contributions from s-channel γ and Z exchange diagrams and from a t-channel neutrino exchange diagram. The latter contributes only to $e_L^- e_R^+ \to W^- W^+$.

The differential cross-section for right-handed electrons is found to be (6,14):

$$\frac{d\sigma_{TT}}{d(\cos\theta)}\bigg|_{e_R^-} = \frac{\pi\alpha^2}{s} \beta^3 \frac{m_Z^4}{(s-m_Z^2)^2} \sin^2\theta$$

$$\frac{d\sigma_{LL}}{d(\cos\theta)}\bigg|_{e_R^-} = \frac{\pi\alpha^2}{32s} \frac{\beta^3}{c_\theta^4} \frac{s^2}{(s-m_Z^2)^2} (5+\beta^2)^2 \sin^2\theta$$

$$\frac{d\sigma_{TL}}{d(\cos\theta)}\bigg|_{e_R^-} = \frac{\pi\alpha^2}{s} \frac{\beta^3}{c_\theta^2} \frac{m_Z^2 s}{(s-m_Z^2)^2} \left(1 + \cos^2\theta + 2\beta \frac{s}{m_Z^2} g_5^Z \cos\theta\right) \quad (17)$$

where $\beta^2 = 1 - 4m_W^2/s$. Other anomalous couplings do not contribute to the forward backward asymmetry in $e_R^- e_L^+ \to W^- W^+$ and they are not considered here.

As can be seen from Eq. 17, there is a term in σ_{TL} that is linear in $\cos\theta$ (the scattering angle in the center of mass). This term gives rise to a forward-backward asymmetry. Although there is a similar term in the differential cross-section for $e_L^- e_R^+ \to W^+ W^-$, in that case one also has a t-channel neutrino exchange diagram that gives rise to a very large forward-backward asymmetry within the minimal standard model. Thus, if we want to isolate the g_5^Z term, it is very important to have right-handed electrons. Since the cross-section for left-handed electrons is several orders of magnitude larger than that for right-handed electrons, it presents a formidable background.

We find (6) that the largest sensitivity to g_5^Z occurs in the forward-backward asymmetry at high center of mass energies. This sensitivity decreases dramatically if there is any contamination of left handed electrons as shown in Figure 1, where we we present the forward-backward asymmetry for an e^+e^- collider with $\sqrt{s} = 500$ GeV. This figure shows the great sensitivity of the ob-

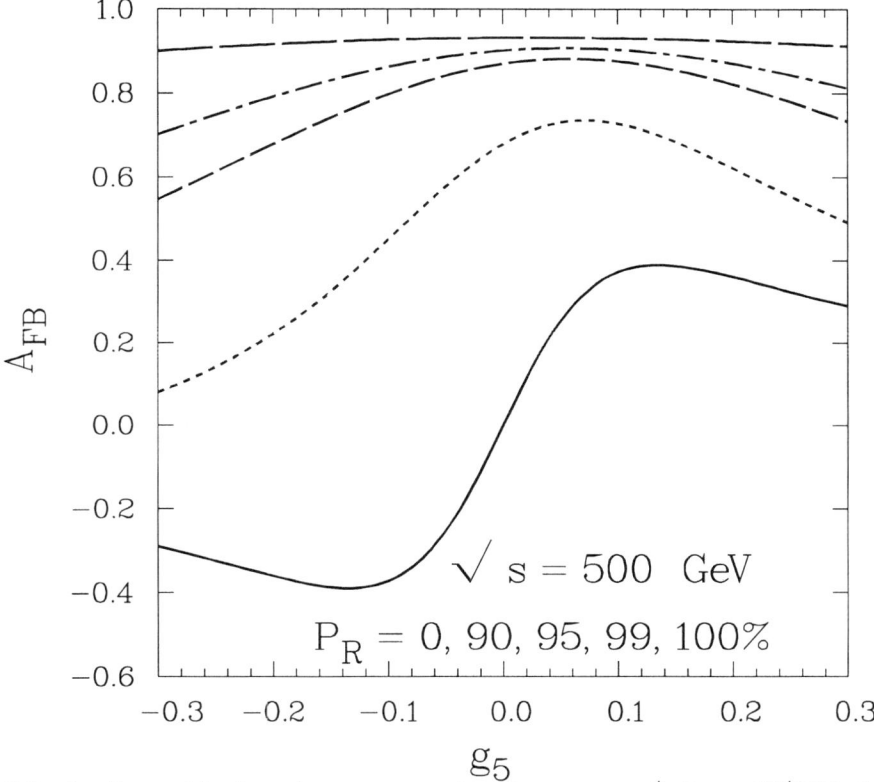

FIG. 1. Forward-backward asymmetry for the process $e^+e^- \to W^+W^-$ for $\sqrt{s} = 500$ GeV. The different curves from upper most to lowest correspond to a fraction of right handed electrons in the beam of 0%, 90%, 95%, 99% and 100%.

servable to the coupling g_5^Z. Unfortunately it also shows how this sensitivity is lost if there is even a small fraction of left handed electrons.

The total cross section is also sensitive to the value of g_5^Z, however, a deviation in the total cross section from the standard model value would not single out the g_5^Z coupling.

High energy $e^-\gamma \to \nu W^- Z$

In this section we explore the possibility of observing the effects of the parity violating operator Eq. 4 via the anomalous four-gauge-boson coupling that it generates. We thus turn our attention to high energy vector-boson fusion experiments. Given the form of the four vector-boson interaction, Eq. 8, we look at processes involving one photon and one Z. There are several possibilities, for example $Z\gamma$ production in high energy e^+e^- or pp colliders. This process, however, suffers from large standard model backgrounds. We consider instead a high energy $e^-\gamma$ collider where we can cleanly identify the process $e^-\gamma \to \nu W^- Z$, and where we can also consider a polarized photon if need be.

To understand the physics, we first use the equivalence theorem to compute the the polarized cross sections for $W\gamma \to wz$, $\sigma(\lambda^W, \lambda^\gamma)$ (6):

$$\sigma_{+-} = \sigma_{-+} = \frac{\pi\alpha^2}{s_\theta^2} \frac{1}{3s}$$

$$\sigma_{++} = \sigma_{--} = \frac{\pi\alpha^2}{s_\theta^2} \frac{1}{3s} \left(|g_5^Z|^2 c_\theta^4 \frac{s^2}{m_W^4} \right)$$

$$\sigma_{L+} = \sigma_{L-} = \frac{\pi\alpha^2}{s_\theta^2} \frac{1}{3s} \left(|g_5^Z|^2 \frac{c_\theta^4}{4} \frac{s^3}{m_W^6} \right) \tag{18}$$

From Eq. 18 we see that the g_5^Z term does not interfere with the lowest order term. This means that we can only construct observables sensitive to g_5^Z that are parity even and can thus be generated by other anomalous couplings. However, it is possible that the cross section is more sensitive to the $|g_5^Z|^2$ term than to those terms proportional to L_{9L}, L_{9R} or L_{10} in very high energy machines. The reason is that the $|g_5^Z|^2$ term is the only one that contributes to the amplitude where all three vector-bosons are longitudinally polarized (this is the source of $\sigma_{L\pm}$ in Eq. 18) and we expect these terms of "enhanced electroweak strength" to dominate at high energies. This is indeed the case, as shown by a numerical simulation (15).

To demonstrate the significant sensitivity of this process to $\hat{\alpha}$, we consider a 2 TeV e^+e^- collider (the $e^-\gamma$ differential cross section is folded with the energy spectrum of the back scattered photon). We make use of the relative enhancement of $\hat{\alpha}$ at higher energies to isolate this coupling with a set of cuts like:

$$|\cos\theta_V| < 0.8, \quad p_T(WZ) > 30 \text{ GeV}, \quad M(WZ) > 0.5 \text{ TeV}. \tag{19}$$

Here the $p_T(WZ)$ cut is optimized to suppress reducible backgrounds from other sources. The numerical results support our earlier conclusion. The interference of g_5^Z with the lowest order term (which vanishes in the effective W approximation) is very small, so it is not possible to single out the g_5^Z term through a parity violating observable. On the other hand, by isolating the high invariant mass region for the WZ pair, we significantly enhance the contribution of g_5^Z with respect to other couplings as shown in Figure 2. In that Figure we show a 3σ significance resulting from the anomalous couplings $\hat{\alpha}$, L_{9L}, and L_{9R}, at $\sqrt{s_{ee}} = 2$ TeV for the cuts of Eq. 19, as a function of the integrated luminosity. We see that the coefficient $\hat{\alpha}$ can be probed here to a level less than 1 $(\Lambda/2\text{ TeV})^2$.

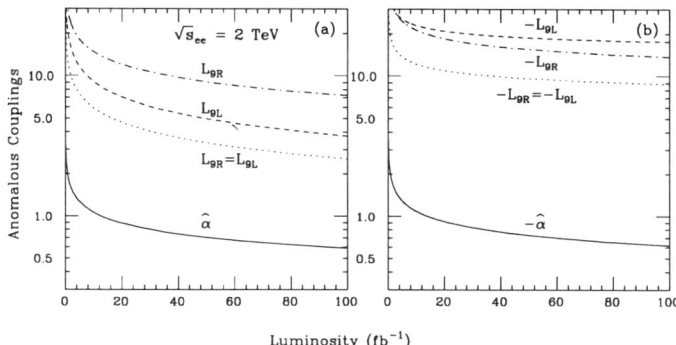

FIG. 2. 3σ sensitivity of an e^+e^- collider at $\sqrt{s_{ee}} = 2$ TeV (operating in the $e^-\gamma$ mode) to $\hat{\alpha}$, L_{9L} and L_{9R} with the cuts Eq. 19. The curves are shown as a function of integrated luminosity, and we set $\Lambda = 2$ TeV.

CONCLUSIONS

Here we summarize the bounds that can be placed on the coupling g_5^Z from all the processes discussed in this talk. We also compare them with the natural size expected for g_5^Z. We see from Table 2 that the current bound at LEP is an order of magnitude better than the bounds from rare decays. An indication of how precise the LEP measurements are is the fact that the LEP bound can only be improved by one order of magnitude in a 2 TeV e^+e^- collider. In principle, g_5^Z can be bound precisely by studying the forward backward asymmetry in $e_L^+ e_R^- \to W^+W^-$, however, it is not clear that it will ever be possible to achieve the high degree of polarization that would be required. From the numbers in Table 2 we conclude that an observation of a non-zero

TABLE 2. Comparison of Bounds on g_5^Z.

Process	Bound on $|g_5^Z|$
Rare Decays	$\mathcal{O}(1)$
Partial Z widths	5×10^{-2}
$A_{fb}(e_L^+ e_R^- \to W^+W^-)$	Potentially very good, needs $P \sim 100\%$
$e^-\gamma \to \nu W^- Z$	5×10^{-3} (in a 2 TeV e^+e^- collider)
Natural size	10^{-4} (with custodial symmetry)
	10^{-2} (without custodial symmetry)

value for g_5^Z would be very strong evidence against a custodial symmetry in the electroweak symmetry breaking sector.

ACKNOWLEDGEMENTS

This work was supported in part by a DOE OJI award under contract number DEFG0292ER40730. I am grateful to S. Dawson for many pleasant collaborations on the matters of this talk. I also thank my collaborators T. Han and K. Cheung.

REFERENCES

1. J. Wudka, these proceedings.
2. Following the work of Gasser and Leutwyler for the case of chiral symmetry breaking in QCD, J. Gasser and H. Leutwyler, *Ann. of Phys.* **158** 142 (1984).
3. T. Appelquist and C. Bernard, *Phys. Rev.* **D22**, 200 (1980); A. Longhitano, *Nucl. Phys.* **B188**, 118 (1981).
4. G. Altarelli, R. Barbieri and F. Caravaglios, *Nucl. Phys.* **B405**, 3 (1993).
5. K. Hagiwara, et. al., *Nucl. Phys.* **B282**, 253 (1987).
6. S. Dawson and G. Valencia, *Phys. Rev.* **D49**, 2188 (1994).
7. S. Dawson and G. Valencia, *Nucl. Phys.* **B348**, 23 (1991).
8. X.-G. He, *Phys. Lett.* **B319** 327 (1993).
9. G. Buchalla and A. Buras, *Nucl. Phys.* **B400**, 225 (1993).
10. S. Dawson and G. Valencia, *Phys. Lett.* **333B**, 207 (1994); *erratum Phys. Lett.* **341B**, 452 (1995).
11. M. Peskin, *Theory of Precision Electroweak Measurements*, Lectures presented at the 17'th SLAC Summer Institute, Stanford, CA (1989).
12. P. Langacker, UPR-0624T; and private communication.
13. The notation and normalization is that of S. Dawson and G. Valencia, hep-ph 9410364 to appear in *Nucl. Phys.*B.
14. C. Ahn, et. al., *Nucl. Phys.* **B309**, 221 (1988).
15. K. Cheung, et. al., *Phys. Rev.* **D51**, 5 (1995).

Effective Lagrangians and Anomalous Couplings

José Wudka*

*Physics Department
University of California, Riverside
Riverside, CA 92521-0413

> The virtual effects of non-Standard Model physics are described using effective lagrangians. Emphasis is given to the estimation magnitude of the effects and to the observability of new physics in processes involving only Standard Model particles.

INTRODUCTION

Despite the successes of the Standard Model it is believed that this model does not describe physics at all scales below the Plank mass. Many possible scenarios of non-Standard Model physics have been explored in the literature and, should experimental evidence for, say a slept-on or a techni-rho be obtained, the guessing game would be over and al efforts will be directed toward the determination of the corresponding supersymmetric or technicolor model.

In the absence of direct evidence (via particle production) of non-Standard Model physics one is lead to the study of the virtual effects that the interactions can generate. The most convenient and general way of studying this type of effects is summarized in the effective lagrangian formalism (1). The basic idea behind this formalism is to assume the existence of new physics at a scale Λ, described by a lagrangian $\mathcal{L}_{\rm new}$. Consider then the effective action obtained by integrating all fields whose mass scale is set by Λ; this object depends on Λ and involves external legs at energies significantly below this scale. It is therefore appropriate to perform a large mass expansion in Λ resulting in an effective lagrangian composed of an infinite series of local operators in the light fields with Λ-dependent coefficients,

$$\mathcal{L}_{\rm eff} = \sum_i \tilde{\alpha}_i(\Lambda) \mathcal{O}_i. \tag{1}$$

This object summarizes the low-energy effects generated by $\mathcal{L}_{\rm new}$ and can be used in all such calculations the only restriction being that all masses and energies should lie below Λ [1].

---

[1] This fact, despite being obvious is sometimes ignored in the literature.

Since this procedure can be followed irrespective of the details of $\mathcal{L}_{\text{new}}$, the same type of expression is obtained for any kind of new physics; the only difference between one theory and another lies in the explicit expression of the coefficients $\tilde{\alpha}_i$ in terms of the couplings of the underlying theory. It is therefore possible to parametrize all type of new physics using these coefficients. In contrast, the operators $\mathcal{O}_i$ are universal and are determined only by the symmetries of the low energy physics (1,4)

Examples of this procedure are quite common. If one studies QED at energies below the electron mass the Euler-Heisenberg lagrangian is obtained (2). If we consider the Standard Model at energies well below the W mass we obtain the QED lagrangian plus a series of operators generating four-fermion interactions, anomalous magnetic couplings, etc. This last example serves to illustrate one important feature of the virtual effects. The electroweak contributions to the anomalous magnetic moment of the muon are only recently being probed [2]. This is so because these are extremely small effects, at the level of 10^{-9} (3). The smallness of this number can be traced back to *(i)* the fact that M_W is much larger than all the scales in the experiment and *(ii)* the fact that the anomalous contributions are loop effects and are correspondingly suppressed. This situation is reproduced when studying non-Standard Model effects: one has to take into consideration the fact that the scale of new physics is significantly above the energies of the experiments and that there are likely to be loop suppression factors.

BUILDING BLOCKS

In practical applications one needs explicit expressions for the operators $\mathcal{O}_i$ in terms of the light fields. In this section I describe the constraints on and variants such construction [3].

Symmetries

The effective operators must reflect some of the symmetries observed at low energies; in particular, all local symmetries must be preserved by the $\mathcal{O}_i$. This is so because a violation of a local symmetry is inconsistent with the assumption that the light vector-bosons have a mass below Λ (5): radiative corrections shift these masses to the cutoff. It has been pointed out (6) that any Lagrangian can be rendered gauge invariant by including the appropriate auxiliary fields; this construction, while perfectly viable, does not specify the transformation properties of the matter fields under the gauge group whence results such as lepton universality are accidental in this approach.

---

[2] In AGS821 at Brookhaven

[3] In the following I will restrict myself to the study of new physics whose scale is not set by the Standard Model.

Global symmetries, in contrast, need not be respected by the effective operators. In fact, the data supporting these symmetries can be used to bound Λ, which in this case represents the scale at which these symmetries are violated.

Particles

The second ingredient needed in the construction of the effective lagrangian is the determination of the low energy excitations. This is straightforward for all but one of the Standard Model particles: the Higgs. Having no detailed information about the processes responsible for the breaking of the symmetries, one is forced to consider various scenarios which differ in the number of light scalars assumed to be present at low energies. I will describe here the two simplest possibilities: either the light excitations coincide precisely with the ones predicted by the Standard Model, or they coincide with the Standard Model spectrum excepting th Higgs.

No light physical scalars

Despite the absence of physical scalars, there is still a need for Goldstone bosons in order to generate masses for the vector bosons. these excitations are most conveniently described using a unitary field U (7). Thus the effective lagrangian is constructed using the Standard Model fields with U replacing the usual scalar doublet.

In this scenario one can define for each operator $\mathcal{O}$ an index (4), $s(\mathcal{O}) = d_\mathcal{O} + f_\mathcal{O}/2$, where $d_\mathcal{O}$ is the number of derivatives in $\mathcal{O}$ and $f_\mathcal{O}$ the number of fermion fields contained in this operator. The terms in $\mathcal{L}_{\text{eff}}$ are then arranged in increasing value of $s(\mathcal{O})$. This ordering is consistent with the loop expansion in the sense that loop contributions generate operators whose indices are larger than those in the vertices of the graphs. This ordering also determines the importance of the various operators; for example, for the operators that do not involve fermions, it corresponds to an ordering in the number of derivatives.

Thus the effective lagrangian takes the form

$$\mathcal{L}_{\text{eff}} = \mathcal{L}_{SM} + \sum_{s(\mathcal{O}_i)=2} \alpha_i^{(2)} \mathcal{O}_i + \sum_{s(\mathcal{O}_i)=4} \alpha_i^{(4)} \mathcal{O}_i + \cdots \quad (2)$$

Examples of such operators with index ≤ 4 are given in Table 1. where W and B denotes, respectively, the $SU(2)_L$ and $U(1)$ gauge-field strengths. The list of effects is not, of course, exhaustive; a complete catalogue can be found in Ref. (14). The phenomenology of this scenario has been studied extensively (15). I have denoted by **TVC** those operators that contribute to the triple-vector-couplings.

Note that other operators contributing to the triple-vector-couplings are subdominant; an illustrative example is $\epsilon_{IJK} W^I_{\mu\nu} W^J_{\nu\rho} W^K_{\rho\mu}$ which contains six derivatives.

TABLE 1. Examples of effective operators when no light scalars are present. Some observable effects are indicated.

Operator	Effects
$\text{tr}\left\{D_\mu U^\dagger\, D^\mu U\right\}$	W and Z masses
$\text{tr}\left\{\tau_3\left(U^\dagger D_\mu U\right)^2\right\}$	ρ-parameter
$W^I_{\mu\nu}B^{\mu\nu}\,\text{tr}\left\{U\tau_3 U^\dagger \tau_I\right\}$	S-parameter; **TVC**
$\text{tr}\left\{\left(D_\mu U^\dagger\, D^\nu U\right)^2\right\}$	**TVC**
$\bar\psi U^\dagger D_\mu U \gamma^\mu \psi$	W, Z-fermion couplings
$\bar\psi \gamma^\mu \psi \bar\psi' \gamma_\mu \psi'$	G_F

Light physical scalars

As I mentioned above I will restrict myself to the case where there is only one light scalar [4]. This scenario satisfies the hypothesis of the decoupling theorem which implies that, up to non-observable renormalization effects [5], all effects generated by the new physics can be expressed as a power series in $1/\Lambda$. In this case the index of an operator equals the canonical dimension of the operator. The effective lagrangian then takes the form

$$\mathcal{L}_{\text{eff}} = \mathcal{L}_{SM} + \frac{1}{\Lambda^2} \sum_{dim=6} \alpha_i \mathcal{O}_i + \cdots \qquad (3)$$

where the terms with higher powers of $1/\Lambda$ are subdominant. I have not included terms of dimension= 5 since all such terms violate the global symmetries of the Standard Model (9). The modifications required to study violations of such symmetries are straightforward and will not be discussed further

Examples of such operators with index = 6 are given in Table 2. Some phenomenological investigations within this scenario can be found in Refs. (9,10) where ϕ denotes the Standard Model doublet.

I will also assume that the underlying theory is weakly coupled. This is based on the requirement that the heavy physics be consistent with a light Higgs. This constraint can be relaxed if a way of protecting the light scalar mass is implemented in the underlying theory (such as supersymmetry); but this leads to a modified low-energy spectrum and will not be considered further.

Note that there is not a one-to-one correspondence between the operators in Tables1 and 2. For example, $\mathcal{O}_{WB}$ and $\mathcal{O}_W$ are of the same order when there is a light Higgs and this is not the case when no light Higgs exists.

---

[4] The formalism can be applied to more complicated situations (8).

[5] Which are of importance, however, when considering naturality issues.

TABLE 2. Examples of effective operators when a light Higgs is present. Some observable effects are indicated.

Operator	Effects
$\mathcal{O}_W = \epsilon_{IJK} W^I_{\mu\nu} W^J_{\nu\rho} W^K_{\rho\mu}$	**TVC**
$\mathcal{O}_{WB} = \phi^\dagger \tau_I \phi W^I_{\mu\nu} B^{\mu\nu}$	S-parameter; **TVC**
$\mathcal{O}^{(3)}_\phi = \left|\phi^\dagger D_\mu \phi\right|^2$	ρ-parameter
$\mathcal{O}^{(1)}_\psi = i \left(\phi^\dagger D_\mu \phi\right) \bar{\psi} \gamma^\mu \psi$	Z, W, H-fermion couplings
$\mathcal{O}_\phi = \frac{1}{3}\left(\phi^\dagger \phi\right)^3$	VEV
$\mathcal{O}_{4\psi} = \bar{\psi}\gamma^\mu \psi \bar{\psi}' \gamma_\mu \psi'$	G_F

ESTIMATES

The expression for the effective lagrangian involves an infinite number of couplings and, because of this, it is apparently of no use in obtaining quantitative estimates. It is fortunate, therefore, that this naive expectation is not realized. The main point is that there exists an ordering of the operators $\mathcal{O}_i$ such that the effects of the higher-order terms become smaller as more terms in this hierarchy are included. This implies that given the precision of an experiment one need only consider a finite number of terms in the effective lagrangian since the effects of the higher-order operators are unobservable.

It is also possible to estimate (or at least bound) the coefficients of the effective operators using general arguments. One can then use these results to provide quantitative estimates of any process and use this to determine its observability at a given experiment. As in the discussion above I need to distinguish between the cases where there is or not a light scalar sector.

No light Higgs

In this situation the effective lagrangian is constructed using the unitary field U, the fermions, and their covariant derivatives. The coefficients of the operators can be estimated using naive dimensional analysis (11). The argument is based on the requirement that the radiative corrections generated by the effective operators to a given coefficient are bounded by the tree-level value of the coefficient. Equivalently, one imagines deriving the renormalization group flow of a given coefficient in the effective lagrangian, and then requires that a change of order one in the renormalization group scale generates a change of order one in the value of the coefficients at low scales.

Naive dimensional analysis generates upper bounds for the coefficients, these are sometimes saturated (as in the case of strong interactions at low energies (12)), but this is not case in general. Applying the above arguments one obtains the generic vertex for the case where there is no light Higgs,

$$c \frac{\Lambda^4}{16\pi^2} \left(\frac{D}{\Lambda}\right)^C \left(\frac{4\pi\psi}{\Lambda^{3/2}}\right)^B U^A \tag{4}$$

where

$$|c| \lesssim 1 \qquad \Lambda = 4\pi v \sim 3\,\text{TeV} \tag{5}$$

As an example consider the operator $c\bar{\ell}\gamma^\mu \left(U^\dagger D_\mu U\right) \ell$ which affects the couplings of the fermions to the vector bosons and is therefore bounded by LEP1: $|c| \lesssim 0.002$. Taking, as suggested by technicolor, $c = (v/\Lambda_{\text{new}})^2$ this bound implies $\Lambda_{\text{new}} \gtrsim 5\,\text{TeV}$

One can also use the effective lagrangian to get the triple-vector-couplings (13) commonly called κ and λ. These are generated by six operators (14), and the naive dimensional analysis estimates are (up to operators involving four derivatives)

$$\begin{array}{ll} \Delta\kappa_\gamma \sim \frac{g^2}{16\pi^2}; & \lambda_\gamma = 0 \\ \Delta\kappa_Z \sim \frac{g^2}{16\pi^2} + \frac{gg'}{16\pi}\Delta T; & \lambda_Z = 0 \end{array} \tag{6}$$

where $\Delta T = $ new physics contributions to the oblique T parameter. There is no contribution to the λ parameters since these arise only from operators containing ≥ 6 derivatives and are correspondingly suppressed.

These estimates are $\sim 3 \times 10^{-3}$, which might get a factor ~ 10 enhancement but may also be suppressed (since naive dimensional analysis provides only upper bounds for the coefficients)

Light Higgs

The dominating effective operators in this case have dimension six and have been catalogued (9,16). The resulting list has 81 operators (for one family of fermions). The presence of such a large number of coefficients makes a general discussion very difficult. Fortunately, this type of general study is not necessary: as I discussed above, the underlying theory is taken to be weakly coupled so that operators generated at tree-level by the underlying dynamics dominate. Such operators can be determined with the mild assumption that the underlying physics is described by a gauge theory (17); from the 81 operators mentioned above, only 12 are tree-level generated.

Operators which are produced only via loops have a generic suppression factor $\sim 1/(16\pi^2)$ and their effects are correspondingly suppressed. It might be thought that the presence of many loops might upset this estimate, but the situation is quite more complicated: should there be a large number of virtual particles the above estimate is indeed modified, but the Higgs mass is also driven to the cutoff and the model no longer corresponds to the light-Higgs scenario (4)

Some of the tree-level-generated operators produce effects which are well-probed (such as non-standard couplings of the Z to the leptons, and the ρ parameter). In this one can easily obtain the (3σ) bounds

$$\mathcal{O}^{(1)}_{\phi\ell} = i\left(\phi^\dagger D_\mu \phi\right) \bar{\ell}\gamma^\mu \ell; \qquad \Lambda_{\text{TeV}} > 2.5/\sqrt{\left|\alpha^{(1)}_{\phi\ell}\right|}$$

$$\mathcal{O}^{(3)}_{\phi\ell} = i\left(\phi^\dagger \tau_I D_\mu \phi\right) \bar{\ell}\tau_I\gamma^\mu \ell; \qquad \Lambda_{\text{TeV}} > 2.5/\sqrt{\left|\alpha^{(3)}_{\phi\ell}\right|}$$

$$\mathcal{O}_{\phi e} = i\left(\phi^\dagger D_\mu \phi\right) \bar{e}\gamma^\mu e; \qquad \Lambda_{\text{TeV}} > 2.7/\sqrt{|\alpha_{\phi e}|}$$

$$\mathcal{O}^{(3)}_{\phi} = i\left(\phi^\dagger D_\mu \phi\right)^2; \qquad \Lambda_{\text{TeV}} > 1.7/\sqrt{\left|\alpha^{(3)}_{\phi}\right|}$$

(7)

where I have taken the coefficients of these operators to be of the form $1/\Lambda^2$, and Λ_{TeV} denotes the value of Λ in TeV units.

None of the tree-level-generated operators contributes to the triple vector couplings. These are generated by the operators

$$\mathcal{O}_W = \epsilon_{IJK} W^I_{\mu\nu} W^J_{\nu\rho} W^K_{\rho\mu}; \quad \mathcal{O}_{\tilde{W}} = \epsilon_{IJK} W^I_{\mu\nu} W^J_{\nu\rho} \tilde{W}^K_{\rho\mu}$$
$$\mathcal{O}_{WB} = \phi^\dagger \tau_I \phi W^I_{\mu\nu} B^{\mu\nu} \qquad \mathcal{O}_{\tilde{W}B} = \phi^\dagger \tau_I \phi \tilde{W}^I_{\mu\nu} B^{\mu\nu}$$

(8)

so that the effective lagrangian describing these non-standard effects is

$$\mathcal{L}_{\text{eff}} = \mathcal{L}_{SM} + \frac{1}{\Lambda^2}\left(g^3 \alpha_W \mathcal{O}_W + gg' \alpha_{WB} \mathcal{O}_{WB} + g^3 \alpha_{\tilde{W}} \mathcal{O}_{\tilde{W}} + gg' \alpha_{\tilde{W}B} \mathcal{O}_{\tilde{W}B}\right)$$

(9)

Since all the above operators are loop generated, it follows that $\alpha_{W,WB,\tilde{W},\tilde{W}b} \sim \frac{1}{16\pi^2}$ which leads to the estimates

$$\lambda = 6\alpha_W \left(\frac{M_W\, g}{\Lambda^2}\right)^2 \sim \left(\frac{0.01}{\Lambda_{\text{TeV}}}\right)^2;$$

$$\Delta\kappa = \alpha_{WB} \left(\frac{2M_W}{\Lambda^2}\right)^2 \sim \left(\frac{0.013}{\Lambda_{\text{TeV}}}\right)^2$$

(10)

and similarly for $\tilde{\lambda}$, $\tilde{\kappa}$.

The bounds from CLEO, AGS821, HERA, LEP1, TeVatron are (or expected) to be ~ 1; but this implies that such measurements of TVC probe physics at $\sim 10^{-2}$ TeV = 10 GeV!

The above numbers are modified slightly in the most optimistic scenarios: should the experimental sensitivity be increased by a factor of ten, and the α be also enhanced by a factor ~ 10, the corresponding experiment would probe physics at the 100GeV scale. It is nonetheless true that a deviation of $O(1)$ from the Standard Model values in the triple vector couplings can only be generated by an enhancement of $\sim 10^4$ which *solely* occurs in these (and maybe other not yet probed) vertices. This is, in my view, an unlikely scenario.

HIGGS REACTIONS

In the previous section it was indicated that there is a set of operators which, being generated at tree level by the underlying dynamics, have the potential of producing the most significant deviations from the Standard Model. Many of these operators generate (non-standard) couplings of the Higgs to the vector-bosons and fermions, and will be probed in future e^+e^- colliders. The said operators also modify the fermion couplings to the gauge bosons (a by-product of gauge invariance), so that the corresponding coefficients are constrained resulting into bounds on the scale of new physics of a few TeV.

The idea which will be developed below is then to assume that the Higgs has been detected and then to use it as a probe for non-standard physics (18). As an example I will concentrate on the processes

$$e^+e^- \to ZH; \quad e^+e^- \to H\bar{\nu}\nu \qquad (11)$$

to which the operators $\mathcal{O}_{\phi\ell}^{(1,3)}$ and $\mathcal{O}_{\phi e}$ defined above (see Table 2 produce significant deviations from the Standard Model results. The generic form of the amplitude for these reactions is

$$\mathcal{A} \sim \mathcal{A}_{SM} \left(1 + \frac{\alpha\, s}{g^2 \Lambda^2} \right) \qquad (12)$$

where α denotes the parameters in $\mathcal{L}_{\text{eff}}$.

Using the Feynman rules obtained from the above operators we obtain the following cross sections [6], where the solid line indicates the Standard Model result and the dotted line the Standard Model plus the effective operators contributions

---

[6] In the plots we included all effects from contributing tree-level–generated dimension 6 operators; for details see Ref. (18).

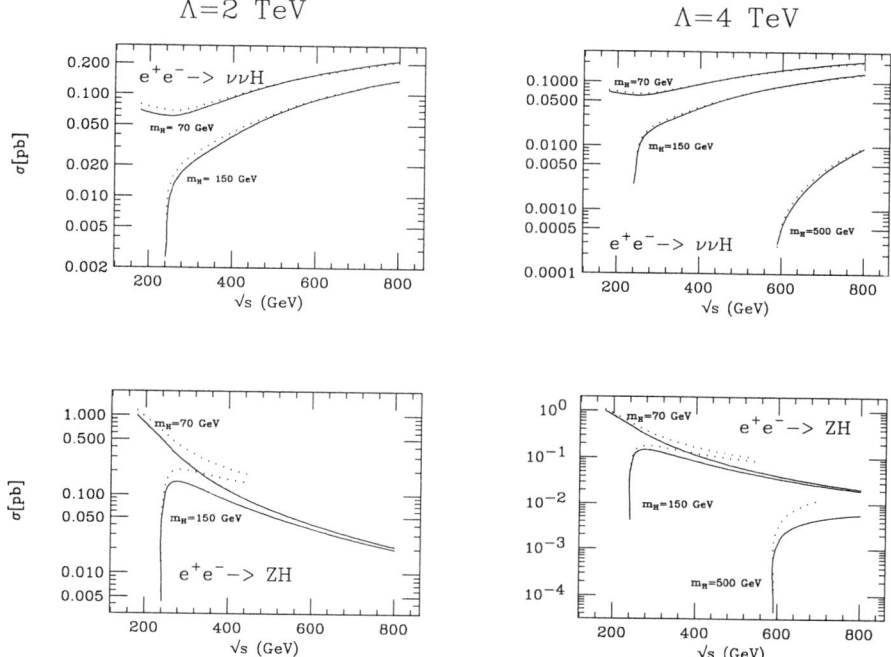

These results show that in the non-standard effects are negligible in the reaction $e^+e^- \to \nu\bar{\nu}H$ while they can be significant in the Bjorken process. Thus one can use the first reaction to determine the properties of the Higgs boson, and the investigate the second searching for new physics effects. To underline the possibility of searching for new physics using the Bjorken process we calculated the statistical significance, shown in the plot below

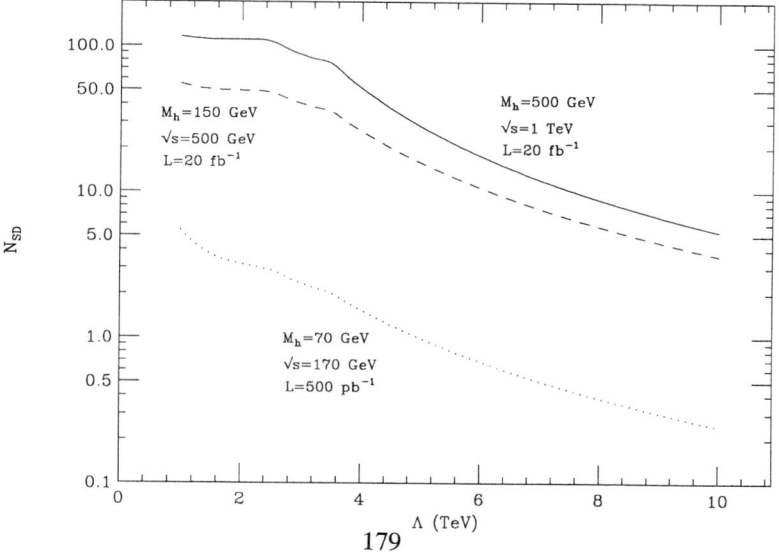

$\mathcal{N}_{SD}$ is defined by

$$\mathcal{N}_{SD} = \frac{|N_{\text{tot}} - N_{SM}|}{\sqrt{N_{\text{tot}}}}. \tag{13}$$

In the above calculation we imposed the constraints from existing data. As can be seen, future colliders will be able to probe new physics into the ~ 10 TeV region using this process.

CONCLUSIONS

I have argued that the effective lagrangian approach is a model and process independent way of studying non-standard effects which provides consistent and reliable estimates of the possible deviations from the Standard Model. The formalism also determines those processes which present the most promising windows into new physics; the sensitivity requirement of a given experiment to observe a process of interest can be determined using this approach.

There are several processes where non-standard effects are expected to be largest. These include Higgs reactions, fermion-fermion scattering and gauge-boson–fermion interactions; the triple vector couplings are not in this category.

REFERENCES

1. S.Weinberg, Physica **A96**, 327 (1979) H. Georgi, Nucl. Phys. **B361**, 339 (1991), Nucl. Phys. **B363**, 301 (1991) J. Polchinski, lectures presented at *TASI 92*, Boulder, CO, Jun 3-28, 1992. M. Einhorn, in *Conference on Unified Symmetry in the Small and in the Large,* Coral Gables, Fl, Jan. 25-27 (1993); talk given at the *Workshop on Physics and Experimentation with Linear e^+e^- Colliders,* Waikoloa, Hawaii, April 26-30, 1993. Univ. of Michigan report UM-TH-93-17.
2. C. Itzykson and J.-B. Zuber, *Quantum Field Theory* (McGraw–Hill, New York, 1980)
3. T. Kinoshita and W.J. Marciano, in: *Quantum Electrodynamics*, edited by T Kinoshita (World Scientific, Singapore, 1990).
4. J. Wudka, Int. J. of Mod. Phys. **A9**, 2301 (1994)
5. M. Veltman, Acta Phys. Pol. **B12**, 437 (1981)
6. C.P. Burgess and D. London, McGill University report MCGILL-92-04, e-Print Archive: hep-ph/9203215 (unpublished).
7. S. Coleman *et. al.*, Phys. Rev. **177**, 2239 (1969). C.G. Callan *et. al.*, Phys. Rev. **177**, 2247 (1969).
8. M.-A. Pérez *et. al.*. U.C. Riverside report UCRHEP-T134 (unpublished)
9. W. Buchmüller and D. Wyler, Nucl. Phys. **B268**, 621 (1986); see also W. Büchmuller *et. al.*, Phys. Lett. **B197**, 379 (1987)

10. A. De Rújula *et. al.*, Nucl. Phys. **B384**, 3 (1992) K. Hagiwara *et. al.*, Phys. Lett. **B318**, 155 (1993). C. Arzt *et. al.*, Phys. Rev. **D49**, 1370 (1994). S. Dawson and G. Valencia, report BNL-60949, e-Print Archive: hep-ph/9410364 (unpublished).
11. S. Weinberg, reference (1). A. Manohar and H. Georgi, Nucl. Phys. **B234**, 189 (1984).
12. J. Gasser and H. Leutwyller, Nucl. Phys. **B250**, 465 (1985).
13. K. Hagiwara *et. al.*, Nucl. Phys. **B282**, 253 (1987).
14. T. Appelquist and G.-H. Wu, Phys. Rev. **D48**, 3235 (1993)
15. See, for example, H. Georgi, Ref. (1). M. Chanowitz *et. al.*, Phys. Rev. **D36**, 1490 (1987). R.D. Peccei and X. Zhang, Nucl. Phys. **B337**, 269 (1990). A. Dobado *et. al.*, Phys. Lett. **B235**, 129 (1990); Z. Phys. **C50**, 465 (1991). M.Golden and L. Randall, Nucl. Phys. **B361**, 3 (1991). B. Holdom, Phys. Lett. **B259**, 329 (1991). J. Bagger *et. al.*, Nucl. Phys. **B339**, 364 (1993). M. J. Herrero and E. Ruiz Morales Nucl. Phys. **B437**, 319 (1995).
16. C.J.C. Burges and H.J. Schnitzer, Nucl. Phys. **B228**, 464 (1983) C.N. Leung *et. al.*, Z. Phys. **C31**, 433 (1986).
17. Arzt *et. al.*, Nucl. Phys. **B433**, 41 (1995)
18. B. Grzadkowski and J. Wudka, report IFT-2/95, UCRHEP-T133 (unpublished)

Low energy constraints on electroweak vector boson self-interactions

K. Hagiwara

Theory Group, KEK, Tsukuba, Ibaraki 305, Japan

Agreements between electroweak precision experiments and the predictions of the Standard Model at the quantum level may imply non-trivial constraints on weak boson self-interactions. I critically examine the universality of the observed electroweak couplings and find that it holds to within 0.2 to 0.5% accuracy, which strongly supports the hypothesis of local gauge symmetry underlying weak-boson interactions. We then assume that non-standard weak boson interactions are characterized by a gauge invariant effective Lagrangian with non-renormalizable higher-dimensional operators. Twelve such operators made of electroweak bosons are obtained in the lowest nontrivial dimension in both the linear and non-linear realization of the symmetry breaking sector. Only a subset of these operators (four in the linear and three in the non-linear Lagrangian) are constrained by the present precision experiments. Although educated order-of-magnitude estimates for the remaining operator terms are possible, low-energy constraints based on such estimates are by no means substitutes for direct measurements at LEP2 and beyond. Finally, the question of whether there already exists quantitative evidence for the standard weak-boson self-couplings in electroweak quantum corrections is critically examined.

---

1. UNIVERSALITY OF THE WEAK-BOSON INTERACTIONS

The Standard Model (SM) of the unified theory of weak and electromagnetic interactions was constructed such that the interactions are both invariant under local $SU(2)_L \times U(1)_Y$ gauge symmetry, and they are of renormalizable form. The observed short-range nature of the weak interactions was then attributed to the weak-gauge-boson masses which are acquired via spontaneous breakdown of the gauge symmetry, $SU(2)_L \times U(1)_Y \to U(1)_{EM}$. This mechanism, named after Higgs, may or may not result in a new massive neutral scalar boson, the Higgs boson, but it should necessarily affect weak-boson scattering amplitudes at high energies ($E \gg m_W$). If the Higgs boson is light ($m_H \lesssim 200$ GeV), as suggested by theories with grand unification of the electroweak and strong gauge interactions, then it can be studied at LEP2, LHC or at a future linear e^+e^- collider of $\sqrt{s} \sim 300 - 400$ GeV. If the Higgs boson has an intermediate mass (200 GeV$\lesssim m_H \lesssim 700$ GeV), it can be studied at the LHC. If, on the other hand, the symmetry breaking is a consequence of a new strong interaction, there may or may not be a Higgs-like scalar resonance, and its detailed study may await a ~ 60 TeV pp collider or a ~ 4 TeV e^+e^- or $\mu^+\mu^-$ collider.

In any case, a complete understanding of electroweak physics is most likely a subject of experiments at future high energy colliders, such as LEP2, LHC, and beyond. In this this report, I would like to review what we have already learned about electroweak physics from precision experiments at LEP1, SLC, the Tevatron, and at low energies. These experiments have recently reached the level of precision that is sensitive to quantum corrections of the SM, and hence they can give us information about physics at very short distances (high energies) that cannot at present be explored directly.

The main theme of this conference is what we have already learned from present experiments and what we will be able to learn in the near future about the self-interactions among the electroweak vector bosons, $W^\pm$, Z and γ. These vector-boson self-couplings are related to the vector boson couplings to quarks and leptons in the SM as a consequence of the local gauge symmetry and the renormalizability of the effective Lagrangian. Local gauge invariance implies that all the gauge-boson couplings should be associated with the covariant derivative,

$$D_\mu = \partial_\mu + ig T^a W^a_\mu + ig' Y B_\mu \qquad (1.1)$$
$$= \partial_\mu + i\frac{g}{\sqrt{2}}(T^+ W^+_\mu + T^- W^-_\mu) + ig_Z(T^3 - s^2 Q)Z_\mu + ieQA_\mu, \qquad (1.2)$$

and renormalizability implies that the gauge-boson couplings to fermions, to bosons and to themselves are universal:

$$\bar{\psi}D_\mu\gamma^\mu\psi \qquad \text{Tr}\{[D_\mu,D_\nu][D^\mu,D^\nu]\} \qquad (D_\mu\phi)^\dagger D^\mu\phi \qquad (1.3)$$

It is this universality of gauge couplings that controls quantum fluctuations at short distances. Because of universality all the quantum fluctuations that have sensitivity to unknown physics at short distances are constrained to have a few universal forms. After identification of these universal terms with known measurements the remaining quantum fluctuations which are sensitive to physics of the weak scale and below can be confronted with other electroweak experiments, independently of unknown short distance physics. This is what a so-called renormalizable field theory can tell us. If it were not for the universality of the gauge couplings, quantum fluctuations at short distances would be uncontrollable, or, equivalently, the high-energy behavior of the weak-boson scattering amplitude would become singular: e.g. the

$W_L W_L \to W_L W_L$ scattering amplitude could in general grow as $(E/m_W)^4$. The degree of sensitivity to short distance physics depends on the actual mass scale of new physics, Λ. The effective Lagrangian which describes physics accurately at and below the weak scale, $v = (\sqrt{2} G_F)^{-1/2} = 246$ GeV, can be written in general as

$$\mathcal{L}_{eff} = \mathcal{L}_{\text{SM}} + \sum_{n \geq 5} \sum_{i} \frac{f_i^{(n)}}{\Lambda^{n-4}} \mathcal{O}_i^{(n)} , \qquad (1.4)$$

where the $\mathcal{O}_i^{(n)}$ are mass-dimension n gauge-invariant operators composed of SM fields. When the scale Λ is very large ($\Lambda \gg v$) the effects of the extra terms (non-renormalizable terms) are small, while they can be significant if $\Lambda \sim v$ and the coefficient f_i is not small.

It is worth noting that the universality of the gauge-boson couplings as depicted in eq.(1.3) is in general violated by higher-dimensional terms in the effective Lagrangian despite local gauge invariance. This can be seen most clearly by examining dimension-six operators of the form

$$\overline{\Psi}_L \gamma^\mu \Psi_L \, \Phi^\dagger D_\mu \Phi , \qquad \overline{\Psi}_R \gamma^\mu \Psi_R \, \Phi^\dagger D_\mu \Phi , \qquad (1.5)$$

where Ψ_L and Ψ_R are appropriate quark or lepton multiplets, and Φ is the SM Higgs doublet. Each operator in (1.5) is separately gauge invariant, and the universality of quark and lepton couplings to the Z boson is violated by these terms. In fact, even within the SM, such an effective interaction is generated in the large m_t limit for the $Z b_L b_L$ vertex, and universality is violated at the effective Lagrangian level. I will show in section 3 that there are many other operators which break the universality of gauge-boson self interactions.

The success of the SM predictions which include quantum corrections while assuming the *absence* of these higher-dimensional terms in the effective Lagrangian of eq.(1.4) suggests that these extra terms are in fact small. I will examine this statement quantitatively in the following sections. We will find in section 3 that some of the operators of mass-dimension six are already constrained to be very small, which suggests, though does not prove, the smallness of other related operators that affect the weak-boson self-interactions. In section 4 I study the quantitative significance of purely SM quantum corrections and examine evidence for standard gauge-boson self-couplings.

Finally, I would like to note that a formal expansion like eq.(1.4) may be useful even when discussing the consequences of a non-gauge theory alternative of the weak bosons [1]. Both the weak bosons and quarks/leptons are composite in such alternative theories, and the natural scale of new physics is m_W itself. There is a priori no reason why the extra terms in the effective Lagrangian are smaller than the terms in $\mathcal{L}_{SM}$. The smallness of the universality violation in such theories has been traditionally explained as a consequence of the underlying current algebra of preon dynamics and vector-boson (W and Z) dominance. On the other hand, there seems to be no reason why the weak-boson self-couplings should obey the gauge-theory universality (1.3). If one blindly calculates quantum corrections by using such non-universal weak-boson self-couplings with a large cut-off scale, Λ, one tends to find stringent constraints on these non-standard couplings as a consequence of the power sensitivity of quantum corrections to the cut-off [2]. One might argue, based on such calculations that weak-boson self-couplings are strongly constrained to have gauge-theory universality even in non-gauge theories. Such constraints disappear, however, if we take the more natural cut-off scale of $\Lambda \sim m_W$. It is more appropriate to state that weak-boson self-couplings in such theories are not constrained through quantum effects simply because we cannot calculate quantum corrections without understanding the dynamics. The great success of the SM predictions at the quantum level, however, makes such alternative scenarios less and less attractive.

2. PRECISION TESTS OF THE UNIVERSALITY OF THE ELECTROWEAK INTERACTIONS

In this section, I review the quantitative significance of the universality of the electroweak gauge-boson couplings.

2.1. Universality within the charged current interactions

First, universality of the charged current couplings to quarks and leptons can be examined by studying the unitarity of the CKM matrix elements.

$$W \longrightarrow \begin{matrix} u \\ d' \end{matrix} = W \longrightarrow \begin{matrix} \nu_\ell \\ \ell \end{matrix} \quad : \text{CKM universality} \qquad (2.1)$$

After the SM radiative corrections are applied, the matrix elements are found in ref. [3]

$$|V_{ud}|^2 = 0.9495 \pm 0.0014, \qquad (2.2a)$$
$$|V_{us}|^2 = 0.0486 \pm 0.0008, \qquad (2.2b)$$
$$|V_{ub}|^2 = 0.00001 \pm 0.000006. \qquad (2.2c)$$

Summing them up, we find

$$|V_{ud}|^2 + |V_{us}|^2 + |V_{ub}|^2 = 0.9981 \pm 0.0016, \qquad (2.3)$$
$$|V_{ud}|^2 + |V_{us}|^2 + |V_{ub}|^2 - 1 = -0.19 \pm 0.16\%. \qquad (2.4)$$

Universality is found to hold to within 0.2% accuracy. It is amusing to note that, if we take the sign and the magnitude of the 1-σ deviation from universality seriously, it is consistent with the expectations of the supersymmetric SM [4].

2.2. Universality among charged and neutral current interactions

The universality of the electroweak couplings of the SM can be conveniently studied by making use of the effective charge form-factors of ref. [5]:

$$\gamma \cdots \gamma \sim \bar{\alpha}(q^2) \tag{2.5}$$

$$\gamma \cdots Z \sim \bar{s}^2(q^2) \tag{2.6}$$

$$Z \cdots Z \sim \bar{g}_Z^2(q^2) \tag{2.7}$$

$$W \cdots W \sim \bar{g}_W^2(q^2) \tag{2.8}$$

We can define [5] the S, T, and U variables of ref. [6] in terms these effective charges,

$$\frac{\bar{s}^2(m_Z^2)\bar{c}^2(m_Z^2)}{\bar{\alpha}(m_Z^2)} - \frac{4\pi}{\bar{g}_Z^2(0)} \equiv \frac{S}{4}, \tag{2.9a}$$

$$\frac{\bar{s}^2(m_Z^2)}{\bar{\alpha}(m_Z^2)} - \frac{4\pi}{\bar{g}_W^2(0)} \equiv \frac{S+U}{4}, \tag{2.9b}$$

$$1 - \frac{\bar{g}_W^2(0)}{m_W^2} \frac{m_Z^2}{\bar{g}_Z^2(0)} \equiv \alpha T, \tag{2.9c}$$

and then it is clear that these variables measure deviations from the naive universality of the electroweak gauge boson couplings. A recent update [7] of the comprehensive analysis of all the electroweak precision data finds e.g. Fig. 1 for the neutral-current charge form-factors as measured at LEP1/SLC. By using these fitted charge form factors Matsumoto finds [7]

$$\left. \begin{array}{l} S = -0.18 + 0.07 \frac{\delta_\alpha}{0.10} \pm 0.20 \\ T = 0.93 \phantom{+0.00\frac{\delta_\alpha}{0.10}} \pm 0.20 \\ U = -0.12 + 0.02 \frac{\delta_\alpha}{0.10} \pm 0.50 \end{array} \right\} \rho_{\text{corr}} = \begin{pmatrix} 1 & 0.84 & -0.08 \\ & 1 & -0.22 \\ & & 1 \end{pmatrix}, \tag{2.10}$$

where the parameter δ_α measures the uncertainty in the SM prediction for the effective charge $1/\bar{\alpha}(m_Z^2)$.

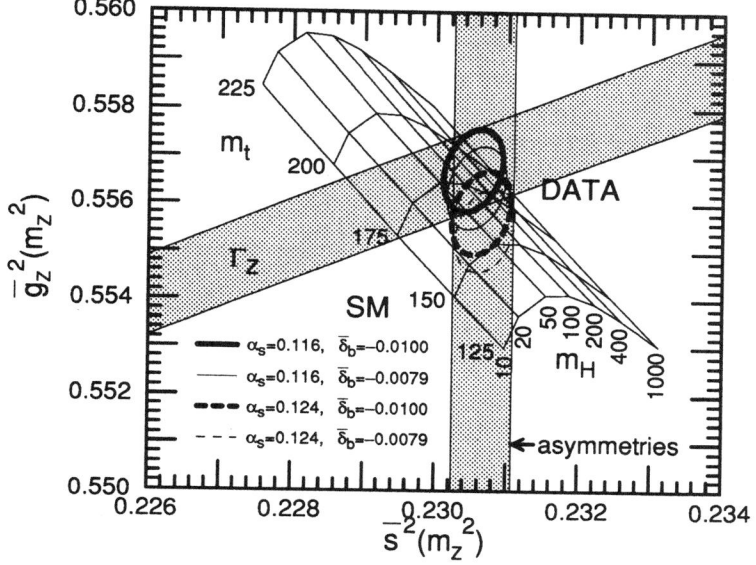

Fig.1. A two-parameter fit to the Z boson parameters in the $(\bar{s}^2(m_Z^2), \bar{g}_Z^2(m_Z^2))$ plane. The $Zb_L b_L$ vertex form-factor, $\bar{\delta}_b(m_Z^2)$, and the QCD coupling, $\alpha_s(m_Z)$, are treated as external parameters. The 1-σ contours are shown for two values of $\alpha_s(m_Z)$, 0.116 (solid lines) and 0.124 (dashed lines), and for two values of $\bar{\delta}_b(m_Z^2)$, -0.0100 (thick lines) and -0.0079 (thin lines). In the SM, $\bar{\delta}_b = -0.0100$ corresponds to $m_t = 175$ GeV, while $\bar{\delta}_b = -0.0079$ corresponds to $m_t = 150$ GeV. Also shown are the SM predictions in the range $125\,\text{GeV} < m_t < 225\,\text{GeV}$ and $10\,\text{GeV} < m_H < 1000\,\text{GeV}$, which are obtained by assuming $\alpha_s(m_Z) = 0.116$ and $\delta_\alpha \equiv 1/\bar{\alpha}(m_Z^2) - 128.72 = 0$.

There recently appeared three new estimates of this quantity,

$$\delta_\alpha \equiv \frac{1}{\bar{\alpha}(m_Z^2)_{\text{SM}}} - 128.72 = \begin{cases} 0.04 \pm 0.10 & \text{Eidelman-Jegerlehner [8]} \\ 0.12 \pm 0.06 & \text{Martin-Zeppenfeld [10]} \\ 0.22 \pm 0.10 & \text{Swartz [9]} \end{cases} \quad (2.11)$$

The estimation of the parameter S is especially sensitive to this uncertainty. I note here that the definition of the effective charge $\bar{\alpha}(q^2)$ in ref. [5] contains both the top-quark and the W-boson contribution [11] to the running of the charge form factors. Consequences of each estimate are, of course, the same whether one counts these corrections as part of the running of the charge form factors or as part of the remaining radiative effects. We find from the fit (2.10) that universality holds well in the combinations S and U, but that the T parameter is found significantly different from zero. In order to test the universality of the couplings, however, we should correct for the SM contributions to these parameters which are included in the definitions of eq.(2.9). The SM contributions are found in ref. [5] as

$$S = -0.23\,[-0.08]\,, \tag{2.12a}$$
$$T = 0.89\,[0.59]\,, \tag{2.12b}$$
$$U = 0.36\,[0.35]\,, \tag{2.12c}$$

for $(m_t, m_H) = (175, 100)\,\text{GeV}$ $[(m_t, m_H) = (175, 1000)\,\text{GeV}]$. New physics contributions to these parameters are then estimated as

$$\Delta S = 0.05\,[-0.12] + 0.07\frac{\delta_\alpha}{0.1} \pm 0.20\,, \tag{2.13a}$$
$$\Delta T = 0.03\,[0.32] \phantom{ + 0.07\frac{\delta_\alpha}{0.1}} \pm 0.20\,, \tag{2.13b}$$
$$\Delta U = -0.48\,[-0.47] + 0.02\frac{\delta_\alpha}{0.1} \pm 0.50\,, \tag{2.13c}$$

respectively. Except for the parameter ΔT they are now all consistent with zero, when the SM contribution is estimated for $[(m_t, m_H) = (175, 1000)\,\text{GeV}]$. If we measure the differences in the effective gauge couplings in eq.(2.9) as deviations from unity as in the case of the CKM universality, we find, after correcting for the SM contributions, that possible new physics contributions to universality violation are constrained as follows:

$$\Delta S\,:0.0 \pm 0.2\%[-0.1 \pm 0.2\%]\,, \tag{2.14a}$$
$$\Delta T\,:0.0 \pm 0.15\%[0.24 \pm 0.15\%]\,, \tag{2.14b}$$
$$\Delta U\,:-0.4 \pm 0.4\%[-0.4 \pm 0.4\%]\,. \tag{2.14c}$$

Three of the four universality tests (the CKM universality of the charged currents, and the S and the T combinations of the differences in the effective charge form-factors) are found to hold at the level of 0.2%, and the last one (the U combination) still holds true at the level of 0.4%. The smallness of non-SM contributions to the universality violation, and especially the near perfect agreement of the observed T value and its SM prediction for $m_t \sim 175$ GeV suggests strongly that the electroweak interactions are gauge interactions, and that non-standard terms in the effective Lagrangian of eq.(1.4) are small. In the next section we study deviations from the SM in an effective Lagrangian with local $SU(2)_L \times U(1)_Y$ gauge invariance.

3. TWO PARAMETRIZATIONS OF NON-STANDARD GAUGE BOSON INTERACTIONS

In this section we examine two parametrizations of the effective Lagrangian of SM fields with spontaneously-broken electroweak gauge invariance and with non-standard interactions among the electroweak bosons. In the first parametrization we use the standard Higgs doublet field as a tool to induce the spontaneous break down of the electroweak gauge symmetry via its vacuum expectation value. The most general Lagrangian consistent with local gauge invariance is then of a renormalizable form at the dimension-four-operator level. Non-standard interactions among the electroweak bosons can then be studied consistently by examining effects of most general operators at the dimension six level [13–15]. In the second parametrization we use the electroweak chiral Lagrangian to parametrize non-standard interactions among weak bosons. There spontaneous symmetry breaking is realised non-linearly, and hence the Lagrangian is non-renormalizable. Consistent perturbative expansion is still possible by examining the low-momentum ($E \ll 4\pi v \sim 1$ TeV) behavior of the theory.

In both languages, the present precision experiments constrain part of the complete set of operators of the given dimension. Unconstrained operators can still affect the weak-boson self-interactions arbitrarily. However, constraints on some of the operators can give us order-of-magnitude estimates for possible deviations from the SM universality of the weak-boson self-interactions.

3.1. Effective Lagrangian with linearly realized symmetry breaking

When the non-SM operators are parameterized with the Higgs doublet field [13], there are four operators of mass dimension six that contribute to the gauge boson propagators [14]. In the notation of ref. [15], they are

$$\mathcal{O}_{DW} = Tr([D_\mu, \hat{W}_{\nu\rho}][D^\mu, \hat{W}^{\nu\rho}]), \tag{3.1}$$

$$\mathcal{O}_{DB} = Tr(\partial_\mu \hat{B}_{\nu\rho} \partial^\mu \hat{B}^{\nu\rho}), \tag{3.2}$$

$$\mathcal{O}_{BW} = \Phi^\dagger \hat{B}_{\mu\nu} \hat{W}^{\mu\nu} \Phi, \tag{3.3}$$

$$\mathcal{O}_{\Phi,1} = (D_\mu \Phi)^\dagger \Phi \, \Phi^\dagger (D^\mu \Phi), \tag{3.4}$$

with the corresponding coupling factors f_i/Λ^2 in the effective Lagrangian of (1.4). Their contributions to the charge form-factors are slightly more involved [15,16] because the operators $\mathcal{O}_{DB}$ and $\mathcal{O}_{DW}$ contribute to the running of the charge form factors. In particular, they contribute to $\bar{\alpha}(m_Z^2)$, which has not been measured accurately. We find

$$\Delta S = -4\pi \frac{v^2}{\Lambda^2} f_{BW}, \tag{3.5a}$$

$$\Delta T = -\frac{1}{2\alpha} \frac{v^2}{\Lambda^2} f_{\Phi,1}, \tag{3.5b}$$

$$\Delta U = 32\pi \frac{m_Z^2 - m_W^2}{\Lambda^2} f_{DW}, \tag{3.5c}$$

$$\Delta \frac{1}{\alpha} \equiv \frac{1}{\bar{\alpha}(m_Z^2)} - \frac{1}{\bar{\alpha}(m_Z^2)_{SM}} = 8\pi \frac{m_Z^2}{\Lambda^2} \left(f_{DW} + f_{DB} \right). \tag{3.5d}$$

Because $\bar{\alpha}(m_Z^2)$ has not been measured accurately, the present experiments poorly constrain the operators $\mathcal{O}_{DW}$ and $\mathcal{O}_{DB}$, but they can be constrained better at LEP2 [16].

In the effective Lagrangian with a Higgs doublet there are eight more operators which consist of the electroweak bosons [15]. Among them, three affect the $WW\gamma$ and WWZ vertices:

$$\mathcal{O}_{WWW} = Tr[\hat{W}_{\mu\nu}\hat{W}^{\nu\rho}\hat{W}_\rho{}^\mu], \tag{3.6}$$

$$\mathcal{O}_W = (D_\mu\Phi)^\dagger \hat{W}^{\mu\nu}(D_\nu\Phi), \tag{3.7}$$

$$\mathcal{O}_B = (D_\mu\Phi)^\dagger \hat{B}^{\mu\nu}(D_\nu\Phi). \tag{3.8}$$

They contribute to the non-standard effective WWZ and $WW\gamma$ couplings [17,18], whose explicit relations are found in refs. [19,15].

Table 1 summarizes contributions of the above twelve gauge-invariant dimension-six operators to two- three- and four-point functions of the electroweak bosons [22]. The upper half of the table contains those vertices without the Higgs field, and lower half gathers those vertices with one or two Higgs fields. A circle indicates presence of an effective interaction. A triangle stands for the interaction which can be absorbed into the redefinition of the SM parameters. Possible effects of these operators in various high-energy processes

	$\mathcal{O}_{WWW}$	$\mathcal{O}_{DW}$	$\mathcal{O}_{DB}$	$\mathcal{O}_{WW}$	$\mathcal{O}_{BB}$	$\mathcal{O}_{BW}$	$\mathcal{O}_W$	$\mathcal{O}_B$	$\mathcal{O}_{\Phi,1}$	$\mathcal{O}_{\Phi,2}$	$\mathcal{O}_{\Phi,3}$	$\mathcal{O}_{\Phi,4}$
WW		o		△								△
ZZ		o	o	△	△	o		o				△
ZA		o	o	△	△	o						
AA		o	o	△	△	o						
WWZ	o	o		△		o	o	o				
WWA	o	o		△		o	o	o				
WWWW	o	o		△			o					
WWZZ	o	o		△			o					
WWZA	o	o		△			o					
WWAA	o	o		△								
HH									o	o	△	△
WWH				o			o					o
ZZH				o	o	o	o	o				o
ZAH				o	o	o	o	o				
AAH				o	o	o						
WWZH				o			o	o				
WWAH				o			o	o				
WWHH				o			o					o
ZZHH				o	o	o	o	o				o
ZAHH				o	o	o	o	o				
AAHH				o	o	o						

TABLE 1. Dimension-six operators and their vertices. Only those vertices up to two Higgs bosons are listed. Vertices indicated by triangles are exactly proportional to those of the standard model.

have been studied in refs. [20]. When a certain type of new physics underlying these dimension six operators is examined, one can obtain relationships between the magnitudes of the coefficients of each operator in the effective Lagrangian [21].

3.2. Effective Lagrangian with non-linearly-realized symmetry breaking

Constraints on new physics contributions to the S, T, and U parameters have immediate consequences when the symmetry breaking sector of the SM is strongly interacting. Higgs-like resonances would then be very heavy, and the effective Lagrangian of the SM becomes that of the chiral Lagrangian of the spontaneous symmetry breaking; $SU(2)_L \times SU(2)_R \to SU(2)_V$. If we allow for $SU(2)_V$ violating effects, there are just three non-standard interaction terms,

$$\mathcal{L}'_1 = \frac{1}{4}\beta_1 v^2 \left[\text{Tr}\left(TV_\mu\right)\right]^2, \tag{3.9a}$$

$$\mathcal{L}_2 = \frac{1}{2}\alpha_1 gg' B_{\mu\nu} \text{Tr}\left[T(W^{\mu\nu})\right], \tag{3.9b}$$

$$\mathcal{L}_8 = \frac{1}{4}\alpha_8 g^2 \left[\text{Tr}\left(TW_{\mu\nu}\right)\right]^2, \tag{3.9c}$$

that contribute to the gauge boson propagators. Here, I adopt the notation of ref. [12]. We find

$$\Delta S = -16\pi\alpha_1, \tag{3.10a}$$

$$\Delta T = 2\frac{\beta_1}{\alpha}, \tag{3.10b}$$

$$\Delta U = -16\pi\alpha_8. \tag{3.10c}$$

The experimental results (2.14) directly constrain these operators through the above identities.

There are eight more C- and P-even operators [23] and one CP-even P-odd operator at the energy-dimension-four level. Many of them give rise to non-standard weak-boson self-couplings [24,12], and hence to the present precision experiments through loop effects [25]. We list all twelve operators in Table 2. The first column lists a set of chiral electroweak operators in the notation of Ref. [12]. The second column contains the corresponding operator in the linear representation of the Higgs field. In general the energy dimension of the linear operator may be greater than the energy dimension of the chiral operator. Here we define those corresponding linear operators of energy dimension greater than six which have not been defined in ref. [15].

$$\mathcal{O}_4^{(8)} = \left[(\mathcal{D}_\mu \Phi)^\dagger (\mathcal{D}_\nu \Phi) + (\mathcal{D}_\nu \Phi)^\dagger (\mathcal{D}_\mu \Phi)\right]^2 \tag{3.11a}$$

$$\mathcal{O}_5^{(8)} = \left[(\mathcal{D}_\mu \Phi)^\dagger (\mathcal{D}^\mu \Phi)\right]^2 \tag{3.11b}$$

$\mathcal{L}_{\text{chiral}}$	$\mathcal{O}^{(n)}_{\text{linear}}$	WW	ZZ	AZ	AA	WWZ	WWA	WWWW	WWZZ	WWZA	WWAA	ZZZZ
$\mathcal{L}'_1 = \frac{\beta_1 v^2}{4}\left[\text{Tr}(TV_\mu)\right]^2$	$\frac{4\beta_1}{v^2}\mathcal{O}_{\Phi,1}$	○										
$\mathcal{L}_1 = \frac{\alpha_1 gg'}{2} B_{\mu\nu}\text{Tr}(TW^{\mu\nu})$	$\frac{4\alpha_1}{v^2}\mathcal{O}_{BW}$	○	○	○	○	○						
$\mathcal{L}_2 = \frac{i\alpha_2 g'}{2} B_{\mu\nu}\text{Tr}(T[V^\mu, V^\nu])$	$\frac{8\alpha_2}{v^2}\mathcal{O}_B$					○	○					
$\mathcal{L}_3 = i\alpha_3 g\,\text{Tr}(W_{\mu\nu}[V^\mu, V^\nu])$	$\frac{8\alpha_3}{v^2}\mathcal{O}_W$					○	○	○	○	○		
$\mathcal{L}_4 = \alpha_4\left[\text{Tr}(V_\mu V_\nu)\right]^2$	$\frac{4\alpha_4}{v^4}\mathcal{O}_4^{(8)}$							○	○			○
$\mathcal{L}_5 = \alpha_5\left[\text{Tr}(V_\mu V^\mu)\right]^2$	$\frac{16\alpha_5}{v^4}\mathcal{O}_5^{(8)}$							○	○			○
$\mathcal{L}_6 = \alpha_6\,\text{Tr}(V_\mu V_\nu)\text{Tr}(TV^\mu)\text{Tr}(TV^\nu)$	$-\frac{64\alpha_6}{v^6}\mathcal{O}_6^{(10)}$								○			○
$\mathcal{L}_7 = \alpha_7\,\text{Tr}(V_\mu V^\mu)\text{Tr}(TV_\nu)\text{Tr}(TV^\nu)$	$-\frac{64\alpha_7}{v^6}\mathcal{O}_7^{(10)}$								○			○
$\mathcal{L}_8 = \frac{\alpha_8 g^2}{4}\left[\text{Tr}(TW_{\mu\nu})\right]^2$	$-\frac{4\alpha_8}{v^4}\mathcal{O}_8^{(8)}$	○	○	○	○	○	○					
$\mathcal{L}_9 = \frac{i\alpha_9 g}{2}\text{Tr}(TW_{\mu\nu})\text{Tr}(T[V^\mu, V^\nu])$	$-\frac{16\alpha_9}{v^4}\mathcal{O}_9^{(8)}$					○	○	○				
$\mathcal{L}_{10} = \frac{\alpha_{10}}{2}\left[\text{Tr}(TV_\mu)\text{Tr}(TV_\nu)\right]^2$	$\frac{128\alpha_{10}}{v^8}\mathcal{O}_{10}^{(12)}$											○
$\mathcal{L}_{11} = \alpha_{11}\,g\,\epsilon^{\mu\nu\rho\sigma}\text{Tr}(TV_\mu)\text{Tr}(V_\nu W_{\rho\sigma})$	$\frac{8\alpha_{11}}{v^4}\mathcal{O}_{11}^{(8)}$			○			○					

TABLE 2. The first column lists chiral operators in the notation of ref.[22] The second column lists the corresponding operator in the linear representation. We employ the notation $\mathcal{O}_i^{(n)}$ where the energy dimension of each operator is denoted by n. The index i is used to distinguish between various operators of the same energy dimension. For convenience the label i is chosen to match label of the corresponding chiral operator. Operators in the second column for which no label n is provided are energy-dimension-six operators by default [14]. The circles show the vertices to which each operator contributes.

$$\mathcal{O}_6^{(10)} = \left[(\mathcal{D}_\mu\Phi)^\dagger(\mathcal{D}_\nu\Phi)\right]\left[\Phi^\dagger(\mathcal{D}^\mu\Phi)\right]\left[\Phi^\dagger(\mathcal{D}^\nu\Phi)\right] \quad (3.11c)$$

$$\mathcal{O}_7^{(10)} = \left[(\mathcal{D}_\mu\Phi)^\dagger(\mathcal{D}^\mu\Phi)\right]\left[\Phi^\dagger(\mathcal{D}_\nu\Phi)\right]\left[\Phi^\dagger(\mathcal{D}^\nu\Phi)\right] \quad (3.11d)$$

$$\mathcal{O}_8^{(8)} = \left[\Phi^\dagger\hat{W}_{\mu\nu}\Phi\right]^2 \quad (3.11e)$$

$$\mathcal{O}_9^{(8)} = \left[\Phi^\dagger\hat{W}_{\mu\nu}\Phi\right]\left[(\mathcal{D}^\mu\Phi)^\dagger(\mathcal{D}^\nu\Phi)\right] \quad (3.11f)$$

$$\mathcal{O}_{10}^{(12)} = \left(\left[\Phi^\dagger (\mathcal{D}_\mu \Phi) \right] \left[\Phi^\dagger (\mathcal{D}_\nu \Phi) \right] \right)^2 \tag{3.11g}$$

$$\mathcal{O}_{11}^{(8)} = i\epsilon^{\mu\nu\rho\sigma} \left[\Phi^\dagger (\mathcal{D}_\mu \Phi) \right] \left[\Phi^\dagger \hat{W}_{\rho\sigma} (\mathcal{D}_\nu \Phi) \right] + \text{h.c.} \tag{3.11h}$$

We employ the notation $\mathcal{O}_i^{(n)}$ where the energy dimension of each operator is denoted by n. The index i is used to distinguish between various operators of the same energy dimension. For convenience the label i is chosen to match the label of the corresponding chiral operator. Finally, the circles in Table 2 denote the vertices to which each operator contributes.

It is worth noting that only four of the twelve chiral operators have corresponding linear operators of energy-dimension-six. Among the remaining eight chiral operators, five have corresponding dimension-eight linear operators, two have dimension-ten partners and the operator $\mathcal{L}_{10}$ corresponds to the dimension-twelve operator $O_{10}^{(12)}$. The energy dimension of the linear operators gives an ordering of their significance when no couplings in the underlying theory are strong. In other words, if the effect of an operator with high energy-dimension in the linear Lagrangian is found to be significant, it signals the presence of strong coupling in the symmetry breaking sector.

4. IS THERE ALREADY INDIRECT EVIDENCE FOR THE STANDARD W SELF-COUPLING?

The success of the SM predictions against precision electroweak experiments at the quantum level suggests a question if there is already evidence for the standard universal gauge-boson self-couplings. It is not so trivial to answer this question, since we should identify which finite portion of the quantum corrections is sensitive to the weak-boson self interactions. Usually, one splits the complete SM radiative corrections in just two pieces, which are separately gauge invariant, the fermionic loop contributions to the gauge-boson self-energies, and the rest. It can then be stated clearly that neither of the corrections alone is consistent with the data, and both contributions are needed to explain the success of the SM radiative corrections [26]. Since the bosonic part of the correction should necessarily contain the weak boson self-interactions, we may already have evidence for universal couplings.

It is not clear to me, however, how much of these finite bosonic correction terms depend on the splitting of the gauge bosons into themselves. For instance, the box diagrams do not contain gauge-boson self-couplings. I therefore split the bosonic corrections into three separately gauge-invariant pieces, 'box'-like, 'vertex'-like and 'propagator'-like pieces by appealing to the S-matrix pinch technique [11]. It is then only the 'vertex'-like and 'propagator'-like pieces which contain the gauge boson self-couplings. Schematically we separate the SM radiative corrections into the following five pieces:

SM radiative correction

$$
\begin{aligned}
&= \text{QED/QCD} &&(A)\\
&+ \text{fermion-loop} &&(B)\\
&+ \text{box} &&(C) \quad (4.1)\\
&+ \text{vertex} &&(D)\\
&+ \text{bosonic-loop} &&(E)
\end{aligned}
$$

Details of this separation for each radiative correction term may be found straightforwardly from the analytic expressions presented in ref. [5]. We find by fitting to recent electroweak data the results of Table 3.

In all the fits, the three precisely known quantities, α, G_F, and m_Z are fixed at their mean measured values. The no-EW entry confronts the tree-level predictions of the SM where only QCD and external QED corrections are applied. In this column $\bar{\alpha}(m_Z^2)$ is calculated by including only contributions from light quarks and leptons with $\delta_\alpha = 0$ for the hadronic uncertainty. It is quite striking to learn [27] that these 'no-EW' predictions agree with experiments at LEP1/SLC very well. It is only the new m_W value [28] and the low energy neutral current experiments which give significantly high χ^2 than the full SM. The next +fermion column gives the result of $A + B$ in eq.(4.1). That the LEP1/SLC data can be fitted well by the no-EW calculation is a consequence of an accidental cancellation between the vertex/box correction to the μ decay matrix elements, (the factor δ_G in the Table), and the T parameter for $m_t \sim 175$ GeV. The predictions of the theory depend on the combination $\delta_G - \alpha T$ when the μ decay constant, G_F, is held fixed, whereas this combination vanishes almost identically when $m_t \sim 175$ GeV. If we include only the fermionic corrections, the T parameter grows from zero to 1.144, while the factor δ_G remains zero [5]. The jump of the χ^2 from 20 to 200 in the LEP1/SLC experiments is a consequence of the absence of this cancellation.

It turned out that the 'box-like' corrections to the μ-decay matrix elements give almost 80% of the total δ_G value. Hence by including only the 'box-like' corrections, the fit improves significantly. This can be seen from the column of +'box', where we give results of A+B+C corrections in eq.(4.1). Up to this stage, no contribution from quantum fluctuations with the weak-boson self-couplings are counted. It is in the next step, the +'vertex' column where I list the results of A+B+C+D corrections, where we can start to see their effects. It turns out that the effects of the remaining 20% correction to δ_G, and the effects in part from the vertex corrections in the Z-decay matrix elements significantly reduce the χ^2 in the LEP1/SLC sector of the experiments, from about 80 down to 20.

TABLE 3.

	data	no-EW	+'fermion'	+'box'	+'vertex'	+'propagator'		
m_t (GeV)		—	175	175	175	175	175	175
m_H (GeV)		—	—	—	—	60	300	1000
S		0.000	-0.066	-0.066	-0.066	-0.283	-0.146	-0.075
T		0.000	1.144	1.144	1.144	0.917	0.768	0.588
U		0.000	0.014	0.014	0.014	0.359	0.354	0.353
$\bar\delta_G$		0.0000	0.0000	0.0043	0.0055	0.0055	0.0055	0.0055
$1/\bar\alpha(m_Z^2)$		128.85	128.86	128.86	128.86	128.72	128.72	128.72
$\bar s^2(m_Z^2)$		0.2312	0.2282	0.2296	0.2300	0.2301	0.2310	0.2317
$\bar g_Z^2(m_Z^2)$		0.5487	0.5582	0.5558	0.5551	0.5565	0.5560	0.5553
$\delta_b(m_Z^2)$		—	—	—	-0.0099	-0.0100	-0.0099	-0.0100
$\bar s^2(0)$		0.2388	0.2360	0.2373	0.2376	0.2386	0.2394	0.2401
$\bar g_Z^2(0)$		0.5487	0.5533	0.5509	0.5502	0.5493	0.5487	0.5480
$\bar g_W^2(0)$		0.4218	0.4271	0.4245	0.4238	0.4245	0.4234	0.4224
Γ_Z(GeV)	2.4974 ± 0.0038	2.4843	2.5354	2.5207	2.4914	2.4972	2.4928	2.4876
		2.4866	2.5377	2.5230	2.4937	2.4995	2.4951	2.4899
		2.4888	2.5400	2.5252	2.4959	2.5017	2.4973	2.4921
σ_h^0(nb)	41.49 ± 0.12	41.50	41.49	41.49	41.48	41.48	41.49	41.49
		41.48	41.47	41.47	41.46	41.46	41.46	41.47
		41.46	41.45	41.45	41.44	41.44	41.44	41.45
R_ℓ	20.795 ± 0.040	20.775	20.822	20.800	20.736	20.734	20.719	20.706
		20.802	20.849	20.827	20.763	20.761	20.746	20.733
		20.829	20.876	20.854	20.790	20.787	20.773	20.759
$A_{FB}^{0,\ell}$	0.0170 ± 0.0016	0.0168	0.0222	0.0196	0.0174	0.0171	0.0157	0.0145
A_τ	0.143 ± 0.010	0.149	0.173	0.162	0.151	0.150	0.143	0.138
A_e	0.135 ± 0.011	0.149	0.173	0.162	0.151	0.150	0.143	0.138
R_b	0.2202 ± 0.0020	0.2183	0.2181	0.2182	0.2157	0.2157	0.2157	0.2157
R_c	0.1583 ± 0.0098	0.1716	0.1719	0.1718	0.1722	0.1722	0.1721	0.1721
$A_{FB}^{0,b}$	0.0967 ± 0.0038	0.1049	0.1214	0.1138	0.1061	0.1053	0.1005	0.0966
$A_{FB}^{0,c}$	0.0760 ± 0.0091	0.0750	0.0878	0.0819	0.0761	0.0755	0.0718	0.0688
A_{LR}	0.1637 ± 0.0075	0.1495	0.1727	0.1620	0.1512	0.1501	0.1434	0.1379
χ^2	($\alpha_s = 0.116$)	24.9	177.9	72.5	21.4	18.2	20.1	32.0
	($\alpha_s = 0.120$)	21.1	191.8	81.0	18.6	17.0	17.0	26.8
	($\alpha_s = 0.124$)	19.1	207.4	91.1	17.4	17.5	15.6	23.3
g_L^2	0.2980 ± 0.0044	0.2905	0.2976	0.2998	0.3015	0.2998	0.2985	0.2973
g_R^2	0.0307 ± 0.0047	0.0304	0.0302	0.0302	0.0293	0.0295	0.0296	0.0297
δ_L^2	-0.0589 ± 0.0237	-0.0591	-0.0596	-0.0641	-0.0634	-0.0634	-0.0634	-0.0634
δ_R^2	0.0206 ± 0.0160	0.0183	0.0181	0.0181	0.0176	0.0177	0.0178	0.0178
χ^2		4.4	0.1	0.3	0.8	0.3	0.2	0.3
s_{eff}^2	0.233 ± 0.008	0.239	0.236	0.235	0.229	0.230	0.231	0.231
ρ_{eff}	1.007 ± 0.028	1.000	1.008	1.016	1.015	1.013	1.012	1.011
χ^2		0.6	0.1	0.1	0.4	0.2	0.1	0.1
Q_W	-71.04 ± 1.81	-74.79	-74.80	-73.03	-72.98	-73.07	-73.17	-73.21
χ^2		4.3	4.3	1.2	1.2	1.3	1.4	1.4
$2C_{1u} - C_{1d}$	0.938 ± 0.264	0.709	0.725	0.729	0.728	0.724	0.720	0.717
$2C_{2u} - C_{2d}$	-0.659 ± 1.228	0.081	0.099	0.102	0.111	0.105	0.100	0.096
χ^2		2.0	1.2	1.1	1.1	1.2	1.4	1.5
m_W	80.24 ± 0.16	79.96	80.46	80.38	80.36	80.43	80.32	80.23
χ^2		3.2	1.9	0.8	0.6	1.4	0.3	0.0
χ^2_{tot}	($\alpha_s = 0.116$)	39.3	185.5	76.1	25.5	22.6	23.4	35.3
	($\alpha_s = 0.120$)	35.6	199.4	84.6	22.6	21.4	20.3	30.1
	($\alpha_s = 0.124$)	33.6	215.0	94.7	21.5	21.9	18.9	26.6

I should therefore conclude that the effect of the 'vertex'-like corrections is significant for the success of the SM at the quantum correction level. Even setting aside the fundamental problem that we could not control quantum fluctuations at short distances if it were not for the universality of the weak-boson self-couplings, it is reassuring to learn that, after cancellation of the short-distance singularity, the remaining finite correction makes the fit even better. I note in passing that the significance of the 'propagator-like' correction term which contains the Higgs-mass dependence of the SM prediction cannot be established at the present level of accuracy, except that adding this last correction worsens the fit when the Higgs boson mass is as large as 1 TeV.

5. CONCLUSIONS

(i) Universality of weak-boson couplings to fermions has been tested at the 0.2% level after inclusion of the SM radiative corrections.

(ii) The most natural explanation of the observed universality of the weak-boson couplings to fermions is local $SU(2)_L \times U(1)_Y$ symmetry which breaks spontaneously to make the W and Z bosons massive.

(iii) Local gauge symmetry implies extended universality of the gauge couplings to include the weak-boson self-interactions.

(iv) The agreement of the SM predictions with precision experiments improves when one includes radiative effects due to 'vertex'-like corrections, which may be regarded as indirect evidence for the universal weak-boson self couplings.

6. ACKNOWLEDGEMENTS

I would like to thank T.Hatsukano, S.Ishihara, S.Matsumoto, R.Szalapski, Y.Yamada and D.Zeppenfeld for fruitful collaboration which made this presentation possible. I would also like to express my sincere gratitude for U.Baur, S.Errede and T.Müller for hospitality and organizing a stimulating conference.

REFERENCES

1. P.D.Hung and J.J.Sakurai, Nucl. Phys. **B143**, 81 (1978);
   J.D.Bjorken, Phys. Rev. **D19**, 335 (1978).
2. M.Suzuki, Phys. Lett. **B153**, 289 (1985);
   H.Neufeld, J.D.Stroughair and D.Schildknecht, Phys. Lett. **B198**, 563 (1987);
   J.J.van der Bij, Phys. Rev. **D35**, 1088 (1987);
   J.A.Grifols, S.Peris and J.Solà, Int. J. Mod. Phys. **A3**, 225 (1988);
   G.L.Kane, J.Vidal and C.-P.Yuan, Phys. Rev. **D39**, 2617 (1989);
   J.J.van der Bij, Phys. Lett. **B296**, 239 (1992).
3. Particle Data Group, L. Montanet *et al.*, Phys. Rev. **D50**, 1173 (1994).
4. R.Barbieri, C.Bouchiat, A.Georges and P.Le Doussal, Nucl. Phys. **B269**, 253 (1986); K.Hagiwara, S.Matsumoto and Y.Yamada, in preparation.
5. K.Hagiwara, D.Haidt, C.S.Kim and S.Matsumoto, Z. Phys. **C64**, 559 (1994).
6. M.E.Peskin and T.Takeuchi, Phys. Rev. Lett. **65**, 964 (1990); Phys. Rev. **D46**, 381 (1992).
7. S.Matsumoto, preprint KEK-TH-418 (1995).

8. S. Eidelman and F. Jegerlehner, Preprint PSI-PR-95-1 (1995).
9. A.D. Martin and D. Zeppenfeld, Preprint MAD/PH/855 (1994).
10. M.L. Swartz, Preprint SLAC-PUB-6710 (1994).
11. J.M.Cornwall and J.Papavassiliou, Phys. Rev. **D40**, 3474 (1989);
    D.C.Kennedy and B.W.Lynn, Nucl. Phys. **B322**, 1 (1989);
    G.Degrassi and A.Sirlin, Nucl. Phys. **B383**, 73 (1992); Phys. Rev. **D46**, 3104 (1992);
    G.Degrassi, B.Kniehl and A.Sirlin, Phys. Rev. **D48**, R3963 (1993);
    J.Papavassiliou and K.Phillippides, Phys. Rev. **D48**, 4255 (1993); preprint NYU-TH-95-03-03;
    J.Papavassiliou and C.Parrinello, Phys. Rev. **D50**, 3059 (1994);
    J.Papavassiliou, Phys. Rev. **D50**, 5958 (1994).
12. W.Buchmüller and D.Wyler, Nucl. Phys. **B268**, 621 (1986).
13. B.Grinstein and M.B.Wise, Phys. Lett. **B265**, 326 (1991).
14. K.Hagiwara, S.Ishihara, R.Szalapski and D.Zeppenfeld, Phys. Lett. **B283**, 353 (1992); Phys. Rev. **D48**, 2182 (1993).
15. K.Hagiwara, S.Matsumoto and R.Szalapski, preprint KEK-TH-440 (1995).
16. K.J.F.Gaemers and G.J.Gounaris, Z. Phys. **C1**, 259 (1979).
17. K.Hagiwara, R.D.Peccei, D.Zeppenfeld and K.Hikasa, Nucl. Phys. **B282**, 253 (1987).
18. A.De Rújula, M.B.Gavela, P.Hernandez and E.Massó, Nucl. Phys. **B384**, 3 (1992).
19. K.Hagiwara, T.Hatsukano, S.Ishihara and R.Szalapski, in preparation.
20. K.Hagiwara, R.Szalapski and D.Zeppenfeld, Phys. Lett. **B318**, 155 (1993);
    K.Hagiwara and M.L.Stong, Z. Phys. **C62**, 99 (1994);
    G.J.Gounaris, J.Layssac and F.M.Renard, preprint PM/95-11;
    G.J.Gounaris and F.M.Renard, preprint PM/95-20.
21. G.J.Gounaris, F.M.Renard and G.Tsirigoti, Phys. Lett. **B338**, 51 (1994); preprint CERN-TH/95-42.
22. T.Appelquist and G.H.Wu, Phys. Rev. **D48**, 3235 (1993).
23. A.Longhitano, Phys. Rev. **D22**, 1166 (1980); Nucl. Phys. **B188**, 118 (1981).
24. B.Holdom, Phys. Lett. **B258**, 156 (1991);
    G.Belanger and F.Boudjema, Phys. Lett. **B288**, 201 (1992);
    A.De Rújula, M.B.Gavela, P.Hernández and E.Massoó, Nucl. Phys. **B384**, 3 (1992);
    P.Hernández and F.Vegas, Phys. Lett. **B307**, 116 (1993).
25. S.Dawson and G.Valencia, preprint BNL-60949 (1994).
26. P.Gambio and A.Sirlin, Phys. Rev. Lett. **73**, 621 (1994).
27. V.A.Novikov, L.B.Okun and M.I.Vysotsky, Mod. Phys. Lett. **A8** (1993) 2529; Erratum A8, 3301 (1993).
28. V.A.Novikov, L.B.Okun, A.N.Rozanov and M.I.Vysotsky, Mod. Phys. Lett. **A9** (1994) 2641;
    Z.Hioki, Phys. Lett. **B340**, 181 (1994).

W and Z Boson Production at HERA

presented for the H1 and ZEUS Collaborations

Roman Walczak

University of Oxford

The HERA experiments, H1 and ZEUS, have each collected approximately 4 pb^{-1} of ep (mainly e$^+$p) integrated luminosity up to 1994. This allows a first look at processes with cross-sections at a picobarn level at the ep centre of mass energy of about 300 GeV. Events possibly containing W or Z bosons were found. Contribution of W and Z bosons to the ep scattering and a sensitivity to the γWW coupling is summarized.

Introduction

Study of W and Z production is a part of a rich physics program of the H1 and ZEUS experiments at the HERA electron-proton collider. Expected cross-sections for the production of real W and Z bosons are about 1 pb with about 50 % uncertainty (1). Measurements of the cross-sections will verify expectations and test some of possible extensions of the Standard Model. A large class of new heavy particles, for example excited fermions, predicted in various models could decay into W or Z bosons. Any difference between the measured cross-section and that predicted within the Standard Model could indicate the existence of such new particles. The cross-section for W production is also sensitive to coupling of the photon to two W's (γWW vertex). This coupling, predicted in the Standard Model but not well measured, might be modified by physics beyond the Standard Model.

Effects due to virtual W and Z bosons can be studied at HERA at previously inaccessible, high Q^2-values (up to $3*10^4$ GeV^2, limited by the event rate). Electroweak parameters like the W propagator mass and the W coupling to fermions at high Q^2 will be measured. Comparisons of measured observables at high Q^2 to extrapolations from low Q^2 may indicate existence of new propagators or substructures of known particles (2).

Each HERA experiment has collected about 4 pb^{-1} of data so far. This corresponds to three lepton beam settings with the proton beam at 820 GeV. About 0.55 pb^{-1} was taken in 1993 with electrons at 26.7 GeV and about

© 1996 American Institute of Physics

0.25 pb^{-1} in 1994 with electrons at 27.5 GeV. The remaining 3.2 pb^{-1} was accumulated in 1994 using a positron beam of 27.5 GeV. The centre of mass energy was about 300 GeV. Both experiments use multipurpose detectors described elsewhere (3,4).

Virtual W and Z bosons

Neutral current (NC) and charged current (CC) deep inelastic scattering (DIS) cross-sections were measured by both experiments at high Q^2 (5,6). Fig. 1 and Table 1 show the cross-sections and their ratios as a function of Q^2 measured by the ZEUS collaboration using a luminosity of 0.540 ± 0.016 pb^{-1}. For the first time one can see that CC and NC cross-sections are similar, demonstrating the equal strengths of the weak and electromagnetic interactions at high Q^2.

Fitting the Q^2 dependence of the CC cross-section with the W boson mass (M_W) as the only free parameter the ZEUS collaboration obtained:
$M_W = 76 \pm 16(\text{stat}) \pm 13(\text{syst})$ GeV.

Using 0.348 ± 0.017 pb^{-1} of e^-p data, the H1 collaboration measured the CC cross-section for transverse momenta (p_t) of the hadron system larger than 25 GeV (corresponds to $Q^2 > 625$ GeV^2):
$\sigma(p_t > 25 \text{ GeV}) = 55 \pm 15(\text{stat}) \pm 6(\text{syst})$ pb.

The cross-section was converted to a neutrino nucleon (νN) cross-section and compared with the low energy neutrino data: see Fig. 2. The deviation from the linear rise of the cross-section due to the W mass is visible. The constraints of this measurements for the W mass are shown in Fig. 3.

For the ratio of the NC to CC cross-sections for $p_t > 25$ GeV of the hadronic systems, the H1 collaboration obtained (8):

$$\frac{\sigma_{e^-p}^{NC}(p_{t_{had}} > 25 GeV)}{\sigma_{e^-p}^{CC}(p_{t_{had}} > 25 GeV)} = 7.2 \pm 2.1(stat) \pm 1.2(syst)$$

Measurement of this ratio with the precision of 1%, which appears a challanging task, corresponds to an error on the W mass of 630 MeV (2).

Production of real W and Z bosons

Recently, detailed calculations of cross-sections for W and Z production were carried out by Baur, Vermaseren and Zeppenfeld (1). In this section we refer to their work. Diagrams illustrating parton level processes contributing to ep→eWX are shown in Fig. 4. A similar set of diagrams describes the ep→eZX reaction. The contributions of diagrams with heavy W or Z propagators are small. For this reason reactions with νWX or νZX in the final state have cross-sections at least ten times smaller than the corresponding ones with the electron instead of the neutrino in the final state. Calculations were done

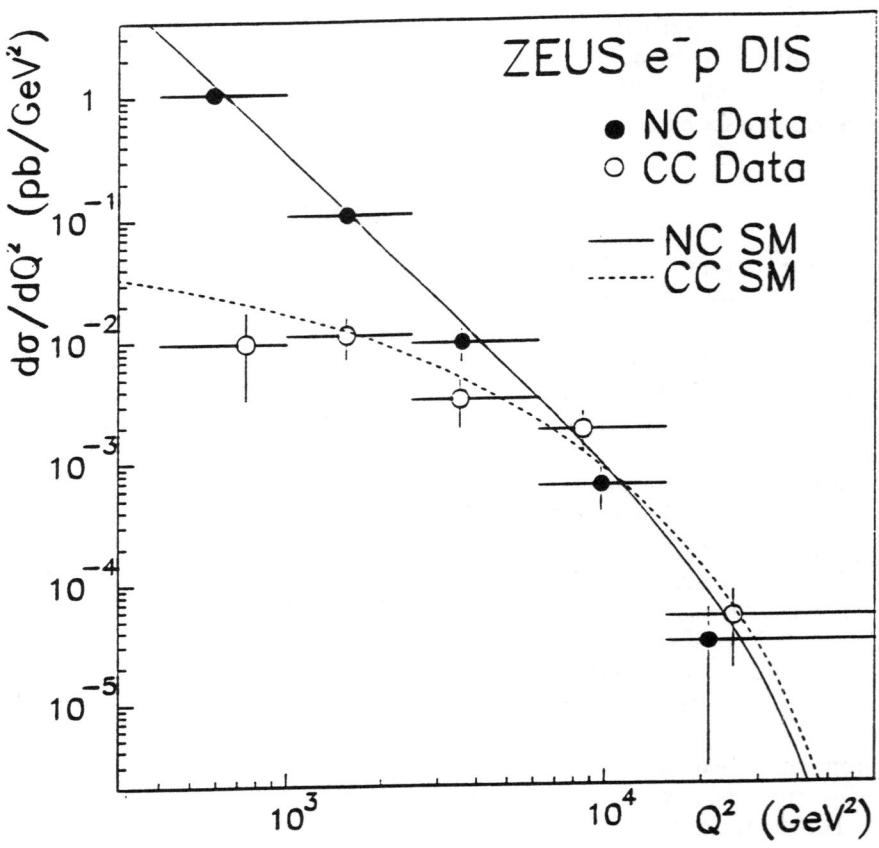

FIG. 1. Differential cross-section for CC and NC e^-p DIS from ZEUS 1993 (5). The points with errors are the data, and the curves are the Standard Model cross-sections. The data are plotted at the average Q^2 of the events in each bin.

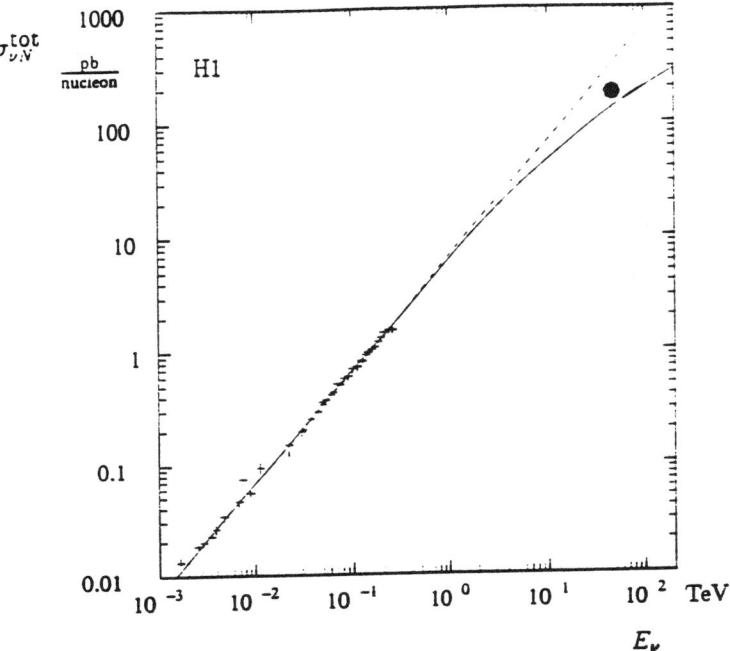

FIG. 2. The energy dependence of the νN cross-section. The crosses represent the low energy neutrino data (7) while the full circle refers to the H1 analysis. The full line represents the predicted cross-section including the W propagator. The dashed line is the linear extrapolation from low energies.

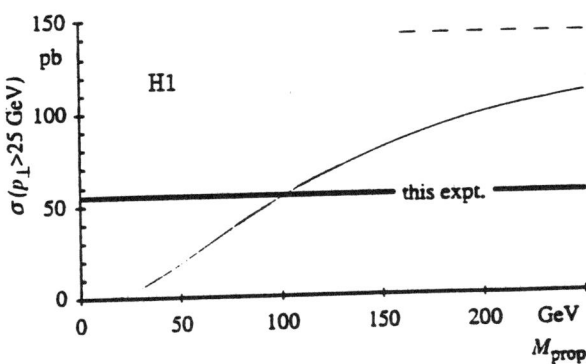

FIG. 3. The CC cross-section predicted as function of the propagator mass M_{prop} (thin solid line). The dashed line indicates the asymptotic case $M_{prop} = \infty$. The shaded region represents the 1σ band of the cross-section measured by the H1 experiment (6).

TABLE 1. Ratios of the NC to CC cross-sections for $Q^2 > Q^2_{min}$ by the ZEUS experiment (5)

Q^2_{min} (GeV^2)	400	1000	2500	6250	15625
$\frac{\sigma_{NC}}{\sigma_{CC}}$ (for $Q^2 > Q^2_{min}$)	$14.7^{+3.4}_{-3.2}$	$4.2^{+1.3}_{-0.9}$	$1.4^{+0.6}_{-0.4}$	$0.4^{+0.3}_{-0.1}$	$0.7^{+1.0}_{-0.5}$

TABLE 2. Theoretical total cross-sections for intermediate boson production (1).

process	cross-section (pb)
ep → eW$^+$X	0.50 - 0.72
ep → eW$^-$X	0.47 - 0.62
ep → eZX	0.34 - 0.43

in the whole phase-space. Regions of the phase-space containig singularities (zero mass quark propagator) or where perturbative methods cannot be applied (elastic or quasielastic proton scattering) were treated separately using approximations and a phenomenological input. It was demonstrated that the total cross-section does not depend on the cuts used to divide the phase-space. It should be noted, however, that the topologies of the events were distorted by the approximations used. For example the W transverse momentum distribution does not change smoothly going from one region of the phase-space to another. Consequently, estimates of experimental acceptances to detect W and Z bosons depend on those cuts.

Calculated total cross-sections for eWX and eZX final states are presented in Table 2. The uncertainties are due to uncertainties of the photon structure function and the variation of the scale at which the photon structure function is evaluated. Additional uncertainties come from the assumption that a virtual photon has the same structure function as the real one and from QCD corrections.

γWW coupling

Production of W bosons depends on the γWW coupling through the diagram (c) in fig. 4. Not all observables are equally sensitive to the contribution of this diagram. For this reason a choice of an observable to estimate the γWW coupling is important for the precision. Transverse momentum distributions of the quark jet from the proton and the ratio of the cross-sections for W and Z production were analysed (9,10). The γWW coupling was parametrized using the κ and λ parameters (11). With 1000 pb^{-1} luminosity, using electron and muon W/Z decay channels only, at 69% C.L. the precision on κ was estimated to be about 0.3-0.4 and on λ about 0.8-0.9. For the same luminosity the precision on κ is higher than the one estimated for the Tevatron but lower on λ.

In principle the γWW vertex can be also studied in ep→ $\nu\gamma$X reaction

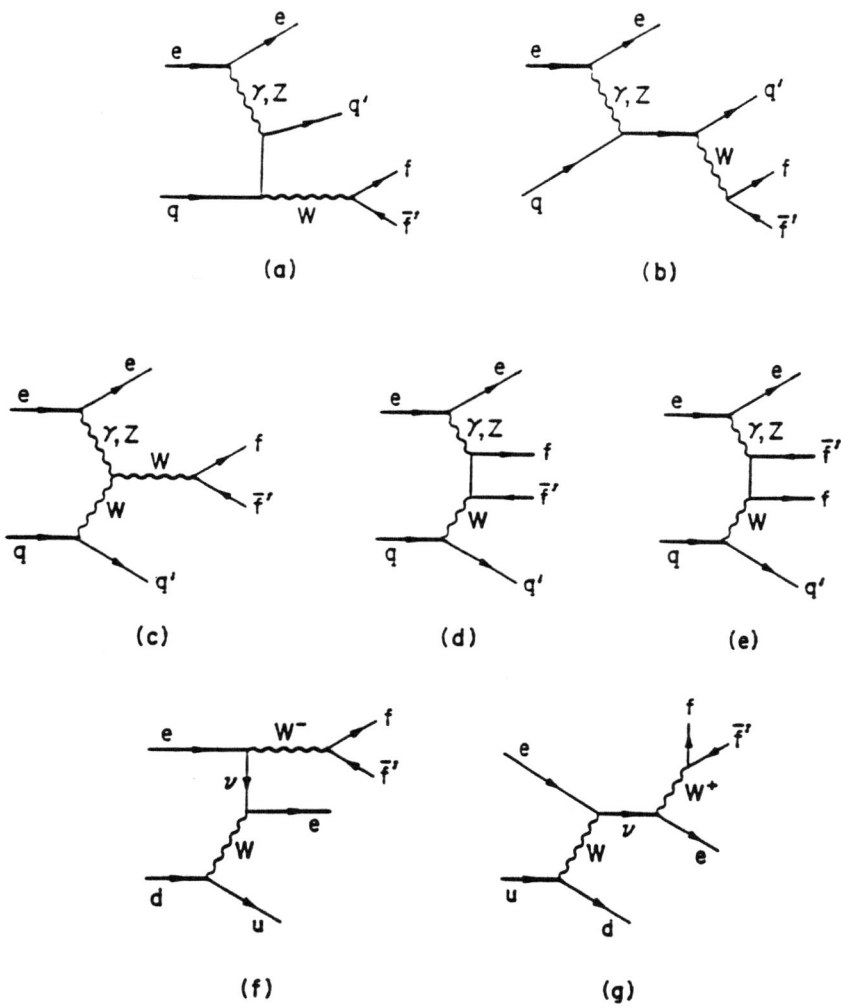

FIG. 4. Feynman graphs for the reaction ep→eWX with W decaying into fermions f.

(17,18). Comparing to the W production process the sensitivity to the γWW coupling is lower.

Search for W in the reaction $e^\pm p \to e^\mp W X$

$W \to e\nu, \mu\nu$

Analysing all 4 pb^{-1} of data, the H1 collaboration found one candidate event corresponding to W^+ decaying into $\mu^+ \nu$ (12), see fig. 5. The muon and the hadronic system consisting of two jets (not resolved by the cone, R=1, jet algorithm) are almost exactly back to back in the plane transverse to the beams. Their transverse momenta are:

$p_t(\mu) = 23.4 \pm 2.4^{+7}_{-5}$ GeV ($\pm stat^{+syst}_{-syst}$)
$p_t(jet1) = 25.3 \pm 3.$ GeV
$p_t(jet2) = 15.2 \pm 1.9$ GeV
and the missing $p_t = 18 \pm 4.8^{+5}_{-7}$ GeV.

Taking the calculated cross-section (1), assuming the full acceptance and demanding that the hadronic system has $p_t > 40$ GeV, the H1 collaboration estimated the probability of observing such an event to be 3%. This event is further discussed by A. Schöning (13). No similar candidate was found in the $e\nu$ channel.

The ZEUS collaboration also looked for W bosons in $e\nu$ and $\mu\nu$ channels. No candidates were found.

$W \to jets$

Search for W bosons in hadronic decays is difficult because of a large QCD background (14). The UA2 collaboration observed hadronic W decays at the CERN $p\bar{p}$ collider (15). The ZEUS collaboration selected a small sample of 55 events, corresponding to 4 pb^{-1} of integrated luminosity, to be used in the search for W and Z bosons. The main requirements, selecting the sample, were: high transverse energy ($E_t > 40$ GeV) of the events, two or more jets of high transverse momenta and away from the beam pipe with polar angles $\theta > 0.19$ (corresponding to pseudorapidity $\eta < 2.35$), measured with respect to the proton beam direction in the ZEUS laboratory frame. Jets were identified using an η, o cone algorithm with the radius R=1. Fig. 6 shows the mass distribution of two highest transverse momentum jets for the sample of the 55 events. Production of W and Z bosons was simulated using the parton level generator EPVEC (1) followed by the JETSET 7.3 (16) code for hadronization and a set of programs describing the ZEUS detector. Fig. 7 shows the mass distribution of the two highest transverse momentum jets for generated events with W (both electric charges) and Z bosons. The overall acceptance (including trigger) for W and Z bosons decaying hadronically is 32% for the W and 39% for the Z. The mass resolution improves for smaller

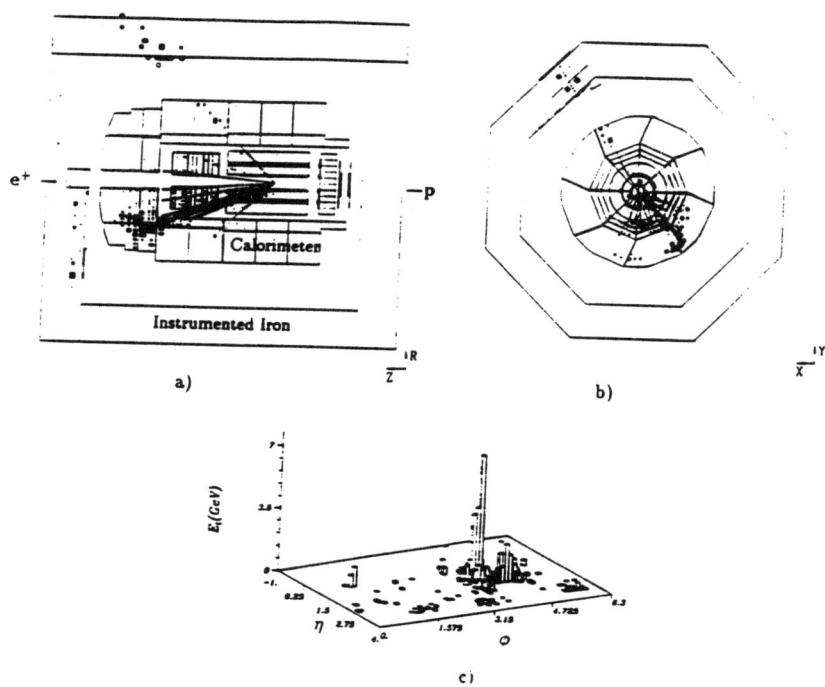

FIG. 5. H1 event display for ep→eW⁺X, $W^+ \to \mu^+\nu$, candidate event: a) R-z view b) R-ϕ view and c) transverse calorimetric energy.

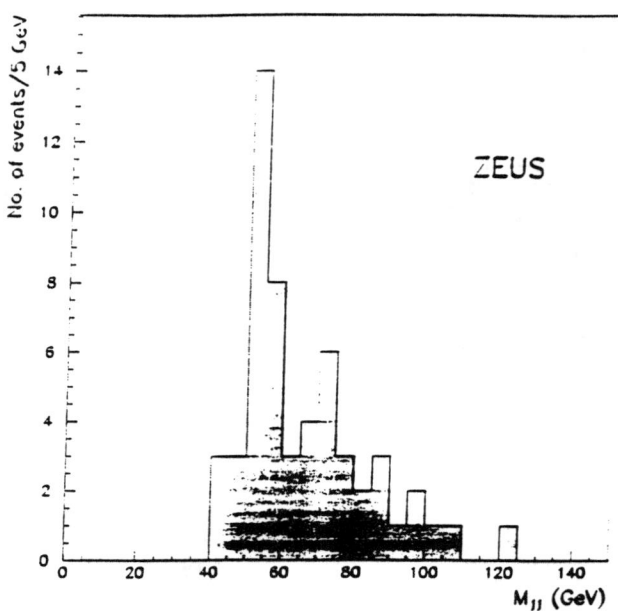

FIG. 6. Mass distribution of two highest transverse momentum jets (not corrected for the detector effects) measured by the ZEUS collaboration.

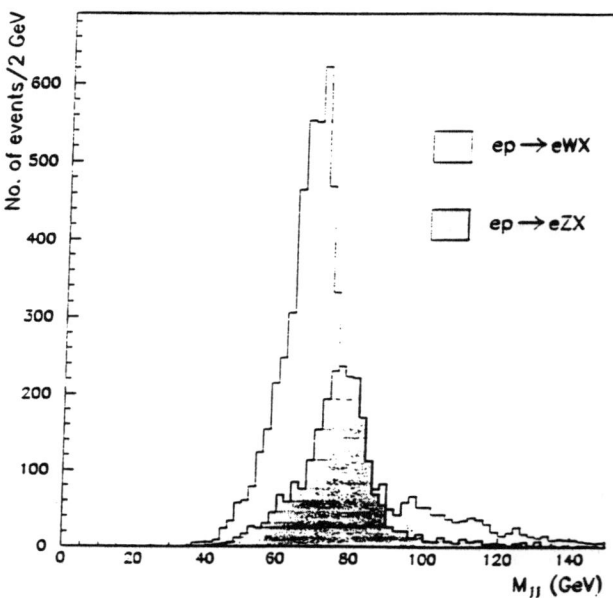

FIG. 7. Monte-Carlo generated mass distribution of two highest transverse momentum jets for events with W (blank) and Z (shaded) bosons (not corrected for the detector effects). The ZEUS detector simulation was used.

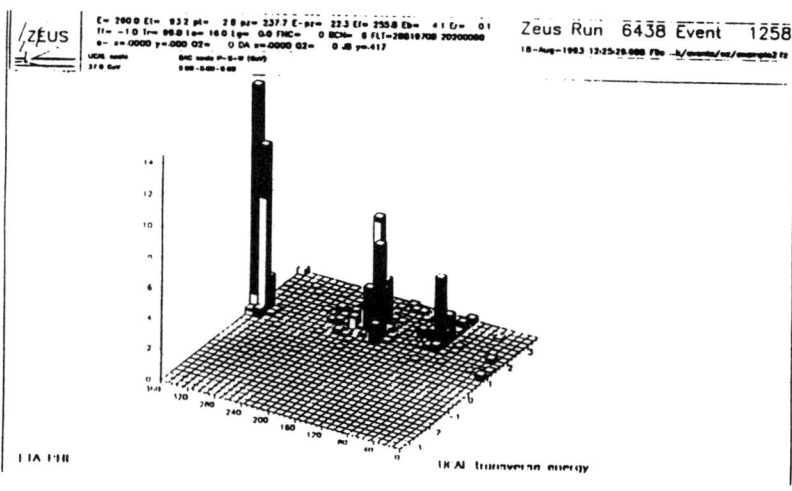

FIG. 8. Transverse energy (in GeV) as a function of pseudorapidity η and azimuthal angle ϕ for ep→eWX or ep→eZX candidate event.

cone radii but this was not yet optimized. Among the 55 events selected one can thus expect about one or two events containing W or Z boson. Fig. 8 shows the transverse energy η, ϕ lego plot of a possible candidate event. A further selection is needed to extract W and Z events from the background. This is under study now.

Conclusions

The NC and CC DIS cross-sections were measured at high Q^2 where contributions of the finite mass W and Z propagators are important. The mass of the W propagator was estimated. The search for W and Z bosons in leptonic and hadronic decay channels has started at HERA. For equal luminosity the sensitivity at HERA to γWW coupling is competitive to the one at the Tevatron or LEP200.

Acknowledgement

We would like to thank U. Baur for his help in installing and understanding the EPVEC (1) code.

REFERENCES

1. U. Baur, J.A.M. Vermaseren, D. Zeppenfeld, Nucl. Phys. B **375**, 3 (1992).
2. Physics at HERA vol. **2**, Proceedings of the Workshop, Hamburg (1991) ed. by W. Buchmüller and G. Ingelman.
3. H1 Collaboration, I. Abt et al., The H1 Detector at HERA, DESY preprint **DESY 93-103** (1993).
4. ZEUS Collaboration, The ZEUS Detector, Status Report 1993, (1993).
5. ZEUS Collaboration, M. Derrick et al., DESY preprint **DESY 95-053** (1995).
6. H1 Collaboration, T. Ahmed et al., Phys. Lett. B **324**, 241 (1994).
7. D. Haidt and H. Pietschmann: Landolt-Börnstein New Series I/10, p. 213, Springer (1988).
8. M. Hapke, Dissertation, DESY **FH1K-94-95**, (1994).
9. U. Baur, D. Zeppenfeld, Nucl. Phys. **325**, 253 (1989).
10. C.S. Kim, Jungil Lee, H.S. Song, Seoul National University preprint, **SNUTP 95-010** (1995).
11. D. Zeppenfeld, this proceedings.
12. H1 Collaboration, T. Ahmed et al., DESY preprint **DESY 94-248** (1994).
13. A. Schöning, this proceedings.
14. H. Baer, J. Ohnemus, D. Zeppenfeld, Z.Phys. C **43**, 675 (1989).
15. UA2 Collaboration, J. Alitti et al., Z.Phys. C **49**, 17 (1991).
16. T. Sjöstrand, CERN preprint, **CERN-TH-6488-92** (1992).
17. T. Helbig and H. Spiesberger in (2).
18. U. Baur and M. A. Doncheski, Phys. Rev. D **46**, 1959 (1992)

Quartic Gauge Boson Couplings

Stephen Godfrey[1]

*Department of Physics, Carleton University,
Ottawa, CANADA K1S 5B6*

Quartic vertices provide a window into one of the most important problems in particle physics; the understanding of electroweak symmetry breaking. I survey the various processes that have been proposed to study quartic gauge boson couplings at future e^+e^-, $e\gamma$, $\gamma\gamma$, e^-e^-, and pp colliders. For the lowest dimension operators that do not include photons, it appears that the LHC will provide the most constraining measurements. However, precision measurements at high energy e^+e^- colliders involving W^+W^- rescattering are also quite sensitive to the effects of a strongly interacting weak interaction. For quartic couplings involving photons, $\gamma\gamma$ collisions appear to be the best place to measure these couplings. Measurements using gauge boson production in $e\gamma$ collisions are almost as precise as the $\gamma\gamma$ processes with $e^+e^- \to VVV$ about an order or magnitude less sensitive.

I. INTRODUCTION

The non-Abelian gauge nature of the standard model predicts, in addition to the trilinear WWZ and $WW\gamma$ couplings (TGV's), quartic gauge boson couplings (QC's). The strength of the couplings is set by the universal gauge couplings of the $SU(2)$ local gauge symmetry. In the standard model there are only three quartic couplings which necessarily involve at least two charged W's; $W^+W^-W^+W^-$, W^+W^-ZZ, and $W^+W^-\gamma\gamma$. Although the $ZZZZ$ vertex is not present in the SM it is present at tree level via Higgs exchange while the $\gamma\gamma ZZ$ vertex is only produced at loop level in the Standard Model.

The trilinear and quartic couplings probe different aspects of the weak interactions. The trilinear couplings test the non-Abelian gauge structure where deviations from the SM can result from integrating out heavy particles in loops (1). In contrast, the quartic couplings can be regarded as a window on electroweak symmetry breaking. Recall that the longitudinal components of the W and Z are Goldstone bosons. The quartic couplings of gauge bosons therefore represent a connection to the scalar sector of the theory. The QC's would arise as a contact interaction manifestation of heavy particle exchange.

It is quite possible that the quartic couplings deviate from their SM values while the TGV's do not. For example, the BESS model is a non-linear realization of symmetry breaking where new structures not present at tree level

---

[1] e-mail: godfrey@physics.carleton.ca

appear in $4W$ couplings (2). There are models with a heavy scalar singlet interacting with the Higgs sector which do not affect the ρ parameter nor the TGV's but do change the $4W$ vertex (3).

Thus, if the mechanism for electroweak symmetry breaking does not reveal itself through the discovery of new particles such as the Higgs boson, supersymmetric particles, or technipions it is quite possible that anomalous quartic couplings could be our first probes into this sector of the electroweak theory.

While considerable effort has been expended to study the trilinear couplings, the quartic couplings are only starting to receive much attention. In this contribution I review recent developments in the study of quartic couplings and attempt to summarize the current status of this subject. In the next section I describe the effective Lagrangians relevant to QC's. I will then describe various processes that have been proposed to study quartic couplings using a wide variety of colliding particles: pp, e^+e^-, $e\gamma$, $\gamma\gamma$, and e^-e^-. In the final section I summarize these results and also add some comments as to where the subject can benefit from further work.

II. EFFECTIVE LAGRANGIANS AND QUARTIC COUPLINGS

The formalism of effective Lagrangians provides a well-defined framework for investigating the physics of anomalous couplings and electroweak symmetry breaking (4–6). The infinite set of terms in $\mathcal{L}_{eff}$ can be organized in an energy expansion where at low energy only a finite number of terms will contribute to a given process. At higher energies more and more terms become important until the whole process breaks down at the scale of new physics. One focuses on the leading operators in the expansion.

Quartic operators can either be associated with trilinear couplings or can be genuinely quartic. The former type is described by:

$$\mathcal{L}^{WW\gamma} = -ie\frac{\lambda_\gamma}{M_W^2}F^{\mu\nu}W^\dagger_{\mu\alpha}W^\alpha_\nu \qquad (1)$$

This operator generates $WW\gamma\gamma$ couplings with strength $e^2\lambda_\gamma$ in addition to $WW\gamma$ couplings. These vertices are not likely to be particularly interesting as the parameter λ_γ will already be constrained from other processes such as $e^+e^- \to WW$ where the TGV contributes but the QC does not appear (1)

We will restrict our discussion to the more interesting, genuinely quartic couplings. We concentrate on the lowest dimension operators that can contribute to a given vertex. We impose custodial $SU(2)$, which is satisfied to high precision by the nearness of the ρ parameter to unity, and $U(1)_{em}$ for the operators involving photons. There are 2 (equivalent) parametrizations which have appeared in the literature. We will begin by describing these parametrizations.

A. General Parametrization

This parametrization was introduced by Bélanger and Boudjema (7,8). There are only two dimension four operators. They do not involve photons since $U(1)_{em}$ requires derivatives which would result in a higher dimension operator. Imposing $SU(2)_C$ the two dimension four operators are given by:

$$\mathcal{L}_4^o = \frac{1}{4} g_o g_W^2 (\vec{W}_\mu \cdot \vec{W}^\mu)^2$$

$$\rightarrow g_o g_W^2 [(W_\mu^+ W^{-\mu})(W_\nu^+ W^{-\nu}) + \frac{1}{c_w^2} W_\mu^+ W_\mu^- Z^\nu Z_\nu + \frac{1}{4c_w^2} Z^\mu Z_\mu Z_\nu Z^\nu]$$

$$\mathcal{L}_4^c = \frac{1}{4} g_c g_W^2 (\vec{W}_\mu \cdot \vec{W}^\nu)(\vec{W}^\mu \cdot \vec{W}_\nu)$$

$$\rightarrow g_c g_W^2 [\frac{1}{2}(W_\mu^+ W^{-\mu} W_\nu^+ W^{-\nu} + W_\mu^+ W^{+\mu} W_\nu^- W^{-\nu})$$

$$+ \frac{1}{c_w^2} W_\mu^+ W^{-\nu} Z^\mu Z_\nu + \frac{1}{4c_w^2} Z^\mu Z_\mu Z_\nu Z^\nu]$$

These operators involve the maximum number of longitudinal modes. These are the most important manifestations of an alternative symmetry breaking scenario. Note that the $ZZZZ$ vertex does not appear in the SM and $W^+W^-W^+W^-$ cannot be probed via 3 boson production in e^+e^-. Photons do not appear in these genuine QC's.

The first operator can be thought of as parametrizing heavy neutral scalar exchange so that we can make the connection:

$$g_o \propto \kappa^2 \left(\frac{M_W^2}{\Lambda^2}\right) \quad (2)$$

where κ is the strength of coupling in the W system. Heavy Higgs exchange, at tree level in the SM, gives κ of order 1. In this case $g_o \simeq 0.2$ corresponds to $M_H = \Lambda \sim 180$ GeV which would most likely be observed directly at a high energy collider invalidating this approach. On the other hand, taking a Higgs mass of 1 TeV yields a contact term of strength $g_0 \simeq 6 \times 10^{-3}$. Thus, to see the effect of a heavy scalar as a deviation to the QC requires very precise measurements.

For the case of scalar exchange $g_o > 0$. In the second operator the $WWZZ$ vertex corresponds to heavy charged scalar exchange so that we could associate it with a triplet of heavy scalars. A specific case of interest is when $g_o = -g_c = g_s < 0$ which could parametrize heavy vector particle exchange which might arise in theories like technicolour. In this case the $4Z$ couplings cancel and the net effect is a rescaling of the SM $4W$ vertex. Bounds on g_s therefore determine the precision with which $4W$ couplings could be measured.

To introduce photons we have to go to dimension 6 operators. We only consider these dim-6 operators since they result in the largest phase space and are therefore most likely to give the largest deviations. Imposing $SU(2)_C$

and $U(1)_{QED}$ and restricting the phenomenological analysis to the C and P conserving operators with a maximum of two photons involves the $\gamma\gamma W^+W^-$ and $\gamma\gamma ZZ$ vertices described by the operators:

$$\mathcal{L}_6^0 = -\frac{\pi\alpha}{4\Lambda^2} a_0 F_{\alpha\beta} F^{\alpha\beta}(\vec{W}_\mu \cdot \vec{W}^\mu) \tag{3}$$

$$\mathcal{L}_6^c = -\frac{\pi\alpha}{4\Lambda^2} a_c F_{\alpha\mu} F^{\alpha\nu}(\vec{W}_\mu \cdot \vec{W}^\nu) \tag{4}$$

Where $\vec{W}_\mu$ is an $SU(2)$ triplet and $F^{\mu\nu}$ and $\vec{W}^{\mu\nu}$ are the $U(1)_{em}$ and $SU(2)$ field strengths respectively. Both operators have contributions from loops but the first can originate from heavy neutral scalar exchange while the second can arise from charged scalars. Note that the $SU(2)$ gauge symmetry predicts that $\gamma\gamma ZZ$ does not appear in the SM. The custodial symmetry imposed on these couplings means that, in leading order in s, they contribute in the same way to the $\gamma\gamma \to WW$ and to $\gamma\gamma \to ZZ$.

There is an additional operator which gives a $W^+W^-Z\gamma$ vertex (9):

$$\mathcal{L}^n = \frac{i\pi\alpha}{4\Lambda^2} a_n \varepsilon_{ijk} W_{\mu\alpha}^i W_\nu^j W^{k\alpha} F^{\mu\nu} \tag{5}$$

The parameter Λ is an unknown "new physics" scale which is often taken to be M_W. This is a little misleading as Λ represents the scale of new physics which one might expect to be $\mathcal{O}(1)$ TeV. One should keep this in mind when gauging the sensitivity of various experiments to the parameters a_i. To facilitate the comparison of different processes I have taken $\Lambda = 1$ TeV, rescaling results where necessary.

$\mathcal{L}_0$ and $\mathcal{L}_c$ affect the value of Δr and therefore contribute to the S and T parameters (10) leading to the rather weak one-sigma constraints (9):

$$\begin{aligned} -700 &< a_0 < 100 \\ -1700 &< a_c < 900. \end{aligned} \tag{6}$$

There are no similar low energy constraints on a_n.

B. Non-Linear Realization

Another widely used effective Lagrangian is the Chiral Lagrangian. It assumes a heavy Higgs boson using a non-linear realization of the Goldstone bosons and assumes a custodial $SU(2)$ (5,11,12).

$$\mathcal{L}_1 = \frac{L_1}{16\pi^2}[Tr(D^\mu \Sigma^\dagger D_\mu \Sigma)]^2 \tag{7}$$

$$\mathcal{L}_2 = \frac{L_2}{16\pi^2}[Tr(D^\mu \Sigma^\dagger D_\nu \Sigma)]^2 \tag{8}$$

where $\Sigma = \exp(iw^i \tau^i/v)$, $v = 246$ GeV, and $D_\mu \Sigma = \partial_\mu \Sigma + \frac{1}{2}igW_\mu^i \tau^i \Sigma - \frac{1}{2}ig' B_\mu \Sigma \tau^3$. In this approach $\mathcal{L}_{1,2}$ would be the most important manifestation of alternative symmetry breaking scenarios in a Higgsless world.

The two approaches are not distinct so that $\mathcal{L}_{1,2}$ is equivalent to $\mathcal{L}_4^{o,c}$ with the mapping:

$$g_{o,c} = \frac{e^2}{16\pi^2} \frac{1}{s_w^2} L_{1,2} \tag{9}$$

Typical models with Goldstone bosons interacting with a scalar, isoscalar resonance like the Higgs boson give $L_i \sim \mathcal{O}(1)$ (13). From precision measurements of the Z^0 widths Dawson and Valencia obtained the weak bounds $-28 \le L_1 + \frac{3}{2}L_2 \le 26$ (14). Imposing perturbative unitarity gives the rough constraints of $|L_1| \le 0.3$ (15). Therefore, the genuine quartic couplings are presently not well constrained by experiment but are limited by perturbative unitarity. To facilate comparison of different processes I have presented results in terms of $L_{1,2}$, rescaling results where necessary (using $\alpha = 1/128$ and $\sin^2_w = 0.23$). I have defined L_s when $L_1 = -L_2$.

Similarly, one can write down operators involving two photons in the Chiral Lagrangian (6):

$$\mathcal{L}_o^{2\gamma} = -\frac{L_o^{2\gamma}}{\Lambda^2} \{ K_o^W g^2 Tr(W_{\mu\nu}W^{\mu\nu}) + K_o^B g'^2 Tr(B_{\mu\nu}B^{\mu\nu}) \tag{10}$$

$$+ K_o^{WB} gg' Tr(W_{\mu\nu}B^{\mu\nu}) \} Tr(D^\alpha \Sigma^\dagger D_\alpha \Sigma) \tag{11}$$

$$\mathcal{L}_c^{2\gamma} = -\frac{L_c^{2\gamma}}{\Lambda^2} \{ K_c^W g^2 Tr(W_{\mu\alpha}W^{\mu\beta}) + K_c^B g'^2 Tr(B_{\mu\alpha}B^{\mu\beta}) \tag{12}$$

$$+ K_o^{WB} gg' Tr(W_{\mu\alpha}B^{\mu\beta}) \} Tr(D^\alpha \Sigma^\dagger D_\beta \Sigma) \tag{13}$$

where $W_{\mu\nu} = \frac{\tau^i}{2}(\partial_\mu W_\nu^i - \partial_\nu W_\mu^i - g\epsilon^{ijk}W_\mu^j W_\nu^k)$ and $B_{\mu\nu} = \frac{1}{2}(\partial_\mu B_\nu - \partial_\nu B_\mu)\tau_3$.

For $\gamma\gamma$ reactions, by making explicit the $U(1)_{QED}$ symmetry, gives the mapping:

$$a_{o,c} = \frac{4e^2}{s_w^2} L_{o,c}^{2\gamma} (K_{o,c}^W + K_{o,c}^B + K_{o,c}^{WB}) \tag{14}$$

III. MEASUREMENT OF QUARTIC COUPLINGS

In this section I survey the various processes that have been proposed to measure quartic couplings.

A. Measurement of the Dimension 4 Operators

1. The Processes $e^+e^- \to W^+W^-Z, ZZZ$

At an 500 GeV e^+e^- collider the W fusion process will be ineffective so that three gauge boson production may be a reasonable substitute for the

TABLE 1. Event rates for various VVV final states in the reaction $e^+e^- \to VVV$ for $\sqrt{s} = 500$ GeV and L=10 fb^{-1}. From Bélanger and Boudjema Ref. (7).

Final State	Events	Comments
WWZ	400	$M_H < 2M_W$ or $M_H > 1$ TeV
	460	$M_H = 200$ GeV
ZZZ	9	$M_H > 1$ TeV
$WW\gamma$	1356	$\theta_{\gamma beam} > 15°$
$ZZ\gamma$	147	$p_{T\gamma} > 20$ GeV
$Z\gamma\gamma$	465	2γ's separated by $15°$

measurement of quartic couplings. In the process $e^+e^- \to VVV$ four W quartic couplings don't contribute so that vertices with at least two neutral vector bosons where one of the neutrals couples to the e^+e^- vertex are likely to be the best tested in e^+e^- collisions. For any model with $SU(2)$, however, $WWWW$ vertices are related to $WWZZ$. $SU(2)$ also predicts a $4Z$ vertex which will contribute to $e^+e^- \to ZZZ$. The event rates for reactions that meet this criteria are shown in Table 1 for $\sqrt{s} = 500$ GeV and assuming an integrated luminosity of L=10 fb^{-1}. The approach used is to look for deviations in the cross sections from their standard model values (7,16).

The process $e^+e^- \to W^+W^-Z$ involves TGV's, QC's and Higgs exchange. In the standard model there is a subtle cancellation between the various contributions. Only anomalous QC's were considered under the assumption that TGV's can be measured better elsewhere and assume the large M_H limit so that Higgs exchange can be neglected. The standard model cross section is 39.88 fb. In their analysis Bélanger and Boudjema included the 67% BR corresponding to the 6 jet and 4 $jet + e^\pm$ or $\mu^\pm$ final states, not including τ's. The signal can be enhanced by using right handed electrons. The cross section is shown in Fig. 1 as a function of L_1. The most dramatic effects are for longitudinal W's with virtually no sensitivity in the TTU mode. The limits obtained on the couplings are based on a 3σ deviation in the total unpolarized cross-section including the 67% BR defined above and only taking statistical errors into account. One obtains the sensitivities (7):

$$-96 < L_1 < 81$$
$$-120 < L_2 < 120$$
$$-81 < L_s < 70 \qquad (15)$$

One could use distributions to distinguish between g_o and g_c. It turns out the the E_Z distribution is especially good at this.

For the process $e^+e^- \to ZZZ$ the only SM contribution is via the Higgs boson so that the SM cross section is very small, $\simeq 1$ fb, making it very sensitive to anomalous couplings. Here Bélanger and Boudjema consider 6 jet and $4jet+ \not{E}$ (not including τ final states) corresponding to 87% of events (7). To use these modes one will need good invariant mass reconstruction to

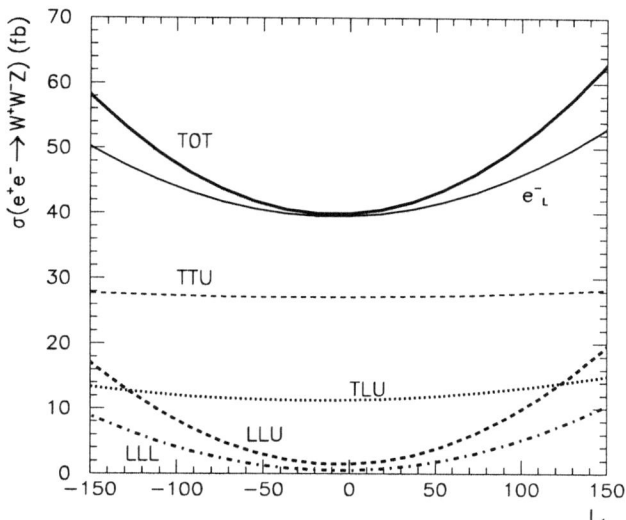

FIG. 1. Cross-section for $e^+e^- \to W^+W^-Z$ as a function of L_1 for $\sqrt{s} = 500$ GeV. Shown are the total unpolarized (TOT) cross-section, with left-handed electrons e_L^- and unpolarized cross-sections for various combinations of vector boson polarizations: T for transverse, L for longitudinal and U for unpolarized. The third label is for the Z polarization. From Bélanger and Boudjema, Ref. (7).

distinguish from the WWZ final states. The largest deviations are seen in the LLU channels. Because the event rate is so small they impose the need for 50 ZZZ events which gives the bounds:

$$-78 < L_1, L_2 < 85. \tag{16}$$

Using a less conservative, naive, 4σ deviation from the SM corresponding to 12 events gives

$$-44 < L_1, L_2 < 48. \tag{17}$$

If deviations were observed, comparing the deviations in the ZZZ mode to those found in the WWZ mode could be used to find the nature of the QC's.

2. The Processes $e^-e^- \to VV'ff'$

$e^-e^- \to VV'ff'$ is another option that has been examined (17–19). It has the advantage of no hadronic background and a low SM cross section due to the cancellation of diagrams. The reactions considered are:

$$\begin{aligned}
e^-e^- &\to e^-e^- Z^0 Z^0 \\
&\to e^- \nu_e Z^0 W^- \\
&\to \nu_e \nu_e W^- W^-
\end{aligned} \tag{18}$$

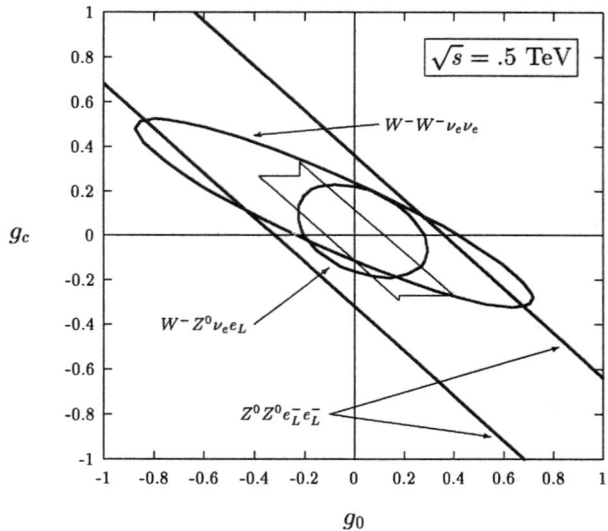

FIG. 2. Contours of observability at 95% C.L. of anomalous QC's g_o and g_c. The measurements are for $\sqrt{s}$ = 500 GeV and L=10 fb^{-1}. The limits which can be obtained under similar conditions in the e^+e^- mode of the same collider are indicated by the thin line. From Cuypers and Kolodziej Ref. (17).

Note that in the last reaction only the combination $g_o + g_c$ ($L_1 + L_2$) can be probed. Cuypers and Kolodziej (17) performed an analysis assuming an integrated luminosity of 10 fb^{-1} and included a 1% systematic error. They used a 10° cut on the primary electrons and included reconstruction efficiencies. They obtained the 95% C.L. contours shown if Fig. 2.

3. W Fusion in pp Collisions

Although WW scattering (5,13,21) is covered by other contributions to these proceedings (22), it is a sufficiently important topic that a few brief comments are included for completeness. If the Goldstone bosons are non-linearly realized then one would expect new strong interactions at ~ 1 TeV responsible for EWSB. This might manifest itself as:

- Longitudinal W states in, for example, technicolour.

- Strong WW interactions in, for example, composite scalars.

Although $W_L W_L$ can be studied in both e^+e^- and pp colliders, because adequate W_L luminosity requires the highest energy possible it is best studied at the higher energy hadron colliders. Isoscalar resonances could be studied in W^+W^- and ZZ scattering, isovector resonances in WZ scattering and non-resonant effects in W^+W^+. The best channel to look for the effect of genuine quartic couplings is the like-sign W pair production, $W^{\pm}W^{\pm}$. Bagger et al. (5) find that $pp \to W_L^+ W_L^+$ scattering at the LHC would be sensitive to $|L_1, L_2| > 1$.

4. $W_L W_L$ Rescattering in e^+e^- and $\gamma\gamma$

In $e^+e^- \to WW$ quartic couplings are studied via the effects of final state interactions (20,21). The rescattering can take place via scalar ($[I,J] = [0,0]$) Higgs like) or vector ($[I,J] = [1,1]$ ρ like) exchange. The W_L's can be related to π's via low energy theorems and chiral perturbation theory. Resonance effects for a ρ like resonance are noticible at a $\sqrt{s} = 500$ GeV collider up to 5 TeV (20). Resonances in the $I = 2$ channel could be studied in e^-e^-.

It may also be possible to study WW rescattering at TeV energies in $\gamma\gamma$ colliders (23,24). Berger and Chanowitz (24) have examined rescattering effects in $\gamma\gamma \to ZZ$ in analogy to $\gamma\gamma \to \pi^0\pi^0$. They concluded that the background overwhelms the signal unless there are strong resonance effects from, for example, an f_{2TC} with mass ~ 3.4 TeV ($N_{TC} = 3$). A very high energy collider of $\sqrt{s_{\gamma\gamma}} = 3.2$ TeV ($\sqrt{s_{e^+e^-}} = 4$ TeV) with high luminosity, of order 100 fb^{-1}, would be needed to see its effects.

B. Measurement of Dimension 6 operators

1. The Processes $e^+e^- \to W^+W^-\gamma$, $ZZ\gamma$, $Z\gamma\gamma$

The process $e^+e^- \to W^+W^-\gamma$ is used to study the $W^+W^-\gamma\gamma$ and $W^+W^-Z\gamma$ couplings. It has the largest cross section of all 3-boson production and is quite sensitive to dimension 6 operators. The largest deviations occur when both W's are longitudinal. Bélanger and Boudjema (7) impose the cuts $P_{T\gamma} > 20$ GeV, $\theta_{e\gamma} > 15°$, and $|\eta_\gamma| < 2$ resulting in a cross section of $\sigma_{WW\gamma} = 135.6$ fb. They used the 79% of the BR that does not include τ's with 45% being $4\,jet + \gamma$. The anomalous QC's contribute significantly to cross-sections with right handed electrons. They obtain the 3σ limits:

$$-62 < a_o < 93$$
$$-110 < a_c < 47 \qquad (19)$$

The different operators give different distributions for E_γ but not for $\theta_{\gamma W}$. $e^+e^- \to ZZ\gamma$ has a SM cross-section of 14.7 fb with the same cuts as above. It turns out that constraints obtained from $e^+e^- \to ZZ\gamma$ are less constraining than the $WW\gamma$ final state and the $e^+e^- \to Z\gamma\gamma$ cross section is very insensitive to anomalous couplings.

Leil and Stirling (25) used the process $e^+e^- \to W^+W^-\gamma$ to study a_n. They imposed the cuts $|\eta_\gamma| \leq 2$, $E_\gamma > 20\%$ to avoid collinear singularities and particle separation of $15°$, obtaining a cross section for $\sqrt{s} = 500$ GeV of $\sigma_{SM} = 123.4$ fb. Using the E_γ spectrum they obtain the additional limit based on L=10 fb^{-1} and requiring 3σ deviations of

$$-610 < a_n < 660. \tag{20}$$

2. The Processes $\gamma\gamma \to W^+W^-$ and $\gamma\gamma \to ZZ$

These reactions are in the pure non-abelian gauge sector of the SM. Both the trilinear and quartic couplings enter. Since the TGV's can be constrained better elsewhere these reactions are ideal tests of the quartic couplings. The $WW\gamma\gamma$ and $ZZ\gamma\gamma$ couplings are related by $SU(2)$ but because they contribute to different observables we can set independent bounds on them.

The process $\gamma\gamma \to W^+W^-$ constitutes the largest cross-section in $\gamma\gamma$ collisions, with a cross-section at 400 GeV of $\sigma = 80$ pb making a $\gamma\gamma$ collider a W-Factory. The angular distributions are shown in Fig. 3. The SM contributions are peaked along the initial photon directions while the anomalous QC's are more central. Even with angular cuts the SM contributions are still large. The photon helicities can be used to separate different contributions. In the $\lambda_1 = \lambda_2$ mode ($J = 0$) the SM does not produce W's of different helicities. This is maintained for a_o so that a_o only contributes to $J = 0$ while a_c contributes to both $J = 0$ and $J = 2$. Because a_o and a_c have the same S-wave amplitudes distinguishing them requires the use of the photon helicity amplitudes. Taking $\cos\theta < 0.7$ the SM cross-section is 17.58 pb so that statistical errors are negligible and the main source of error is systematics. For L=10 fb^{-1} and assuming $\Delta\sigma/\sigma = 3\%$ Bélanger and Boudjema (8) obtain:

$$-7.8 < a_o < 3.1 \quad J_Z = 0$$
$$-16 < a_c < 0.56 \quad J_Z = 0$$
$$-3.1 < a_c < 3.1 \quad J_Z = 2 \tag{21}$$

Ratios could also have been used which eliminates the need to measure the $\gamma\gamma$ luminosity. Using angular distributions could give additional information.

The process $\gamma\gamma \to ZZ$ is attractive as the SM background is very small. $SU(2)$ relates the $ZZ\gamma\gamma$ vertex to the $WW\gamma\gamma$ vertex so combining this and the previous reaction is an ideal way of testing for $SU(2)$ symmetric QC's. As before, a_o contributes to $J_z = 0$ while a_c contributes to both the $J_z = 2$ and $J_z = 0$ channels making it possible to distinguish the 2 quartic couplings. The $J_Z = 0$ and $J_Z = 2$ channels can be distinguished using polarization and angular distribution information.

Unfortunately, in the original analysis of these process it was assumed that the SM cross section was zero. A subsequent 1-loop calculation by Jikia (26) found that, although small, the SM cross section was not neglible and was dominated by the transverse modes. Nevertheless it is believed that properly including the SM contribution will still result in useful bounds, in much the

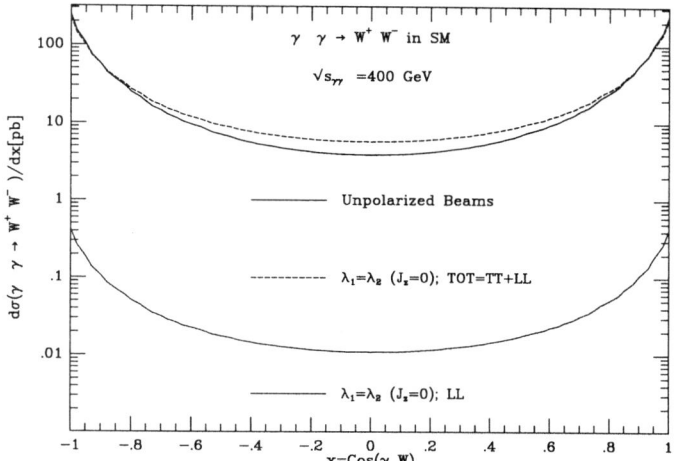

FIG. 3. W angular distribution in the process $\gamma\gamma \to W^+W^-$ at $\sqrt{s} = 400$ GeV for different initial photon helicities. From Bélanger and Boudjema, Ref. (8).

same way as the $\gamma\gamma \to W^+W^-$ case (6). In the absence of a detailed analysis we describe the estimate of Baillargeon et al. (6). The SM $Z_L Z_L$ contribution in the heavy Higgs mass limit is quite small at all energies, ~ 1 fb. Therefore to obtain a crude estimate as to how the limits are changed it is sufficient to include the SM $Z_T Z_T$ contribution that is not affected by anomalous QC's. The limits are based on the total cross section only. One could exploit the fact that the TT cross section is relatively insensitive to the J_Z of the initial two photons to construct an asymmetry such as $\sigma(J_Z = 0) - \sigma(J_Z = 2)$ to reduce the SM background. This, of course assumes that the new physics does not contribute equally to the two J_Z. Baillargeon et al. include only the visible, unambiguous ZZ signal with one Z decaying hadronically and the other leptonically with the cut $\cos\theta_Z < 0.866$. The criteria of observability was based on requiring 3σ statistical deviation from the SM cross-section.

$$\begin{array}{lll} |a_o| < 2 & |a_c| < 5 & (\sqrt{s_{ee}} = 500 \text{ GeV L=10 fb}^{-1}) \\ |a_o| < 0.3 & |a_c| < 0.7 & (\sqrt{s_{ee}} = 1 \text{ TeV L=60 fb}^{-1}) \end{array} \qquad (22)$$

3. The Processes $\gamma\gamma \to W^+W^-Z$ and $\gamma\gamma \to W^+W^-\gamma$

Éboli et al., (27) have studied the the processes $\gamma\gamma \to W^+W^-Z$ and $\gamma\gamma \to W^+W^-\gamma$. They found that the constraints from the first reaction on a_n is as

TABLE 2. Sensitivities of a_o, a_c and a_n to $e\gamma \to VV'f$ corresponding to 3σ deviations, varying one coupling at a time. For events containing a photon in the final state the cut $p_{T\gamma} > 15$ GeV was used to eliminate collinear divergences. From Éboli et al. Ref. (9).

Final State	a_o	a_c	a_n
WWe	$-33 < a_o < 5.6$	$-230 < a_c < 220$	$-700 < a_n < 700$
$Z\gamma e$	$-150 < a_o < 150$	$-200 < a_c < 220$	—
ZZe	$-4.5 < a_o < 4.4$	$-15 < a_c < 15$	—
$W\gamma\nu$	$-87 < a_o < 84$	$-89 < a_c < 170$	—
$WZ\nu$	—	—	$-190 < a_n < 120$

restrictive as the contraint obtained in $e\gamma$ collisions. The limits on a_c are an order of magnitude better than those coming from the e^+e^- mode and are comparable to the limits that can be obtained in the $e\gamma$ mode. However, they are a factor of 2 weaker that those obtained from $\gamma\gamma \to W^+W^-$. The limits on a_o are slightly better than those obtained in the e^+e^- mode but an order of magnitude worse than those obtained in $e\gamma$ or $\gamma\gamma \to W^+W^-$.

4. The Processes $e\gamma \to VV'f$

A number of authors have studied the effects of anomalous QC's on the reactions $e\gamma \to VV'f$ (9,15,28). The cross sections are summarized in Fig. 4 which gives the cross sections in $e\gamma$ collisions as a function of $\sqrt{s}$ (29). The WWe and ZZe final states are most sensitive to a_o and a_c although WWe is insensitive to a_n. The cross section $\sigma(WWe)$ is an order of magnitude larger than $\sigma(WW\gamma)$ due to t-channel photon exchange. Likewise, for ZZe t-channel photon exchange is introduced by the $ZZ\gamma\gamma$ vertex, not present in the SM, making it a very sensitive process. The results from an analysis of Éboli et al. (9) for the sensitivies of $e\gamma \to VV'f$ to the various QC's, are summarized in Table 2. Their results are based on 3σ effects based on statistics for 10 fb^{-1} integrated luminosity. The conclusion of these studies is that they are not quite as good as those coming from $\gamma\gamma$ reactions for the study of QC's except for a_n which requires the smaller phase space process $\gamma\gamma \to W^+W^-Z^0$ in $\gamma\gamma$ collisions.

IV. SUMMARY

One of the most important problems in particle physics is the understanding of electroweak symmetry breaking. If the Higgs boson is "heavy" and electroweak symmetry breaking is non-linearly realized then the quartic vertices will provide a window into EWSB. I have surveyed various processes that have been proposed to study quartic couplings. For dimension 4 operators $e^+e^- \to W^+W^-Z$, $\to ZZZ$, $e^-e^- \to VV'ff'$, $pp \to WW + X$,

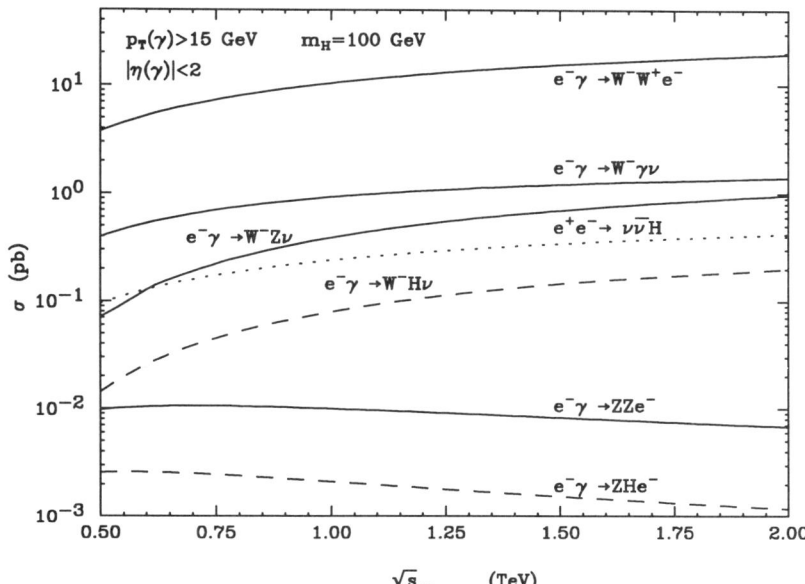

FIG. 4. Cross sections for $e\gamma$ processes as a function of $\sqrt{s}$ with the acceptance cuts $p_T(\gamma) > 15$ GeV and $|\eta(\gamma)| < 2$. From K. Cheung Ref. (29).

and WW rescattering in $e^+e^- \to W^+W^-$ have been considered and for dimension 6 operators $e^+e^- \to W^+W^-\gamma$, $ZZ\gamma$, $\gamma\gamma \to W^+W^-$, ZZ, and $e\gamma \to WW\gamma$, $ZZ\gamma$, $WZ\nu$.

For the dimension 4 operators it appears that the LHC will provide the most constraining measurements. However, is far from clear whether the LHC will be able to disentangle this sector of the weak interaction. It is possible, then, that precision measurements at high energy e^+e^- colliders through W^+W^- rescattering could be our first glimpse of a strongly interacting weak interaction.

For the dimension 6 quartic couplings involving photons, $\gamma\gamma$ collisions appear to be the best place to measure these couplings. They can be measured at least an order of magnitude more precisely than using 3-boson production in e^+e^-. Measurements using gauge boson production in $e\gamma$ collisions are almost as precise as the $\gamma\gamma$ processes.

The study of quartic couplings are still in the preliminary stages. It would be useful for the purposes of comparing different processes that a consistent parametrization of the vertices be adopted and that the different processes be analysed in a consistent way. The most dramatic effect of QC's is when all vector bosons are longitudinal. Therefore, an important next step is to include the decays of the W's and Z's into fermions and their reconstruction. After all, it is the fermions which are observed not the gauge bosons themselves. This would enable more sophisticated polarization studies that would simulate the experimental separation of W's and Z's and the separation of longitudinal and transverse gauge bosons.

ACKNOWLEDGMENTS

The author is most grateful to Genevieve Bélanger for many helpful communications in preparing this review and a careful reading of the manuscript. The author thanks Genevieve Bélanger, Fawzi Boudjema, Kingman Cheung, and Frank Cuypers for graciously supplying him with the figures included here. The author thanks the organizers of TGV95 for their kind invitation to attend a most enjoyable meeting and the Deans of Research and Science at Carleton University for financial support to attend the meeting. This research was supported in part by the Natural Sciences and Engineering Research Council of Canada.

REFERENCES

1. A comprehensive review on trilinear gauge boson couplings is given by H. Aihara et al., To appear in *Electroweak Symmetry Breaking and Beyond the Standard Model*, eds. T. Barklow, S. Dawson, H. Haber and J. Siegrist (World Scientific).
2. R. Casalbuoni, S. de Curtis, D. Dominici, and R. Gatto, Nucl. Phys. **B282**, 235 (1987); G. Cvetič and R. Kögerler, Nucl. Phys. **B363**, 401 (1991).
3. A. Hill and J.J. van der Bij, Phys. Rev. **D36**, 3463 (1987).
4. For a recent review of effective Lagrangians see: F. Boudjema, Proceedings of the *Workshop on Physics and Experiments with Linear e^+e^- Colliders*,eds. F.A. Harris et al., Waikoloa Hawaii, April 26-30, 1993 (World Scientific, 1994) p.712.
5. J. Bagger, S. Dawson, and G. Valencia, Nucl. Phys. **B399**, 364 (1993).
6. M. Baillargeon, G. Bélanger, and F. Boudjema, *Proceedings of Two-Photon Physics from Daphne to LEP200 and Beyond*, eds. F. Kapusta and J. Parisi (World Scientific, 1994) p.267, (hep-ph/9405359).
7. G. Bélanger and F. Boudjema, Phys. Lett. **B288**, 201 (1992).
8. G. Bélanger and F. Boudjema, Phys. Lett. **B288**, 210 (1992).
9. O.J. P.Éboli, M.C. Gonzalez-Garcia and S.F. Novaes, Nucl. Phys. **B411**, 381 (1994).
10. M.E. Peskin and T. Takeuchi, Phys. Rev. **D46**, 381 (1992).
11. T. Appelquist and C. Bernard, Phys. Rev. **D22**, 200 (1980).
12. A. Longhitano, Nucl. Phys. **B188**, 118 (1981).
13. S. Dawson, *Proceedings of the Beyond the Standard Model III Workshop*, eds S. Godfrey and P. Kalyniak (World Scientific, 1993)p. 188.
14. S. Dawson and G. Valencia, Nucl. Phys. **B439**, 3 (1995).
15. K. Cheung, S. Dawson, T. Han, and G. Valencia, Phys. Rev. **D51**, 5 (1995).
16. C. Grosse-Knetter and D. Schildknecht, Phys. Lett. **B302**, 309 (1993).
17. F. Cuypers and K. Kolodziej, Phys. Lett. **B344**, 365 (1995).
18. C. Bilchak, M. Kuroda and D. Schildknecht, Nucl. Phys. **B299**, 7 (1988).
19. M. Kuroda, F.M. Renard and D. Schildnecht, Z. Phys. **C40**, 575 (1988).

20. T.L. Barklow, these proceedings; T.L. Barklow, in *Physics and Experiments with Linear Colliders - Volume I*, ed. R. Orava *et al.*, (World Scientific, 1992) p. 423.
21. For a recent review see T. Han, Proceedings of the *Workshop on Physics and Experiments with Linear e^+e^- Colliders*,eds. F.A. Harris *et al.*, Waikoloa Hawaii, April 26-30, 1993 (World Scientific, 1994) p.270.
22. R.S. Chivukula, these proceedings. See also R.S. Chivukula *et al.*, To appear in *Electroweak Symmetry Breaking and Beyond the Standard Model*, eds. T. Barklow, S. Dawson, H. Haber and J. Siegrist (World Scientific).
23. K. Cheung, Phys. Lett. **B323**, 85 (1994).
24. M.S. Berger and M.S. Chanowitz, Proceedings of the *Workshop on Gamma-gamma Colliders*, March 28-31 1994, Berkelely California, (hep-ph/9406413); See also J.F. Donoghue and T. Torma, Nucl. Phys. **B424**, 399 (1994).
25. G. Abu Leil and W.J. Stirling, J. Phys. G **21**, 517 (1995), (hep-ph/9406317).
26. G.V. Jikia, Phys. Lett. **B298**, 224 (1993); Nucl. Phys. **B405**, 24 (1993).
27. O.J. P.Éboli, M.B. Magro, P.G. Mercadante, S.F. Novaes, hep-ph/9503432.
28. R. Rosenfeld, Northeastern University report NUB-3086/94-Th, hep-ph/9403356.
29. K. Cheung, Nucl. Phys. **B403**, 572 (1993).

Exact and Approximate Radiation Amplitude Zeros — Phenomenological Aspects

Tao Han

Davis Institute for High Energy Physics
Department of Physics, University of California, Davis 95616

We review the phenomenological aspects of the exact and approximate Radiation Amplitude Zeros (RAZ) and discuss the prospects of searches for these zeros at current and future collider experiments.

I. EXACT AND APPROXIMATE RADIATION AMPLITUDE ZEROS

More than 15 years ago, the pioneer studies on vector-boson pair production (1–3) revealed a surprise: the angular distribution for $f_1 \bar{f}_2 \to W^- \gamma$ develops a pronounced zero (3,2) at

$$\cos\theta = (Q_{f_1} + Q_{f_2})/(Q_{f_1} - Q_{f_2}), \tag{1}$$

where θ is the W^- scattering angle with respect to the incident fermion (f_1) direction in the center of mass (c.m.) frame, and Q_{f_i} the electric charge of fermion f_i. Figure 1 demonstrates this unusual angular distribution for $e^- \nu, d\bar{u} \to W^- \gamma$ processes, in which the zero occurs at $\cos\theta = 1, -1/3$, respectively. The authors of Ref. (3) stated in the abstract that "... We can offer no explanation for this behavior".

In fact, it is not difficult to see what is happening for some simple cases. Take $d\bar{u} \to W^- \gamma$ as an example. There are three Feynman diagrams to contribute at the Born level: a t-channel diagram with an amplitude proportional to Q_u/t, a u-channel diagram proportional to Q_d/u, and an s-channel diagram proportional to $Q_{W^-}/(s - M_W^2)$, where $t = (p_d - p_W)^2 = -\frac{1}{2}(s - M_W^2)(1 - \cos\theta)$. Notice the charge relation in the Standard Model $Q_d - Q_u = Q_{W^-}$, and the kinematic relation $s - M_W^2 = -t - u$, one can easily cast the amplitude into the form

$$\mathcal{M} \sim \left(\frac{Q_u}{t} + \frac{Q_d}{u}\right) F(\sigma_i, \lambda_i, p_i), \tag{2}$$

where $F(\sigma_i, \lambda_i, p_i)$ denotes a reduced matrix element as a function of the fermion helicity σ_i, vector-boson polarization λ_i and the external momenta p_i. We see immediately that this amplitude develops a zero at a special angle determined by Eq. 1.

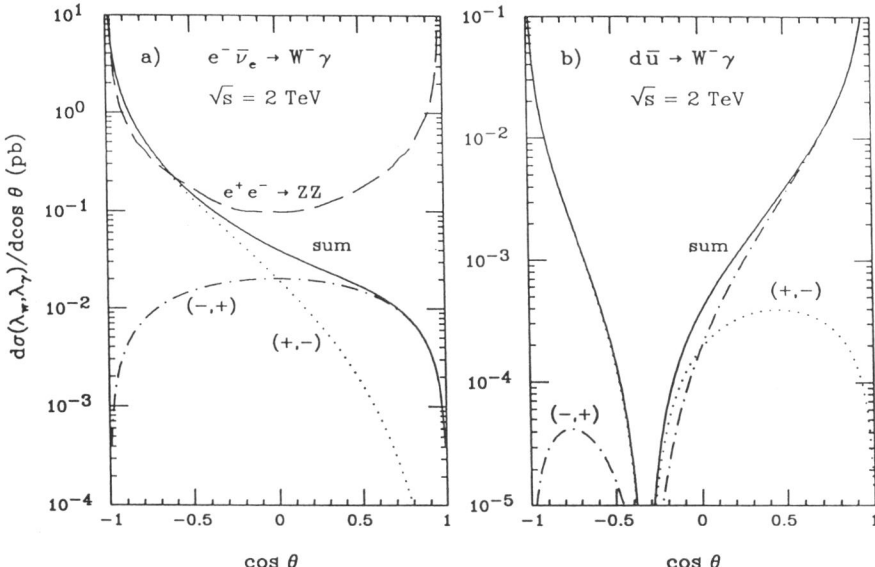

FIG. 1. Differential cross sections $d\sigma(\lambda_w, \lambda_\gamma)/d\cos\theta$ for a). $e^-\bar{\nu}_e \to W^-\gamma$ and b). $d\bar{u} \to W^-\gamma$, where θ is the polar angle between W^- and the incident fermion (e^- or d) in the c.m. frame. For comparison, the differential cross section for $e^+e^- \to ZZ$, in which there is no RAZ, has been included in a).

Not long after this discovery, several groups (4–6) further examined this interesting feature. It was found that in gauge theories, any tree-level 4-particle (spin ≤ 1) Feynman amplitudes with one or more massless gauge particles can be factorized into two factors, one of which contains the dependence of internal quantum numbers (such as charges) and the other contains the dependence of spin and polarization indices (4). This factorization is a special case for a more general theorem (5), which states that *for a tree-level n-particle (spin ≤ 1) amplitude with one photon, the amplitude develops a zero when the factor $Q_i/p_i \cdot q$ are equal for $i = 1, 2...n-1$, where Q_i and p_i are the charge and momentum for the ith particle, respectively. and q the photon momentum.* There is certainly a deeper explanation for this phenomena, having something to do with the relationship between the internal gauge symmetry and the space-time symmetry. One can find a very nice discussion in Bob Brown's talk at this conference (7), or from the classical papers on this subject (5). Following the literature, we will call those zeros Radiation Amplitude Zeros (RAZ) (8).

It should be noted, however, that

- not all of the RAZ occur in physical region — in fact, most of them do not. The above theorem can be translated into an intuitive necessary condition for RAZ to occur in physical region: *along with a massless gauge boson, the other particles involved in the process must have the same sign of electric charges.* We will call this condition the "same-sign rule".

- although loop diagrams (and bubbles) do not significantly alter the nature of RAZ (9), higher order real emissions spoil the RAZ (10-12). It was suggested (11) that one can regain the Born-level kinematics by vetoing additional final state particles, thus recovering an "approximate" zero in practice.

It is natural to ask what may happen in a theory with a spontaneously broken gauge symmetry, such as the Standard Model (SM). It is conceivable that the radiation of a Z-boson may have some similarity to that of a photon. For the case of $d\bar{u} \to W^- Z$, the amplitude can be written as (13)

$$\mathcal{M} \sim X F_X(\sigma_i, \lambda_i, p_i) + Y F_Y(\sigma_i, \lambda_i, p_i), \qquad (3)$$

where X and Y are combinations of coupling factors

$$X = \frac{s}{2}\left(\frac{g_-^{f_1}}{u} + \frac{g_-^{f_2}}{t}\right), \quad Y = g_-^{f_1} \frac{M_Z^2 s}{2u(s - M_W^2)}, \qquad (4)$$

with the left-handed neutral current couplings $g_-^{f_1} - g_-^{f_2} = Q_W \cot\theta_w$, and $F_{X,Y}(\sigma_i, \lambda_i, p_i)$ contain the spin dependent part and is roughly proportional to the product of the vector-boson wave functions $\epsilon_w^* \cdot \epsilon_z^*$. It is obvious that without the Y-term, the helicity amplitudes would factorize. In this case, all amplitudes would simultaneously vanish for $g_-^{f_1}/u + g_-^{f_2}/t = 0$, analogous to the $W\gamma$ case in Eq. 2. Since Y is directly proportional to M_Z^2, one may naively expect full factorization when $M_Z^2 \ll s$. In fact, in the high energy limit, only three helicity amplitudes remain non-zero:

$$\mathcal{M}(\lambda_w = \pm, \lambda_z = \mp) \longrightarrow \frac{1}{\sin\theta}(\lambda_w - \cos\theta)\left[(g_-^{f_1} - g_-^{f_2})\cos\theta - (g_-^{f_1} + g_-^{f_2})\right],$$

$$\mathcal{M}(\lambda_w = 0, \lambda_z = 0) \longrightarrow \frac{1}{2}\sin\theta \frac{M_Z}{M_W}(g_-^{f_2} - g_-^{f_1}). \qquad (5)$$

While the dominant amplitudes $\mathcal{M}(\pm, \mp)$ fully factorize in the high energy limit, $\mathcal{M}(0,0)$ behaves differently. This can be traced to the special energy-dependence of the polarization vectors for longitudinal vector bosons, $\epsilon_v \sim \sqrt{s}/M_V$. Since the Y-term in Eq. 3 goes like $(M_Z^2/s)\epsilon_w^* \cdot \epsilon_z^*$, the $\mathcal{M}(0,0)$ amplitude remains finite at high energies.

The combined effect of the zero in $\mathcal{M}(\pm, \mp)$ and the relatively small contributions from the remaining helicity amplitudes results in an approximate zero for the $f_1 \bar{f}_2 \to W^\pm Z$ differential cross section at

$$\cos\theta \simeq (g_-^{f_1} + g_-^{f_2})/(g_-^{f_1} - g_-^{f_2}) \simeq \begin{cases} \frac{1}{3}\tan^2\theta_w \simeq 0.1 & \text{for } d\bar{u} \to W^- Z, \\ -\tan^2\theta_w \simeq -0.3 & \text{for } e^-\bar{\nu}_e \to W^- Z. \end{cases}$$

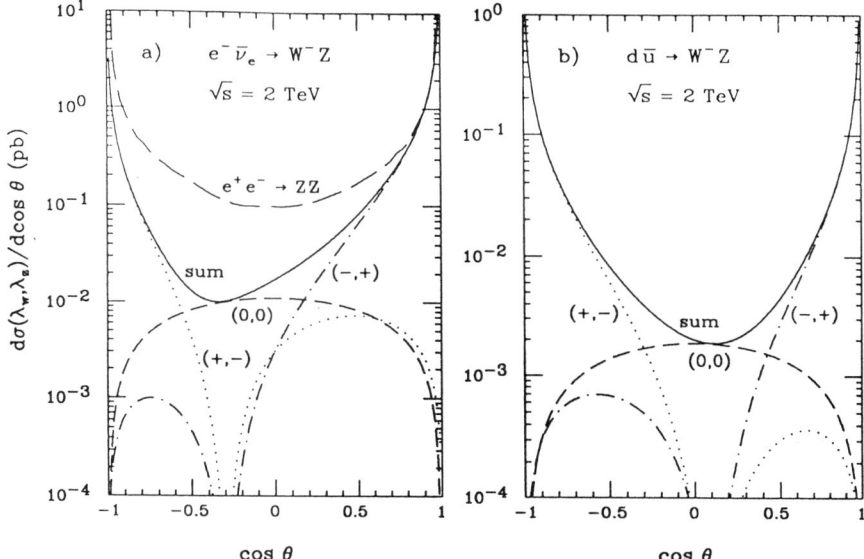

FIG. 2. Differential cross sections $d\sigma(\lambda_w, \lambda_z)/d\cos\theta$ for a). $e^-\bar{\nu}_e \to W^-Z$ and b). $d\bar{u} \to W^-Z$, where θ is the polar angle between W^- and the incident fermion (e^- or d) in the c.m. frame. For comparison, the differential cross section for $e^+e^- \to ZZ$, in which there is no RAZ, has been included in a).

This is illustrated in Fig. 2 where the differential cross sections are shown for $e^-\bar{\nu}_e \to W^-Z$ and $d\bar{u} \to W^-Z$ for $(\lambda_w, \lambda_z) = (\pm, \mp)$ and $(0,0)$, as well as the unpolarized cross section, which is obtained by summing over all W- and Z-boson helicity combinations (solid line). For both reactions, the total differential cross section displays a pronounced minimum at the location of the zero in $\mathcal{M}(\pm, \mp)$. Due to the $1/\sin\theta$ behaviour of $\mathcal{M}(\pm, \mp)$, the $(+, -)$ and $(-, +)$ amplitudes dominate outside of the region of the zero. In order to demonstrate the influence of the zero in $\mathcal{M}(\pm, \mp)$ on the total angular differential cross section, the $\cos\theta$ distribution for $e^+e^- \to ZZ$ has been included in Fig. 2a). The zero in the $(\pm, \mp)$ amplitudes causes the minimum in the WZ case to be much more pronounced than the minimum in $e^+e^- \to ZZ$.

It is important to note that the RAZ are the direct results from subtle gauge cancellation. Non-standard couplings, such as those Δg_1, $\Delta\kappa$ and λ (14) spoil these cancellations and eliminate the (approximate) zeros. This can be seen from the additional contributions to the SM amplitudes, for the $W\gamma$ process,

$$\Delta\mathcal{M}_{W\gamma}(\pm, \pm) = \frac{F}{2}\sin\theta\left[\Delta\kappa + \frac{\lambda}{r_w}\right], \quad (6)$$

$$\Delta\mathcal{M}_{W\gamma}(0, \pm) = \frac{F}{2}\frac{(1+\lambda_\gamma\cos\theta)}{\sqrt{2r_w}}\left[\Delta\kappa + \lambda\right], \quad (7)$$

where $F = V_{f_1 f_2} e^2 / \sqrt{2} \sin \theta_w$ and $r_v = M_V^2/s$; the corresponding contributions to the WZ production amplitudes are

$$\Delta \mathcal{M}_{wz}(\pm, \pm) = \frac{F}{2} \frac{Q_W \cot \theta_w}{1 - r_w} \beta \sin \theta \left[\Delta g_1 + \Delta \kappa + \frac{\lambda}{r_w} \right], \tag{8}$$

$$\Delta \mathcal{M}_{wz}(0, 0) = \frac{F}{2} \frac{Q_W \cot \theta_w}{1 - r_w} \frac{\beta \sin \theta}{\sqrt{2 r_w 2 r_z}} 2 \left[\Delta g_1 (1 + r_w) + \Delta \kappa\, r_z \right], \tag{9}$$

$$\Delta \mathcal{M}_{wz}(\pm, 0) = \frac{F}{2} \frac{Q_W \cot \theta_w}{1 - r_w} \frac{\beta(1 - \lambda_w \cos \theta)}{\sqrt{2 r_z}} \left[2 \Delta g_1 + \lambda \frac{r_z}{r_w} \right], \tag{10}$$

$$\Delta \mathcal{M}_{wz}(0, \pm) = \frac{F}{2} \frac{Q_W \cot \theta_w}{1 - r_w} \frac{\beta(1 + \lambda_z \cos \theta)}{\sqrt{2 r_w}} \left[\Delta g_1 + \Delta \kappa + \lambda \right], \tag{11}$$

where $\beta = [(1 - r_w - r_z)^2 - 4 r_w r_z]^{1/2}$. Due to angular momentum conservation, the $(\pm, \mp)$ amplitudes which dominate in the SM do not receive any contributions from the anomalous couplings. The amplitude zeros in these two helicity configurations for both $W\gamma$ and WZ channels thus remain exact. All other helicity amplitudes are modified in the presence of non-standard $WW\gamma/WWZ$ couplings. At high energies the anomalous contributions grow proportional to $\sqrt{s}$ (s) for $\Delta \kappa$ (Δg_1 and λ) and eventually dominate the cross section. The nature of the RAZ is thus sensitive to new physics in the vector-boson sector.

II. PROSPECTS OF EXPERIMENTAL SEARCHES FOR RAZ

Clearly, the radiation amplitude zeros (RAZ) are a very interesting feature of gauge theories and it would be desirable to experimentally observe this distinctive phenomena. However, we emphasize that studying these "zeros" is not to search for "nothing". Rather, we would hope to find new physics in the vector-boson sector and the amplitudes near RAZ are especially sensitive to the deviation from the SM. This is the motivation to examine the feasibility of experimental searches for RAZ.

A. $W\gamma$ Production at Hadron Colliders: $p\bar{p}, pp \to W^\pm \gamma \to l^\pm \nu \gamma$

The successful $p\bar{p}$ collider experiments at the Fermilab Tevatron may provide suitable environment for searching for RAZ and for testing the anomalous gauge boson couplings (15). However, it is non-trivial to carry out the searches for the RAZ experimentally. The problems, both theoretical and experimental, include:

1. *reconstruction of the $q\bar{q}$ c.m. frame:* it is impossible to non-ambiguously reconstruct the parton c.m. frame to define the scattering angle to

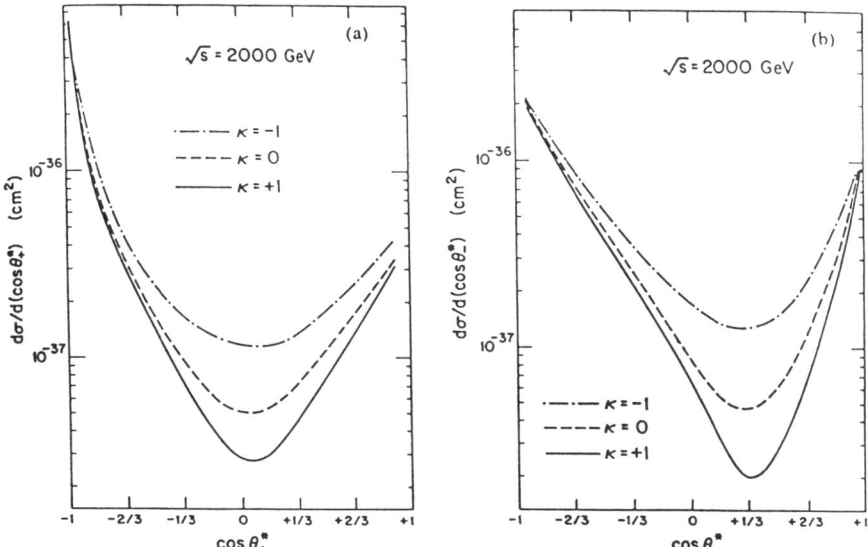

FIG. 3. Differential cross sections a). $d\sigma/d\cos\theta^*_+$ b). $d\sigma/d\cos\theta^*_-$ for $p\bar{p} \to W^-\gamma$, $W^- \to e^-\bar{\nu}_e$. Note that θ^* here is the polar angle between γ and p in the c.m. frame. Effects from an anomalous coupling κ are also shown, where $\kappa = 1$ corresponds to the SM results. Acceptance cuts are described in Ref. (17).

obtain $d\sigma/d\cos\theta$ since the reconstruction of the neutrino momentum (p_ν) from constraint $(p_l + p_\nu)^2 = M_W^2$ is subject to a two-fold ambiguity (16,17).

2. *z-axis along the incident fermion moving direction:* in hadron colliders, there are two types of parton-level contributions to the same final state: $d_1\bar{u}_2 \to W^-\gamma$ and $\bar{u}_1 d_2 \to W^-\gamma$. Since the polar angle θ is defined with respect to incident fermion moving direction $\vec{p}_d$, it is then impossible to non-ambiguously identify the direction of z-axis (along the d-quark). In $p\bar{p}$ collisions at Tevatron energies, due to the valence quark dominance, the contribution from $d_1\bar{u}_2$ is much larger than that from $\bar{u}_1 d_2$, so that one can simply assign the z-axis along the proton direction. However, in pp collisions, those contributions are equal, making the z-axis identification intrinsically impossible.

3. *higher order corrections:* the RAZ in $d\bar{u} \to W^-\gamma$ is exact only for the $2 \to 2$ Born-level process. Additional jets from higher order QCD radiation (10–12) will spoil the subtle cancellation and thus fill up the zero. One has to reject (or veto) the additional jets to recover the Born-level kinematics (11).

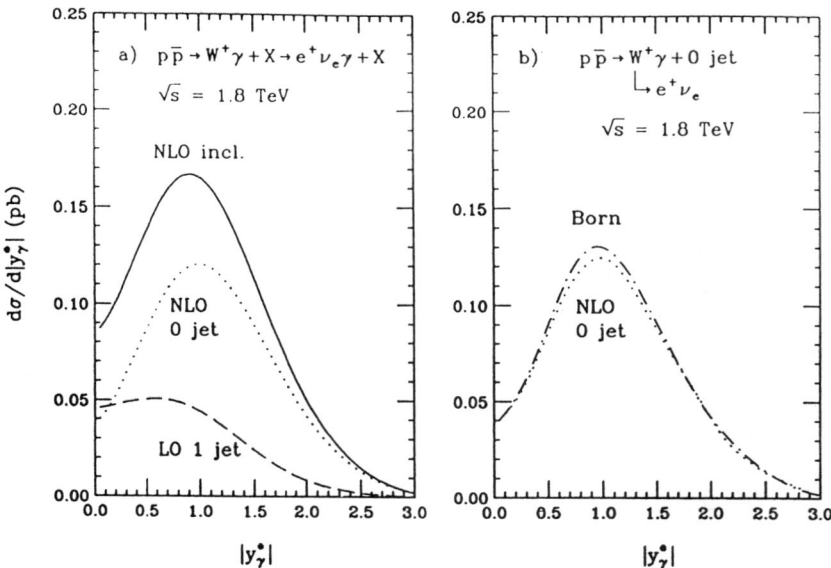

FIG. 4. The differential cross section for the photon rapidity in the reconstructed center of mass frame for the reaction $p\bar{p} \to W^+\gamma \to e^+\nu_e\gamma$ at $\sqrt{s} = 1.8$ TeV in the SM. a) The inclusive NLO differential cross section (solid line) is shown, together with the $\mathcal{O}(\alpha_s)$ 0-jet (dotted line) and the (LO) 1-jet (dashed line) exclusive differential cross sections. b) The NLO $W\gamma$+0 jet exclusive differential cross section (dotted line) is compared with the Born-level LO differential cross section (dot-dashed line). A jet is defined as $p_T^j > 10$ GeV and $|\eta^j| < 2.5$. Other cuts imposed are described in Ref. (11).

4. *W^- radiative decay:* for the channel $d\bar{u} \to W^-\gamma \to e^-\bar{\nu}_e\gamma$, a single W^- (Drell-Yan) production with subsequent radiative decay $d\bar{u} \to W^- \to e^-\bar{\nu}_e\gamma$ gives the same final state but different kinematical structure. Those events should be kept separated. This could be achieved by imposing a transverse mass cut (17–19) slightly above M_W, $M_T(l^\pm\nu,\gamma) > 90$ GeV.

5. *backgrounds:* the most severe background for the $W^-\gamma$ final state seems to be the misidentification of a photon from a jet $j \to \gamma$, due to the much larger production rate for W^-j. Good γ-j discrimination factor is needed to successfully identify the signal.

The first attempt to realistically study the RAZ at the Tevatron was carried out in Ref. (17). Due to the two-fold ambiguity in constructing the neutrino momentum, the authors studied two polar angle distributions $\cos\theta_+^*$ and $\cos\theta_-^*$, corresponding to the two solutions for $\cos\theta_\nu^* > \cos\theta_e^*$ and $\cos\theta_\nu^* < \cos\theta_e^*$, respectively. Although one is unable to tell the correct p_ν solution on an even-by-event basis, it is seen from Fig. 3 that $\cos\theta_-^*$ reflects

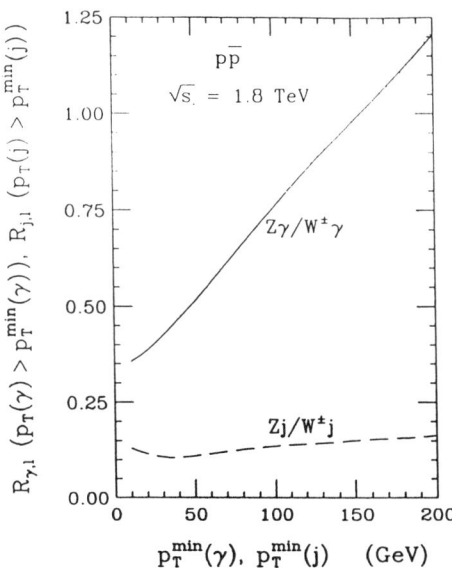

FIG. 5. The ratio of integrated cross sections as a function of the minimum transverse momentum of the photon, $p_T^{min}(\gamma)$, at the Tevatron. The dashed line shows the corresponding ratio of Zj to $W^{\pm}j$ cross sections for comparison.

the zero location better. This can be understood in terms of the $V - A$ coupling. Namely, e^- ($\bar{\nu}_e$) prefers to move in the forward (backward) direction so that $\cos\theta_\nu^* < \cos\theta_e^*$, which corresponds to the $\cos\theta_-^*$ solution. Fig. 3 also demonstrates the anomalous coupling effects that tend to fill up the dip.

The RAZ in Eq. 1 corresponds to the photon rapidity in c.m. frame

$$y_\gamma^* = \frac{1}{2}\ln\frac{1+\cos\theta_\gamma}{1-\cos\theta_\gamma} = \frac{1}{2}\ln(-\frac{Q_2}{Q_1}), \tag{12}$$

which gives $y_\gamma^* \simeq \pm 0.35$ for $W^{\mp}\gamma$ channel. As a direct reflection of the RAZ, the photon rapidity spectrum in the c.m. frame develops a clear dip in the central region after summing over the two solutions for p_ν (18,19). A problem arises when we include QCD radiative corrections (12). Although moderate at the Tevatron energies, the QCD corrections tend to fill up the dip and to increase the cross section in a similar way as the anomalous couplings (11). Figure 4 shows the differential cross section for the photon rapidity in the reconstructed center of mass frame for the reaction $p\bar{p} \to W^+\gamma \to e^+\nu_e\gamma$ at $\sqrt{s} = 1.8$ TeV in the SM. The inclusive next-to-leading-order (NLO) differential cross section [solid line in a)] is seen to be significantly larger than the Born-level leading-order (LO) approximation [dot-dashed line in b)] and tend to fill in the dip near zero. However, the NLO $W\gamma + 0$ jet exclusive differential cross section (dotted line) is comparable to the Born-level LO result. This important observation implies that if we study the 0-jet exclusive

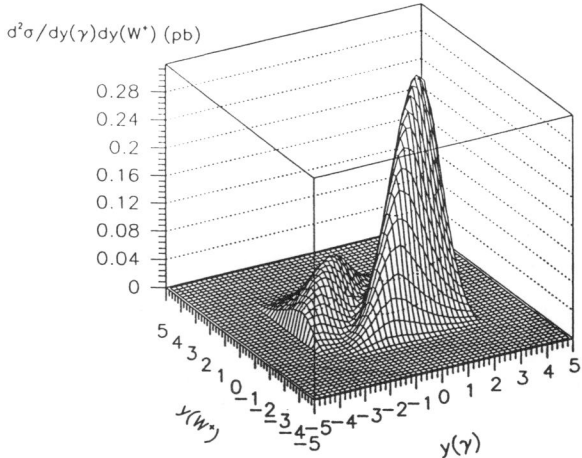

FIG. 6. The double differential distribution $d^2\sigma/dy_\gamma dy_W$ for $p\bar{p} \to W^+\gamma \to \ell^+\nu\gamma$, $\ell = e, \mu$, in the Born approximation at the Tevatron (1.8 TeV). The cuts imposed are described in Ref. (21).

process $p\bar{p} \to W\gamma +$ 0-jet $\to e\nu_e +$ 0-jet, namely, if we veto the extra jet(s) from higher order QCD processes, we recover most of the feature in the Born level and thus regain the sensitivity to study the anomalous couplings.

It is noted that the RAZ for $p\bar{p} \to W^\pm \gamma \to e^\pm \nu_e$ occur in the central region $\cos\theta = \pm\frac{1}{3}$ (and $y_\gamma^* \simeq 0$ averagely). Therefore, deviations from the SM will largely happen in high transverse momentum $p_T(\gamma)$ region. This feature has been carefully examined in a recent paper (20). It is shown that, as a function of a cutoff on the photon transverse momenta $p_T^{min}(\gamma)$, the ratio of integrated cross sections $R_{\gamma,l} = \sigma(\gamma Z)/\sigma(\gamma W)$ for $Z\gamma$ process (which has no RAZ) and for $W\gamma$ process (which has a RAZ) is a clear indication of a zero behavior, as shown in Fig. 5. We see that in high $p_T(\gamma)$ region, the rate for $W\gamma$ process is significantly smaller than that of $Z\gamma$ process. In contrast, the ratio versus a cutoff on a jet transverse momentum $p_T^{min}(j)$ for Zj and Wj production is flat over a large $p_T(j)$ range. The advantage of looking at the cross section ratio versus $p_T^{min}(\gamma)$ is to have avoided the c.m. frame and z-axis ambiguities, while the price to pay is to lose the information about the exact RAZ location.

Some more interesting observation has been made recently in Ref. (21). Recall the rapidity in the lab frame (y) is a sum of that in c.m. frame (y^*) and a term reflecting the c.m. frame motion

$$y = y^* + \frac{1}{2}\ln\left(\frac{x_1}{x_2}\right), \tag{13}$$

where $x_{1,2}$ are the parton momentum fractions.

If we take the rapidity difference between the photon and the W, then the difference is invariant under the longitudinal boost. Therefore, the rapidity

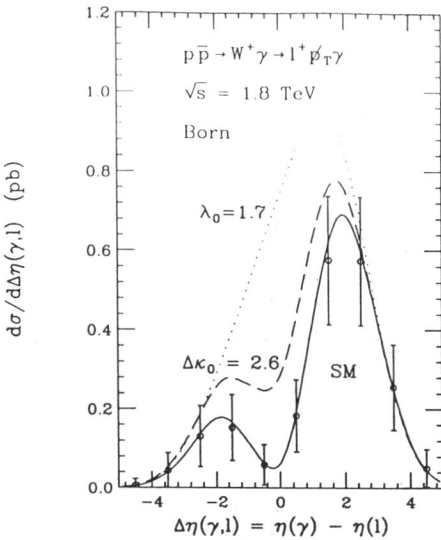

FIG. 7. The pseudorapidity difference distribution, $d\sigma/d\Delta\eta(\gamma,\ell)$, for $p\bar{p} \to W^+\gamma$, $W^\pm \to \ell^+\nu$ with $\ell = e$, μ, at the Tevatron in the Born approximation for anomalous $WW\gamma$ couplings. The curves are for the SM (solid), $\Delta\kappa_0 = 2.6$ (dashed), and $\lambda_0 = 1.7$ (dotted). Only one coupling is varied at a time. The error bars indicate the expected statistical uncertainties for an integrated luminosity of 22 pb^{-1}. The cuts imposed are described in Ref. (21).

correlation between W-γ in the c.m. frame is preserved in the lab frame

$$\Delta y = y_\gamma - y_W = y_\gamma^* - y_W^* \simeq -0.4. \qquad (14)$$

We thus have a chance to avoid the frame ambiguity if we choose the variable in such a clever way. Fig. 6 demonstrates the rapidity correlation in the lab frame (21). We see an impressive "valley" for the rapidity correlation, given by $y_\gamma - y_W \simeq -0.4$. In order to implement this idea more realistically, we must use the final state momentum of $l^\pm$, rather than that of W. Fortunately, based on helicity arguments, the charged lepton in $W\gamma$ process goes dominantly along the W moving direction, so that W-γ rapidity correlation is largely preserved,

$$\Delta\eta(\gamma,l) = \eta(\gamma) - \eta(l) \simeq -0.3. \qquad (15)$$

One could directly study the rapidity difference, which would have the advantage for higher statistics than the double differential cross section. This is shown in Fig. 7. The curve for the pseudorapidity difference in the SM (solid) presents a clear dip at -0.3. The authors of Ref. (21) have also estimated the error bars for expected statistical uncertainties with an integrated luminosity of 22 pb^{-1}. Since the CDF/D0 collaborations have accumulated about 100 pb^{-1} each (at the time of writing), one can anticipate that an experimental study along this line may first observe the clear dip reflecting the RAZ. The effects from anomalous couplings are also demonstrated in the figure.

Finally, two remarks are in order. First, we have thus far concentrated on Tevatron energies. At the LHC, due to the more severe problems regarding the z-axis definition and much larger QCD corrections to the Born amplitudes, the conclusions in studying the RAZ seem rather pessimistic. Secondly, we have not discussed much about the background issue. It turns out that if we could achieve a $j \to \gamma$ misidentification factor at a level of 10^{-3}, the background may not be too severe (21,15).

B. W Radiative Decay: $W^\pm \to l^\pm \nu \gamma$, $q\bar{q}'\gamma$

It was shown (22) that the W-radiative decay, $W \to f_1 \bar{f}_2 \gamma$, also presents a RAZ. Refs. (17-19) studied the process $p\bar{p} \to W^\pm \to e^\pm \nu \gamma$ at the Tevatron. This process develops a zero at the kinematical boundary $\cos\theta_{l\gamma} = -1$ in W-rest frame. It can be effectively separated from the $W\gamma$ associated production by imposing a transverse mass cut, $M_T(l^\pm \nu, \gamma) < 90$ GeV; and it also has larger statistics. However, the RAZ is less pronounced due to the single-zero behavior (17,22) and the difficulty for W-rest frame reconstruction. It is therefore less sensitive to anomalous couplings.

Ref. (23) discussed the zero in the hadronic decay process $W^- \to d\bar{u}\gamma$. In this case, it is a double-zero as usual and the W-rest frame reconstruction may be relatively easier. However, the event identification may be difficult in hadron collider experiments; and it will suffer from low statistics for $e^+e^- \to W^+W^-$ with a radiative W decay.

C. WZ Production: $p\bar{p}, pp \to W^\pm Z \to l^\pm \nu l^+ l^-$

As discussed earlier, the Born amplitude for $q_1 \bar{q}_2 \to W^\pm Z$ develops a zero at high energies (13) at $\cos\theta \simeq \pm\frac{1}{3}\tan^2\theta_W \approx \pm 0.1$. Following the proposal of studying the rapidity correlation analogous to $\Delta\eta(\gamma, \ell)$ for $p\bar{p} \to W^+\gamma$ process in Fig. 7, one can examine the rapidity correlation (24) via $\Delta y(Z, \ell_1) = y(Z) - y(l_1)$ where l_1 is the charged lepton from W decay. Figure 8 shows the differential cross section $d\sigma/\Delta y$. There is a dip near 0.1 as predicted in the SM (solid curve), although it will not be easy to convincingly establish the effect due to the less pronounced dip for this channel and limited number of $W^\pm Z \to \ell_1^\pm \nu_1 \ell_2^+ \ell_2^-$ events expected (see the estimated statistical error bars in the figure for 10 fb^{-1} luminosity). At the LHC energies, the zero is further washed out due to larger QCD radiative corrections and the z-axis ambiguity.

It is amusing to note (13) that $e\nu_e$ or $\mu\nu_\mu$ collisions above the WZ threshold would in principle provide a clean environment for event reconstruction. The location of the zero at $\cos\theta \approx \pm 0.3$ is ideal for experimental studies of the

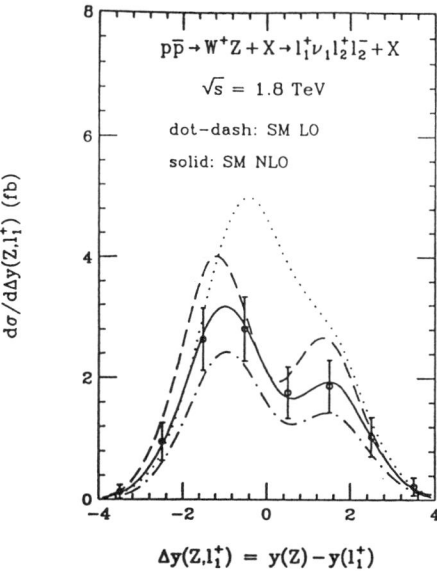

FIG. 8. The differential cross section for the rapidity difference $\Delta y(Z, \ell_1)$ for $p\bar{p} \to W^+Z + X \to \ell_1^+ \nu_1 \ell_2^+ \ell_2^- + X$ at $\sqrt{s} = 1.8$ TeV. The solid and dot-dashed curves show the inclusive NLO and the LO SM prediction, respectively. The dashed and dotted lines give the results for $\Delta\kappa^0 = +1$ and $\Delta\kappa^0 = -1$. The error bars associated with the solid curves indicate the expected statistical uncertainties for an integrated luminosity of 10 fb^{-1}. The cuts imposed are described in Ref. (24).

$W^{\pm}Z$ final state, unlike the case for $e^-\bar{\nu}_e \to W^-\gamma$ where the zero is located at the kinematical boundary ($\cos\theta = 1$) resulting a single-zero.

D. $qq' \to qq'\gamma$ And $eq \to eq\gamma$

Certain single photon radiation processes in quark scattering, such as

$$uu \to uu\gamma, \quad u\bar{d} \to u\bar{d}\gamma, \quad dd \to dd\gamma, \quad d\bar{u} \to d\bar{u}\gamma \qquad (16)$$

present a RAZ (25) at

$$\cos\theta_\gamma = (Q_2 - Q_1)/(Q_2 + Q_1), \qquad (17)$$

where Q_1 and Q_2 are the electric charges for initial state quarks and θ_γ the photon scattering angle with respect to the incident quark. But some other processes such as

$$u\bar{u} \to u\bar{u}\gamma, \quad ud \to ud\gamma, \quad d\bar{d} \to d\bar{d}\gamma, \quad du \to du\gamma \qquad (18)$$

do not. The cause for the difference is the "same-sign rule", as stated earlier. The locations of the zeros are clearly sensitive to the fractional charges of the quarks (25), although there are no triple vector-boson self-interactions involved. However, after convoluting with the hadron structure functions, the RAZ becomes a dip. At low energies where the valence quarks dominate, there is a good chance one could find the RAZ in these processes. Our experimental colleagues may consider to re-examine the low energy data, such as that at CERN ISR pp collider ($\sqrt{s} \sim$ 30 - 60 GeV), for this purpose. At higher energies, such as at the Fermilab Tevatron (26), the QCD multiple-jet processes would completely swamp the RAZ signal.

It is straightforward to calculate the processes $e^{\pm}p \to e^{\pm}X\gamma$ (27,28) by simply replacing one of the quarks by $e^{\pm}$ in processes 16,18. Once again, at low energies where the valence quarks dominate, it is possible to examine the dip resulted from the RAZ. At HERA energies, however, the RAZ effects in e^-p collisions seem to be largely washed out, while it is claimed to be more promising in e^+p collisions (28), again due to the argument of the "same-sign rule". Inclusion of more realistic experimental simulation may further worsen this situation.

III. RAZ IN THEORIES BEYOND THE STANDARD MODEL

The RAZ is a general feature in gauge theories. There are in fact many more processes beyond the SM in which the RAZ occur. The RAZ theorem has been generalized to supersymmetric theories with massless gaugino emission (29) and RAZ have been found in the exact supersymmetric limit for processes (30) such as

$$d\bar{u} \to \tilde{W}\tilde{\gamma}, \quad \gamma e \to \tilde{W}\tilde{\nu}_L \quad etc.. \tag{19}$$

In this limit, the RAZ locate at the same places as those for the SM partners.

The RAZ is also found in charged Higgs boson production $p\bar{p} \to H^{\pm}\gamma$ (31), although the small Yukawa coupling of $H^{\pm}$ to light fermions would make this process unobservable. A more promising process is for the decay $H^+ \to t\bar{b}\gamma$ (32) if kinematically accessible. Similarly, the RAZ effects in radiative decays of other charged scalar particles such as lepto-quarks are also studied (33).

IV. SUMMARY

Certain tree-level processes involving massless gauge bosons and charged particles present radiation amplitude zeros (RAZ). With higher order radiative corrections and in a more realistic experimental environment, those zeros are always approximate or become dips. In the SM with a spontaneously broken gauge symmetry, WZ final state develops approximate zeros at high energies. In general, the nature of those zeros is sensitive to the gauge couplings of vector bosons and that of fermions as well. Studying these RAZ experimentally may thus provide probes to physics beyond the SM.

Progress has been made in studying the RAZ both theoretically and experimentally, e.g., $p\bar{p} \to W^{\pm}\gamma \to l^{\pm}\nu\gamma$ and $W^{\pm}Z \to l^{\pm}\nu l^{+}l^{-}$ at Fermilab Tevatron energies. Other processes such as $qq' \to qq'\gamma$ at low energies, and $e^+p \to e^+X\gamma$ at HERA should be examined at a level with realistic experimental acceptance to draw further conclusion. It is clearly challenging to experimentally observe those "approximate" zeros. Hopefully one day, we would be able not only to observe the RAZ, but also in so doing to find some hints on new physics in the vector-boson sector.

ACKNOWLEDGEMENT

I would like to thank Uli Baur, Steve Errede and Thomas Müller for their invitation and their nice organization of such a stimulating conference. I am also grateful to Bob Brown for illuminating discussions during the conference. This work was supported in part by the U.S. Department of Energy under Contract No. DE-FG03-91ER40674.

REFERENCES

1. R. W. Brown and K. O. Mikaelian, Phys. Rev. **D19**, 922 (1979).
2. R. W. Brown, K. O. Mikaelian and D. Sahdev, Phys. Rev. **D20**, 1164 (1979).
3. K. O. Mikaelian, M. A. Samuel and D. Sahdev, Phys. Rev. Lett. **43**, 746 (1979).
4. D. Zhu, Phys. Rev. **D 22**, 2266 (1980); C. J. Goebel, F. Halzen and J. P. Leveille, Phys. Rev. **D 23**, 2682 (1981).
5. S. J. Brodsky and R. W. Brown, Phys. Rev. Lett. **49**, 966 (1982); R. W. Brown, K. L. Kowalski and S. J. Brodsky, Phys. Rev. **D 28**, 624 (1983); R. W. Brown and K. L. Kowalski, Phys. Rev. **D 29**, 2100 (1984).
6. M. A. Samuel, A. Sen, G. S. Sylvester and M. L. Laursen, Phys. Rev. **D 29**, 994 (1984).
7. R. W. Brown, in these proceedings.
8. For a review on this subject, see, e.g., J. Reid and G. Tupper, in *Cape Town 1990, Proceedings, Phase structure of Strongly Interacting Matter*, p.317.
9. M. L. Laursen, M. A. Samuel, A. Sen and G. Tupper, Nucl. Phys. **B226**, 429 (1983); M. L. Laursen, M. A. Samuel and A. Sen, Phys. Rev. **D 28**, 650 (1983).
10. G. Tupper, Phys. Lett. **B 156**, 400 (1985).
11. U. Baur, T. Han and J. Ohnemus, Phys. Rev. **D48**, 5140 (1993).
12. J. Ohnemus, Phys. Rev. **D50**, 1931 (1994).
13. U. Baur, T. Han, and J. Ohnemus Phys. Rev. Lett. **72**, 3941 (1994).
14. K. Hagiwara, R. Peccei, D. Zeppenfeld, and K. Hikasa, Nucl. Phys. **B282**, 253 (1987); also see other relevant talks in these proceedings.
15. See the experimental talks given by H. Aihara, T. Fuess, H. Johari, G. Landsberg, D. Neuberger, B. Wagner, C. Wendt, J. Womersley, L. Zhang, in these proceedings.

16. J. Stroughair and C. Bilchak, Z. Phys. **C 26**, 415 (1984); J. Gunion, Z. Kunszt, and M. Soldate, Phys. Lett. **B 163**, 389 (1985); J. Gunion and M. Soldate, Phys. Rev. **D 34**, 826 (1986); W. J. Stirling *et al.*, Phys. Lett. **B 163**, 261 (1985).
17. J. Cortes, K. Hagiwara, and F. Herzog, Nucl. Phys. **B 278**, 26 (1986).
18. U. Baur and D. Zeppenfeld, Nucl. Phys. **B 308**, 127 (1988).
19. U. Baur and E. L. Berger, Phys. Rev. **D41**, 1476 (1990).
20. U. Baur, S. Errede and J. Ohnemus, Phys. Rev. **D48**, 4103 (1993).
21. U. Baur, S. Errede, G. Landsberg, Phys. Rev. **D50**, 1917 (1994).
22. T. R. Grose and K. O. Mikaelian, Phys. Rev. **D23**, 123 (1981).
23. M. A. Samuel and G. Tupper, Prog. Theor. Phys. **74**, 1352 (1985); F. Boudjema, C. Hamzaoui, M. A. Samuel and J. Woodside, Phys. Rev. Lett. **63**, 1906 (1989); J. H. Reid, G. Tupper and M. van Zijl, Phys. Lett. **B 218**, 473 (1989).
24. U. Baur, T. Han and J. Ohnemus, Phys. Rev. **D51**, 3381 (1995).
25. K. Hagiwara, F. Halzen and F. Herzog, Phys. Lett. **B135**, 324 (1984); J. Reid and M. Samuel, Phys. Rev. **D39**, 2046 (1989).
26. M. D. Doncheski and F. Halzen, Z. Phys. **C52**, 673 (1991).
27. C. L. Bilchak, J. Phys. **G11**, 1117 (1985).
28. G. Couture, Phys. Rev. **D39**, 2527 (1989); J. Reid, G. Li and M. A. Samuel, Phys. Rev. **D41**, 1675 (1990).
29. R. W. Brown and K. L. Kowalski, Phys. Lett. **B144**, 235 (1984).
30. V. Barger, R. Robinett, W.-Y. Keung and R. J. N. Phillips, Phys. Lett. **B131**, 372 (1983); R. Robinett, Phys. Rev. **D30**, 688 (1984); *ibid.* **D31**, 1657 (1985).
31. X. G. He and H. Lew, Mod. Phys. Lett. **3**, 1199 (1988).
32. J. Reid, G. Tupper, G. Li and M. A. Samuel, Phys. Lett. **B241**, 105 (1990); B. Mukhopadhyaya, M. A. Samuel, J. Woodside, J. Reid, G. Tupper, Phys. Lett. **B247**, 607 (1990).
33. N. G. Deshpande, X. G. He and S. Oh, Phys. Rev. **D51**, 2295 (1995).

Strong $W_L W_L$ Scattering[†]

R. Sekhar Chivukula*

*Department of Physics
Boston University
590 Commonwealth Avenue
Boston MA 02215

[†] BUHEP-95-17
hep-ph/9505202

I describe theories of a strongly-interacting electroweak symmetry breaking sector and discuss the expected size of anomalous weak-boson couplings in these models.

SIGNATURES OF ELECTROWEAK SYMMETRY BREAKING IN WW SCATTERING

The physics of electroweak symmetry breaking must appear at energies of order a TeV or lower. To see this, consider a thought experiment (1), the scattering of longitudinally polarized W^+ and W^-:

$$\text{(diagrams)} \tag{1}$$

Using the Feynman-rules of the electroweak gauge theory we can calculate $W_L^+ W_L^-$ scattering at tree level. We find that this amplitude grows like E_{cm}^2:

$$\mathcal{A} = \frac{g^2 s}{8 M_W^2} (1 + \cos\theta^*) \quad , \tag{2}$$

plus terms that do not grow with s. Projecting onto the s-wave state, we find

$$\mathcal{A}^{l=0} = \frac{g^2 s}{128 \pi M_W^2} \sim \left(\frac{\sqrt{s}}{2.5 \text{ TeV}}\right)^2 \quad . \tag{3}$$

Unitarity implies that the dynamics associated with EWSB has to appear before an energy scale of around 2.5 TeV (1) (2). There are three possibilities:

- There may be additional particles with masses less than or of order of a TeV, or

- the W and Z interactions may become strong at energies of order a TeV, or

© 1996 American Institute of Physics

- both of the above.

It is important to note that the amplitude calculated above *universal* (3) (4): the calculation depended only on the *gauge structure* of the standard model and on the relationship

$$\rho = \frac{M_W}{M_Z \cos\theta_W} \approx 1 \quad , \tag{4}$$

and will hold *regardless* of the dynamics responsible for electroweak symmetry breaking. Therefore, in order to understand the dynamics of electroweak symmetry breaking, it will be necessary to characterize the physics which cuts-off the growth in the longitudinal gauge boson scattering amplitudes.

The universality of the scattering amplitudes for longitudinal gauge boson scattering can also be seen as a consequence of the fact that the longitudinal components of the gauge-bosons are the "eaten" Goldstone Bosons of $SU(2) \times U(1)$ breaking. More formally, the "Equivalence Theorem" (5) (3) states that any amplitude involving the scattering of longitudinal gauge-bosons is equal to the same amplitude involving the corresponding Goldstone Bosons (which would be present in the ungauged theory), up to corrections of order $(M_W/E)^2$. The low-energy scattering amplitudes of Goldstone Bosons, however, are determined by the low energy theorems of chiral dynamics (*c.f.* PCAC in QCD) and are determined by the symmetry structure of the theory. In a theory in which $\rho = 1$, the symmetry structure of the electroweak symmetry breaking sector is naturally $SU(2) \times SU(2)$ (6), and the scattering of the longitudinal components of the W and Z are determined by analogs of Weinberg's low-energy theorems in QCD (4).

THEORIES OF ELECTROWEAK SYMMETRY BREAKING

The Standard One-Doublet Higgs Model

In the standard one-doublet Higgs model one introduces a fundamental scalar doublet of $SU(2)_W$:

$$\phi = \begin{pmatrix} \phi^+ \\ \phi^0 \end{pmatrix} \quad , \tag{5}$$

which has a potential of the form

$$V(\phi) = \lambda \left(\phi^\dagger \phi - \frac{v^2}{2} \right)^2 \quad . \tag{6}$$

In the potential, v^2 is assumed to be positive in order to favor the generation of a non-zero vacuum expectation value for ϕ. This vacuum expectation value breaks the electroweak symmetry, giving mass to the W and Z. When symmetry breaking takes place, the four degrees of freedom in ϕ divide up.

Three of them become the longitudinal components, W_L and Z_L, of the gauge bosons, and the fourth, commonly called H (for Higgs particle), is left over

$$\phi = \Omega \begin{pmatrix} 0 \\ \frac{H+v}{\sqrt{2}} \end{pmatrix} . \tag{7}$$

Here, Ω is an $SU(2)$ matrix. If we make an $SU(2)_W$ gauge transformation until Ω is the identity, we arrive at unitary gauge.

The exchange of the Higgs boson contributes to $W_L W_L$ scattering. In the limit in which E_{cm} is large compared to the masses of the particles in the process, the leading contribution (in energy) from Higgs boson exchange exactly cancels the bad high-energy behavior in $W_L^+ W_L^-$ scattering

$$\rightarrow \mathcal{A} = -\frac{g^2 s}{8 M_W^2}(1 + \cos\theta^*) , \tag{8}$$

plus terms which do not grow with energy.

At tree-level the Higgs boson has a mass given by $m_H^2 = 2\lambda v^2$. In order for this theory to give rise to strong W and Z interactions, it would be necessary that the Higgs boson be heavy and, therefore, that λ be large.

This explanation of electroweak symmetry breaking is unsatisfactory for a number of reasons. For one thing, this model does not give a dynamical explanation of electroweak symmetry breaking. For another, when embedded in theories with additional dynamics at higher energy scales, these theories are technically unnatural (7).

Perhaps most unsatisfactory, however, is that theories of fundamental scalars are probably "trivial" (8), *i.e.*, it is not possible to construct an interacting theory of scalars in four dimensions that is valid to arbitrarily short distance scales. In quantum field theories, fluctuations in the vacuum screen charge – the vacuum acts as a dielectric medium. Therefore there is an effective coupling constant which depends on the energy scale (μ) at which it is measured. The variation of the coupling with scale is summarized by the β–function of the theory

$$\beta(\lambda) = \mu \frac{d\lambda}{d\mu} . \tag{9}$$

The only coupling in the Higgs sector of the standard model is the Higgs self-coupling λ. In perturbation theory, the β-function is calculated to be

$$\rightarrow \beta = \frac{3\lambda^2}{2\pi^2} . \tag{10}$$

Using this β–function, one can compute the behavior of the coupling constant as a function of the scale[1]. One finds that the coupling at a scale μ is related

---

[1] Since these expressions were computed in perturbation theory, they are only valid when $\lambda(\mu)$ is sufficiently small. We will return to the issue of strong coupling below.

to the coupling at some higher scale Λ by

$$\frac{1}{\lambda(\mu)} = \frac{1}{\lambda(\Lambda)} + \frac{3}{2\pi^2} \log \frac{\Lambda}{\mu} \quad . \tag{11}$$

In order for the Higgs potential to be stable, $\lambda(\Lambda)$ has to be positive. This implies that

$$\frac{1}{\lambda(\mu)} \geq \frac{3}{2\pi^2} \log \frac{\Lambda}{\mu} \quad . \tag{12}$$

Thus, we have the bound

$$\lambda(\mu) \leq \frac{2\pi^2}{3 \log \left(\frac{\Lambda}{\mu}\right)} \quad . \tag{13}$$

If this theory is to make sense to arbitrarily short distances, and hence arbitrarily high energies, we should take Λ to ∞ while holding μ fixed at about 1 TeV. In this limit we see that the bound on λ goes to zero. In the continuum limit, this theory is trivial; it is free field theory.

The inequality above can be translated into an upper bound on the mass of the Higgs boson (9). From the bound above, we have

$$\frac{\Lambda}{\mu} \leq \exp \left(\frac{2\pi^2}{3\lambda(\mu)} \right) \quad , \tag{14}$$

but

$$m_H^2 \sim 2v^2 \lambda(m_H) \quad , \tag{15}$$

thus

$$\Lambda \leq m_H \exp \left(\frac{4\pi^2 v^2}{3 m_H^2} \right) \quad . \tag{16}$$

For a given Higgs boson mass, there is a *finite* cutoff energy at which the description of the theory as a fundamental scalar doublet stops making sense. This means that the standard one-doublet Higgs model can only be regarded as an *effective* theory valid below this cutoff.

The theory of a relatively light weakly coupled Higgs boson, can be self-consistent to a very high energy. For example, if the theory is to make sense up to a typical GUT scale energy, 10^{16} GeV, then the Higgs boson mass has to be less than about 170 GeV (10). In this sense, although a theory with a light Higgs boson does not really answer any of the interesting questions (*e.g.*, it does not explain *why* $SU(2)_W \times U(1)_Y$ breaking occurs), the theory does manage to postpone the issue up to higher energies.

The theory of a heavy Higgs boson (*i.e.* with a mass of about 1 TeV), however, does not really make sense. Since we have computed the β-function in perturbation theory, our answer is only reliable at energy scales at which $\lambda(\mu)$ (as well as the Higgs boson mass) is small. Fortunately, non-perturbative lattice calculations are available. Early estimates (11) indicated that if the theory was to make sense up to 4 TeV, the mass of the Higgs boson had to be less than about 640 GeV. More recent results (12) imply that this bound may be relaxed somewhat; one might be able to get away with an 800 GeV Higgs boson, but the Higgs boson mass is certainly bounded by a value of this order of magnitude. The triviality limits on the mass of the Higgs boson imply that it is not possible for the W_L and Z_L scattering amplitudes in the standard model to truly become large at energies well below the cutoff. This result is especially interesting because it implies that if nothing shows up below energies of the order 700–800 GeV, then something truly "non-trivial" is going on. We just have to find it.

Technicolor

In models with fundamental scalars, electroweak symmetry breaking can be accommodated if the parameters in the potential (which presumably arise from additional physics at higher energies) are suitably chosen. By contrast, technicolor theories strive to explain electroweak symmetry breaking in terms of physics operating at an energy scale of order a TeV. In technicolor theories, electroweak symmetry breaking is the result of chiral symmetry breaking in an asymptotically-free, strongly-interacting gauge theory with massless fermions. Unlike theories with fundamental scalars, these theories are technically natural: just as the scale Λ_{QCD} arises in QCD by dimensional transmutation, so too does the weak scale v in technicolor theories. Accordingly, it can be exponentially smaller than the GUT or Planck scales. Furthermore, asymptotically-free non-abelian gauge theories may be fully consistent quantum field theories.

In the simplest technicolor theory (13) one introduces a (massless) left-handed weak-doublet of "technifermions", and the corresponding right-handed weak-singlets, which transform as N's of a strong $SU(N)_{TC}$ technicolor gauge group. In analogy to the (approximate) chiral $SU(2)_L \times SU(2)_R$ symmetry on quarks in QCD, the strong technicolor interactions respect an $SU(2)_L \times SU(2)_R$ global chiral symmetry on the technifermions. When the technicolor interactions become strong, the chiral symmetry is broken to the diagonal subgroup, $SU(2)_{L+R}$, producing three Nambu-Goldstone bosons which become, via the Higgs mechanism, the longitudinal degrees of freedom of the W_L and Z_L. Because the left-handed and right-handed techni-fermions carry different electroweak quantum numbers, the electroweak interactions break to electromagnetism. If the f-constant of the theory, the analog of f_π in QCD, is chosen to be $v \approx 246$ GeV, then the W mass has its observed value. Furthermore (6), the remaining $SU(2)_{L+R}$ custodial symmetry insures that, to lowest order in the hypercharge coupling, $M_W = M_Z \cos\theta_W$.

In addition to the "eaten" Nambu-Goldstone bosons, such a theory will give rise to various resonances, the analogs of the ρ, ω, and possibly the σ, in

QCD. In general, the growth of the W_L and Z_L scattering amplitudes are cut off by exchange of these heavy resonances,

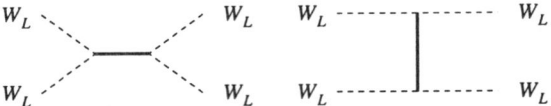

just as in QCD the growth of pion–pion scattering amplitudes are cut off by QCD resonances. Scaling from QCD, we expect that the masses of the various resonance will be of order a TeV. Unlike the situation in models with only fundamental scalars in the symmetry breaking sector, the scattering of longitudinal W and Z bosons can truly be strong.

In figure 1, we show the data for the scattering amplitude of $\pi^+\pi^0$ at low-energies, as well as the corresponding low-energy theorem. We see that while the growth of the scattering amplitude begins close to the low-energy theorem prediction, it is significantly enhanced (and unitarized) by the presence of the ρ-resonance. We expect a similar behavior in technicolor theories, with the energy scale enhanced by a factor of $v/f_\pi \approx 2600$.

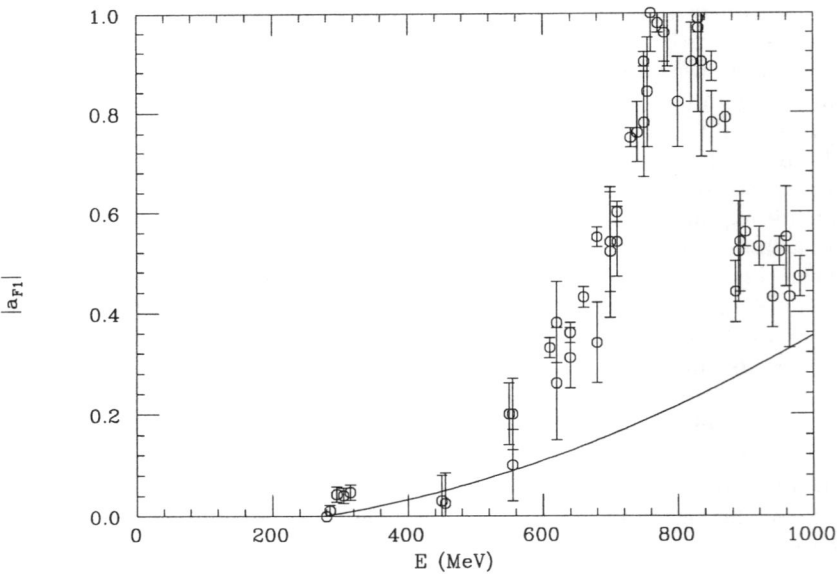

FIG. 1. Data (14) and low-energy theorem prediction for the spin-1/isospin-1 pion scattering amplitude.

The most direct signal for technicolor, therefore, is an enhancement in the production of WZ pairs at high invariant-mass, coming from the production of the technicolor analog of the ρ-meson in QCD (15) (3). If the technirho resonance(s) are too heavy to be observed at the LHC, there may be an enhancement in the isospin-2 $W^+W^+ + W^-W^-$ channel which is large enough to be observed (16) (17). Detecting technicolor at the LHC is likely to be quite challenging, however. Recent estimates (18) (17) of the luminosity required to detect a technicolor at the LHC indicate that it would be necessary to accumulate of order 100 fb^{-1}, and that this would result in a signal of only a few tens of events (over a background of comparable size!).

Inelastic Channels in WW-Scattering

Up to now, we have assumed that the *only* "light" particles in the electroweak symmetry breaking sector are the longitudinal components of the W and Z. In a theory of this sort, the behaviors described above are generic: the growth in the $W_L W_L$ scattering amplitude may be cut-off by light, narrow resonances (such as in the weakly-coupled standard model) or by heavy, broad resonances (such as would be expected in the simplest technicolor model). However, if the global symmetry structure of the theory is larger than $SU(2) \times SU(2)$, there may be additional (pseudo-)Goldstone bosons. These additional particles give rise to *inelastic* channels for vector-boson scattering, and may have dramatic consequences for the behavior of the theory.

Consider a technicolor model with a global $SU(N_f)_L \times SU(N_f)_R$ chiral symmetry which breaks spontaneously to the vectorial $SU(N_f)$ subgroup, breaking the weak interactions and producing $N_f^2 - 1$ Goldstone (or pseudo-Goldstone) bosons. The low energy theorem for the $SU(N_f)$ singlet, spin singlet scattering amplitude of these bosons is (19)

$$a_{singlet} = \frac{N_f N_d s}{32\pi v^2}, \qquad (17)$$

where N_d is the number of technifermion doublets. In analogy to the analysis of the $W_L^+ W_L^-$ scattering amplitude given at the beginning of this talk, we see that as N_f and N_d increase, *i.e.* as the number of inelastic channels in $W_L W_L$ scattering grow, the scale by which the dynamics of EWSB must appear *decreases* (20) (21). For example, in the one family technicolor model (22) $N_f = 8$ and $N_d = 4$. In this model $a_{singlet}$ would exceed unitarity at 440 GeV, and we expect that new physics must appear at the energy scale or lower.

Mitch Golden and I studied the phenomenology of a model of electroweak-symmetry breaking with many inelastic channels in a toy-model based on a scalar $O(N)$ theory (21). We showed that, although the new physics occurs at relatively low energies, this new low-energy physics can be hard to detect. The presence of the large numbers of inelastic channels can result in *elastic* W and Z scattering amplitudes that are small and structureless at all energies, i.e. lacking any discernible resonances (see Fig. 2). Nonetheless, the theory can be strongly interacting and the *total* W and Z cross sections large: most of the cross section is for the production of particles other than the W or Z. In such a model, discovering the electroweak symmetry breaking sector will depend on

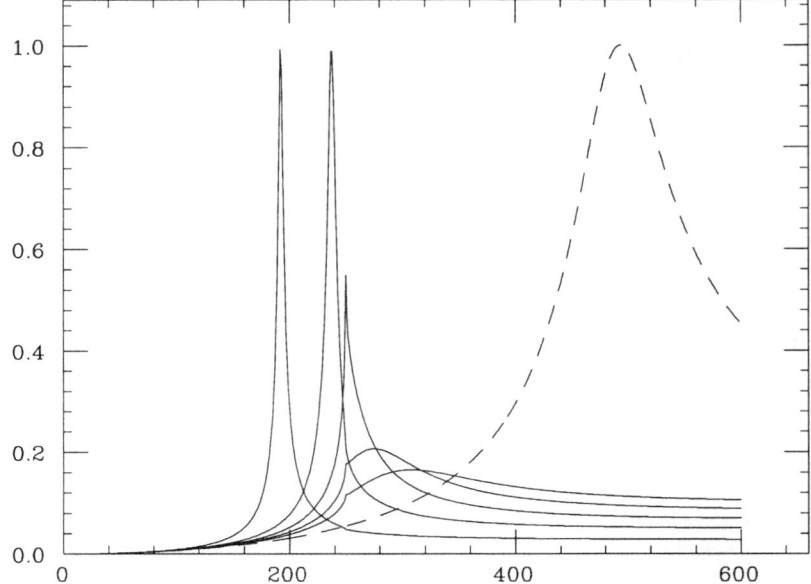

FIG. 2. The absolute value of the (weak) isospin-0 $W_L W_L$ scattering amplitude in a toy $O(N)$ model of electroweak symmetry breaking (21). The model contains 32 pseudo-Goldstone bosons, and the different solid curves show the change in the amplitude as the mass of the pseudo goldstone bosons is adjusted. The right-most nearly structureless amplitude corresponds to the case where the "Higgs" in this model is strongly coupled, but can decay to paris of pseudos in addition to pairs of weak gauge bosons. The dashed-line corresponds to the same scattering amplitude in the standard model with a 500 GeV Higgs boson.

the observation of the other particles and our ability to associate them with symmetry breaking. This implies that we must keep an open mind about the experimental signatures of the electroweak symmetry breaking sector and that we cannot rely solely on two gauge boson final states.

WHAT DOES THIS IMPLY FOR ANOMALOUS WEAK-BOSON SELF-INTERACTIONS?

It would seem natural that in a theory of a strongly interacting electroweak symmetry breaking sector, like technicolor, there could be large corrections to the electroweak gauge-boson self-couplings. For example, one would expect that the coupling of one longitudinal gauge-boson to two transverse gauge-bosons would acquire a form-factor similar to the electromagnetic form-factor of the pion in QCD.

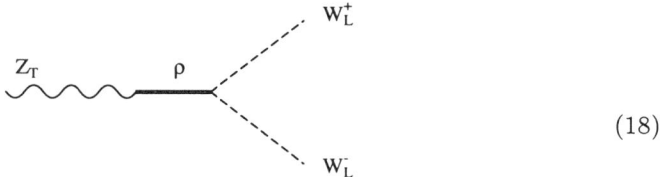

(18)

As discussed by Wudka at this conference (23), one can use dimensional analysis to estimate the size of the corrections to the weak-boson self-couplings (24)

$$\Delta g_1, \ \Delta\kappa = \mathcal{O}\left(\frac{g^2}{16\pi^2}\right) , \qquad (19)$$

and

$$\lambda = \mathcal{O}\left(\frac{g^4}{(16\pi^2)^2}\right) , \qquad (20)$$

Using these estimates, we see that deviations in κ and g_1 are expected to be of order 10^{-3}, while λ is expected to be of order 10^{-5} or 10^{-6}.

I would like to emphasize here that these dimensional estimates have a simple physical interpretation in terms of the form-factor picture that I discussed above (see also the discussion of Willenbrock at this conference (25)). As in QCD, we expect that the scale of variation of the form factor is given by the mass of the lowest-lying resonance in in the appropriate channel, namely by the mass of a vector meson. Furthermore, in the limit that M_W, $M_Z \to 0$, we know that the vector bosons must couple to a conserved current and that the vector-boson self couplings must be of canonical form (26). Therefore, we can estimate that the size of the anomalous couplings κ and g must be

$$\Delta g_1, \ \Delta\kappa = \mathcal{O}\left(\frac{M_W^2}{M_{\rho TC}^2}\right) \qquad (21)$$

and, given that λ is the coefficient of a *dimension-6* operator and the normalization chosen in (24),

$$\lambda = \mathcal{O}\left(\frac{M_W^4}{M_{\rho TC}^4}\right) . \qquad (22)$$

Afficianados of dimensional analysis (27) will see immediately that these two estimates are, in fact, consistent since the dimensional analysis estimate of the lightest-resonance mass in models of electroweak symmetry breaking are of order $4\pi v$.

What are the prospects for the experimental detection of deviations of this size? Baur, Han, and Ohnemus (28) have recently considered this issue for a variety of colliders. The prospects are discouraging. For example, for the LHC with an integrated luminosity of 100 fb^{-1}, they find that one may be able to probe to the level of 10^{-2} for Δg and λ. This is not sufficient to be sure to probe effects of the size predicted above.

On the other hand, one might wonder if the analysis of the effects of anomalous weak-boson self-interactions is consistent with the results given in the previous section. From the analysis given above, we see that a Δg to order 10^{-2} would arise in a model with a technirho of mass approximately 1 TeV. This is consistent with the analysis of (18): in both cases one is looking for the effects of technirho mesons on the production of WZ pairs!

CONCLUSIONS

A strongly interacting symmetry breaking sector will result in one or more resonances which are either:

- Heavy (with masses of order a TeV) and broad (in the case that elastic W and Z scattering dominates). Detection will require an integrated luminosity of order 100 fb^{-1} or more at the LHC.

- Light and broad (in the case that inelastic channels are important). In this case detection will hinge on observing particles other than the W_L and Z_L and identifying them as being associated with EWSB.

In the first case, the lightest resonances in the electroweak symmetry breaking sector are expected to be the technivector mesons, the analogs of the ρ and ω in QCD. The masses of these resonances are expected to be of order a TeV, and one expects an enhancement of WZ and/or WW production at energies of this order of magnitude. One may think of the "tail" of the technirho as giving rise to anomalous weak-boson self-interactions. The expect size of the resulting anomalous gauge boson vertices is small, with $\Delta\kappa$ and Δg of order 10^{-3} and λ of order 10^{-5} or 10^{-6}.

ACKNOWLEDGEMENTS

I gratefully acknowledge the support of NSF Presidential Young Investigator Award and a DOE Outstanding Junior Investigator Award.

This work was supported in part by the National Science Foundation under grant PHY-9057173 and by the Department of Energy under contract DE-FG02-91ER40676.

REFERENCES

1. B. Lee, C. Quigg, and H. Thacker, Phys. Rev. Lett. **38**, 883 (1977).
2. M. Veltman, Acta. Phys. Polon. **B8**, 475 (1977).
3. M. Chanowitz and M.K. Gaillard, Nucl. Phys. **B261**, 379 (1985).
4. M.Chanowitz, M. Golden and H. Georgi, Phys. Rev. Lett. **57**, 2344 (1987) and Phys. Rev. **D36**, 1490 (1987).
5. J. Cornwall, D. Levin and G. Tiktiopoulos, Phys. Rev. **D10**, 1145 (1974); C. Vayonakis, Lett. Nuovo Cimento **17**, 383 (1976).
6. M. Weinstein, Phys. Rev **8**, 2511 (1973).
7. G. 't Hooft, in *Recent Developments in Gauge Theories*, G. 't Hooft, et. al., eds., Plenum Press, New York NY 1980.
8. K. G. Wilson, Phys. Rev. **B4**, 3184 (1971); K. G. Wilson and J. Kogut, Phys. Rep. **12**, 76 (1974).
9. R. Dashen and H. Neuberger, Phys. Rev. Lett. **50**, 1897 (1983).
10. L. Maiani, G. Parisi and R. Petronzio, Nucl. Phys. **B136**, 115 (1978).
11. M. Lüscher and P. Weisz, Nucl. Phys. **B318**, 705 (1989); J. Kuti, L. Lin and Y. Shen, Phys. Rev. Lett. **61**, 678 (1988); A. Hasenfratz et. al., Phys. Lett. **B199**, 531 (1987); A. Hasenfratz et. al., Nucl. Phys. **B317**, 81 (1989); G. Bhanot et. al., Nucl. Phys. **B353**, 551 (1991) and **375**, 503 (1992) E.
12. U. M. Heller, H. Neuberger, and P. Vranas, Nucl. Phys. **B399**, 271 (1993); K. Jansen, J. Kuti, and C. Liu, Phys. Lett. **B309**, 119 (1993).
13. S. Weinberg, Phys. Rev. **D19**, 1277 (1979); L. Susskind, Phys. Rev. **D20**, 2619 (1979); E. Farhi and L. Susskind, Phys. Rep. **74**, 277 (1981).
14. J. F. Donoghue, C. Ramirez, and G. Valencia, Phys Rev **D38**, 2195 (1988).
15. E. Eichten, I. Hinchliffe, K. Lane, and C. Quigg, Rev. Mod. Phys. **56**, 579 (1984).
16. M. Berger and M. Chanowitz Phys. Lett. **B263**, 509 (1991).
17. J. Bagger, et. al., Phys. Rev. **D49**, 1246 (1994).
18. M. Chanowitz and W. Kilgore, Phys. Lett. B **322**, 147 (1994) and **347**, 387 (1995).
19. R. Cahn and M. Suzuki, Phys. Rev. Lett. **67**, 169 (1991).
20. M. Soldate and R. Sundrum, Nucl. Phys. **B340**, 1 (1990).
21. R. S. Chivukula and M. Golden, Phys. Lett. **B267**, 233 (1991).
22. E. Farhi and L. Susskind, Phys. Rev. **D20**, 3404 (1979).
23. J. Wudka, these proceedings.
24. K. Hagiwara et. al., Nucl. Phys. **B282**, 253 (1987).
25. S. Willenbrock, these proceedings.
26. S. Weinberg and E. Witten, Phys. Lett. **B96**, 59 (1980).
27. S. Weinberg, Physica **96A**, 327 (1979); A. Manohar and H. Georgi, Nucl. Phys. **B234**, 189 (1984); H. Georgi, Weak Interactions and Modern Particle Theory, The Benjamin/Cummings Publishing Company, Inc., Menlo Park, CA, 1984.
28. U. Baur, T. Han, and J. Ohnemus, Phys. Rev. **D51**, 3381 (1995).

Triple Gauge Couplings: Does Lep-1 obviate LEP-2?

Pilar Hernández

Lyman Laboratory
Harvard University
Cambridge, MA 02138

> I discuss the constraints on anomalous gauge boson couplings from present data and compare them with future direct measurement at LEP-200. The conclusion is that LEP-200 is unlikely to reveal new physics in this sector.

INTRODUCTION

A few years ago, in a most controversial work (1) (2), we studied the indirect constraints imposed on anomalous triple gauge boson interactions, by the data from the first two years of LEP running and other low energy experiments. We concluded that direct measurement of these interactions at LEP-200 could hardly improve on those constraints. A new analysis of the present data, including 94' data from LEP, reinforces the conclusion of our previous work. In this paper, I will briefly review the analysis of (1) (2), focussing on the most polemic points, and present the results of an actualized analysis.

The basic result is easy to understand. The key observation is that anomalous gauge interactions not only affect present measurements through quantum corrections at one loop (5), but most importantly by the relations that $SU(2) \times U(1)$ gauge invariance imposes between these interactions and fermion-boson interactions and/or boson self-energies. The latter ones are measured by present data at the 1% level, so this translates into similar constraints on anomalous triple gauge boson interactions (TGV's) through gauge symmetry. LEP-200 on the other hand, will only be able to measure these interactions to the 10% level.

ANOMALOUS GAUGE COUPLINGS

In order to make this argument precise, we first need to parametrize a generic type of new physics. We use an effective lagrangian parametrization (6). The only assumption here is that the new physics is caracterized by an energy scale Λ distinct from the M_z scale. Typically, $\Lambda \gg M_z$. In this situation, the physics at $E \sim M_z$, can be described by an effective theory obtained after integrating out the degrees of freedom relevant at the scale Λ.

No trace is left of the new physics other than a tower of new interactions of light particles, which are described by higher dimensional operators, weighted by appropiate powers of Λ. The unkown couplings of these higher dimensional operators parametrize the new physics:

$$\mathcal{L}_{eff} = \mathcal{L}_{light} + \sum_i \alpha_i \frac{\mathcal{O}^{d_i}}{\Lambda^{d_i-4}} \tag{1}$$

Well-known are the uses of effective lagrangians to study deviations from the SM (6). This parametrization is simple and has some important properties,

- Generality: if we allow all possible higher dimensional operators, we can in principle parametrize a general class of theories.

- Independence of the process: the q^2 dependence is well reproduced for $q^2 < \Lambda^2$. This is important since we want to compute theoretical predictions for processes at different q^2.

- Predictability: if the energy of the experiment $E << \Lambda$ and if the strength of the higher dimensional operators is that determined by their cutoff dependence (naturality assumption: $\alpha_i \sim O(1)$), we can truncate the infinite series and consider, for instance, only the leading effects $O(E/\Lambda)$. This reduces the set of parameters to a small number and thus gives predictive power whenever we have more observables than parameters.

Still, we have to decide which are the gauge symmetries G that we must impose on the effective lagrangian. This will determine also the form of $\mathcal{L}_{light}$, since, once the particle content is known, the low energy lagrangian contains all the possible interactions allowed by symmetries. At this point we may distinguish two situations.

Decoupling New Physics The low energy theory is renormalizable (light Higgs), i.e. the limit $\Lambda \to \infty$ of eq. 1 exists. In this case, $SU(2) \times U(1)_Y \in G$, and all the higher dimensional operators must be symmetric under $SU(2) \times U(1)_Y$. Possible underlying theories of this type include extended gauge groups, SUSY, extended technicolor, L-R models, GUTS, etc.

A list of all the next-to-leading operators $\sim O(E^2/\Lambda^2)$ that preserve $SU(2) \times U(1)$ and CP was given in ref. (7). Of the full list, we only consider those giving rise to anomalous triple gauge vertices (TGV's). They are listed in Table 1. The so called blind directions produce direct effects on the triple couplings, δg_Z, $\delta \kappa_\gamma$, $\delta \kappa_Z$, λ_γ, λ_z, but do not produce any tree level effect on present observables. Their effects start at one loop. The fermionic operators only have indirect effects through the renormalization of the input parameters, which we choose to be those best measured α, G_F and M_z. Their effects on present observables are thus tree level. Finally, the bosonic operators produce both direct and indirect effects and induce also tree level effects on present experiments.

Bosonic Operators	Fermionic Operators	Blind Directions
$\mathcal{O}_{WB} \equiv \Phi^\dagger \vec{\sigma}\, \Phi \vec{W}_{\mu\nu} B^{\mu\nu}$	$\mathcal{O}_{e\mu} \equiv \vec{J}_\rho(L_e)\vec{J}^\rho(L_\mu)$	$\mathcal{O}_W \equiv 1/3!\vec{W}_\mu^\nu \times \vec{W}_\nu^\lambda \cdot \vec{W}_\lambda^\mu$
$\mathcal{O}_\Phi \equiv J_\rho(\Phi)J^\rho(\Phi)$	$\mathcal{O}_e \equiv \vec{J}_\rho(\Phi)\vec{J}^\rho(L_e)$	$\mathcal{O}_{B\Phi} \equiv iB^{\mu\nu}(D_\mu\Phi)^\dagger D_\nu\Phi$
$\mathcal{O}_{DW} \equiv [D^\rho \vec{W}_{\mu\nu}]^\dagger [D_\rho \vec{W}^{\mu\nu}]$	$\mathcal{O}_\mu \equiv \vec{J}_\rho(\Phi)\vec{J}^\rho(L_\mu)$	$\mathcal{O}_{W\Phi} \equiv i\vec{W}^{\mu\nu}(D_\mu\Phi)^\dagger \vec{\sigma} D_\nu\Phi$

TABLE 1. Effective operators parametrizing decoupling new physics. With $J_\rho(\Phi) \equiv i\frac{1}{2}\Phi^\dagger D_\rho\Phi + h.c.$, $\vec{J}_\rho(\Phi) \equiv i\frac{1}{2}\Phi^\dagger \vec{\sigma} D_\rho\Phi + h.c.$, $\vec{J}_\rho(L_l) \equiv \frac{1}{2}\bar{L}_l\gamma_\rho\vec{\sigma}L_l$.

Non-Decoupling New Physics The low energy is non-renormalizable. In this case, the limit $\Lambda \to \infty$ does not exist, or in other words, the scale Λ cannot be arbitratily large and must be related to the scale M_z. It is still possible that the ratio M_z/Λ is small enough so that an effective lagrangian treatment at $E \sim M_z$ is justified (6). The most interesting possibility of this kind is that of a dynamical symmetry breaking sector, e.g. technicolor.

The most general effective lagrangian to describe the spontaneous symmetry breaking pattern $G \to H$ has a non-linearly realized (12) symmetry group G and a linearly realized H. Some assumptions are needed to incorporate the successes of the SM to describe the physics up to $E \leq M_z$. The minimal of them being:

$$SU(2) \times U(1) \quad \in G$$
$$U(1)_{em} \quad \in H$$
$$\text{3 light goldstone bosons}$$

Only two possibilities satisfy the previous assumptions (13):

$$G = SU(2)_L \times SU(2)_R \times U(1)$$
$$H = SU(2)_C \times U(1)$$

or

$$G = SU(2)_L \times U(1)$$
$$H = U(1)$$

In the first case, $\rho = 1$, while in the second case $\rho - 1 \sim O(1)$ naturally. Only a fine tunning can give $\rho - 1 \sim 1\%$, as observed. For this reason, we only consider the first possibility.

The fermion interactions of the effective lagrangian are assumed to be the same than those of the SM [1], while the leading boson interactions are contained in operators of the form,

$$\mathcal{O}_i \sim \beta_i v^2 \Lambda^2 Tr[(\frac{D_\mu}{\Lambda})^{n_d} U^{n_g} (\frac{gW^{\mu\nu}}{\Lambda^2})^{n_2} (\frac{g'B^{\mu\nu}}{\Lambda^2})^{n_1}] \qquad (2)$$

---

[1] In models of dynamical symmetry breaking anomalous fermion-boson interactions are naturally smaller, since they occur at higher order loops or are proportional to the yukawa couplings.

Bosonic Operators	Blind Directions
$\mathcal{L}_1 \equiv gg'\beta_1 B^{\mu\nu} Tr[TW_{\mu\nu}]$	$\mathcal{L}_2 \equiv ig'\beta_2 B^{\mu\nu} Tr[T[V_\mu, V_\nu]]$
$\mathcal{L}'_1 \equiv g'^2 \beta'_1 v^2 (Tr[TV_\mu])^2$	$\mathcal{L}_3 \equiv ig\beta_3 Tr[W^{\mu\nu}[V_\mu, V_\nu]]$
$\mathcal{L}_8 \equiv g^2 \beta_8 (Tr[TW_{\mu\nu}])^2$	$\mathcal{L}_9 \equiv ig\beta_9 Tr[TW_{\mu\nu}]Tr[T[V_\mu, V_\nu]]$

TABLE 2. Effective operators parametrizing non-decoupling new physics. With $T \equiv U\sigma_3 U^\dagger$, $V_\mu \equiv (D_\mu U)U^\dagger$ and $W^{\mu\nu} \equiv \frac{\vec{\sigma}}{2}\vec{W}_{\mu\nu}$.

A power counting analysis of loop corrections (11) gives,

$$\beta_i \sim \beta_i^{tree\ level} + O(\frac{\Lambda^2}{4\pi v^2}) \qquad (3)$$

The choice $\Lambda \sim 4\pi v$, sets the natural strength of the couplings $\beta_i \sim 1$. This also implies that an effective lagrangian is possible for $E \sim v < \Lambda$, where the sum of all the operators eq. 2 can be safely truncated.

The lowest order bosonic lagrangian is described by standard kinetic terms for the transverse components of the gauge fields together with,

$$\mathcal{L}^o = \frac{v^2}{2} Tr(D^\mu U^\dagger D_\mu U) \qquad (4)$$

whose coefficient is determined by the scale of symmetry breaking $v \sim 246 GeV$. $U = exp(i\pi_a \sigma_a/v)$ and its covariant derivative,

$$D_\mu U = \partial_\mu + ig\frac{\vec{\sigma}}{2}\vec{W}_\mu U - ig' U \frac{\sigma_3}{2} B_\mu \qquad (5)$$

To next-to-leading order $\sim E^2/\Lambda^2$ the list was given in ref. (8)[2]. Those that modify the triple gauge coulings are listed in Table 2.

- Generality versus Symmetry

The need for imposing gauge invariance has been often called a theoretical prejudice, restricting the generality of new physics that we should look for in the experiment. The reason being that gauge symmetry is broken at low energies in any case. This lost of generality is however only apparent. In reality, imposing $SU(2) \times U(1)$ symmetry is the most economic way of looking for the relevant degreees of freedom. In Fig. 1, we show a typical plane of two anomalous directions in the gauge couplings. The arrow points in the direction where gauge symmetry is exact. The contour is the constraint obtained from precission measurements. If we look for new physics in any non-symmetric direction, by considering its observable effects through radiative corrections, we will find generically that these are quadratically divergent. If we then

---

[2] We do not consider a very interesting operator which violates C and P but preserves CP. See (9) (10).

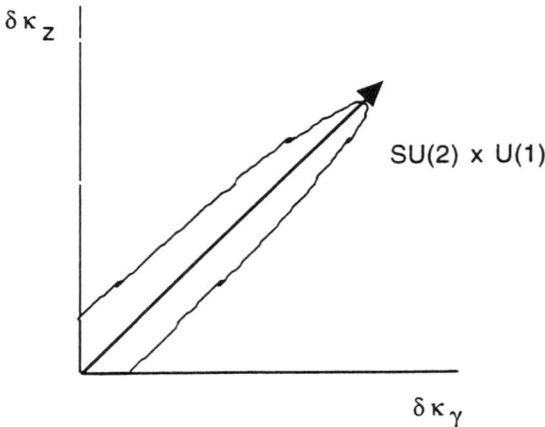

FIG. 1. Experimental constraints in a $SU(2) \times U(1)$ symmetric theory.

interpret the cutoff as the physical scale associated with the anomalous gauge couplings, we will find a very strong limit on this scale (5). Again a lot of confussion surrounds the interpretation of these quadratic divergences. It is well known that although the coefficients of the logarithmic divergences can be computed exactly in the effective theory, the quadratic corrections cannot. This however, does not mean that these divergences are not physical. They are indeed. They tell us that a counterterm must be present, which will have an effect at tree level on the precission measurements and whose natural scale will be that determined by the divergence. In other words, the constraint derived by considering the quadratically divergent term is a good order of magnitude estimate.

On the other hand, when new physics points in a symmetric direction, two situations can occur. In the first case, the anomalous triple gauge couplings are related to other couplings by the symmetry, and these cases will be typically constrained at the 1% level aswell. Else, they can be genuine "blind directions", when no anomalous couplings occur in fermion-boson interactions or boson self-energies. In this case, the gauge symmetry ensures there are no quadratic divergences. Consecuently, these directions are less constrained by present data and determine the small window, where LEP-200 could give some information.

PRESENT CONSTRAINTS ON TGV'S

For the non-blind directions, a tree level analysis is enough. The effects on present experiments enter through corrections to the boson self-energies and/or to the fermion boson couplings. The reader is referred to refs. (1), (2) for the explicit formulae.

We express the results in terms of the dimensionless parameters,

$$\epsilon_i \equiv \alpha_i \frac{v^2}{\Lambda^2} \qquad (6)$$

for the operators of the first and second columns of table 1. The results for the operators in the first column of table 2 are expressed in terms of the β_i couplings.

For blind directions, the one-loop calculations have been computed in refs. (1)- (3). Recently also in ref. (4). For details, we refer to the original references. The only point I would like to discuss here is the calculation of the effects of blind directions in the non-linear lagrangian, i.e. operators $\mathcal{L}_2$, $\mathcal{L}_3$ and $\mathcal{L}_9$ in table .

As expected, from the non-decoupling nature of the new physics in this case, there are quadratic divergences in the one-loop calculation (for blind directions in the linearly realized effective lagrangian, the corrections are at most logarithmic). There are two possible interpretations of these terms, both leading to the same conclusion. The first is just to consider them as physical with $\Lambda \sim 4\pi v$. The other possibility is to renormalize them. This however requires adding one or more counterterm in non-blind directions, of the same order in the expansion E/Λ as the original direction, i.e. a combination of $\mathcal{L}_1$, $\mathcal{L}'_1$ and $\mathcal{L}_8$. Now, from eq. 3, we see that the natural size of the coupling of this counterterm is set precisely by the quadratic divergences we have computed. Thus, we conclude that typically, within an order of magnitude accuracy, the quadratic divergences give the right physics.

One logical possibility is that the naturality assumption eq. 3 breaks down. In this situation quadratic divergences may give a completely wrong answer, but in this case even the use of the truncated effective lagrangian comes into question and the predictibility of the method is lost. We will assume this is not the case, for selfconsistency.

For the actual analysis, we use the following set of experimental measurements:

LEP (94' Glasgow)	$\Gamma_Z, \sigma_h, R_l, A^l_{FB}, A^b_{FB}, P_\tau$ (15)
Parity Violation in Cesium	Q_W (16)
W Mass	M_W (17)
Deep Inelastic Neutrino Scattering	R_ν (19)

This set is optimal in the sense that the experimental correlations are minimal and we neglect them. We also assume there are no accidental cancellations between the different anomalous directions.

The SM predictions have been obtained with ZFITTER motecarlo ref. (14). We leave m_t as a free parameter and include the window of $M_H \in$ ($50 GeV, 1 TeV$) into the theoretical error. For each operator, we construct,

$$\chi_i^2 = \sum_{j=1}^{N_{obs}} \frac{(X_j^{exp} - X_j^{SM}(m_t) - \epsilon_\alpha \delta X_{\alpha j})^2}{(2\Delta_j^{exp} + \Delta_j^H + \Delta_j^{th})^2} \qquad (7)$$

where X_j^{exp} and X_j^{SM} are respectively the measured value of the observable j and the theoretical prediction. $\delta X_{\alpha j}$ is the correction produced in the observable j by the operator α. Finally, $\Delta_j^{exp}, \Delta_j^H$ and Δ_j^{th} are respectively the experimental error (we work at the 2σ confidence level), the theoretical error due to the Higgs mass window and the theoretical error from ZFITTER.

The constraints are obtained from the 2σ contours,

$$\chi_i^2(m_t, \epsilon_i) = \chi_i^2(min) + 1 \qquad (8)$$

The numerical results are the following,

- Decoupling Physics (Table 1)
  Non-Blind Directions $|\epsilon_i| \leq 0.1 - 1\ \%$
  Blind Directions $|\epsilon_i| \leq 1$

- Non-Decoupling SB Sector (Table 2)
  Non-Blind Directions $|16\pi^2 \beta_i| \leq 1$
  Blind Directions $|16\pi^2 \beta_i| \leq 1 - 5$

FUTURE CONSTRAINTS ON TGV'S AT LEP200

A simple analysis to estimate the future constraints at LEP-200 on the operators of tables 1 and 2 was carried out in (1), (2). The study is based on the differential cross sections, which is the most sensitive observable, taking into account the small statistics expected. We did a montecarlo simulation of 10^4 W pairs at $E_{cm} = 200 GeV$, which corresponds to an integrated luminosity of $500 fb^{-1}$. We then performed a 2σ χ^2-test on the anomalous differential cross section, normalized to its standard value. Again we assumed there are no accidental cancellations between the different operators and considered one at a time.

Fig. 2 shows the χ^2 contours for non-blind directions. The region outside the vertical arrows would be excluded by LEP200. We have superimposed the present constraints. In all cases, non-blind directions are constrained already enough to safely conclude that LEP-2 will not be sensitive to anomalous deviations of the gauge couplings in these directions. The results for the chiral operators $\mathcal{L}_1$ and $\mathcal{L}_1'$ can be read using the relations in unitary gauge,

$$\mathcal{L}_1 \to -\frac{2gg'\beta_1}{v^2} O_{WB}$$

$$\mathcal{L}_1' \to -\frac{16g'^2 \beta_1'}{v^2} O_\Phi$$

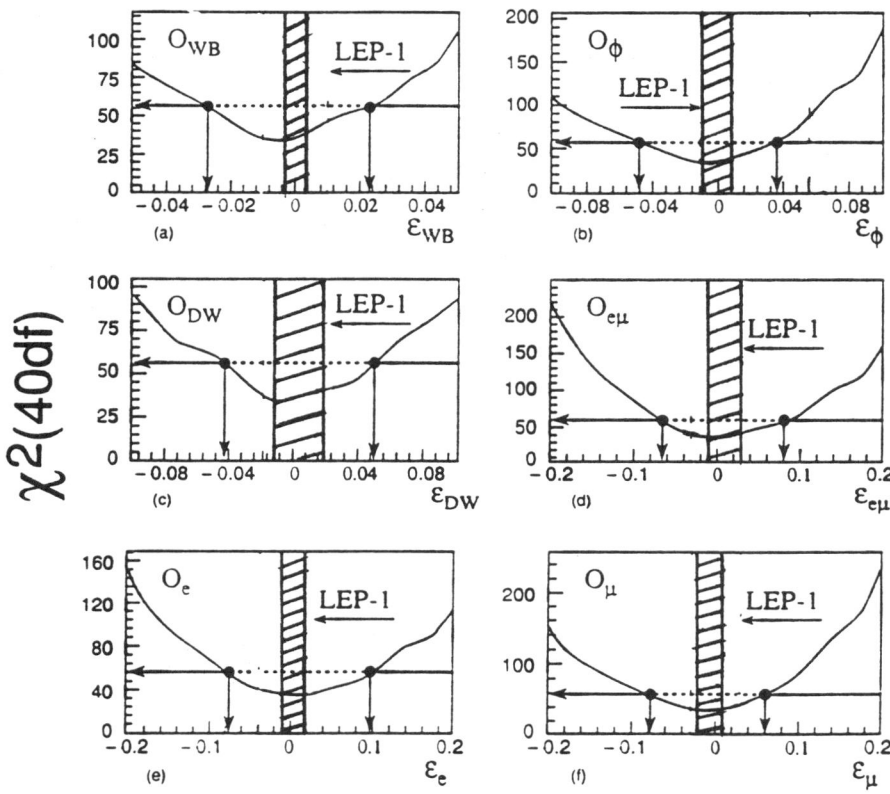

FIG. 2. χ^2 contours measuring the significance of the effect of non-blind operators at LEP200. The arrows signal the 2σ confidence level values. The vertical bands are the regions allowed from present data.

The results for blind directions, in the case of decoupling physics (light Higgs) are however more promising. Fig. shows the 2σ allowed horizontal bands from LEP-200, as compared to the present 2σ contours. While new physics pointing in the direction of $O_{B\Phi}$ is presently excluded, bounds on anomalous O_W and $O_{W\Phi}$ could be improved by a factor 3-5 at LEP200.

On the other hand, blind directions for a non-decoupling symmetry breaking sector are much better constrained at present and there is no window for LEP-2. See (2), (4).

More sophisticated analysis including the information for polarized cross sections do not improve our results significantly and definitely does not change the conclusions. In the most optimistic situation only a 20

CONCLUSIONS

We have seen that a effective lagrangian parametrization of generic new physics affecting the triple gauge coupling interactions leads naturally to simultaneous anomalies in fermion-boson interactions and boson self-energies, which are strongly constrained by present data to be smaller than 1% (18). These relations are imposed by gauge invariance $SU(2) \times U(1)$ in the case of decoupling new physics and by custodial symmetry in the case of a non-decoupling symmetry breaking sector. The genuine anomalous gauge coupling interactions are represented by the so called blind directions. For the decoupling physics there are three,

$$\begin{aligned} O_W &\rightarrow & \lambda_\gamma = \lambda_z \\ O_{B\Phi} &\rightarrow & \kappa_z = -\tfrac{s^2}{c^2}\kappa_\gamma \\ O_{W\Phi} &\rightarrow & \kappa_\gamma = scg_Z = -\tfrac{c^2}{s^2}\kappa_z \end{aligned}$$

For non-decoupling symmetry breaking sector the relevant directions are,

$$\begin{aligned} \mathcal{L}_9 &\rightarrow & \kappa_\gamma = \kappa_z \\ \mathcal{L}_2 &\rightarrow & \kappa_z = -\tfrac{s^2}{c^2}\kappa_\gamma \\ \mathcal{L}_3 &\rightarrow & \kappa_\gamma = scg_Z = -\tfrac{c^2}{s^2}\kappa_z \end{aligned}$$

As expected, non-blind directions have been found to be too tightly constrained by present data to expect any new information from LEP-200. In the case, of a non-decoupling sector, new physics cannot point exclusively in blind directions, since they generate quadratic divergences that must be renormalized by counterterms pointing in non-blind directions. Taking into account the natural size of these counterterms, we conclude that no effect will be found at LEP-200 either. Finally, the only window left for LEP-200 are the blind directions O_W and $O_{W\Phi}$. Although it is possible that new physics points exclusively in these directions, no model or symmetry is known that would produce such effects.

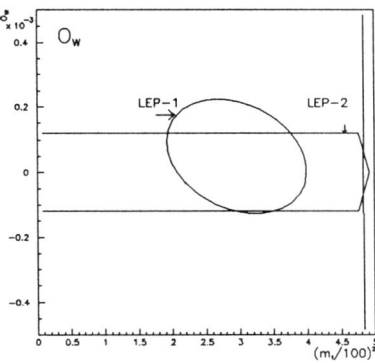
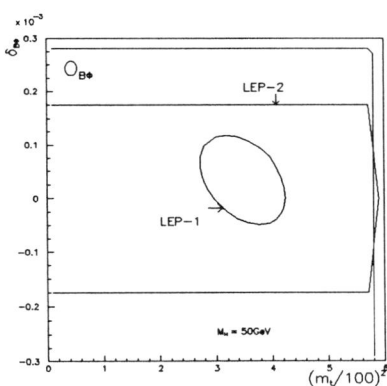
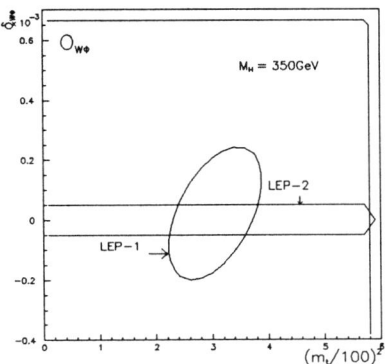

FIG. 3. 2σ confidence level contours allowed by present data in the blind directions of decoupling new physics, compared with LEP200 exclusion horizontal bands. $\delta_{W\Phi} \equiv (\alpha^3/4\pi)^{1/2}/(4sc^2)\epsilon_{W\Phi}$ $\delta_{B\Phi} \equiv (\alpha^3/4\pi)^{1/2}/(4s^2c)\epsilon_{B\Phi}$ and $\delta_W \equiv (\alpha^3/4\pi)^{1/2}/(8s^3)\epsilon_W$

Although the prospects for LEP-200 in finding anomalies in the gauge sector look very unpromising, other measurements from LEP-200 can be important. M_W is expected to be measured with a precission 1%, improving considerably present precission tests of the SM. Also, the analysis presented here, which is based on an effective lagrangian parametrization, has nothing to say about the possibility of new resonances appearing in the energy range of LEP-200.

ACKNOWLEDGMENTS

I thank my collaborators in the work presented here A. De Rújula, M.B. Gavela, E. Massó and F.J. Vegas. This work has been partially supported by grant NSF-PHYS-92-18167.

REFERENCES

1. A. De Rújula et al., Nucl. Phys. B **384**, 3-58 (1992).
2. P. Hernández, F.J. Vegas, Phys. Lett. B , 5298 (1978).
3. K. Hagiwara et al., Phys. Lett. **B283** (1992) 353.
4. S. Dawson and G. Valencia, hep-ph 9410364 to appear in Nucl. Phys. **B**.
5. See for instance G.L. Kane, J. Vidal and C.P.Yuan, Phys. Rev. **D39** (1989) 2617.
6. For a review and references see H. Georgi, Ann. Rev. Nucl. and Part. Sci. **43**, 209 (1993).
7. W. Buchmuller and D. Wyler, Nucl. Phys. **B268** (1986) 621; C. Burges and H.J. Schnitzer, Nucl. Phys. **B228**(1983) 464; C. Leung, S.T. Love and S. Rao, Z. Phys. **C31** (1986) 433.
8. A. Longhitano, Nucl. Phys. **B188** 118 (1981); T. Appelquist and C. Bernard, Phys. Rev. **D22**, 200 (1980).
9. G. Valencia, these proceedings.
10. T. Appelquist and G.H. Wu, Phys. Rev. **D51** (1995)240.
11. A. Manohar, H.Georgi, Nucl. Phys. bf B234(1984) 189.
12. S.Coleman, J. Wess and B. Zumino, Phys.Rev. **177**(1969)2239; C.G.Callan, S. Coleman, J. Wess and B. Zumino, hys.Rev. **177**(1969)2247.
13. M. Chanowitz, H. Georgi, M. Golden, Phys. Rev. **D36** (1987)1490.
14. D.Bardin et al., CERN-TH.6443/92 and references thererin.
15. D.Schaile, CERN-PPE-94-162.
16. M.C. Noecker et al. Phys.Rev.Lett. **61** (19988) 310; S.A. Blundell et al. Phys. Rev. Lett. **65** (1990) 1411; V.A.Dzuba et al. Phys.Lett **141A**(1989)147.
17. J. Alliti et al. UA2 Col, Phys.Lett **B241**(1990)150. F. Abe et al. CDF Col. Phys.Rev. **D43** (1991) 2070.
18. D. Schaile, these procedings.
19. H. Abramowicz et al. CDHS collaboration. Phys.Rev.Lett **57** 9(1986) 298; J.V.Allaby et al. CHARM Coll, Phys.Lett. **B177**(1986)446.
20. S. Katsanevas et al. DELPHI 92-166 PHYS 250(1992); D. Treille, CERN-PPE 93-54. M. Bilenky et al. BI-TP 92/44 (1993).

Understanding Something About Nothing: Radiation Zeros

Robert W. Brown[1]

Physics Department
Case Western Reserve University
Cleveland Ohio 44106

Radiation symmetry is briefly reviewed, along with its historical, experimental, computational, and theoretical relevance. A sketch of the proof of a theorem for radiation zeros is used to highlight the connection between gauge-boson couplings and Poincare transformations. It is emphasized that while mostly bad things happen to good zeros, the weak-boson self-couplings continue to be intimately tied to the best examples of exact or approximate zeros.

INTRODUCTION

There are radiation zeros all over the place. See Fig. 1. Almost all Born amplitudes for the radiation of photons and gluons and other massless gauge bosons have such zeros. One reason no one notices them, however, is that only a small fraction occur in physical regions of scattering or decay. See Fig. 2 later on.

The conditions for physical "null zones" are sufficiently restrictive that the first examples of radiation zeros (the Mikaelian-Samuel-Sahdev zeros (1)), were discovered only when radiative weak-boson production was analyzed. (Striking dips in the angular distributions of $\bar{\nu}e \to W\gamma$ and $q\bar{q} \to W\gamma$ corresponding to the zeros had been seen earlier. (2)) The zero in the $q\bar{q}$ channel has proven to be an important signature in testing the trilinear $WW\gamma$ couplings, as wonderful progress is drawing tighter and tighter bounds. So far, all are converging to the standard model predictions. (3) The progress has been accelerated by theoretical support by Baur, Errede, and Landsberg (4), showing that laboratory rapidity correlations involving the photon and the charged decay lepton display a pronounced dip if the radiation zero exists. In a very nice talk, Tao Han has shown us the details (and plots) at this conference. (5)

---

[1] Certain work behind this paper has been supported in part by the NSF and the CWRU industrial problem solving group.

FIG. 1. Even in the early days of QED, zeros could have been found.

Additional history. Permit me to digress for a moment. Our original work (6,2) on the production of electroweak pairs in proton collisions

$$p\bar{p},\ pp \to WW, ZZ, WZ, W\gamma \tag{1}$$

took place in the late 70's. This was presented as an alternative to electron colliders, betting that the necessary energy threshold would first be reached by proton machines. And the theoretical discovery of the first radiation zero came in the midst of looking at the corresponding Born calculations. So it is gratifying, after all these years, to be at the present conference where many experimental results, including those pertaining to the $W\gamma$ zero, have now been achieved.

In view of the role of the $W\gamma$ channel, I have been given the opportunity to review the theory of the radiation zero, and to connect it to another signature of interest, "approximate" zeros. Owing to the fact that out of all the people from all the old collaborations from CWRU, Oklahoma State, Wisconsin, and SLAC, I happen to be the only one present means I have some responsibility to mention the early work.[2] This is also a welcome chance to talk about a theoretical development that, because of the lack of examples, is not well known, but perhaps should be. Continuing experimental progress provides a good incentive.

---

[2] I was asked by a young CWRU graduate student whether this meant all the others had passed away.

In particular, I would like to emphasize a picture that emerges in the proof of general radiation zero theorems. The emission, or absorption, of a gauge boson can be viewed as a local Lorentz transformation (or better, a Poincare transformation) of the particle doing the emitting or absorbing. This may be of importance in future model building, especially since our only successful theories to date, gauge theories, lead to universal forms for this transformation.

RADIATION INTERFERENCE THEOREMS

It seems to me that the experimental, computational and theoretical consequences of radiation zeros and their generalizations are easiest to describe if we first look at an archetypal theorem for their existence and location. The classical limit of such zeros will also be evident. The primary example of the general set of radiation zero theorems found by Brodsky, Kowalski, and meeski (7,8) is for the emission of single photons.

Single-photon theorem

The theorem is the following. Consider the quantum amplitude M_γ for the emission of a single photon with momentum q. Besides the photon, assume there are n other particle legs (k particles in and $n-k$ particles plus γ out, say). The theorem is that the tree amplitude approximation vanishes, independent of anybody's spin, for common charge-to-light-cone-energy ratios, viz.

$$M_\gamma(tree) = 0$$
$$\text{if } \frac{Q_i}{p_i \cdot q} = \text{same, all } i \qquad (2)$$

where the ith particle has electric charge Q_i and four-momentum p_i. The stipulations are that all couplings must be "gauge-theoretic". That is, the photon couplings to the particles must be as prescribed by local gauge theory, and any derivative couplings among the particles themselves must be gauge covariant. Also, scalar, spinor, and vector particles can be accommodated (spins ≤ 1). We return at the end to the question of higher spins.

Physical null zones

The factors $\frac{Q_i}{p_i \cdot q}$ come from the coupling and particle propagator denominator. It is easy to go into the complex plane to make them equal, and hence zeros are all over the (complex) place. But for them to be equal in physical phase space, the first and obvious requirement is that all charges must have the same sign (since $p \cdot q \geq 0$). This knocks out many reactions. By the way, there are only $n-2$ independent equations. Interestingly and importantly, the last ratio is automatically the same if the rest are equal, by virtue of charge and momentum conservation.[3]

In the garden variety reactions, we could look, for example, at electron-electron bremsstrahlung, $e^-e^- \to e^-e^-\gamma$. The Born amplitude vanishes if the photon is at right angles to the c.m. beam direction, and the final electrons have the same energy. The null zone is two-dimensional, and the reason this radiation zero was not noticed in radiative corrections calculations is that the final-state phase space is sufficiently high dimensional.

Relevance

EXPERIMENTAL The null zone conditions explain why it took so long to discover radiation zeros. *We had to wait for fractionally charged quarks and weak bosons in order to get three things: Same-sign fermion-antifermion pairs, a process well-approximated by a Born amplitude, and only three particles plus the photon so the null zone was simple.* The amplitude for $q\bar{q} \to W\gamma$ vanishes at a photon scattering angle determined by the quark charges. (The $\nu e \to W\gamma$ amplitude vanishes at the edge of phase space and is easily misinterpreted as a helicity constraint.) The corresponding null zone in radiative W decay (9) is a line in the Dalitz plot.

The first is still the best example. Again, I can refer to Tao Han's talk, but the relevance is that only for the very restrictive couplings coming out of gauge theory does the zero occur. In view of the results of many calculations with anomalous couplings, and the proof (see below) of the pertinent theorem, I do not think it would be hard to put together another theorem. Accepting the present particle content of our world, *only in the standard electroweak theory will there be a zero in the $W\gamma$ reaction*. In any other theory including composite models the rapidity dip will get filled in. We say such experiments test gauge theories and test the electromagnetic properties of weak bosons. :)

What about other photonic zeros? The aforementioned zero in electron-electron bremsstrahlung is less interesting as a test, and involves more particles. There has been work on electron-quark and quark-antiquark bremsstrahlung, as tests of quark properties, but these lead to jet identification experiments, along with the more complicated phase space. (10–13) We will come back to the subject of more tests of weak-boson properties in a bit. :|

But as noted most zeros are unphysical. :(

COMPUTATIONAL As a by-product of the general proof of radiation zeros, we learn how to rewrite the set of Feynman tree diagrams as a smaller number of factored terms, separately gauge-invariant. It is possible to combine

---

[3] What I'm saying is, if you have 10% rotten apples and 10% rotten oranges, then the overall percentage of rotten fruit is the same 10%.

the CALKUL photon polarization vectors, for which the fermionic degrees of freedom are used as the bases, and very much simplify the amplitude analysis.

The radiation symmetry may be used, as gauge invariance is used, to check the increasingly laborious calculations used in higher-order perturbation theory studies. There is an analogous non-Abelian radiative symmetry that exists to check QCD jet calculations, for example.

THEORETICAL The mechanism in the radiation zero phenomena that has rather uniquely shown itself is the relation between a photon coupling and local Poincare invariance. We repeat that only gauge theory couplings lead to universal space-time transformations, and it is of value to sketch the proof in order to see how this connection is exposed and what universality means.

PROOF HIGHLIGHTS

We can describe the proof fairly succinctly in terms of photon attachments in a minute. But first let us lay some groundwork.

Vertex source graphs

Define a "source" graph T_G with which we will generate the complete tree radiation amplitude by the attachment of a photon in all possible ways. After simplification of the product of vertex and propagator for the attachment of a photon to a given leg, the complete tree for a source graph V_G made up of a single vertex (no internal lines) is

$$M_\gamma(V_G) = \sum \frac{Q_i J_i}{p_i \cdot q} \qquad (3)$$

where we see the $Q/p \cdot q$ factors emerge from the coupling and propagator denominator. The coefficient "vertex current" J_i is the result of inserting the current j_i into the ith leg, with

$$j = j_{conv} + j_{spin} + j_{cont} + j_{YM} \qquad (4)$$

The convection current is $p \cdot \epsilon$,[4] the spin current is a first-order momentum-space Lorentz transformation of the wave function (e.g., $\frac{i}{4}\sigma^{\alpha\beta}\omega_{\alpha\beta}$ for a spinor), the contact current is the corresponding transformation of single derivative couplings and the Yang-Mills current has the form of $\omega_{\mu\nu} \times$ (the q terms in the Yang-Mills vertex). The first-order Lorentz parameter tensor is

$$\omega_{\mu\nu} = q_\mu \epsilon_\nu - \epsilon_\mu q_\nu \qquad (5)$$

---

[4] Sign changes in going from final to initial legs are left understood throughout these formulas.

where ϵ is the photon polarization vector.

The key identity is that the sum over the vertex currents is zero

$$\sum J_i = 0 \qquad (6)$$

because the sum over the convection currents vanishes by momentum conservation, the sum over the combination of spin currents and contact currents vanishes by Lorentz invariance, and the sum over the Yang-Mills current vanishes by the Bianchi identity. Equal $Q/p \cdot q$ factors can be pulled out of the sum Eq.(3) and the complete radiative process for a vertex source graph vanishes by Eq.(6).

Allowed couplings

The vertex graph could have involved any number of particles, but the universal forms of the various currents j_i are preserved only if there are important restrictions on any derivative couplings present. There can be no derivatives of Dirac fields; single derivatives of scalar fields are allowed; single derivatives of vector fields and double derivatives of scalar fields are allowed but only in Yang-Mills trilinear form (reminding us of the relationship between the longitudinal vector boson and the Goldstone bosons in spontaneous symmetry breaking). And the photon couplings also must follow the gauge algorithm: All derivatives are replaced by covariant derivatives.

Connection to space-time symmetry

A succinct way of looking at the previous result is that the attachment of a photon generates transformations. The convection current corresponds to a (first-order) displacement of that leg's wave function, the spin current to its Lorentz transformation, and the contact current to a Lorentz transformation of the derivative coupling. For common $Q/p \cdot q$, all these transformations correspond to the same universal element and they cancel by invariance. The photon attachment is a first-order Poincare transformation on the vertex amplitude, but by the invariance of that amplitude, it becomes "unattached" in the null zone.

General source graphs

To finish the proof, consider general tree source graphs T_G with (fixed) internal lines. In tacking a photon in all possible ways onto the source graph, the new ingredient is internal line attachment. But there are Ward-like identities[5] that simplify the problem and are of the generic form (D denotes propagators)

---

[5] For a scalar line, the identity is the same as the Ward identity.

$$D(p-q)\Gamma D(p) + \text{(seagulls, if any)} = D(p-q)j\frac{Q}{p\cdot q} + \frac{Q}{p\cdot q}jD(p) \qquad (7)$$

Note that we get exactly the same kind of $\frac{Q}{p\cdot q}j$ factor for each internal leg of each vertex that we had for the external legs. Furthermore, in the null zone the internal ratio $\frac{Q}{p\cdot q}$ will equal the common external ratio by the same rotten fruit calculation described earlier. These identities have reduced the problem to a sum of vertex-source problems, and by the earlier arguments, the null zone applies to all trees.

Gauge invariant reorganization of Feynman diagrams

The result of applying the decomposition identities and the contracted forms of the external leg attachments is a new reduced and rearranged amplitude, a sum over vertices

$$M_\gamma(T_G) = \sum M_\gamma(V_G) R(V_G) \qquad (8)$$

with the vertex attachments separated out from the rest of the graph factors R. Each term in the sum is separately gauge invariant. The radiation zero is evident from the fact that each factor $M_\gamma(V)$ vanishes in the null zone.

Alternate theorem. Notice that charge conservation $\sum Q_i = 0$ is dual to Eq.(6) and implies

$$M_\gamma(T_G) = 0 \quad \text{if} \quad \frac{J_i}{p_i\cdot q} = \text{same} \qquad (9)$$

This alternate interference theorem refers to zeros that, by contrast, are spin-dependent but independent of charge.

Radiation symmetry and representation

Both theorems can be stated as symmetries. (14) They correspond to invariance under either replacement,

$$\frac{Q_i}{p_i\cdot q} \to \frac{Q_i}{p_i\cdot q} + C \quad \text{or} \quad \frac{J_i}{p_i\cdot q} \to \frac{J_i}{p_i\cdot q} + C' \qquad (10)$$

By the appropriate choice of C and C', all zeros can be made explicit in a "radiation representation" for the vertex amplitudes:

$$M_\gamma(V_G) = \sum p_i\cdot q \left(\frac{Q_i}{p_i\cdot q} - \frac{Q_j}{p_j\cdot q}\right)\left(\frac{J_i}{p_i\cdot q} - \frac{J_k}{p_k\cdot q}\right) \qquad (11)$$

for $i \neq j, k$. This reduces $M_\gamma(T_G)$ to a single factored form for the simplest case, $n = 3$, in agreement with the original work by Goebel, Halzen and Leveille. (15,16) Their beguiling statement for the original $q\bar{q} \to W^\pm\gamma$ is that the three Feynman amplitudes reduce to the sum of two Abelian (QED-like) diagrams multiplied by a factor in which the zero is explicit.

MULTI-PHOTONS AND GENERAL DECOUPLING THEOREM

What about neutrals amongst the n particles, including other photons? The quick answer from an examination of the $Q/p \cdot q$ factor is the right one. All neutral particles must be massless and travel in the same direction as the photon. Strictly, in the proof we must have the corresponding current J_i vanish in the null zone, and this is what is found.[6]

In fact, the multiphoton zeros and the connection to local Poincare transformations follow from a decoupling theorem (17,18) for the scattering of a system of particles immersed in an external electromagnetic plane wave. On the way to the theorem it is first shown that for a particle coupled to an external electromagnetic plane wave the wave functions for spins 0, 1/2, 1 all can be written in the form

$$\Psi(x) = ULT\chi(x) \tag{12}$$

where χ is the free solution. The U, L, T are local gauge, Lorentz, and displacement transformations, respectively, whose first-order terms are exactly the ones we have been talking about.

The identity used to show that these are indeed solutions ends up being a cornerstone to the whole radiation zero business. It is

$$(UT)^{-1} D^\mu UT = \Lambda^{\mu\nu} \partial_\nu \tag{13}$$

in which D is the covariant derivative (hiding the plane wave inside it) and Λ is the finite Lorentz transformation (little group element, actually) corresponding to ye olde ω. *It is observed that the plane wave has been swallowed up into a local space-time symmetry of the equation of motion.* The Λ terms go away by invariance, and isn't this familiar by now?

To finish the story of the theorem, consider the tree amplitude for the scattering of a system of particles with no external field. If we turn on an external electromagnetic field, the internal and external legs of the tree amplitude are altered according to the Fourier transforms of the ULT factors. In the null zone, all factors collapse to unity from charge conservation, Lorentz invariance, and momentum conservation. I have not done justice to the whole story, but the bottom line is that, like Perseus, the system of particles can be invisible to an external plane wave in special regions of phase space.

WHEN BAD THINGS HAPPEN TO GOOD ZEROS

Like some other recently well-publicized evidence, measurements of radiation zeros are destined to be contaminated, compromised and ultimately

---

[6] For very special cases, like Compton amplitudes, there are no radiation zeros, physical or non-physical, if the null zone corresponds to forward scattering of massless neutral vector particles.

FIG. 2. But most zeros are not in the physical phase space.

corrupted. See Fig. 2. To start with, higher-order corrections will not vanish in the null zone. The internal factors $Q/p \cdot q$ in closed loops correspond to momenta p that is integrated over; these factors are certainly not fixed by the outside legs. *It is interesting, however, that radiation symmetry Eqs.(10) still hold for the complete quantum amplitude to all orders in perturbation theory.*

Anomalous photon couplings spoil the zeros in lowest order. Many speakers at this conference have focused on limits that can be set on the κ and λ parameters for both $WW\gamma$ and WWZ trilinear couplings. As indicated by the earlier remarks, the $W\gamma$ zero provides litmus tests for these parameters. Any deviations from the gauge theory values lead to momentum dependence in the photon attachment currents such that the first-order transformations can no longer be universal.

Higher derivatives than the ones announced as "gauge-theoretic" produce terms in the J currents that are higher order in q. These terms have *a priori* no mechanism, no additional symmetry to effect their cancellation in a sum over all the legs of a given vertex. The Yang-Mills $\mathcal{O}(q^2)$ is the one exception that proves the rule.

Generalization to a Larger Gauge Sector

There are zeros associated with any gauge group when the corresponding massless gauge bosons are emitted. The "charges" now are the Clebsch-Gordan coefficients coming from the attachment of a boson belonging to the adjoint representation of the gauge group. The bad things here are two-fold. In QCD, color charges are averaged or summed over in hadronic reactions. The zeros are for the most part washed out, even when perturbation theory is applicable (in deep inelastic reactions, and so forth). In thinking of the weak bosons themselves, electroweak symmetry is broken. *Radiation zeros require the internal symmetry to remain good: The charges "Q_i" must be conserved.* Of course, the weak-boson masses are far from vanishing; a nonvanishing q^2 itself ruins radiation interference.

It is not surprising that in a supersymmetric or extended supersymmetric world, the photonic zeros would have partner photino zeros and "sphotino" zeros. There are "szeros" and "xeros" in the supersymmetrically extended gauge sector. (19,20) The well-known problem is that SUSY is experimentally evasive and hardly unbroken.

RECENT WORK

An "approximate zero" in WZ production by fermion-antifermion annihilation has been proposed recently by Baur, Han and Ohnemus (21,5) as another test of the self-couplings. At high energies, the sensitivity to the WWZ vertex is not so terribly different than that of the $W\gamma$ channel on the $WW\gamma$ vertex. It was briefly noted in the early, detailed paper on radiation zeros (8) that the dip structure found (2) in the WZ angular distributions corresponds to an approximate radiation interference zero. Baur *et al.* show that even in the high-energy limit there is a nonvanishing longitudinal helicity amplitude. Even if the c.m. energy is large compared with the masses, the couplings still refer to a broken symmetry theory and the zero will remain approximate at high energy. The approximate zero corresponds to broken radiation symmetry, nevertheless, and the origin and mechanism for the partial cancelations are the same as for the exact zeros.

Some day, experiments on multiboson zeros, exact or approximate, will make sense. Recent discussions of multiphoton/gluon reactions (22,23) could be extended to the broken symmetry weak-boson production channels. In WZZ production, for example, quadrilinear couplings come into play, along with a multi-Z approximate zero. But whether these couplings can be probed, even in the next generation of colliders, is very much open to question.

RIGHT AND LEFT

What is right

It may seem reasonable to say that the radiation zero phenomena are well understood. Among other things, they are another property of gauge theories. We have the following nice connections.

Classical interference. The zeros are the generalization of the well-known vanishing of classical nonrelativistic electric and magnetic dipole radiation occurring for equal charge/mass ratios (indeed, the low-energy limit of the null zone conditions!) and gyromagnetic g-factors. The null zone is exactly the same as that for the completely destructive interference of radiation by charge lines (a classical convection current calculation (8)) and is preserved by the fully relativistic quantum Born approximation for gauge theories.

Gauge couplings as transformations. Proving radiation symmetry theorems has brought forth the fact that gauge boson couplings to particles, including self-couplings, can be interpreted as transformations of the particles in both internal and external space. This is understandable since the gauge bosons belong to adjoint representations of both the Lie gauge groups and the Poincare space-time group. Only for the gauge couplings, however, do we get the universal Poincare generator representation by the spin and contact currents, which are necessary for radiation symmetry. For the SUSY and extended SUSY cases, we generate universal local supersymmetry and chiral transformations, respectively.

The long and the short of it. We have noted that the gauge-theory Born amplitudes have the same null zone as the classical radiation patterns. In the short-distance limit, gauge theories can be renormalized, corresponding to good high-energy behavior for the Born amplitudes. Thus, only for couplings that correspond to $g = 2$, for example, do both the short and long distance behaviors fall into these special categories. (Brodsky and Schmidt (24) emphasize the magic of $g = 2$, including its implications for photoabsorption sum rules.) We have this connection between the small and large distance scales.

What is left

Still, there are to me, at least, some loose ends.

Spin barrier. It has not been possible to find a theory of photon couplings, for instance, to particles with spins greater than one that preserves radiation symmetry. I at least have failed thus far in attempts to use supermultiplets in the gauge and matter sectors motivated by supersymmetry and string theories. Passarino tells how gravitons spoil radiation symmetry, and how their radiation does not seem to have any analogous zeros. (25) One way of describing this theoretical wall is in terms of a power series in photon momentum. (modulo the $Q/p \cdot q$ factor). The zeroth and first order terms are controlled by translation and Lorentz symmetry, respectively. The isolated instance of second-order terms in the Yang-Mills source vertex is controlled by the Bianchi identity, reminiscent of a curved-space-time symmetry. Higher spins lead to additional second-order and higher-order terms for which there must be new symmetries controlling them.

Mixing internal and external spaces. The null zone is defined by equations mixing internal charges and phase space. The original $W\gamma$ zero occurs at angles given in terms of quark charges. Radiation symmetry can be rewritten $Q_i \to Q_i + Cp_i \cdot q$. The mixing together of internal and external parameters suggests a look at ideas such as those behind Kalusza-Klein theories where a fifth or higher dimension is defined in terms of charges and gauge fields. The radiation symmetry could be part of the larger space-time symmetry, and radiation zeros a decoupling of the extra coordinate(s).

ACKNOWLEDGMENT

Many thanks go to Ulrich Baur, Tao Han and Thomas Müller for their help and hospitality. I am grateful to Ken Kowalski for discussions on these matters through the years.

REFERENCES

1. K.O. Mikaelian, M.A. Samuel, and D. Sahdev, Phys. Rev. Lett. **43**, 746 (1979).
2. R.W. Brown, D. Sahdev, and K.O. Mikaelian, Phys. Rev. D **20**, 1164 (1979).
3. The progress in the experimental study of trilinear couplings through the production of all electroweak pairs at Fermilab has been the subject of talks by Aihara, Fuess, Johari, Landsberg, Neuberger, Wagner, and Zhang at this conference. See those papers for references.
4. U. Baur, S. Errede, and G. Landsberg, Phys. Rev. D, to be published.
5. Tao Han, this conference, and references therefrom.
6. R.W. Brown and K.O. Mikaelian, Phys. Rev. D **19**, 922 (1979).
7. S.J. Brodsky and R.W. Brown, Phys. Rev. Lett. **49**, 966 (1982). See also M.A. Samuel, Phys. Rev. D **27**, 2724, (1983).
8. R.W. Brown, K.L. Kowalski and S.J. Brodsky, Phys. Rev. D **28**, 624 (1983).
9. T.R. Grose and K.O. Mikaelian, Phys. Rev. D **23**, 123 (1981).
10. C.L. Bilchak, J. Phys. G: Nucl. Phys. **11**, 1117 (1985).
11. G. Couture, Phys. Rev. D **39**, 2527 (1987).
12. J. Reid, G. Li and M.A. Samuel, Phys. Rev. D **41**, 1675 (1990).
13. M.A. Doncheski and F. Halzen, Z. Phys. C **52**, 673 (1992).
14. R.W. Brown, Proc. Europhys. Study Conf. (Erice, Italy, 1983), ed. H. Newman (Plenum, New York, 1985).
15. C.J. Goebel, F. Halzen, and J.P. Leveille, Phys. Rev. D **23**, 2682 (1981). (1992).
16. Dongpei Zhu, Phys. Rev. D **22**, 2266 (1980).
17. R.W. Brown and K.L. Kowalski, Phys. Rev. Lett. **51**, 2355 (1983).
18. R.W. Brown and K.L. Kowalski, Phys. Rev. D **30**, 2602 (1984).
19. R.W. Brown and K.L. Kowalski, Phys. Lett. **144B**, 235 (1984).
20. D. DeLaney, E. Gates and O. Tornqvist, Phys. Lett. **186B**, 91 (1987).
21. U. Baur, T. Han and J. Ohnemus, Phys. Rev. Lett. **72**, 3941 (1994).
22. F.K. Diakonos et al., Phys. Lett. **303B**, 177 (1993).
23. R.W. Brown, M.E. Convery and M.A. Samuel, Phys. Rev. D **49**, 2290 (1994).
24. S.J. Brodsky and I. Schmidt, SLAC-PUB-95-6761 (1995).
25. G. Passarino, Nucl. Phys. **B241**, 48 (1984).

Measuring WWZ and $WW\gamma$ Couplings at LEP II

J. Busenitz*

*Department of Physics and Astronomy
The University of Alabama
Tuscaloosa, AL 35487

> We present an overview of the experimental study of WWZ and $WW\gamma$ couplings with the process $e^+e^- \to W^+W^-$ at LEP II. Estimates of the sensitivity at LEP II to these couplings, obtained from simulation studies carried out for the L3 detector, are given.

INTRODUCTION

Starting in the second half of 1995, the center-of-mass energy at LEP will be raised above the Z resonance with the goal of crossing W^+W^- production threshold around mid-1996. In the following 3-4 years, it is planned that LEP will deliver an integrated luminosity of 500 pb^{-1} per experiment in the 175-210 GeV energy range. This phase of LEP operation is commonly referred to as LEP II or LEP 200.

Assuming the Standard Model, approximately 8000 W^+W^- final states will be produced per experiment for $\sqrt{s}$=176 GeV and an integrated luminosity pf 500 pb^{-1}. As shown by Figure 1, the production of W pairs depends on the WWZ and $WW\gamma$ gauge boson couplings. These trilinear gauge vertices are a striking feature of the Standard Model and figure prominently in its high-energy behavior. To date, however, direct measurements of the WWZ and $WW\gamma$ couplings have not been very precise. From the high statistics alone, one would expect that measurements at LEP II will result in a significant improvement in the precision of our direct knowledge of these couplings. The attraction of measuring WWZ and $WW\gamma$ couplings at LEP II is heightened by the events being clean and suffering from relatively low backgrounds, as is characteristic of e^+e^- collider experiments, and by the fact that all four major LEP experiments (ALEPH, DELPHI, L3, OPAL) have competitive capabilities for measuring these couplings. Before discussing in detail the measurement of WWV ($V = Z, \gamma$) couplings with W^+W^- events, we note that the potential at LEP II for measuring WWV couplings from single W production (1) and quartic couplings $(\gamma/Z)WW\gamma$ from $e^+e^- \to W^+W^-\gamma$ events (2) has also been studied, and that estimates have been made of the sensitivity at LEP II to $ZZ\gamma$ and $Z\gamma\gamma$ couplings (3).

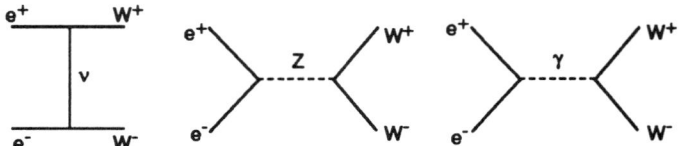

FIG. 1. Lowest-order Feynman diagrams for $e^+e^- \to W^+W^-$

OVERVIEW OF MEASUREMENT OF WWV COUPLINGS IN $e^+e^- \to W^+W^-$

Important quantities to measure

In order to extract complete information on WWV couplings from the study of $e^+e^- \to W^+W^-$, we must measure the dependence of the cross section on $\sqrt{s}$, the energies and momenta at which the W's are produced and the W helicities. For the sake of simplicity, we shall assume that the W^+W^- center-of-mass frame coincides with the laboratory frame, which amounts to ignoring initial-state radiation (ISR), and we shall also neglect the finite width of the W. Assuming that the bulk of running at LEP II will be done with unpolarized beams, the cross section will not depend on the W azimuthal production angles. The energy-momentum variables on which the cross section depends then reduce to the center-of-mass energy and the polar angle of one of the W's. The helicities of the W's cannot be measured directly, but since the W decay amplitudes depend on W helicity, information about helicity can be obtained on a statistical basis by measuring the dependence of the cross section on the W decay angles.

Therefore, the important quantities to measure are the center-of-mass energy, the polar angle at which one of the W's is produced, and the decay angles for each W. Our definition of the angles is shown schematically in Figure 2. It is conventional to take the polar angle at which one of the W's is produced to be the angle θ_W at which the W^- is produced with respect to the e^- beam direction. It is also conventional to take the W decay angles, (θ_-, ϕ_-) for the W^- and (θ_+, ϕ_+) for the W^+, to be those of the down-type daughter. Here we follow (4) in in choice of coordinate system for measuring the decay angles: the z-axis coincides with the W^- direction of flight and the x-axis is parallel to the vector product $\vec{p}_{W^-} \times \vec{p}_{e^-}$.

How well can the important quantities be measured?

We address this question in two steps. First, we consider whether or not enough information is available to measure the quantity, modulo a discrete ambiguity and typical experimental smearing. Second, we indicate the expected

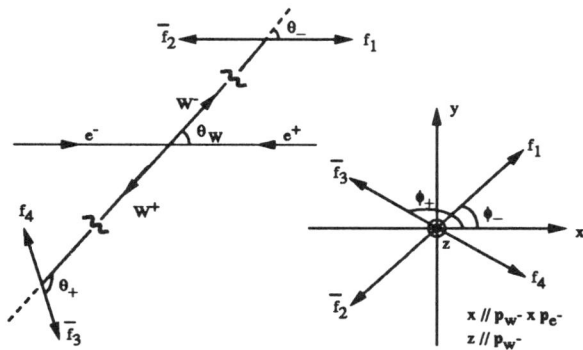

FIG. 2. Definition of W production and decay angles

resolutions on quantities under circumstances in which they are measurable. The effects of ISR are briefly discussed at the end.

In general, by charge and momentum conservation, reconstruction of the momentum of one of the W's and determination of the sign of one of the W's suffices to determine the momenta and signs of both W's. In that case θ_W is determined unambiguously. If the momenta of the W's can be reconstructed but their signs cannot be determined, then as can be seen from Figure 2, there is a twofold ambiguity in θ_W: $\theta_W \leftrightarrow \pi - \theta_W$. Measurement of the decay angles, (θ_+, ϕ_+) or (θ_-, ϕ_-), requires reconstruction of the momentum of the parent W, reconstruction of the momentum of at least one of the two daughters, and tagging of at least one of the daughters so that the momentum of the down–type daughter can be determined. For reconstructing momenta, one can take advantage of total four–momentum conservation, knowledge of the W mass (to within finite width effects), and knowledge of the τ mass in the case that W leptonic decay into the τ channel has occurred. If the momenta of the W decay products are measured but the down–type daughter is not tagged, there is a twofold ambiguity in the decay angles: $(\theta, \phi) \leftrightarrow (\pi - \theta, \phi + \pi)$.

To be more specific on whether or not the angular variables can be measured, we must take into account how the W's decay. We consider the main classes one by one.

Both W's decay hadronically (46% of the events). Each W decays into two quarks, which are manifest experimentally as two or more hadronic jets. The procedure of combining jets to reconstruct the parent W can be aided by the constraint that, to within finite W width effects and experimental resolution, the invariant mass of the jet daughters is the W mass. Assuming that the daughter jets are well-reconstructed and combined correctly (see following section for estimate of how efficiently this can be done), the momenta of the W's are well–determined. To measure θ_W unambiguously, one also needs to determine the charge of at least one of the W's. Studies based on the DELPHI

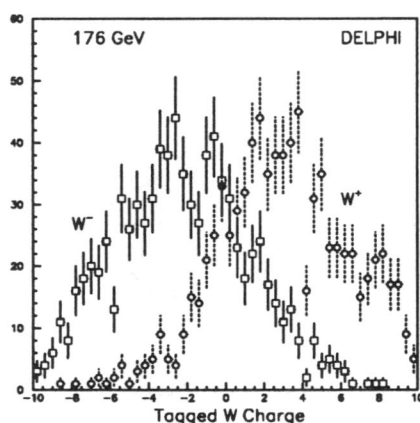

FIG. 3. Tagged W charge calculated from the daughter jet constituents.

detector have been carried out to determine how reliably the W charge can be determined from its daughter jet constituents. Figure 3, reproduced from (6), shows the result for well–reconstructed W's. With appropriate cuts on the tagged W charge, it appears that θ_W could be measured reliably in some fraction of events. In order to measure (θ_+, ϕ_+) and (θ_-, ϕ_-) the momenta of the primary daughters have to be reconstructed and at least one of the primary daughter quarks has to be tagged. In the predominant case where the W decays to two jets, the momenta of the primary daughters can be reconstructed straightforwardly. For decays to 3 or more jets, there may be some ambiguity in the primary quark daughter momenta due to jet combinatorics. Tagging the daughter quarks as up–type or down–type does not seem experimentally feasible. In that case, (θ_+, ϕ_+) and (θ_-, ϕ_-) can be measured only to within a twofold ambiguity.

One W decays hadronically while the other decays leptonically into the e or μ channel (29% of the events). The momentum of the W decaying hadronically can be reconstructed from the hadronic jets, and the charge of the W decaying leptonically is directly tagged by measuring the charge of the e or μ. For charged leptons produced at angles well away from the beamline, charge measurement by the LEP experiments is very reliable overall for the typical lepton momenta at LEP II energies. Thus the momentum of the W^-, and consequently θ_W, is measured reliably and unambiguously. For the leptonic decay, the charged lepton is the down–type daughter. Since the parent momentum can be determined despite the undetected neutrino, (θ, ϕ) can be obtained. For the hadronic decay, (θ, ϕ) can be measured to within a twofold ambiguity for the reasons indicated in the previous paragraph.

One W decays hadronically while the other decays leptonically into the τ channel (14% of the events). The LEP experiments are able to correctly tag the τ charge from its decay products in most cases, and therefore θ_W can be measured. The decay angles (θ, ϕ) for the hadronic decay can be measured to within the usual twofold ambiguity. Reconstructing the angle of the τ in its parent W rest frame is complicated by undetected neutrinos, one in the case the τ decays hadronically and two in the case it decays leptonically. In the case that the τ decays hadronically, there are 6 unknown, namely, the three-momentum of the τ neutrino from W decay and the three-momentum of the neutrino from τ decay, and 6 constraints (total four-momentum conservation, the W mass, and the τ mass). The τ momentum can in principle be determined up to some ambiguity, but studies are needed to understand how severe the ambiguity is. For leptonic τ decay, there are more unknowns than constraints.

Both W's decay leptonically to e/μ (5% of the events). It has been shown (7) that the momenta of the parent W's can be reconstructed up to a twofold ambiguity using the standard constraints and that the direction of the W^- for the one possibility typically differs little from its direction for the other possibility. In other words, θ_W can be measured with a very mild twofold ambiguity. For a each of the two slightly different possibilities for θ_W, the decay angles (θ_+, ϕ_+) and (θ_-, ϕ_-) can be calculated unambiguously.

Both W's decay leptonically with at least one decay occurring to the τ channel (5% of the events). We mention this class for completeness. Since it comprises only 5% of the events, its potential impact on the measurement of couplings is limited at best. To measure any of the production and decay angles without severe smearing/ambiguities seems unlikely since there are at least three neutrinos produced among the W and τ decays.

Of the event classes just described, the so-called semileptonic WW events where one W decays hadronically and the other decays leptonically to the e or μ channel is the most promising single class for measuring WWV couplings. It comprises a large fraction of events and is the class for which the production and decay angles can be best determined.

The second part of the question concerns the experimental resolutions on the angles in the case that they can be measured. A preliminary estimate has been made for the semileptonic WW channel assuming the L3 detector performance. The angular variables were determined by making a constrained kinematic fit to the measured jet and charged lepton momenta, the constraints being total four-momentum conservation and that the invariant mass of the jets and the invariant mass of the charged lepton and neutrino be the W mass allowing for the finite W width. For events with a good fit, the angular resolutions were found to be described by two-component gaussians. The widths σ_1 and σ_2 and their respective relative normalizations R_1 and R_2 are given in Table 1 for $\cos\theta_W$, $\cos\theta_\ell$, and ϕ_ℓ where θ_ℓ and ϕ_ℓ are the decay angles for the leptonic decay. The resolutions are adequate given the angular scales over which the cross sections vary appreciably.

TABLE 1. Angular resolutions for L3 (preliminary)

Variable	σ_1	R_1	σ_2	R_2
$\cos\theta_W$	0.011	47%	0.032	53%
$\cos\theta_\ell$	0.010	25%	0.052	75%
ϕ_ℓ	0.021	55%	0.079	45%

In the above, we have neglected the effects of ISR. This radiation is typically emitted in the far forward direction and consequently goes undetected. Since the average energy of ISR is a few GeV at LEP II for W pair production, it may be argued that its neglect would not lead to serious errors in the determination of angles. However, if one is using a kinematic fit to determine these angles, there is the danger that neglect of ISR would lead to significant inefficiency due to events failing the fit. Also, neglecting ISR precludes the possibility of checking experimentally that the event generator and detector simulation one is using models the effects of ISR correctly. Explicitly taking into account ISR nominally introduces at least one unknown (corresponding to the assumption that the photon is emitted along the beam direction) and this would lead to more severe ambiguities in the determination of the angles, especially for events in which there are two or more neutrinos produced.

Event Selection–Efficiencies and Backgrounds

Event selection must be applied not only to distinguish WW events from other Standard Model processes but also to distinguish between the various WW event classes since the procedure for determination of the relevant angular quantities varies from one class to another. Below we review the results of selection studies for three categories, (a) hadronic WW events, ie. events in which both W's decay to quarks, (b) semileptonic WW events in which one W decays to hadrons and the other decays to the e or μ channel, and (c) leptonic WW events in which both W's decay to the e or μ channels.

Hadronic WW events: These events, which are predominantly 4–jet events with little missing energy, are to be distinguished from other Standard Model processes such as $e^+e^- \to Z/gamma \to$ jets and $e^+e^- \to ZZ \to$ jets as well as from semileptonic WW events. In a simulation study (8) done for the DELPHI detector, the selection efficiency was found to be about 50% using an algorithm which imposed a 4–jet interpretation on the events and required a good kinematic refit of the measured jet quantities for the hypothesis of $e^+e^- \to W^+W^-$. We are unaware of any estimates of the surviving backgrounds; for center–of–mass energies below ZZ threshold, at least, we expect them to be relatively small.

Semileptonic WW events: The experimental signature of these events is an isolated high energy electron or muon, a large amount of missing energy due to the undetected neutrino, and hadronic jets from the decay of the opposite W. Background processes with the potential for producing events with a similar signature include two–photon processes, $e^+e^- \to Z/\gamma(\gamma) \to$ jets(γ), $e^+e^- \to Ze^+e^- \to$ jets e^+e^-, single W production, $e^+e^- \to ZZ \to$ jets $\ell^+\ell^-$, semileptonic WW events in which the leptonic decay goes through the τ channel, and hadronic WW events. The notation (γ) refers to photons from initial-state radiation.

We have carried out a study of semileptonic WW event selection for the L3 detector with backgrounds $e^+e^- \to Z/\gamma(\gamma) \to$ jets(γ), $e^+e^- \to ZZ \to$ jets $\ell^+\ell^-$, and semileptonic WW events in the τ channel explicitly taken into account. A combination of cuts that was found to result in a relatively high efficiency for selecting semileptonic WW events while strongly suppressing the background is the following: (i) more than 16 clusters in the electromagnetic and hadronic calorimeters, (ii) visible energy greater than 80 GeV, (iii) $E_\nu > 20$ GeV and $|\cos\theta_\nu| < 0.966$ where the neutrino momentum was reconstructed using four–momentum conservation (no kinematic fit), (iv) an isolated, positively identified electron or muon with momentum greater than 20 GeV and polar angle at least 22 degrees from the beamline, and (v) invariant mass of the charged lepton and neutrino in the range 60–200 GeV. The last cut is important for suppressing the semileptonic WW events in the τ channel. With these cuts, the selection efficiency is approximately 72% at 176 GeV and 67% at 190 GeV. The signal–to–background ratio at both energies is about 20. Backgrounds we did not consider but which may be comparable in magnitude to the ones we did are single W production and $e^+e^- \to Ze^+e^- \to$ jets e^+e^- (6).

Leptonic WW events: These events have a simple topology: two energetic oppositely–charged leptons and a large amount of missing energy. After cuts to select events containing two positively identified leptons of opposite charge and momentum greater than 20 GeV, the surviving backgrounds are leptonic WW events in which at least one of the W's has decayed into the τ channel and the τ has decayed leptonically, $e^+e^- \to Z/\gamma(\gamma) \to \ell^+\ell^-(\gamma)$, and (for center–of–mass energies greater than approximately 190 GeV) $e^+e^- \to ZZ \to \ell^+\ell^-\nu\bar{\nu}$. In the context of L3, we have found that after making additional cuts to suppress these backgrounds, the resulting selection efficiency is about 65% (53%) and the signal–to–background ratio is approximately 10 (7.5) for $\sqrt{s}=180$ GeV (200 GeV).

Extracting couplings from data

The most general effective Lagrangian for $WW\gamma$ and WWZ interactions which is invariant under Lorentz and electromagnetic gauge transformations contains 13 independent couplings. In the notation of (4), these couplings are g_1^Z, κ_V, λ_V, g_4^V, g_5^V, $\tilde{\kappa}_V$, and $\tilde{\lambda}_V$ ($V = \gamma, Z$). The couplings g_1^Z, κ_V, λ_V conserve C, P, and CP separately while the remaining couplings violate C, P, and/or CP. The Standard Model values are 1 for g_1^Z and κ_V and are 0 for all other couplings.

If one attempts to measure the values of these couplings from the WW data while allowing the parameters to vary simultaneously, not surprisingly the errors are large. Motivated by the outstanding success of the Standard Model to date and various theoretical arguments, one may assume that the Lagrangian contains additional symmetries, in which case the number of independent free parameters is reduced. Studies of the effective Lagrangian under various assumptions of discrete and local symmetries are reported in (5,9–11).

Several different approaches for extracting a given set of couplings from the data have already been investigated. Common to two of these approaches are fits to the angular distributions. In one approach (5), the angular distributions are fitted (binned maximum likelihood) for the differential cross sections $d\sigma^{TT}/d\cos\theta_W$, $d\sigma^{TL}/d\cos\theta_W$, and $d\sigma^{LL}/d\cos\theta_W$ and for the $W^\pm$ spin density matrix elements; here TT, TL, and LL refer to the polarizations of the produced W's. The couplings are then fitted to the differential cross sections and spin density matrix elements. In the other approach (12) based on making fits to the angular distributions, the free couplings are fit directly (unbinned maximum likelihood) to the angular distributions. In terms of the uncertainties on the fitted couplings, the two approaches are comparable for cases in which there is only one independent coupling while the latter approach gives somewhat smaller errors overall if there are two or more independent couplings. An advantage of the former approach is that the fit results for the differential cross section and spin density matrix elements obtained in the first step are model independent.

A third approach to extract the values of the couplings from the data involves "optimal variables" (13). In this approach, it is assumed that deviations of WW production cross section from the Standard Model prediction are linear in the anomalous couplings. (This is a reasonable assumption if the anomalous couplings are small because the largest contributions to the cross section from anomalous couplings are likely to arise from interference with Standard Model amplitudes.) Then one can identify observables whose average values can be related to the anomalous couplings and which can be shown to lead to the smallest possible statistical error for the given luminosity. Further studies are needed to understand the potential of this method under conditions where detector acceptances and resolutions are taken into account.

ESTIMATES OF SENSITIVITY TO WWV COUPLINGS AT LEP II

We have made an estimate of the sensitivity of LEP II to WWV couplings in the semileptonic WW channel assuming a specific form for the effective Lagrangian and taking into account experimental inefficiencies and resolutions per the L3 detector.

For the effective Lagrangian, we followed (9) in assuming that only dimension–6 operators are present, that C, P, and CP are conserved, and that the Lagrangian is SU(2)⊗U(1)–symmetric; additionally we have assumed the operator coefficients f_B and f_W are equal. This results in only two free couplings, which we take to be λ_γ and $\Delta\kappa_\gamma \equiv \kappa_\gamma - 1$. The remaining couplings are then defined as $\lambda_Z = \lambda_\gamma$, $\Delta g_1^Z = \Delta\kappa_\gamma/2\cos^2\theta$ (θ is the Weinberg angle), and $\Delta\kappa_Z = \Delta\kappa_\gamma - \Delta g_1^Z$.

The program LEPWW was used to generate WW events for various values of $\Delta\kappa_\gamma$ and λ_γ with initial–state radiation turned on. The generated quark and charged lepton momenta were smeared parametrically according to L3 resolutions. The events were then subjected to a set of selection cuts equivalent to that described in the previous section. The angular variables were calculated directly from the smeared momenta ignoring the presence of initial–state radiation. The surviving events were binned in $\cos\theta_W$, $\cos\theta_\ell$, ϕ_ℓ, $\cos\theta_j$, and ϕ_j where the subscript ℓ refers to the decay to leptons and j to the decay to hadrons. For a particular choice of angular binning, the event samples generated at different values of $\Delta\kappa_\gamma$ and λ_γ were analyzed to determine the coefficients of a parameterization for the effective cross section:

$$\sigma_i = \sigma_i^{SM} + a_i \Delta\kappa_\gamma + b_i(\Delta\kappa_\gamma)^2 + c_i \lambda_\gamma + d_i(\lambda_\gamma)^2 + e_i \Delta\kappa_\gamma \lambda_\gamma \quad (1)$$

The index i refers to the ith angular bin, σ_i is the effective cross section for the ith bin, σ_i^{SM} is the corresponding cross section according to the Standard Model, and the coefficients are $a_i,...,e_i$. For the event statistics used, the maximum number of bins for any combination of angular binning was less than 600 in order to control statistical fluctuations. Also, in calculating the coefficients, the e^+ and μ^+ samples were combined and e^- and μ^- samples were combined. No attempt was made to include the effects of backgrounds since they are comparatively small and are thus not likely to degrade the sensitivity to anomalous couplings provided, of course, they are understood.

To estimate the errors in the measurement of $\Delta\kappa_\gamma$ and λ_γ for a given choice of angular binning, we generated 1000 independent event samples. For each sample, the number of events in the ith bin was chosen according with Poisson statistics with mean $(500\text{ pb}^{-1})\sigma_i^{SM}$. We then carried out a maximum likelihood fit of equation (1), multiplied by the luminosity, to determine $\Delta\kappa_\gamma$ and λ_γ. For the choices of angular binning we discuss here, the distribution of the fitted values of $\Delta\kappa_\gamma$ and λ_γ for the 1000 samples is well–approximated by a two–dimensional gaussian, hence we may represent the error contours by ellipses calculated from the correlation matrix for the 1000 fit results.

Figure 4 shows the estimated errors at the 95% C.L. for a center–of–mass energy of 176 GeV for three different choices of binning. Using the notation $(n_1, n_2, n_3, n_4, n_5)$ to denote the number of equal–sized divisions made in $\cos\theta_W$, $\cos\theta_\ell$, ϕ_ℓ, $\cos\theta_j$, and ϕ_j, respectively, the dashed contour corresponds to the binning (8,8,8,1,1), the solid line corresponds to (4,4,4,3,3) with no quark tagging, and the dotted line corresponds to (4,4,4,3,3) with reliable quark tagging assumed. Comparing the dashed and dotted contours in Figure 4, it is evident that, even though it was necessary to subdivide the data

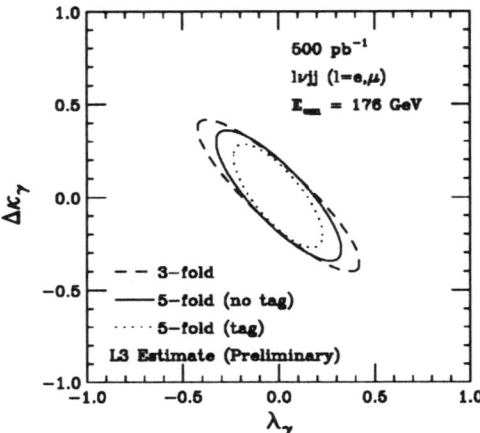

FIG. 4. 95%–C.L. error contours for fit to distributions differential in $\cos\theta_W$, $\cos\theta_\ell$, and ϕ_ℓ(dashed contour), differential in $\cos\theta_W$, $\cos\theta_\ell$, ϕ_ℓ, $\cos\theta_j$, and ϕ_j with untagged quarks (solid contour), and differential in $\cos\theta_W$, $\cos\theta_\ell$, ϕ_ℓ, $\cos\theta_j$, and ϕ_j with tagged quarks (dotted contour)

in $\cos\theta_W$, $\cos\theta_\ell$, and ϕ_ℓ more coarsely when we also subdivided it in $\cos\theta_j$ and ϕ_j, there is nonetheless a clear gain in sensitivity realized by taking into account the correlations between all 5 angles rather than just 3 of them. The importance of exploiting angular correlations to increase sensitivity to WWV couplings has already been pointed out in previous studies (9,5,12). While we believe that it will not be possible to reliably tag the quarks in hadronic W decay, we have included the dotted contour to indicate the gain in sensitivity that would be realized if it were possible. Also borne out by Figure 4 is that the errors on $\Delta\kappa_\gamma$ and λ_γ are strongly correlated at LEP II energies. Figure 5 shows the error contours for two different center–of–mass energies, 176 GeV and 190 GeV, obtained for (4,4,4,3,3) binning with no quark tagging. The significant reduction in errors with an increase of 14 GeV in center–of–mass energy is due to the fact that anomalous WWV couplings strengthen rapidly with increasing energy. Estimates of sensitivity to anomalous couplings are also reported in (5,12) for a range of assumptions about the structure of the effective Lagrangian. We lack space to adequately summarize the results of these studies, but we strongly recommend that those interested in the subject consult these papers. We would note in particular that the hadronic WW channel is also considered in (12). From the results presented, it is evident that inclusion of the hadronic channel yields would yield a modest gain in sensitivity relative to using semileptonic WW events alone.

In making our sensitivity estimates, the reduction of the WW production cross section by ISR, the effects of detector acceptance and resolution on event reconstruction, and the impact of event selection have been approximately

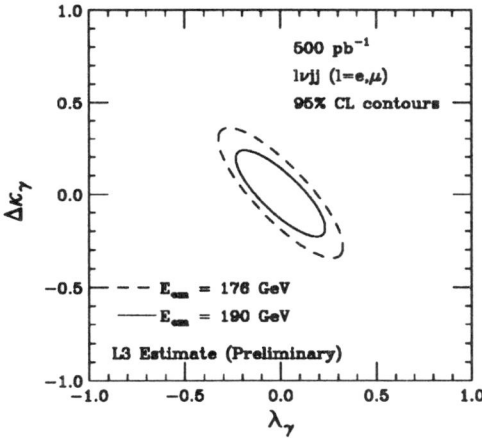

FIG. 5. 95%–C.L. error contours for 176 GeV (dashed) and 190 GeV (solid) for fit to distribution differential in all five angular variables with quarks untagged

TABLE 2. 1-σ errors on $\Delta\kappa_\gamma$ and λ_γ from one–parameter fits at 176 (190) GeV

	$\sigma_{\Delta\kappa}$	σ_λ
this study	0.080 (0.055)	0.075 (0.056)
other studies	0.046 (0.033) (5)	0.055 (0.032) (12)

taken into account. For the estimates made by Sekulin (12) and Bilenky et al. (5), such effects were not included, however, the approaches to fitting the angular distributions for the couplings are superior to ours. Thus, for a particular choice of effective Lagrangian, the sensitivity estimate obtained from our approach and the estimates obtained in the other approaches roughly bracket the range in which the "most realistic" sensitivity estimate lies. To get an indication of this range, we have carried out single–parameter fits to angular data binned according to (4,4,4,3,3) without quark tagging. The single–parameter fits differ from the two–parameter fits described earlier only in that Equation (1) was used with either $\Delta\kappa_\gamma$ or λ_γ fixed to zero. The one-standard–deviation errors are given in Table 2 for $\sqrt{s}$ of 176 GeV (190 GeV) along with the corresponding results obtained in the previous studies for the same relationship among couplings.

CONCLUDING REMARKS

Much work has already been done in exploring strategies for extracting information on WWV couplings from the LEP II data and estimating the precision that can be obtained. In further preparation, full detector simulation and realistic event reconstruction for WW events and backgrounds are required, how to handle radiative effects, especially ISR, must be addressed, and the feasibility of using other channels, e.g. semileptonic WW events in the τ channel, to improve the measurement precision needs to be studied. A workshop on LEP II has recently been convened, for which one of the main topics is how LEP II can be best exploited to measure WWV couplings, so we anticipate much additional progress during the coming year. Based on present indications, we can look forward to having a large amount of data on $e^+e^- \to W^+W^-$ within a few years that can be used to significantly advance our empirical knowledge of WWZ and $WW\gamma$ interactions.

ACKNOWLEDGEMENTS

We would like to thank Martin Pohl for useful discussions on estimating the experimental sensitivity to WWV couplings at LEP II. The results presented on semileptonic and leptonic WW event selection with the L3 detector were obtained in collaboration with Roger McNeil. This work was supported by the U.S. Department of Energy.

REFERENCES

1. C.G. Papadopoulos, Phys. Lett. B 333 (1994) 202.
2. Ghadir Abu Leil and W.J. Stirling, Phys. Rev. D 49 (1994) 3751.
3. H. Aihara et al., Fermilab preprint PUB-95/031.
4. K. Hagiwara et al., Nucl. Phys. B282 (1987) 253.
5. M. Bilenky et al., Nucl. Phys. B409 (1993) 22.
6. Ralph Eichler, Report of the Triple Gauge Boson Vertex Working Group, LEPC, November 1992.
7. J.B. Hansen, LEP II Workshop Working Group on Three Gauge Couplings, Meeting #1, 17 January 1995.
8. DELPHI Note 92-166 PHYS 250.
9. K. Hagiwara et al., Phys. Rev. D 48 (1993) 2182.
10. A. DeRujula et al., Nucl. Phys. B384 (1992) 3.
11. P. Hernandes and J. Vegas, Phys. Lett. B 307 (1992) 116.
12. R. Sekulin, Phys. Lett. B 338 (1994) 369.
13. M. Diehl and O. Nachtmann, Z. Phys. C62 (1994) 397.

Testing Vector Boson Self-Interactions in Future Tevatron Experiments

Christopher Wendt

University of Wisconsin, Madison WI, 53706

The prospects are summarized for studying the self-couplings of electroweak vector bosons in future Tevatron experiments. For studies which have already been done by CDF and D0 in the 1992-95 Tevatron runs, the sensitivities are extrapolated to future higher luminosity. New approaches will also become feasible with higher luminosity. The various channels are compared.

INTRODUCTION

The CDF and D0 collaborations have reported bounds on $WW\gamma$ and WWZ couplings, using data from $p\bar{p}$ collisions at $\sqrt{s} = 1.8\,\text{TeV}$ taken in Tevatron runs Ia (1992-93) and Ib (1994-95) (1,2). The integrated luminosity used in the studies to date has been of order $20\,\text{pb}^{-1}$. By the end of run Ib, the integrated luminosity available to each experiment is expected to be approximately $100\,\text{pb}^{-1}$, and future Tevatron runs may allow exposures of $1000\,\text{pb}^{-1}$ or even $10000\,\text{pb}^{-1}$.

In this talk we will consider the analyses that have been done so far, and extrapolate the sensitivity of each to higher luminosity. These include $W\gamma$ and $Z\gamma$ production; searches for WW and WZ pairs in the lepton(s) plus jets channel at high transverse momentum; and WW production in the pure leptonic channel. In addition, the availability of larger integrated luminosity will allow new analyses such as measurement of WWZ couplings from WZ production with pure leptonic decays. The expected sensitivities of the various techniques will be compared.

Much of the work described here was done in the context of the recent DPF Long-Range Planning Study, and more details can be found in reference (3).

In all of the analyses to be considered, the sensitivity to couplings is largely due to their effects on differential cross sections for diboson production at high values of subprocess energy $\sqrt{\hat{s}}$. In general, each cross section is a quadratic function of the couplings, and the minimum cross section is obtained for couplings very close to the standard model values. Therefore, for practical purposes the yield is equal to that given by the standard model plus a term proportional to $\Delta\xi^2$, where $\Delta\xi$ represents the deviation of any coupling from its standard model value. A measurement of the yield therefore places bounds on the deviations $\Delta\xi$.

The way that the sensitivity scales with luminosity can be different for each analysis. For example, in an analysis where the backgrounds are small but the predicted standard model signal is still too small for observation with the available integrated luminosity (L), the experimental limits on $\Delta\xi$ will scale like $L^{-1/2}$. (This assumes that the standard model is the correct model, and that the experimental cuts are not varied with L.) If instead we considered a channel with a significant standard model signal, again with little background, we would find that the experimental limits on $\Delta\xi$ scaled like $L^{-1/4}$. Real cases are more complicated than suggested by such arguments, but this just adds to the differences in scaling behavior for different analyses. Because of this, the analyses with the best sensitivity to date are not guaranteed to remain so with higher luminosity.

EXTRAPOLATION OF PAST MEASUREMENTS

$W\gamma$, $Z\gamma$

The CDF and D0 experiments have both studied $W\gamma$ and $Z\gamma$ production (2). In the standard model, the photon P_T distribution is predicted to fall off rapidly at high P_T. In the case of anomalous couplings, the distribution would be enhanced at high P_T, and therefore a fit to the observed P_T distribution gives bounds on the couplings.

The method has been applied to simulated data at high luminosity (3) with the results shown in Figures 1 and 2. In these simulations, it was assumed that the observed data follow the standard model prediction. The limits are those expected from the $W \to e\nu$ decay channel, with electron and photon coverage in the region $|\eta| < 2$. Experimental efficiencies were taken from the CDF run Ia analysis.

For the $W\gamma$ analysis (Figure 1), an integrated luminosity of $1\,\text{fb}^{-1}$ (50 times as large as that used so far) would give sensitivities to $WW\gamma$ couplings $|\Delta\kappa| \sim 0.4$ and $|\lambda| \sim 0.12$ at 95% CL, corresponding to an improvement of a factor 5 with respect to current results. Another factor of 10 in integrated luminosity would yield an additional factor of 2 in sensitivity.

For the $Z\gamma$ analysis (Figure 2), the projections may be compared with current 95% CL limits on $ZZ\gamma$ couplings, which are $|h_{30}^Z| \lesssim 2$ and $|h_{40}^Z| \lesssim 0.5$ for a form factor scale $\Lambda_{FF} = 500\,\text{GeV}$. However, the limits in this process are very sensitive to the assumed value of Λ_{FF}, and the apparent improvement by a factor between 10 and 100 at high luminosity is substantially an artifact of the different choice $\Lambda_{FF} = 1500\,\text{GeV}$. At constant Λ_{FF}, an increase from $1\,\text{fb}^{-1}$ to $10\,\text{fb}^{-1}$ improves sensitivity by a factor of two, similar to the $W\gamma$ case.

WW, WZ in lepton(s) plus jets channel

As shown by CDF, good sensitivity to $WW\gamma$ and WWZ couplings can be obtained from searches for WW or WZ production, in the channel where

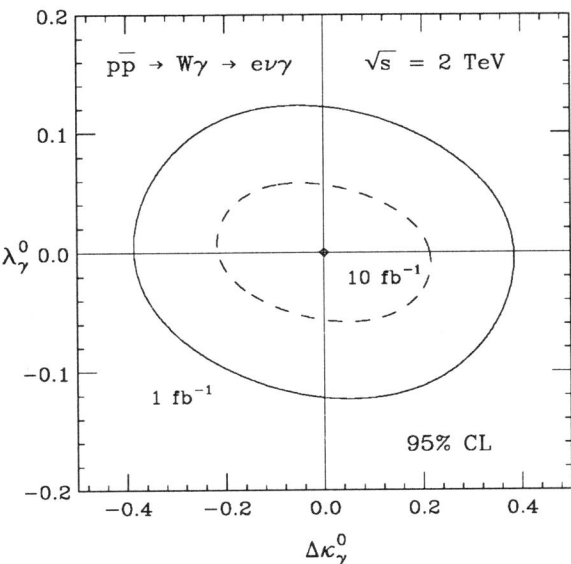

FIG. 1. Expected 95% CL bounds on $WW\gamma$ couplings from the process $p\bar{p} \to W\gamma$, with $W \to e\nu$ decays. Other couplings are held at standard model values. From ref. (3).

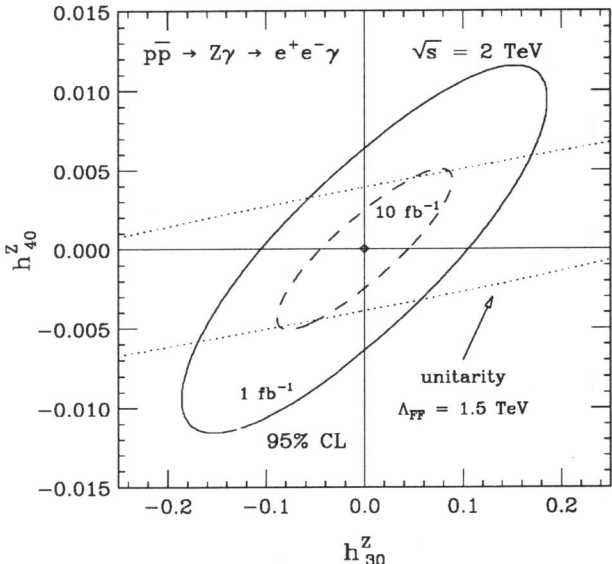

FIG. 2. Expected 95% CL bounds (solid and dashed) on $ZZ\gamma$ couplings from the $Z\gamma$ process, with form factor scale $\Lambda_{FF} = 1500$ GeV. The region outside the dotted curve is excluded by unitarity requirements (7). From ref. (3).

one boson decays into leptons and the other decays into jets (1). In this method, the large background from QCD W + jets production is suppressed by selecting event candidates with very high boson transverse momentum. As a result of this cut, the predicted number of WW or WZ events in the standard model is essentially zero, as is the background. However, the cut leaves intact a large part of the high-P_T region where the cross section would be enhanced by anomalous values of the $WW\gamma$ or WWZ couplings. The resulting sensitivity to the couplings is better than a similar search in the pure leptonic decay mode, because of the good acceptance for jets and the more favorable branching fractions. For the run Ia analysis, with integrated luminosity 20 pb^{-1}, the optimal boson P_T cut was found to be $P_T^W > 130$ GeV or $P_T^Z > 100$ GeV. The 95% CL bounds obtained were $|\Delta\kappa| \lesssim 1.2$ and $|\lambda| \lesssim 0.8$, assuming a form factor scale of 1000 GeV. Extrapolating this method to higher integrated luminosity, one observes that if the cut were left at the same place, the QCD background (and its uncertainty) would mask signals for $|\lambda| \lesssim 0.3$, even for very large statistical samples. Sensitivity better than this can be obtained by focussing on higher P_T, that is, by re-optimizing the boson P_T cut for each luminosity. Even better, a fit can be made to the complete boson P_T spectrum, taking into account explicitly the background and its uncertainty.

The sensitivity obtained from a simple optimized boson P_T cut was compared with that from a fit to the P_T spectrum in bins of width 50 GeV, assuming in the latter case that the background spectrum could be calculated with 20% systematic uncertainty in each bin. The fit to the complete spectrum gave only very slightly better results than the optimized simple P_T cut method. For the simple P_T cut method, the optimal cut was found to be 150, 200, or 250 GeV for integrated luminosity of 100, 1000 or 10000 pb^{-1} respectively.

Figure 3 shows the projected sensitivity for luminosities ranging from 100 pb^{-1} to 50 fb^{-1}, using the P_T spectrum fit method. Cuts and efficiencies were taken from the CDF run Ia analysis, except that electron and muon coverage was assumed to extend over the pseudorapidity range $|\eta| < 2$. Unlike the CDF run Ia analysis, simulated events with extra jets satisfying $P_T > 50$ GeV were rejected. This cut was imposed to reduce background from top quark production and to reduce the effects of QCD corrections to WW or WZ production. However, neither of these was found to be a very large effect. The QCD W+jets background was calculated with the VECBOS tree-level matrix element Monte Carlo program (4), scaled to match observed CDF data. The ultimate sensitivity to the couplings shown is in the range 0.1-0.2 for 10 fb^{-1}, corresponding to an improvement of a factor 5-10 compared to current results. This improvement is similar to that expected for the $W\gamma$ analysis.

From simulation, it was determined that detection efficiency for the $W \rightarrow jj$ or $Z \rightarrow jj$ decay suffers by a few percent for boson $P_T \gtrsim 250$ GeV, and strongly (more than 50%) for boson $P_T \gtrsim 330$ GeV. The loss is due to merging of the two jets, and degraded two-jet mass resolution. Fortunately, even for the highest luminosities considered, the sensitivity comes mostly from events with

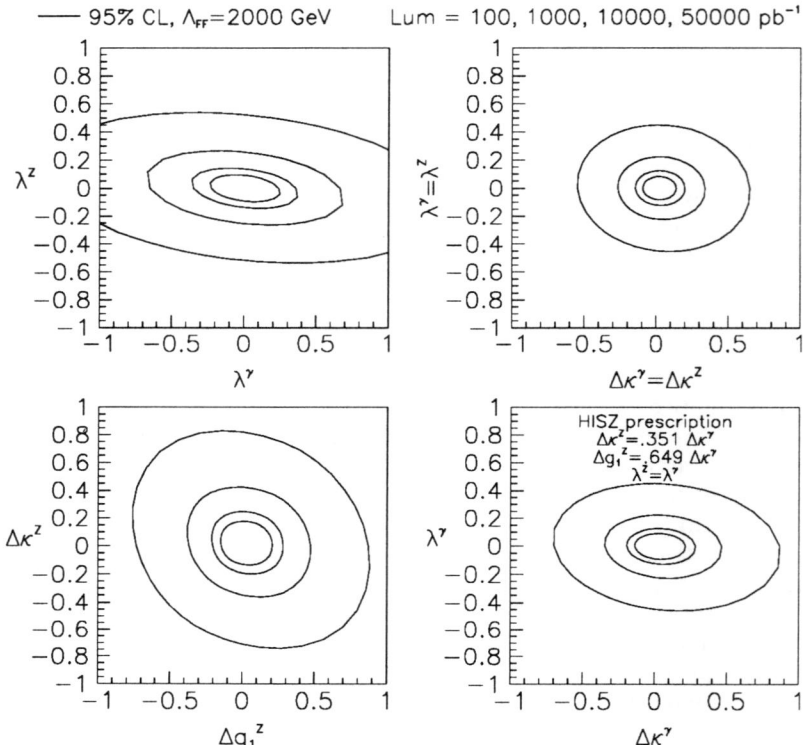

FIG. 3. Projected 95% CL bounds on $WW\gamma$ and WWZ couplings from WW, WZ production with decay into lepton(s) plus jets. For each plot, all couplings other than those shown are held at standard model values. The four curves in each plot correspond to integrated luminosities of $100\,\text{pb}^{-1}$ (outer), $1000\,\text{pb}^{-1}$, $10000\,\text{pb}^{-1}$, and $50000\,\text{pb}^{-1}$ (inner). The assumed form factor scale is $\Lambda_{FF} = 2000\,\text{GeV}$.

boson $P_T \lesssim 300\,\text{GeV}$ where this effect is fairly weak.

WW in pure leptonic channel

Both CDF and D0 have studied the WW production process in the pure leptonic decay channel (1), and the D0 group has extrapolated their expected sensitivity to higher luminosity (5). Based on the sample analyzed so far ($14\,\text{pb}^{-1}$), they obtained 95%CL limits $|\Delta\kappa| < 2.5$ and $|\lambda| < 1.9$ for $\Lambda_{FF} = 1000\,\text{GeV}$. The backgrounds from fake leptons and from top quark production are relatively easy to control. A large part of the background from top quark production is suppressed by a cut $|\vec{P}_T^{\text{had}}| < 30\,\text{GeV}$, where $\vec{P}_T^{\text{had}}$ is the vector sum of calorimeter energy transverse to the beam axis, omitting that deposited by leptons from the W decays. This cut has a rejection factor of 6 for top quark events, and an efficiency of $91\% \pm 5\%$ for WW events.

With a sample of size $1\,\text{fb}^{-1}$, the D0 group expects to observe approximately 50 $WW \to l\nu l\nu$ events, assuming standard model couplings. This will allow a good measurement of the cross section, and would correspond to limits on the couplings $|\lambda| \lesssim 0.3$ and $|\Delta\kappa| \lesssim 0.4$ at 95%CL with $\Lambda_{FF} = 1000\,\text{GeV}$. This sensitivity is comparable to that expected in the lepton(s) plus jets channel for a similar luminosity. Relatively speaking, the increased luminosity improves sensitivity of this channel more than the others, so that it will be more competitive with them than it is now.

Comparison of channels

The three methods discussed above for measuring $WW\gamma$ and WWZ couplings have comparable overall projected sensitivities. The methods differ however in their relative sensitivity to $WW\gamma$ and WWZ couplings, and in the dependence on sensitivity to the form-factor scale Λ_{FF}. They are therefore complementary. The $ZZ\gamma$ and $Z\gamma\gamma$ couplings are probed only in one process ($Z\gamma$ production).

The difference between $WW\gamma$ and WWZ sensitivities is trivial in the case of the $W\gamma$ process, which is wholly insensitive to WWZ couplings. Only if some relation is assumed between $WW\gamma$ and WWZ couplings, e.g. the HISZ relations (6), can the $W\gamma$ process provide information on WWZ couplings. In contrast, WW production is sensitive to both $WW\gamma$ and WWZ couplings, but again the bounds obtained will depend on the assumed relation between the two. The lepton(s) plus jets method admits both WW and WZ production, and also concentrates on very high $\sqrt{\hat{s}}$; for both these reasons its relative sensitivity to $WW\gamma$ and WWZ differs from WW studies in the pure leptonic channel. In fact the method has been found to be more sensitive to WWZ couplings than $WW\gamma$ couplings, but not so much as the pure leptonic WZ channel (below).

The assumed form factor scale (Λ_{FF}) enters into the comparison because the methods are most sensitive to the couplings $\Delta\xi$ at a fairly high energy scale

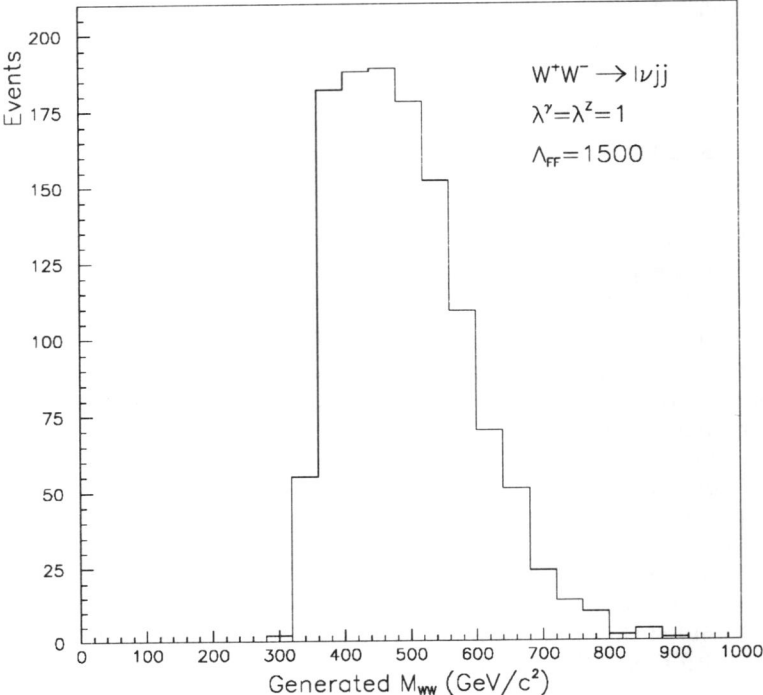

FIG. 4. Distribution of subprocess energy for simulated WW events that pass all cuts in the CDF run Ia analysis of the lepton plus jets mode. Anomalous couplings $\lambda^\gamma = \lambda^Z = 1$ were assumed ($\Lambda_{FF} = 1500$ GeV), but other values for the couplings give very similar results.

$\sqrt{\hat{s}}$, which is different for each method, whereas the limits are conventionally quoted on the couplings evaluated at a common $\sqrt{\hat{s}} = 0$. To extrapolate from one scale to another it is necessary to assume a particular dependence of the couplings on $\sqrt{\hat{s}}$, which by convention takes a standard form that avoids unitarity violations at high $\sqrt{\hat{s}}$ (7):

$$\Delta\xi(\sqrt{\hat{s}}) = \frac{\Delta\xi(0)}{(1 + \hat{s}/\Lambda_{FF}^2)^2}$$

With this form, anomalous couplings vary only slowly below the unknown scale Λ_{FF} but are rapidly suppressed above it.

In the $W\gamma$ process, most of the sensitivity comes from hard photons with $P_T^\gamma \gtrsim 100\,\text{GeV}$, corresponding to $\sqrt{\hat{s}} \gtrsim 230\,\text{GeV}$. Because of this, in the run Ia CDF analysis the sensitivity was roughly constant for $\Lambda_{FF} \gtrsim 700\,\text{GeV}$ but was degraded for smaller Λ_{FF}, worsening by a factor of two for $\Lambda_{FF} \sim 350\,\text{GeV}$.

The Λ_{FF} dependence in the lepton(s) plus jets method (WW,WZ pair production) is stronger than in the $W\gamma$ process. Here the requirement of very high boson P_T, which is necessary to suppress backgrounds, results in sensitivity to the couplings' values at a higher energy scale. Figure 4 shows the $\sqrt{\hat{s}}$ distribution for simulated WW events which would pass the cuts used in the run Ia CDF analysis. This simulation assumed anomalous $WW\gamma$ and WWZ couplings $\lambda = 1$ and $\Lambda_{FF} = 1500\,\text{GeV}$, but the concentration of events around $\sqrt{\hat{s}} = 500\,\text{GeV}$ is primarily set by the cuts and does not shift by much as the couplings are varied. For future high luminosity runs the region of sensitivity will move from 500 GeV to even higher $\sqrt{\hat{s}}$. Extrapolating expected bounds from this scale to $\sqrt{\hat{s}} = 0$, the projected sensitivity as a function of Λ_{FF} is shown in Figure 5 for an integrated luminosity of $10\,\text{fb}^{-1}$. Significant loss of sensitivity is seen for $\Lambda_{FF} \lesssim 800\,\text{GeV}$.

One concludes that although sensitivity in the $W\gamma$ process is roughly comparable to WW, WZ in the lepton(s) plus jets mode for $\Lambda_{FF} \gtrsim 1000\,\text{GeV}$, if deviations from standard model are restricted to the energy scale 500 GeV or smaller then they could potentially be observed in $W\gamma$ but missed in WW, WZ. The dependence on Λ_{FF} for the pure leptonic WW method is intermediate between these two cases.

NEW PROSPECTS AT HIGH LUMINOSITY

WZ in pure leptonic channel

The set of methods sensitive to $WW\gamma$ and WWZ couplings is rounded out by WZ pair production with pure leptonic decays. In the past, the small cross section and branching ratios for this channel have made it unobservable, but with an integrated luminosity of $10\,\text{fb}^{-1}$ one expects about 400 events (before cuts and efficiencies). In addition about 60 ZZ pairs with pure leptonic decay are expected. Such samples will be sufficient for rough measurement of the cross section, and in the WZ case they will allow more detailed studies.

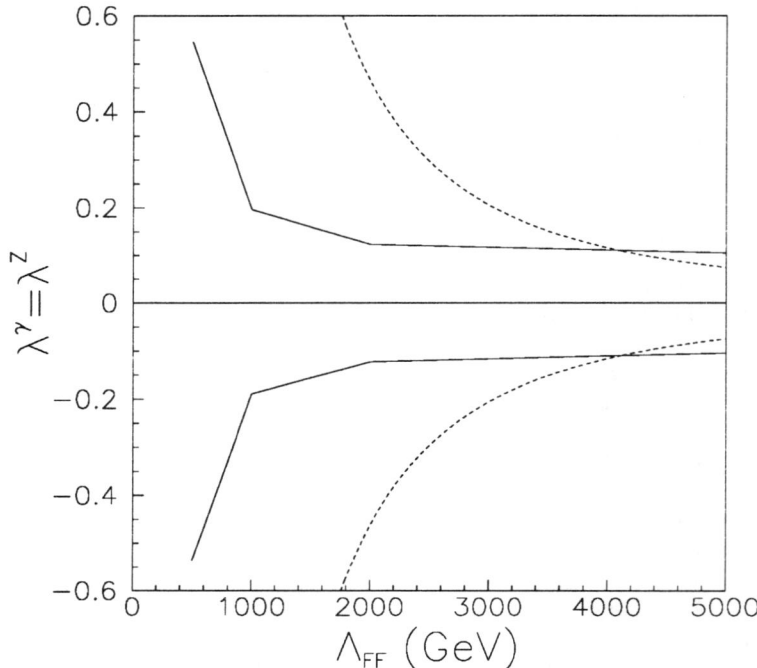

FIG. 5. Sensitivity of the WW, WZ lepton(s) plus jets channel for the anomalous coupling λ, as a function of the assumed form factor scale Λ_{FF}. The solid lines show upper and lower 95% CL limits expected for integrated luminosity of $10\,\text{fb}^{-1}$. The dotted curve shows the unitarity bound (7).

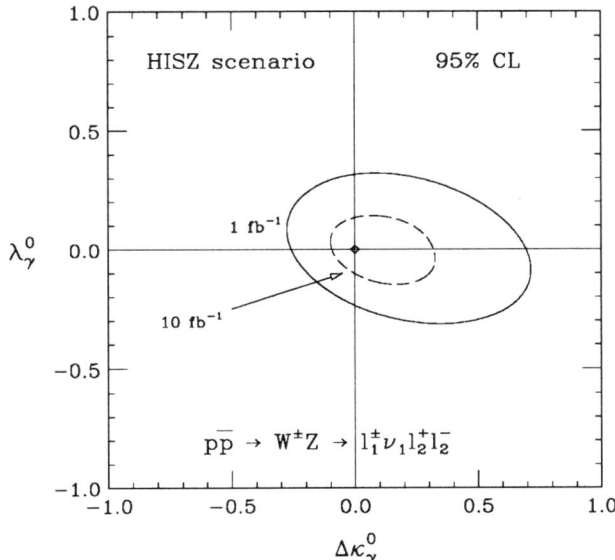

FIG. 6. Expected 95% CL bounds on $WW\gamma$, WWZ couplings from WZ production with pure leptonic decay. The couplings λ^γ, λ^Z, g_1^Z, $\Delta\kappa^\gamma$ and $\Delta\kappa^Z$ are assumed to be connected by the HISZ relations (6). From ref. (8).

The particular advantage of studying the $WZ \to l\nu ll$ channel is that it is sensitive only to the WWZ vertex, and therefore can provide information on this vertex without any assumptions on the relation between $WW\gamma$ and WWZ couplings. In addition, backgrounds are expected to be small. Figure 6 shows the expected future bounds on couplings from this channel (8). The detector model assumed in this study was somewhat more optimistic than in the others, enhancing the sensitivity by perhaps 30%. Even so, although in the past the channel has not been competitive, it is clear that at higher luminosity its sensitivity will approach that of the other channels. The dependence on the form factor scale Λ_{FF} should be similar to the purely leptonic WW process, *i.e.* less severe than WW, WZ decaying into lepton(s) plus jets.

$Z\gamma$, WZ, ZZ with invisible Z decay

In principle, the processes with invisible Z decay ($Z \to \nu\bar{\nu}$) have an advantage over their counterparts with pure leptonic Z decay, namely the larger branching fraction. The Z decay is "observed" via missing transverse momentum in all these processes, and thus one of the main backgrounds is expected to arise from jet energy mismeasurement in γ plus jet, W plus jet, and Z plus jet events. Careful studies of these backgrounds, in the context of real detectors, and their effect on sensitivity to anomalous couplings have yet to be done. It has been pointed out that they can be substantially reduced by a requirement of high transverse momentum for the other (observed) boson in the event (9).

In the case of $WZ \to l\nu\nu\bar{\nu}$, the event signature is a single lepton, missing transverse momentum, and transverse mass larger than the W mass. Therefore this process is afflicted with an additional irreducible background from the high-mass tail of Drell-Yan single W production. A study has been done to assess the sensitivity of this WZ channel to the WWZ couplings, taking account of the Drell-Yan background but not that from fake missing transverse momentum. The latter has been found to be substantially smaller than the high-mass Drell-Yan continuum in the CDF experiment (10). The bounds on the couplings were determined using fits to simulated spectra of transverse mass above the W mass. Lepton acceptance and identification efficiencies were assumed to be typical of the CDF experiment. With integrated luminosity of $10\,\text{fb}^{-1}$, expected bounds are $|\Delta\kappa^Z| < 0.5$ and $|\lambda^Z| < 0.5$ for $\Lambda_{FF} = 1000\,\text{GeV}$ and assuming the HISZ relations (6) among the couplings λ^Z, Δg_1^Z and $\Delta\kappa^Z$. This sensitivity is much poorer than that from pure leptonic WZ decay (Figure 6), which is $-0.04 < \Delta\kappa^Z < 0.11$ and $|\lambda^Z| < 0.15$ for similar assumptions.

Radiation zeroes in $W\gamma$, WZ

In both $W\gamma$ and WZ production, the standard model exhibits strong cross section dips at certain angles. These result in corresponding dips in the distribution of Δy, where Δy is the rapidity difference between the γ or Z and

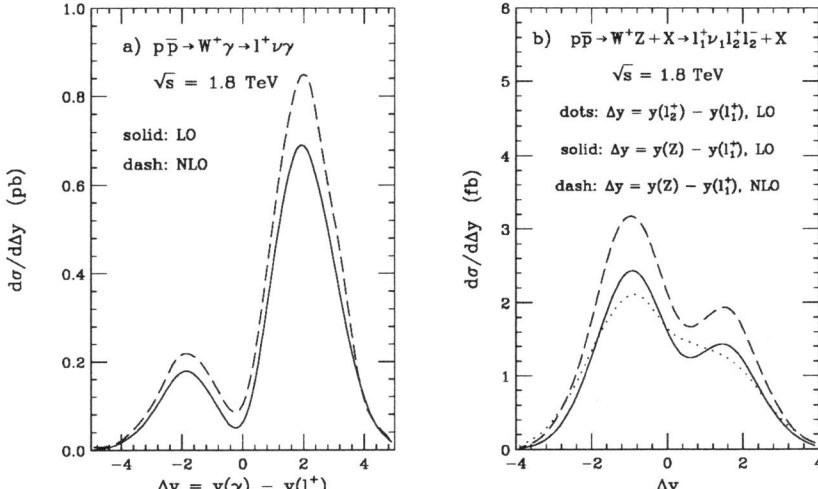

FIG. 7. Distribution of Δy in $W\gamma$ and WZ production, where Δy is the difference between rapidity of the e^+ or μ^+ from W^+ decay and the rapidity of the other boson (γ or Z). From refs. (8,11).

the charged lepton from the decay $W^+ \to l^+ \nu$ (11,12,8). These are shown in Figure 7. For an integrated luminosity of $1\,\text{fb}^{-1}$, the $W\gamma$ distribution of Figure 7a will be mapped out by approximately 1000 events (after cuts). Significant results may already be obtained with the 100 events expected by the end of run Ib. In the WZ case (Figure 7b), integrated luminosity of $10\,\text{fb}^{-1}$ will be required to map out the distribution with about 50 events (after cuts).

The radiation zeroes are an interesting prediction of the standard model, and their observation would be very important. However, the sensitivity of these features to anomalous couplings is not as great as the other methods already described.

SUMMARY

With integrated luminosity in the range $1-10\,\text{fb}^{-1}$, future Tevatron experiments will probe $WW\gamma$ and WWZ couplings with a precision in the range $0.1-0.3$. The experiments will also be able to search for anomalous $Z\gamma\gamma$ and $ZZ\gamma$ couplings with a sensitivity more than an order of magnitude better than in the past. These statements rest largely on extrapolation of results from mature analyses at CDF and D0 with lower integrated luminosity.

The sensitivity to $WW\gamma$ and WWZ couplings will be shared among many different channels. Comparable sensitivities are expected from $W\gamma$; WW or WZ decaying to lepton(s) plus jets; pure leptonic WW decays; and pure leptonic WZ decays. This is somewhat different from the current situation, where the best limits come from $W\gamma$ and from WW, WZ in the lepton(s) plus jets mode. It is important to study all these channels, not only to improve statistical power by combining results, but also because the channels are somewhat complementary to each other, each with its own strengths and weaknesses.

REFERENCES

1. T. A. Fuess, these proceedings.
2. H. Aihara, these proceedings.
3. H. Aihara et al., "Anomalous Gauge Boson Interactions," University of Wisconsin preprint number MAD/PH/871, to be published in *Electroweak Symmetry Breaking and Beyond the Standard Model*, eds. T. Barklow et al..
4. F. A. Berends et al., Nucl. Phys. **B357**, 32 (1991).
5. T. Diehl, private communication.
6. K. Hagiwara et al., Phys. Rev. **D41**, 2113 (1990).
7. U. Baur and D. Zeppenfeld, Phys. Lett. **B201**, 383 (1988).
8. U. Baur, J. Ohnemus, and T. Han, Phys. Rev. **D51**, 3381 (1995).
9. U. Baur and E. L. Berger, Phys. Rev. **D47**, 4889 (1993).
10. F. Abe et al., Phys. Rev. Lett. **74**, 341 (1995).
11. U. Baur, S. Errede, and G. Landsberg, Phys. Rev. **D50**, 1917 (1994).
12. S. Frixione, P. Nason, and S. Ridolfi, Nucl. Phys. **B383**, 3 (1992).

Probing Tri-linear Couplings at the LHC

John Womersley*

*Fermi National Accelerator Laboratory[1]
Batavia, Illinois 60510

> Prospects for measuring the tri-linear couplings of gauge bosons at the LHC are reviewed. Studies carried out by the ATLAS collaboration, and for the 1994 DPF Long Range Planning Workshop, are described. It will be possible to probe WWV anomalous couplings with a precision of order $10^{-1} - 10^{-3}$ if the form factor scale $\Lambda_{FF} > 2\,\text{TeV}$. This just reaches the interesting region where one may hope to see deviations from the standard model given present limits on the scale of new physics. The $Z\gamma V$ limits are much more sensitive to Λ_{FF}; for $\Lambda_{FF} > 1.5\,\text{TeV}$, $h_3^Z < \mathcal{O}(10^{-3})$ and $h_4^Z < \mathcal{O}(10^{-5})$. It will be possible to see the amplitude zero in $W\gamma$ and WZ production at the LHC but this physics is probably better explored at the Tevatron.

STATUS OF THE LHC

The CERN council unanimously approved the LHC project on December 16, 1994. While the full, high-luminosity project is approved, the machine may be brought into operation in two stages depending on the resources available: firstly with a centre of mass energy of approximately 10 TeV in 2004, using two-thirds of the dipole magnets, and then the full 14 TeV in 2008. Financial contributions from non-member states are to be used to accelerate the construction of the machine; the expected level of contributions would enable the full LHC to be available in 2004. The timescale, staging and non-member states' contributions will be reviewed in 1997. CERN has agreed to operate under very stringent financial constraints during the full period of LHC construction.

LHC DETECTORS

There are two large, general-purpose pp collider detectors approved for LHC: ATLAS (1) and CMS (2). Both collaborations completed Technical

---

[1] operated by the Universities Research Association, Inc., for the U.S. Department of Energy.

Proposals for their detectors in December 1994.

ATLAS uses a tracking system employing silicon pixels, silicon strip detectors, and a transition radiation tracker, all contained within a superconducting solenoid. The charged track resolution is $\Delta p_T/p_T = 30\%$ at $p_T = 500\,\text{GeV}/c$. The tracker is surrounded by an electromagnetic calorimeter using a lead-liquid argon accordion design; the EM calorimeter covers $|\eta| < 3$ (with trigger coverage of $|\eta| < 2.5$) and has a resolution of $\Delta E/E = 10\%/\sqrt{E} \oplus 0.7\%$. The hadronic calorimeter uses scintillator tiles in the barrel, and liquid argon in the endcaps; its resolution is $\Delta E/E = 50\%\sqrt{E} \oplus 3\%$. Separate forward calorimeters cover the region $3 < |\eta| < 5$ with a resolution $\Delta E/E = 100\%\sqrt{E} \oplus 10\%$. Surrounding the calorimeters is the muon system. Muon trajectories are measured using three layers of chambers (MDT's and CSC's) in a spectrometer using a large air-core toroid magnet. The resulting muon momentum measurement is $\Delta p_T/p_T = 8\%$ at $p_T = 1\,\text{TeV}/c$ and $\Delta p_T/p_T = 2\%$ at $p_T = 100\,\text{GeV}/c$. Muons may be triggered on over the range $|\eta| < 2.2$.

The CMS tracking system is based on silicon pixels, silicon strip detectors, and microstrip gas chambers. The charged track resolution is $\Delta p_T/p_T = 5\%$ at $p_T = 1\,\text{TeV}/c$ and $\Delta p_T/p_T = 1\%$ at $p_T = 100\,\text{GeV}/c$. CMS has chosen a precision electromagnetic calorimeter using lead tungstate ($PbWO_4$) crystals, covering $|\eta| < 3$ (with trigger coverage of $|\eta| < 2.6$). Its resolution is $\Delta E/E = 2\%\sqrt{E} \oplus 0.5\%$. The surrounding hadronic calorimeter uses scintillator tiles in the barrel and endcaps; its resolution is $\Delta E/E = 65\%\sqrt{E} \oplus 5\%$. The region $3 < |\eta| < 5$ is covered by forward calorimeters using parallel-plate chambers or quartz fibers and having a resolution of about $\Delta E/E = 130\%\sqrt{E} \oplus 10\%$. The calorimeters are contained in a 4 tesla superconducing coil which provides the magnetic field for charged particle tracking. Muon trajectories outside the coil are measured in four layers of chambers (drift tubes and CSC's) embedded in the iron return yoke. The resulting muon momentum measurement covers the range $|\eta| < 2.4$ with a resolution $\Delta p_T/p_T = 5\%$ at $p_T = 1\,\text{TeV}/c$ and $\Delta p_T/p_T = 1\%$ at $p_T = 100\,\text{GeV}/c$. The muon trigger extends over $|\eta| < 2.1$.

Significant contributions to both detectors are planned to be made by U.S. groups. For ATLAS, these groups involve about 200 physicists and engineers from 27 U.S. institutions; for CMS, about 300 physicists and engineers from 37 U.S. institutions. Contributions to ATLAS include one half to one third of the silicon pixels, one third to one quarter of the silicon strips, and the barrel transition radiation tracker; readout for the full liquid argon calorimeter, the EM section of the forward calorimeters, and about one third of the scintillator tile calorimeter; the endcap muon system, and contributions to the level 1 and level 2 triggers. For CMS, the list includes the forward silicon pixels, and one quarter of the forward MSGC's; the barrel hadron calorimeter system and EM calorimeter front-end; the endcap muon system, and contributions to the level 1 and level 2 triggers.

DIBOSON PRODUCTION AT HADRON COLLIDERS

The trilinear WWV and $Z\gamma V$ couplings ($V = Z, \gamma$) may be probed at hadron colliders using diboson final states. The generic set of Feynman diagrams contributing to diboson production is shown in Fig. 1.

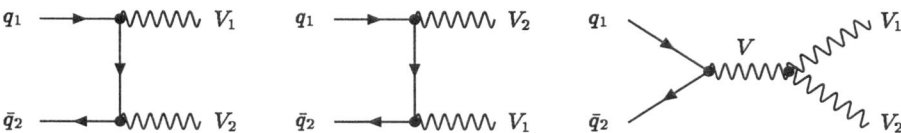

Figure 1: *Generic Feynman diagrams contributing to di-boson production in hadronic collisions. V, V_1, $V_2 = W, \gamma, Z$.*

Following the notation used elsewhere in these proceedings, we shall parametrize the CP-conserving WWV anomalous couplings in terms of $\Delta\kappa_V$ and λ_V, where $\kappa_V = 1$ and $\lambda_V = 0$ in the Standard Model for $V = Z, \gamma$. In general, we would expect anomalous couplings of order m_W^2/Λ^2 if Λ is the scale for new physics, so if $\Lambda \sim 1$ TeV then $\Delta\kappa_V, \lambda_V \sim 0.01$. The $Z\gamma V$ anomalous couplings are parametrized in terms of h_3^V and h_4^V, where $h_3^V = h_4^V = 0$ in the Standard Model and deviations are expected to be $\mathcal{O}(m_W^4/\Lambda^4)$.

To maintain unitarity, the observed anomalous couplings must be modified by a form factor; so (for example)

$$\Delta\kappa_V(q^2) = \frac{\Delta\kappa_V^0}{(1+q^2/\Lambda_{FF}^2)^n} \qquad (1)$$

where Λ_{FF} is the form factor scale and $n = 2$ for $\Delta\kappa, \lambda$ and $n = 3, 4$ for h_3^V, h_4^V.

ATLAS STUDIES

The ATLAS collaboration have studied (3) the sensitivity of their proposed detector to anomalous couplings in the $W\gamma$ and WZ modes; the W^+W^- signal is swamped by $t\bar{t}$ background. A form factor scale $\Lambda_{FF} = 10$ TeV was used.

For the $W\gamma$ final state, events were assumed to be triggered using a high-p_T lepton plus high-p_T photon candidate. The background includes contributions

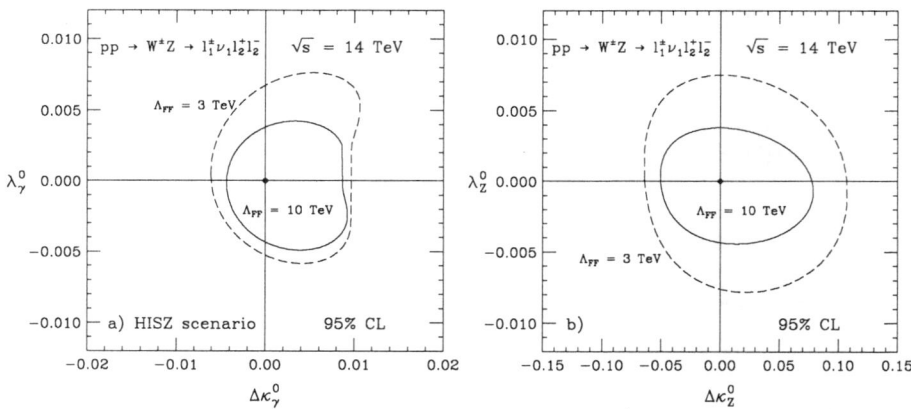

Figure 2: 95% CL sensitivity limits from $W^\pm Z \to \ell_1^\pm \nu_1 \ell_2^+ \ell_2^-$ at the LHC a) in the HISZ scenario (5) and b) if only $\Delta\kappa_Z$ and λ_Z are allowed to deviate from the Standard Model.

from events with a real lepton and a real photon (e.g. $b\bar{b}\gamma$, $t\bar{t}\gamma$, and $Z\gamma$); a fake lepton but a real photon (e.g. γ + jet); and a fake photon with a real lepton (e.g. W + jet, $b\bar{b}$, and $t\bar{t}$). Rejection factors of 10^4 against jets faking photons and 10^5 against jets faking electrons were assumed. It is claimed that there are essentially no fake muons in ATLAS. To reduce backgrounds, events were selected with $p_T^\gamma > 100\,\text{GeV}/c$, $p_T^\ell > 40\,\text{GeV}/c$, and $|\eta^\ell| < 2.5$. Events with jets were also vetoed, to further reduce backgrounds and to lessen the importance of higher-order QCD corrections. In an integrated luminosity of $10^5\,\text{pb}^{-1}$, 7500 events remain, with a signal to background ratio of 3:1. The p_T^γ distribution is then fitted in the region where the standard model prediction is 15 events (above about 600 GeV/c, yielding limits of $|\Delta\kappa_\gamma| < 0.04$ and $|\lambda_\gamma| < 0.0025$ (95% C.L.).

Similar techniques were used for the WZ state. The trigger was three high-p_T leptons, and the backgrounds are from $Zb\bar{b}$, Z + jet, $b\bar{b}$ and $t\bar{t}$ processes. Events were selected with $p_T^\ell > 25\,\text{GeV}/c$, $|\eta^\ell| < 2.5$, $|m_{\ell_1 \ell_2} - m_Z| < 10\,\text{GeV}/c^2$, and $m_T(\ell^3, E_T^{miss}) > 40\,\text{GeV}/c^2$; a jet veto was also imposed. In $10^5\,\text{pb}^{-1}$, 4000 events then remain, with a signal to background ratio of 2:1. The p_T^Z distribution is again fitted in the region where the standard model prediction is 15 events (above about 380 GeV/c), yielding limits of $|\Delta\kappa_Z| < 0.07$ and $|\lambda_Z| < 0.005$ (95% C.L.).

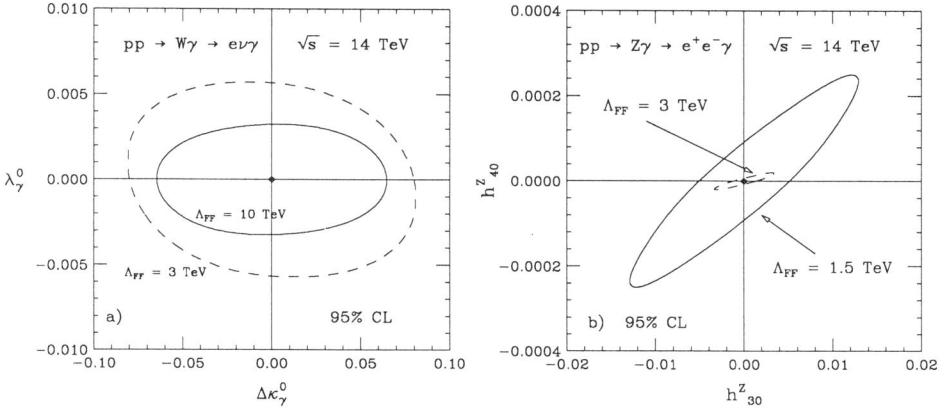

Figure 3: 95% CL sensitivity limits for a) $WW\gamma$ couplings from $W\gamma$ production and b) $ZZ\gamma$ couplings from $Z\gamma$ production at the LHC. Results are displayed for an integrated luminosity of 100 fb^{-1} and two different form factor scales.

DPF WORKING GROUP STUDIES

Studies (4) have also been carried out for the 1994 DPF Long Range Planning Workshop [2].

For WZ states, the $eee\nu$ signal only was considered, and it was required that $p_T^\ell > 25\,\text{GeV}/c$, and $E_T^{miss} > 50\,\text{GeV}$. A binned likelihood fit to the p_T^Z distribution then yields limits on $\Delta\kappa_Z$ and λ_Z which are shown in Fig 2.

For $W\gamma$ and $Z\gamma$ states, a combination of ATLAS resolutions and CDF efficiencies was assumed. It was required that $p_T^\ell > 40\,\text{GeV}/c$, $p_T^\gamma > 25\,\text{GeV}/c$, $E_T^{miss} > 25\,\text{GeV}$ ($W\gamma$ only), and $m(\ell\ell\gamma) > 110\,\text{GeV}/c^2$ ($Z\gamma$ only). A separation of $\Delta R > 0.7$ between the lepton and photon was required, and events with any jet with E_T above 50 GeV were vetoed. A binned likelihood fit to the p_T^γ distributions then yields limits on $\Delta\kappa_\gamma$, λ_γ, h_3^Z and h_4^Z which are shown in Fig 3.

The limits obtained in all the above studies are summarised in Table 1.

---

[2] by the subgroup on Anomalous Gauge Boson Interactions within the "Electroweak Symmetry Breaking — Beyond the Standard Model" group of the 1994 Long Range Planning Workshop of the Division of Particles and Fields of the American Physical Society.

Table 1: Expected 95% CL limits on anomalous WWV, $V = \gamma, Z$, and $ZZ\gamma$ couplings from experiments at the LHC (pp collisions at $\sqrt{s} = 14$ TeV; $\int \mathcal{L} dt = 100\,\text{fb}^{-1}$). Only one of the independent couplings is assumed to deviate from the SM at a time. The limits obtained for $Z\gamma\gamma$ couplings almost coincide with those found for h_3^Z and h_4^Z.

channel	study	limit $\Lambda_{FF} = 3$ TeV	limit $\Lambda_{FF} = 10$ TeV
$pp \to W^{\pm}\gamma \to e^{\pm}\nu\gamma$	DPF	$\|\Delta\kappa_\gamma^0\| < 0.080$	$\|\Delta\kappa_\gamma^0\| < 0.065$
		$\|\lambda_\gamma^0\| < 0.0057$	$\|\lambda_\gamma^0\| < 0.0032$
	ATLAS		$\|\Delta\kappa_\gamma^0\| < 0.04$
			$\|\lambda_\gamma^0\| < 0.0025$
$pp \to W^{\pm}Z \to \ell\nu\ell\ell$	DPF	$-0.0060 < \Delta\kappa_\gamma^0 < 0.0097$	$-0.0043 < \Delta\kappa_\gamma^0 < 0.0086$
$\ell = e, \mu$, HISZ (5)		$-0.0053 < \lambda_\gamma^0 < 0.0067$	$-0.0043 < \lambda_\gamma^0 < 0.0038$
$pp \to W^{\pm}Z \to \ell\nu\ell\ell$	DPF	$-0.064 < \Delta\kappa_Z^0 < 0.107$	$-0.050 < \Delta\kappa_Z^0 < 0.078$
$\ell = e, \mu$, $\Delta g_1^Z = 0$		$-0.0076 < \lambda_Z^0 < 0.0075$	$-0.0043 < \lambda_Z^0 < 0.0038$
	ATLAS		$\|\Delta\kappa_Z^0\| < 0.07$
			$\|\lambda_Z^0\| < 0.005$

channel	study	limit $\Lambda_{FF} = 1.5$ TeV	limit $\Lambda_{FF} = 3$ TeV
$pp \to Z\gamma \to e^+e^-\gamma$	DPF	$\|h_{30}^Z\| < 0.0051$	$\|h_{30}^Z\| < 0.0013$
		$\|h_{40}^Z\| < 9.2 \cdot 10^{-5}$	$\|h_{40}^Z\| < 6.8 \cdot 10^{-6}$

AMPLITUDE ZERO IN $W\gamma, WZ$

It is found (6) that the standard model amplitude for the processes $q\bar{q} \to W\gamma, WZ$ exhibit an approximate amplitude zero which leads to a dip in the distribution of rapidity difference $\Delta y = y_\gamma - y_\ell$ ($W\gamma$) or $y_Z - y_{\ell^+}$ (WZ) near $\Delta y = 0$. In Fig. 4 the distributions of Δy are shown for the LHC. While the dip is still apparent, the LHC is much less sensitive to this behaviour than is the Tevatron, owing to the dominance of gluon-gluon scattering at the LHC.

CONCLUSIONS

At the LHC it will be possible to probe WWV anomalous couplings with a precision of order $10^{-1} - 10^{-3}$ if the form factor scale $\Lambda_{FF} > 2$ TeV. This just reaches the interesting region where one may hope to see deviations from the

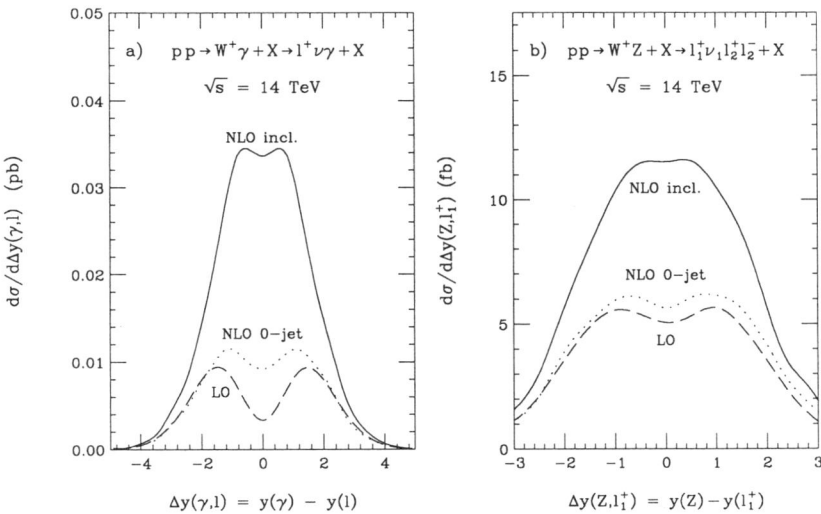

Figure 4: Rapidity difference distributions in the SM at the LHC. a) The photon lepton rapidity difference spectrum in $pp \to \ell^+ \displaystyle{\not}p_T \gamma$. b) The $y(Z) - y(\ell_1^+)$ distribution in $pp \to W^+Z$.

standard model given present limits on the scale of new physics. The $Z\gamma V$ limits are much more sensitive to Λ_{FF}; for $\Lambda_{FF} > 1.5\,\text{TeV}$, $h_3^Z < \mathcal{O}(10^{-3})$ and $h_4^Z < \mathcal{O}(10^{-5})$.

It will be possible to see the amplitude zero in $W\gamma$ and WZ production at the LHC but this physics is probably better explored at the Tevatron.

ACKNOWLEDGMENTS

The author would like to thank U. Baur and S. Errede for their help in the preparation of this document, and for making available Figs. 2–4.

REFERENCES

1. ATLAS collaboration, Technical Proposal, CERN/LHCC 94-43 (LHCC/P2), December 1994.
2. CMS collaboration, Technical Proposal, CERN/LHCC 94-38 (LHCC/P1), December 1994.
3. D. Fouchez, ATLAS internal note PHYS-NO-160 (1994), unpublished.
4. H. Aihara *et al.*, FERMILAB-Pub-95/031.
5. K. Hagiwara, S. Ishihara, R. Szalapski, and D. Zeppenfeld, Phys. Lett. **B283** (1992) 353; Phys. Rev. **D48** (1993) 2182.
6. R.W.Brown *et al.*, Phys. Rev. **D20** (1979) 1164; K.O.Mikaelian *et al.*, Phys. Rev. Lett. **43** (1979) 746; U. Baur, T. Han and J. Ohnemus, Phys. Rev. Lett. **72** (1994) 3941.

Studies of WWγ and WWZ Couplings at Future e^+e^- Linear Colliders

Timothy L. Barklow

Stanford Linear Accelerator Center
Stanford University, Stanford, California 94309

Techniques for studying the $WW\gamma$ and WWZ trilinear vector boson couplings at future e^+e^- linear colliders are reviewed. We also investigate how strong longitudinal W boson scattering can be studied through the process $e^+e^- \to W^+W^-$. Quantitative estimates of the trilinear coupling sensitivities at center of mass energies of 0.5 TeV and 1.5 TeV are included.

INTRODUCTION

A new era in experimental particle physics has opened recently with the direct measurements of trilinear vector boson couplings by the CDF (1) and D0 (2) collaborations at the Tevatron. In the next few years experiments at the Tevatron with the Main Injector and at LEP II will measure these couplings with greater accuracy. In the longer term it is expected that future colliders such as the LHC and the next e^+e^- linear collider (NLC) will improve the precision of trilinear vector boson coupling measurements by an order of magnitude or more over the Main Injector and LEP II results.

The theoretical issues surrounding the study of of trilinear vector boson couplings at the LHC and NLC have been discussed extensively in the literature (3). In this paper we concentrate on the experimental techniques used to measure directly the $WW\gamma$ and WWZ trilinear vector boson couplings at the NLC. We shall deal solely with the measurement of these couplings through the process $e^+e^- \to W^+W^-$. And, since the experimental techniques are closely related, we shall also review how strong longitudinal W boson scattering can be studied using $e^+e^- \to W^+W^-$.

We assume two stages for the center of mass energy and luminosity of the NLC (4). In the initial stage the center of mass energy is 500 GeV and the design luminosity is 0.8×10^{34} cm^{-1} s^{-1}. In a later stage the center of mass energy reaches 1500 GeV with a design luminosity of 1.9×10^{34} cm^{-1} s^{-1}. We will assume 10^7 seconds at the design luminosity for our integrated luminosity. Initial state electron polarization will play an important role in our analysis, and we assume that 90% polarization will be available at these center of mass energies and luminosities.

We will use the trilinear vector boson coupling formalism of Ref. (5) in our discussion. For example, we will be referring later to the parameters κ_γ, κ_Z, λ_γ, λ_Z, and g_1^Z, which are defined in Eq. 2.1 of Ref. (5)

ANALYSIS OF W^+W^- FINAL STATE

We perform a full final state helicity analysis on $e^+e^- \to W^+W^-$ in order to obtain the greatest sensitivity to anomalous trilinear vector boson couplings and to strong longitudinal W boson scattering effects. The final state topology with one W decaying to an electron or muon (plus neutrino) and the other decaying hadronically is best for such an analysis. Other event topologies will not be considered.

The W^- production angle Θ is defined to be the angle between the initial state e^- and the W^- in the e^+e^- rest frame. We define θ^* and ϕ^* to be the polar and azimuthal angles of the lepton in the rest frame of the W which decays leptonically. $\overline{\theta}^*$ and $\overline{\phi}^*$ are the polar and azimuthal angles of the quark jets in the rest frame of the W which decays hadronically. We assume no quark flavor tagging, and so we must average over the quark and antiquark directions.

Multi-Differential Cross Section Formalism

Consider the topology where the W^- decays leptonically and the W^+ decays hadronically. The multi-differential cross section in the narrow width approximation is then given by (5)

$$\frac{d\sigma(\kappa_\gamma, \lambda_\gamma, \kappa_Z, \lambda_Z; \cos\Theta, \cos\theta^*, \phi^*, \cos\overline{\theta}^*, \overline{\phi}^*, P_e)}{d\cos\Theta\ d\cos\theta^*\ d\phi^*\ d\cos\overline{\theta}^*\ d\overline{\phi}^*} =$$

$$\frac{9\beta B_{lh}}{8192\pi^3 s} \sum_{\lambda,\lambda',\overline{\lambda},\overline{\lambda}'} \mathcal{Q}_{\lambda'\overline{\lambda}'}^{\lambda\overline{\lambda}}(\kappa_\gamma, \lambda_\gamma, \kappa_Z, \lambda_Z; \cos\Theta, P_e) \mathcal{D}_{\lambda'}^{\lambda}(\cos\theta^*, \phi^*) \overline{\mathcal{H}}_{\overline{\lambda}'}^{\overline{\lambda}}(\cos\overline{\theta}^*, \overline{\phi}^*)$$

(1)

where

$$\mathcal{Q}_{\lambda'\overline{\lambda}'}^{\lambda\overline{\lambda}} \equiv (1+P_e) \sum_{\overline{\sigma}} \mathcal{M}_1(+, \overline{\sigma}, \lambda, \overline{\lambda}; \Theta) \mathcal{M}_1^*(+, \overline{\sigma}, \lambda', \overline{\lambda}'; \Theta)$$
$$+ (1-P_e) \sum_{\overline{\sigma}} \mathcal{M}_1(-, \overline{\sigma}, \lambda, \overline{\lambda}; \Theta) \mathcal{M}_1^*(-, \overline{\sigma}, \lambda', \overline{\lambda}'; \Theta) \qquad (2)$$

and

$$\overline{\mathcal{H}}_{\overline{\lambda}'}^{\overline{\lambda}}(\cos\overline{\theta}^*, \overline{\phi}^*) \equiv \left[\overline{\mathcal{D}}_{\overline{\lambda}'}^{\overline{\lambda}}(\cos\overline{\theta}^*, \overline{\phi}^*) + \overline{\mathcal{D}}_{\overline{\lambda}'}^{\overline{\lambda}}(-\cos\overline{\theta}^*, \overline{\phi}^* + \pi)\right] \quad . \qquad (3)$$

Here P_e is the initial state electron polarization and $\mathcal{M}_1$ is the W^+W^- production amplitude as given in Ref. (5). The variables $\lambda, \lambda', \overline{\lambda}, \overline{\lambda}'$ are indices for the W^+W^- polarizations where $\lambda = +, -, 0$ denotes right-handed transverse, left-handed transverse, and longitudinal polarization, respectively.

The W^- leptonic decay tensor $\mathcal{D}^\lambda_{\lambda'}(\cos\theta^*, \phi^*)$ is given explicitly by

$$\mathcal{D}^+_+ = \frac{1}{2}(1-\cos\theta^*)^2, \qquad \mathcal{D}^-_- = \frac{1}{2}(1+\cos\theta^*)^2, \qquad \mathcal{D}^0_0 = \sin^2\theta^*,$$

$$\mathcal{D}^+_0 = \frac{1}{\sqrt{2}}(1-\cos\theta^*)\sin\theta^* e^{i\phi^*},$$

$$\mathcal{D}^-_0 = -\frac{1}{\sqrt{2}}(1+\cos\theta^*)\sin\theta^* e^{-i\phi^*},$$

$$\mathcal{D}^+_- = \frac{1}{2}\sin^2\theta^* e^{2i\phi^*}, \tag{4}$$

and $\mathcal{D}^{\lambda'}_\lambda = (\mathcal{D}^\lambda_{\lambda'})^*$. The W^+ hadronic decay tensor $\overline{\mathcal{H}}^{\overline{\lambda}}_{\overline{\lambda}'}(\cos\overline{\theta}^*, \overline{\phi}^*)$ is given by

$$\overline{\mathcal{H}}^+_+ = 1 + \cos^2\overline{\theta}^*, \qquad \overline{\mathcal{H}}^-_- = 1 + \cos^2\overline{\theta}^*, \qquad \overline{\mathcal{H}}^0_0 = 2\sin^2\overline{\theta}^*,$$

$$\overline{\mathcal{H}}^+_0 = \sqrt{2}\cos\theta^* \sin\overline{\theta}^* e^{-i\overline{\phi}^*}$$

$$\overline{\mathcal{H}}^-_0 = -\sqrt{2}\cos\theta^* \sin\overline{\theta}^* e^{i\overline{\phi}^*}$$

$$\overline{\mathcal{H}}^+_- = \sin^2\overline{\theta}^* e^{-2i\overline{\phi}^*} \tag{5}$$

The expressions for the leptonic decay tensor elements demonstrate how the lepton polar decay angle θ^* helps separate right and left-handed transversely polarized W^- bosons from each other and from longitudinally polarized W^- bosons. Furthermore, comparing $\mathcal{D}^0_0$ and $\mathcal{D}^+_-$ we see that the azimuthal angle ϕ^* is of value in disentangling longitudinal polarizations from the inteference between right and left-handed transverse polarizations.

The analyzing power of the hadronic decay tensor is somewhat degraded due to the averaging of the quark and antiquark directions. For example, right and left-handed transversely polarized W^+ bosons are indistinguishable. Nevertheless, the quark polar decay angle is still of value in distinguishing transverse polarizations from longitudinal polarizations, and the quark azimuthal angle continues to separate longitudinal polarizations from the interference between right and left-handed transverse polarizations. From the above considerations we choose to use the following five measured variables in our maximum likelihood fits: $\cos\Theta, \cos\theta^*, \phi^*, \cos\overline{\theta}^*$, and $\overline{\phi}^*$. These variables are reconstructed from detected quantites such as the energy and angles of the lepton and quark jets.

Reconstruction of $\cos\Theta, \cos\theta^*, \phi^*, \cos\bar{\theta}^*, \bar{\phi}^*$

We shall now describe in detail how the measured variables $\cos\Theta$, $\cos\theta^*$, ϕ^*, $\cos\bar{\theta}^*$, and $\bar{\phi}^*$ are reconstructed. Our detected quantities are (all laboratory frame):

$$|\vec{P}_l| = \text{lepton momentum},$$
$$\theta_l = \text{polar angle of lepton},$$
$$\phi_l = \text{azimuthal angle of lepton},$$
$$\beta_j = |\vec{P}_j|/E_j \text{ where } (E_j, \vec{P}_j) \text{ is the four-vector for jet}\#j, \; j=1,2$$
$$\theta_j = \text{polar angle of jet}\#j, \; j=1,2$$
$$\phi_j = \text{azimuthal angle of jet}\#j, \; j=1,2$$
$$\kappa = |\vec{P}_2|/E_1. \tag{6}$$

Note that, with the exception of $|\vec{P}_l|$ all quantities are unitless, which helps reduce the systematic error associated with their measurement. The quantity $|\vec{P}_l|$ has units of GeV, but it is straightforward to measure the momentum of an isolated e^- or μ^- very well.

If we use only jet information then we can obtain estimates for the W^- polar and azimuthal angles. We denote these estimates by Θ_0 and Φ_0, respectively, in order to distinguish them from our final estimates of Θ and Φ. The quantities Θ_0 and Φ_0 are given by:

$$\cos\Theta_0 = \chi\left(\frac{\beta_1\cos\theta_1 + \kappa\cos\theta_2}{[\beta_1^2 + \kappa^2 + 2\kappa\beta_1\hat{n}_1\cdot\hat{n}_2]^{\frac{1}{2}}}\right) \tag{7}$$

$$\Phi_0 = \tan^{-1}\left[\frac{\beta_1\sin\theta_1\sin\phi_1 + \kappa\sin\theta_2\sin\phi_2}{\beta_1\sin\theta_1\cos\phi_1 + \kappa\sin\theta_2\cos\phi_2}\right] \tag{8}$$

where

$$\chi = \begin{cases} -1 & \text{if the } W^- \text{ decays leptonically} \\ +1 & \text{if the } W^+ \text{ decays leptonically} \end{cases} \tag{9}$$

and

$$\hat{n}_i = (\sin\theta_i\cos\phi_i, \sin\theta_i\sin\phi_i, \cos\theta_i) \text{ for } i=1,2. \tag{10}$$

Next we define the variable α, which is used in conjunction with the lepton momentum to obtain the best estimate for $\cos\Theta$ and Φ:

$$\alpha = \tan^{-1}\left[\frac{\sin\Theta_0\Psi_0}{\cos\theta_l\sin\Theta_0\Omega_0 - \sin\theta_l\cos\Theta_0}\right] \tag{11}$$

where

310

$$\Omega_0 = \cos\phi_l \cos\Phi_0 + \sin\phi_l \sin\Phi_0 \qquad (12)$$

$$\Psi_0 = -\sin\phi_l \cos\Phi_0 + \cos\phi_l \sin\Phi_0 \; . \qquad (13)$$

Define θ_T to be the angle between the lepton and the W^- in the lab frame. The variable θ_T depends only on the lepton momentum:

$$\cos\theta_T = \frac{\chi}{2\beta_W |\vec{P}_l|}\left[\frac{s - 2\sqrt{s}E_l + M_{W_0}^2 + m_l^2}{E_b} - 2(\sqrt{s} - E_l)\right] \qquad (14)$$

where β_W is the velocity of the W^- in the lab frame, M_{W_0} is the nominal mass of the $W^\pm$, E_l is the lepton energy, m_l is the lepton mass, and s is the square of the e^+e^- center of mass energy. Our best estimate of Φ, the azimuthal angle of the W^- in the lab frame, is then given by:

$$\Phi = \tan^{-1}\left(\frac{\sin\theta_T(\cos\theta_l \sin\phi_l \cos\alpha + \cos\phi_l \sin\alpha) + \sin\theta_l \sin\phi_l \cos\theta_T}{\sin\theta_T(\cos\theta_l \cos\phi_l \cos\alpha - \sin\phi_l \sin\alpha) + \sin\theta_l \cos\phi_l \cos\theta_T}\right)$$
$$(15)$$

With these definitions in hand we can write expressions for the measured variables:

$$\cos\Theta = -\sin\theta_l \sin\theta_T \cos\alpha + \cos\theta_l \cos\theta_T \qquad (16)$$

$$\cos\theta^* = \frac{\cos\theta_T + \chi\beta_W}{1 + \chi\beta_W \cos\theta_T} \qquad (17)$$

$$\phi^* = \tan^{-1}\left(\frac{\sin\theta_l \Psi}{\cos\Theta \sin\theta_l \Omega - \sin\Theta \cos\theta_l}\right) \qquad (18)$$

$$\cos\overline{\theta}^* = \frac{-\chi\beta_W + \beta_1\omega}{[(1 - \chi\beta_W\beta_1\omega)^2 - (1 - \beta_1^2)(1 - \beta_W^2)]^{\frac{1}{2}}} \qquad (19)$$

$$\overline{\phi}^* = \tan^{-1}\left[\frac{\sin\theta_1 \overline{\Psi}}{\cos\Theta \sin\theta_1 \overline{\Omega} - \sin\Theta \cos\theta_1}\right] \qquad (20)$$

where

$$\omega = \sin\Theta \sin\theta_1 \overline{\Omega} + \cos\Theta \cos\theta_1 \qquad (21)$$

$$\Omega = \cos\Phi \cos\phi_l + \sin\Phi \sin\phi_l \qquad (22)$$

$$\Psi = -\sin\Phi \cos\phi_l + \cos\Phi \sin\phi_l \qquad (23)$$

$$\overline{\Omega} = \cos\Phi \cos\phi_1 + \sin\Phi \sin\phi_1 \qquad (24)$$

$$\overline{\Psi} = -\sin\Phi \cos\phi_1 + \cos\Phi \sin\phi_1 \qquad (25)$$

For the above expressions to be valid the W^+ and W^- must be produced with $M_{WW} = \sqrt{s}$ and $M_W = M_{W_0}$ where M_{WW} denotes the invariant mass of the W^+W^- pair and M_W is the mass of either the W^+ or W^-. The dotted histograms in Figs.1a and 1b show the distributions for M_{WW} and M_W, respectively, before any cuts are applied. The spread in M_{WW} is due to both beamstrahlung and bremsstrahlung.

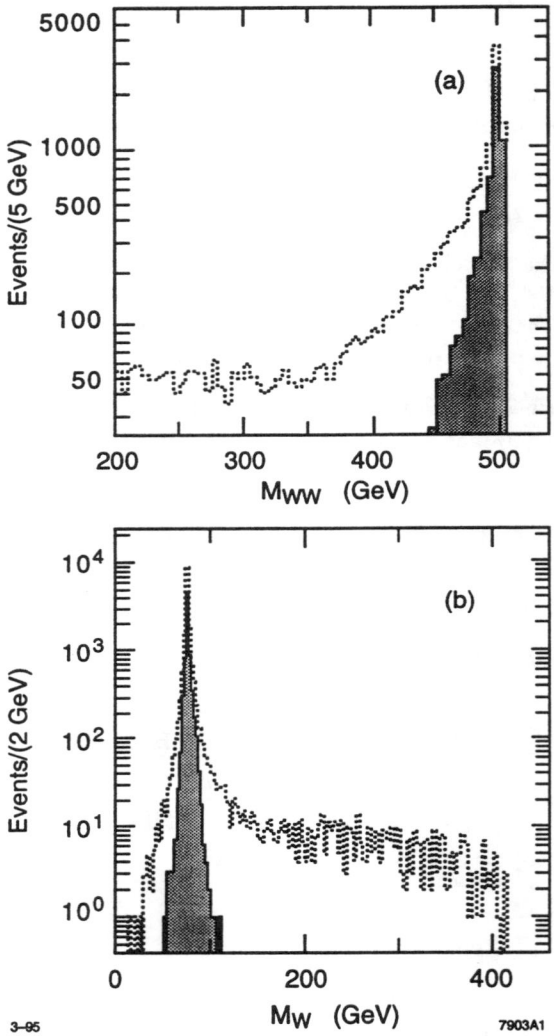

FIG. 1. Distributions for the W^+W^- mass (a) and the $W^\pm$ mass (b) for $\sqrt{s} = 500$ GeV. The dotted histograms are before any cuts, and the shaded histograms are after the cut given by Eq. (31) has been applied. Although no analysis cuts were applied to the Monte Carlo sample in creating the dotted histograms, the Monte Carlo sample itself was generated under the restriction $|\cos\Theta_{true}| < 0.9$.

Experimental Cuts

It is desirable for many reasons to form a W^+W^- event sample with $M_{WW} \approx \sqrt{s}$ and $M_W \approx M_{W_0}$. We have just mentioned the issue of variable reconstruction. Another reason is that if events with $M_{WW} < 475$ GeV are not removed then we would be forced to rely on the measured multi-differential beamstrahlung spectrum with its concomitant systematic error. Also, if events with M_W far from the pole mass M_{W_0} are removed then we need only consider the three amplitudes associated with on-shell W^+W^- production when performing our maximum likelihood fits.

We impose two cuts to obtain a background free sample of on-shell W^+W^- events with $M_{WW} = \sqrt{s}$. First, we require that $|\cos\Theta| < 0.9$ in order to ensure that the event is well within the detector (assumed hermetic for $|\cos\theta| < 0.98$). For the second cut we form a χ^2 variable which tests the consistency of the event with the production of a pair of on-shell W^+W^- bosons with $M_{WW} = \sqrt{s}$. In order to construct this χ^2 variable we reconstruct the mass of the leptonically decaying W (M_{W_1}) and the mass of the hadronically decaying W (M_{W_2}).

The variable M_{W_2} is reconstructed by imposing 4 energy-momentum constraints and solving for the 4 unknowns $\{\vec{P}_\nu, M_{W_2}\}$, where $\vec{P}_\nu$ is the momentum of the neutrino from the leptonically decaying W. The energy-momentum constraints are given by:

$$\begin{aligned}
a_x M_{W_2} + P_{l_x} + P_{\nu_x} &= 0 \\
a_y M_{W_2} + P_{l_y} + P_{\nu_y} &= 0 \\
a_z M_{W_2} + P_{l_z} + P_{\nu_z} &= 0 \\
b M_{W_2} + E_l + \sqrt{\vec{P}_\nu^2} &= \sqrt{s}
\end{aligned} \qquad (26)$$

where

$$\vec{a} = \frac{\vec{\beta}_H}{\sqrt{1-\beta_H^2}}, \quad b = \frac{1}{\sqrt{1-\beta_H^2}} \qquad (27)$$

and $\vec{\beta}_H$ is the velocity of the hadronically decaying W in the lab system. The variable M_{W_2} is given explicitly by

$$M_{W_2} = \frac{1}{\sqrt{1-\beta_H^2}}\left[G + \xi\sqrt{G^2 - (s - 2\sqrt{s}E_l + m_l^2)(1-\beta_H^2)}\right] \qquad (28)$$

where $G = \sqrt{s} - E_l + \vec{P}_l \cdot \vec{\beta}_H$ and $\xi = \pm 1$ reflects the two-fold ambiguity in the solution. The solution $\xi = -1$ always gives the correct solution for center-of-mass energies less than 3 TeV.

We define

$$\chi^2 \equiv \frac{(M_{W_1} - M_{W_0})^2}{\Gamma_W^2} + \frac{(M_{W_2} - M_{W_0})^2}{\Gamma_W^2} \tag{29}$$

where

$$M_{W_1} = \left[(E_l + E_\nu)^2 - (\vec{P}_l + \vec{P}_\nu)^2\right]^{1/2} . \tag{30}$$

Our second cut is then

$$\chi^2 < 30 . \tag{31}$$

We shall refer to this cut as the χ^2 cut.

The shaded histograms in Fig.1a and 1b show the distributions for M_{WW} and M_W respectively after the χ^2 cut has been applied. Before the χ^2 cut the mean values and standard deviations of M_{WW} and M_W are $\langle M_{WW}\rangle = 451$ GeV, $\sigma_{M_{WW}} = 74$ GeV, $\langle M_W\rangle = 87$ GeV, and $\sigma_{M_W} = 36$ GeV. After the cut the mean values and standard deviations are $\langle M_{WW}\rangle = 493$ GeV, $\sigma_{M_{WW}} = 13$ GeV, $\langle M_W\rangle = 80$ GeV, and $\sigma_{M_W} = 2.9$ GeV. The efficiency of the χ^2 cut is 45%.

The χ^2 cut also rejects background from W^+W^- events in which one W boson decays hadronically and the other decays to a tau lepton which subsequently decays to an electron or muon. In addition, this cut is very effective in rejecting events that are not $e^+e^- \to W^+W^-$.

MAXIMUM LIKELIHOOD FORMALISM

We use the maximum likelihood method to fit for trilinear vector boson coupling parameters such as $\Delta\kappa_\gamma$ and λ_γ. Our maximum likelihood function L is given by

$$L = \prod_1^N f(\vec{x}_i, \vec{a}) \tag{32}$$

where N is the number of events, f is the probability density function, $\vec{x}_i$ are the measured variables $\vec{x}$ for event i, and $\vec{a}$ are the fit parameters. The probability density function is normalized to unity:

$$\int d\vec{x}\, f(\vec{x}, \vec{a}) = 1 \tag{33}$$

The measured variables $\vec{x}$ are $\vec{x} = (\cos\Theta, \cos\theta^*, \phi^*, \cos\overline{\theta}^*, \overline{\phi}^*)$. The fit parameters are $\vec{a} = (\Delta\kappa_\gamma, \lambda_\gamma)$, for example. The covariance matrix V for the $\vec{a}$ that maximizes L is given by

$$(V^{-1})_{ij} = N \int d\vec{x}\, \frac{1}{f}\left(\frac{\partial f}{\partial a_i}\right)\left(\frac{\partial f}{\partial a_j}\right) . \tag{34}$$

If we have an unnormalized probability density function $\mu(\vec{x}, \vec{a})$ then we write

$$f = \frac{\mu}{\Xi}, \quad \Xi(\vec{a}) = \int d\vec{x}\, \mu(\vec{x}, \vec{a}) \quad . \tag{35}$$

In order to account for detector resolution, experimental cuts, initial state radiation, initial state electron polarization, and distinct final state event topologies we use the following expression for the unnormalized probability density function μ:

$$\mu_{kl}(\vec{x}_l, \vec{a}) = \int d\vec{x}'_l\, d\vec{q}_l\, d\vec{z}\, r_l(\vec{x}_l, \vec{x}'_l, \vec{q}_l, \vec{z}\,) \eta_l(\vec{x}'_l, \vec{q}_l, \vec{z}\,) t_{kl}(\vec{x}'_l, \vec{q}_l, \vec{z}\,, \vec{a}) h_l(\vec{z}\,) \quad , \tag{36}$$

where k is the initial state electron polarization index and l is the final state event topology index. Here $\vec{x}'$ denotes the true values of the measured variables, and $\vec{q} = (q^2, \bar{q}^2)$, where q^2 and $\bar{q}^2$ are the invariant masses squared of the leptonically decaying W and hadronically decaying W respectively. We define $\vec{z} = (z_1, z_2)$ where $z_1 = E_{e^-}/E_b$, $z_2 = E_{e^+}/E_b$, E_b is the beam energy and $E_{e^\pm}$ are the electron and positron energies following initial state bremsstrahlung and beamstrahlung. The resolution function is $r(\vec{x}, \vec{x}', \vec{q}, \vec{z}\,)$, $\eta(\vec{x}', \vec{q}, \vec{z}\,)$ is the detection efficiency function, $t(\vec{x}', \vec{q}, \vec{z}\,, \vec{a})$ is the multi-differential cross section, and $h(\vec{z}\,)$ is the multi-differential luminosity.

With initial state polarizations and distinct final state event topologies, the expression for $\left(V^{-1}\right)_{ij}$ is somewhat complicated unless we make some additional defintions. Define

$$\Xi_{kl}(\vec{a}) = \int d\vec{x}_l\, \mu_{kl}(\vec{x}_l, \vec{a}) \tag{37}$$

and

$$\lambda_k = \frac{\mathcal{L}_k}{\mathcal{L}} \tag{38}$$

where $\mathcal{L}_k$ is the luminosity at polarization k and $\mathcal{L}$ is the total luminosity. Note that the systematic error for λ_k will be much smaller than the systematic error for $\mathcal{L}_k$. Next define

$$\Lambda(\vec{a}) = \sum_{k,l} \lambda_k \Xi_{kl} \tag{39}$$

$$\alpha_{kl}(\vec{a}) = \frac{\lambda_k \Xi_{kl}}{\Lambda} \tag{40}$$

$$\zeta_{ikl}(\vec{a}) = \frac{1}{\Xi_{kl}} \frac{\partial \Xi_{kl}}{\partial a_i} - \frac{1}{\Lambda} \frac{\partial \Lambda}{\partial a_i} \tag{41}$$

$$\psi_{ikl}(\vec{x}_l, \vec{a}) = \frac{1}{\mu_{kl}} \frac{\partial \mu_{kl}}{\partial a_i} - \frac{1}{\Xi_{kl}} \frac{\partial \Xi_{kl}}{\partial a_i} \tag{42}$$

$$\omega_{ijl}(\vec{x}_l, \vec{a}) = \frac{1}{\Lambda} \sum_k \lambda_k \mu_{kl} \psi_{ikl} \psi_{jkl} \quad . \tag{43}$$

The inverse of the covariance matrix is then given by

$$(V^{-1})_{ij} = N \sum_l \left[\Phi_{ijl} + \Omega_{ijl} \right] \tag{44}$$

where

$$\Phi_{ijl} = \sum_k \alpha_{kl} \zeta_{ikl} \zeta_{jkl} \tag{45}$$

$$\Omega_{ijl} = \int d\vec{x}_l \, \omega_{ijl} \quad . \tag{46}$$

RESULTS

We will now plot 95% confidence level contours for some fit parameters based on the covariance matrix calculation described above. In making these plots our expression for $\mu_{kl}(\vec{x}_l, \vec{a})$ has been simplified. The errors on our reconstructed quantities $\cos\Theta$, $\cos\theta^*$, ϕ^*, $\cos\overline{\theta}^*$, $\overline{\phi}^*$ are small enough that the resolution function r_l can be approximated by a delta function. Also, the imposition of our χ^2 cut, Eq. (31), allows us to approximate the efficiency function η_l by a delta function at $q^2 = M_{W_0}^2$, $\overline{q}^2 = M_{W_0}^2$, $z_1 = 1$, and $z_2 = 1$.

As a result of these considerations we can write

$$\mu_{kl}(\vec{x}_l, \vec{a}) = R_T \eta_\chi t_{kl}(\vec{x}_l, \vec{a}) \tag{47}$$

where $t_{kl}(\vec{x}_l, \vec{a})$ is now the narrow width multi-differential cross section Eq. (1) with initial state electron polarization $P_e(k)$ at the nominal $\sqrt{s}$. The factor $R_T = 1.4$ is the ratio of the total W^+W^- cross section calculated with beamstrahlung, bremsstrahlung, and finite W width effects included to the total W^+W^- cross section calculated in the narrow width approximation at the nominal $\sqrt{s}$. The factor $\eta_\chi = 0.45$ is the detection efficiency of the χ^2 cut applied to a Monte Carlo event sample generated with beamstrahlung, bremsstrahlung, and finite W width effects included.

Limits on $\Delta\kappa_V$ and λ_V

We start with a two parameter fit of $\Delta\kappa_\gamma$ and λ_γ where the couplings $\Delta\kappa_Z$, λ_Z, and Δg_1^Z are assumed to be related to $\Delta\kappa_\gamma$ and λ_γ according to the following prescription:

$$\Delta g_1^Z = \frac{1}{2\cos^2\theta_W} \Delta\kappa_\gamma \tag{48}$$

$$\Delta\kappa_Z = \frac{1}{2}(1 - \tan^2\theta_W) \Delta\kappa_\gamma \tag{49}$$

$$\lambda_Z = \lambda_\gamma \quad . \tag{50}$$

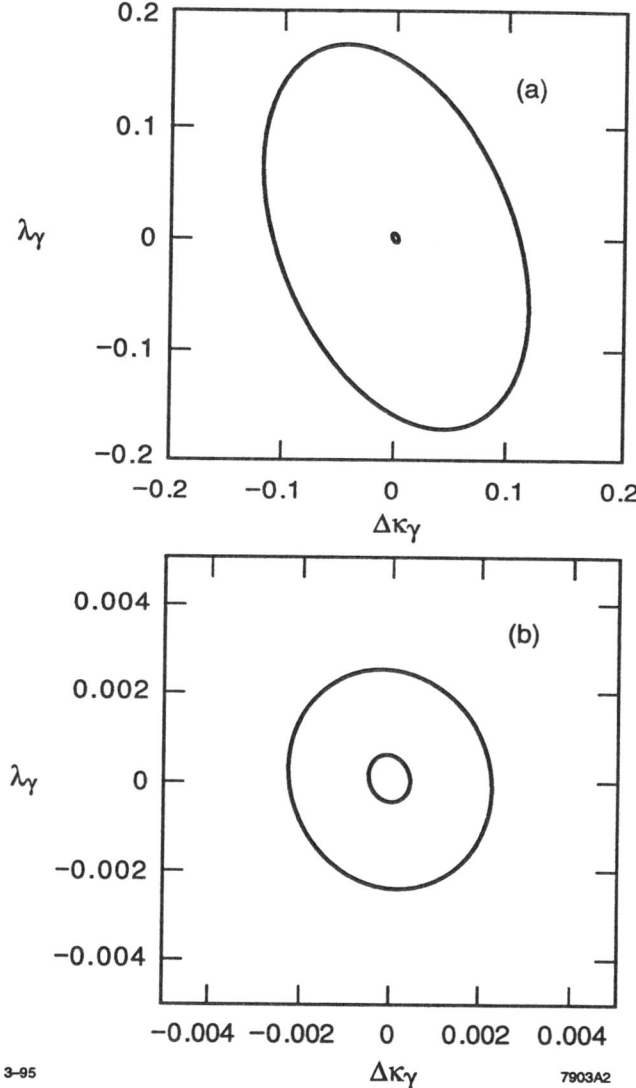

FIG. 2. 95% confidence level contours for $\Delta\kappa_\gamma$ and λ_γ in the HISZ scenario. The outer contour in (a) is for $\sqrt{s} = 190$ GeV and 0.5 fb^{-1} luminosity. The inner contour in (a) and the outer contour in (b) is for $\sqrt{s} = 500$ GeV with 80 fb^{-1} luminosity. The inner contour in (b) is for $\sqrt{s} = 1500$ GeV with 190 fb^{-1} luminosity. In each case the initial state electron polarization is 0%.

This theoretically motivated scheme is called the HISZ scenario (6).

We perform our fit of $\Delta\kappa_\gamma$ and λ_γ at center of mass energies of 190, 500, and 1500 GeV assuming 0% initial state electron polarization. The luminosity at $\sqrt{s} = 190$ GeV is assumed to be 0.5 fb^{-1}. The outer contour in Fig.2a is the 95% confidence level contour for $\sqrt{s} = 190$ GeV. The 95% confidence level contour for $\sqrt{s} = 500$ GeV is given by the inner contour in Fig.2a and the outer contour in Fig.2b. The improvement in precision in going from LEP II to a $\sqrt{s} = 500$ GeV NLC is dramatic. The inner contour in Fig.2b is the 95% confidence level contour for $\sqrt{s} = 1500$ GeV.

Instead of assuming relationships between the $\Delta\kappa_V$ and λ_V parameters in our maximum likelihood fits, we can allow $\Delta\kappa_\gamma$, $\Delta\kappa_Z$, λ_γ, and λ_Z to be indepedent parameters and fit for all four parameters simultaneously. The outer contours in Figs.3a and 3b are two-dimensional projections of the four-dimensional 95% confidence level ellipsoid for a simultaneous fit of $\Delta\kappa_\gamma, \Delta\kappa_Z, \lambda_\gamma$ and λ_Z at $\sqrt{s} = 500$ GeV with 0% initial state electron polarization. We see that there is a significant correlation between $\Delta\kappa_\gamma$ and $\Delta\kappa_Z$ and between λ_γ and λ_Z in the outer contours. In addition, the area encompassed by the outer contour in Fig.3b is significantly larger than the area encompassed by the outer contour in Fig.2b.

If however 90% initial state electron polarization is assumed, then we obtain the inner contours shown in Figs.3a and 3b. Initial state electron polarization has removed the correlation between the fit parameters. Also, comparing the outer contour of Fig.2b with the inner contour of Fig.3b we see that initial state polarization has restored the trilinear coupling resolution of our four parameter fit to that of our two parameter fit. In effect, our knowledge of the well-measured $e^+e^-\gamma$ and e^+e^-Z couplings is being used to disentangle the $W^+W^-\gamma$ coupling parameters from the W^+W^-Z coupling parameters.

Strong $W_L W_L$ Scattering Effects

A final state helicity analysis of $e^+e^- \rightarrow W^+W^-$ can provide information about the $J = 1$ partial wave in the scattering process $W_L W_L \rightarrow W_L W_L$ where W_L denotes a longitudinally polarized W boson (7). In particular, such an analysis is quite sensitive to a vector resonance. Strong $W_L W_L$ scattering effects are incorporated by multiplying the standard model amplitude for $e^+e^- \rightarrow W_L W_L$ by the complex form factor F_T where

$$F_T = \exp[\frac{1}{\pi}\int_0^\infty ds'\delta(s', M_\rho, \Gamma_\rho)\{\frac{1}{s'-s-i\epsilon} - \frac{1}{s'}\}], \qquad (51)$$

$$\delta(s) = \frac{1}{96\pi}\frac{s}{v^2} + \frac{3\pi}{8}\left[tanh(\frac{s-M_\rho^2}{M_\rho\Gamma_\rho})+1\right], \qquad (52)$$

$v = 240$ GeV, M_ρ is the techni-rho mass and Γ_ρ is the techni-rho width. The variable F_T can be thought of as the techni-pion form factor in analogy with

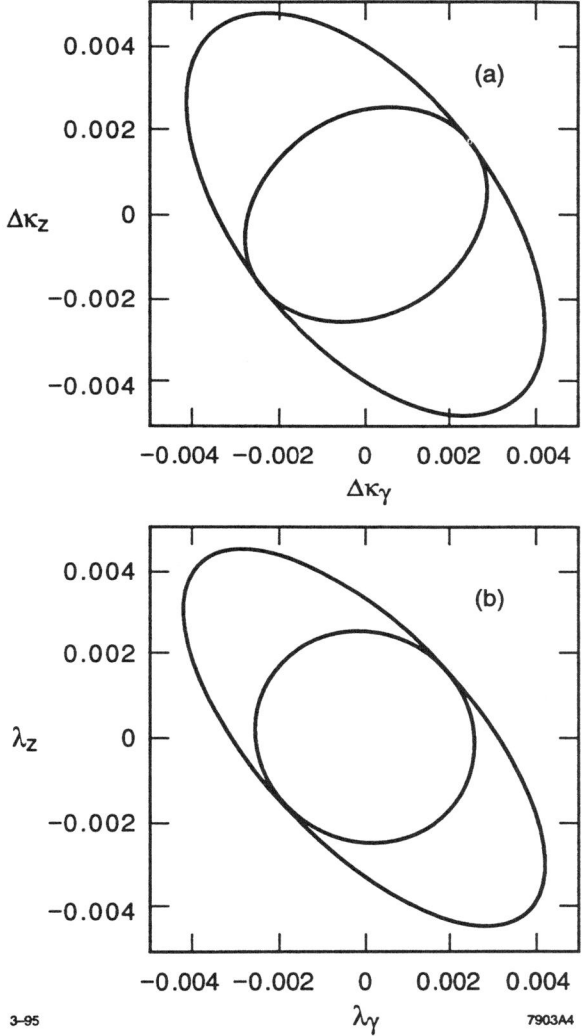

FIG. 3. Two-dimensional projections of the four-dimensional 95% confidence level ellipsoid for a simultaneous fit of $\Delta\kappa_\gamma, \Delta\kappa_Z, \lambda_\gamma$ and λ_Z at $\sqrt{s} = 500$ GeV with 80 fb^{-1}. The outer contours are for 0% initial state electron polarization and the inner contours are for 90% initial state electron polarization.

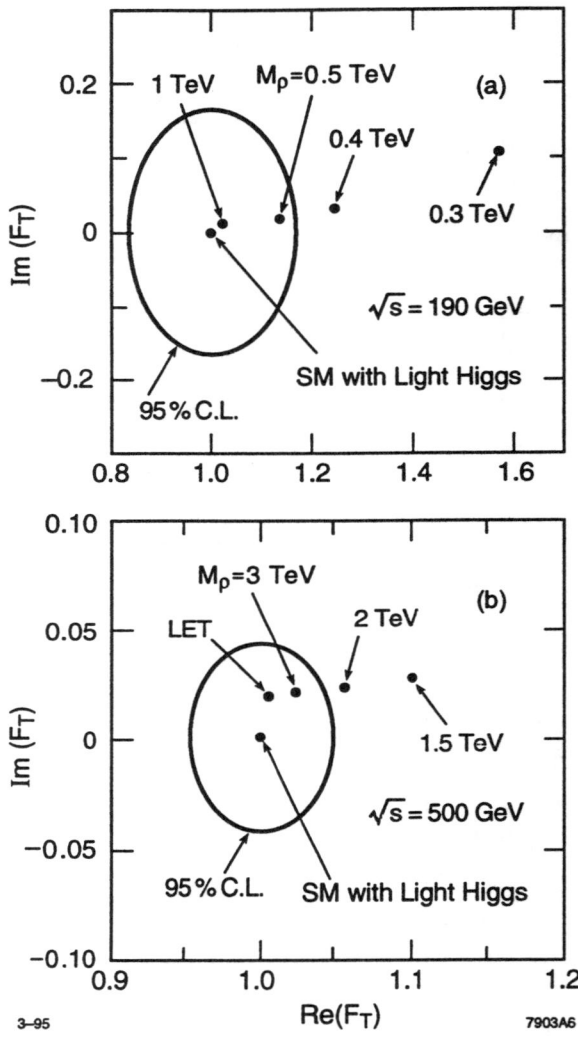

FIG. 4. 95% confidence level contours for the real and imaginary parts of F_T at (a) $\sqrt{s} = 190$ GeV with 0.5 fb^{-1} and at (b) $\sqrt{s} = 500$ GeV with 80 fb^{-1}. The initial state electron polarization is 0%. The values of F_T for various techni-rho masses are indicated in the figures.

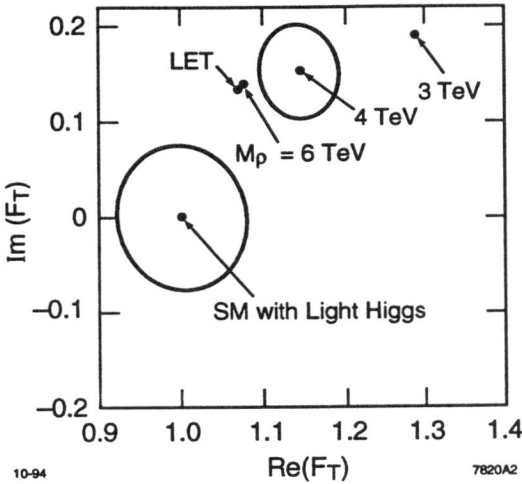

FIG. 5. Confidence level contours for the real and imaginary parts of F_T at $\sqrt{s} = 1500$ GeV with 190 fb^{-1}. The initial state electron polarization is 0%. The contour about the light Higgs value of $F_T = (1,0)$ is 95% confidence level and the contour about the $M_\rho = 4$ TeV point is 68% confidence level.

the rho-dominated pion form factor for $e^+e^- \rightarrow \pi^+\pi^-$. Note that for an infinite techni-rho mass $\delta(s)$ becomes

$$\delta(s) = \frac{1}{96\pi}\frac{s}{v^2}, \qquad (53)$$

reflecting the low energy theorem (LET) amplitude for longitudinal gauge boson scattering.

The maximum likelihood formalism that we used to fit for $\Delta\kappa_\gamma$ and λ_γ can just as well be used to fit for the real and imaginary parts of F_T. Fig.4a shows the 95% confidence level contour for the real and imaginary parts of F_T at $\sqrt{s} = 190$ GeV and 0.5 fb^{-1}, and Fig.4b shows the contour at $\sqrt{s} = 500$ GeV and 80 fb^{-1}. Initial state electron polarization does not play a role in these fits since the form factor F_T is common to both the $W^+W^-\gamma$ and W^+W^-Z amplitudes. We see that LEP II can exclude techni-rho masses up to about 480 GeV. The NLC at $\sqrt{s} = 500$ GeV can exclude techni-rho masses up to about 2.5 TeV and can discover techni-rho resonances with masses up to about 1.5 TeV.

Figure 5 contains confidence level contours for the real and imaginary parts of F_T at $\sqrt{s} = 1500$ GeV with 190 fb^{-1}. Shown are the 95% confidence level contour about the light Higgs value of F_T, as well as the 68% confidence level (i.e., 1σ probability) contour about the value of F_T for a 4 TeV techni-rho. We see that even the non-resonant LET point is well outside the light Higgs 95% confidence level region. In fact, the LET point intersects the 99.99945%

confidence level contour about the light Higgs point, corresponding to a 4.5σ signal. The 6 TeV and 4 TeV techni-ρ points correspond to 4.8σ and 6.5σ signals, respectively.

CONCLUSION

In conclusion, the process $e^+e^- \to W^+W^-$ can be used to measure with high accuracy the $W^+W^-\gamma$ and W^+W^-Z trilinear vector boson couplings. The NLC can set 95% confidence level limits on $\Delta\kappa_\gamma, \Delta\kappa_Z, \lambda_\gamma, \lambda_Z$ of about 3×10^{-3} at $\sqrt{s} = 500$ GeV and of about 5×10^{-4} at $\sqrt{s} = 1500$ GeV. With 90% initial state electron polarization these limits remain valid when all four parameters are fitted simultaneously.

In addition, the process $e^+e^- \to W^+W^-$ is an effective probe of strong electroweak symmetry breaking. Techni-rho resonances with masses up to 1.5 TeV can be discovered with one year of design luminosity at $\sqrt{s} = 500$ GeV, while techni-rho resonances of arbitrarily large mass can be disovered at $\sqrt{s} = 1500$ GeV with one year of design luminosity.

REFERENCES

1. F. Abe et al. (CDF Collaboration), FERMILAB-Pub-94/236-E (preprint, August 1994); F. Abe et al. (CDF Collaboration), FERMILAB-Pub-95/036-E (preprint, March 1995).
2. J. Ellison (D0 Collaboration), FERMILAB-Conf-94/329-E (preprint, November 1994); S. Abachi et al. (D0 Collaboration), FERMILAB-Pub-95/044-E (preprint, March 1995).
3. H. Aihara et al., FERMILAB-Pub-95/031 (preprint, March 1995) and references therein.
4. R.D.Ruth, SLAC-PUB-6751 (preprint, March 1995), to appear in "DPF Working Groups' Reports" (1995).
5. K. Hagiwara, R.D. Peccei, D. Zeppenfeld, and K. Hikasa, *Nucl. Phys.* **B282** 253 (1987).
6. K. Hagiwara, S. Ishihara, R. Szalapski, and D. Zeppenfeld, *Phys. Lett.* **B283** 353 (1992); *Phys. Rev.* **D48**, 2182(1993).
7. M.E.Peskin, in *Physics in Collision 4*, ed. A.Seiden (Editions Frontieres, Gif-sur-Yvette, 1984).
8. J. Bagger, S. Dawson, and G. Valencia, *Nucl. Phys.* **B399** 364 (1993).

Anomalous Gluon Self-Interactions and $t\bar{t}$ Production

Elizabeth H. Simmons* and Peter Cho**

*Department of Physics, Boston University
590 Commonwealth Avenue, Boston MA 02215
**Lauritsen Laboratory, California Institute of Technology
Pasadena, CA 91125

> Strong-interaction physics that lies beyond the standard model may conveniently be described by an effective Lagrangian. The only genuinely gluonic CP-conserving term at dimension six is the three-gluon-field-strength operator G^3. This operator, which alters the 3-gluon and 4-gluon vertices form their standard model forms, turns out to be difficult to detect in final states containing light jets. Its effects on top quark pair production hold the greatest promise of visibility.

INTRODUCTION

The hallmark of a non-abelian gauge theory is the self-interaction of its gauge fields. Most of this conference has been devoted to the electroweak vector bosons; this talk[1] will focus, instead, on the self-interactions of the gluons. Any experimental indication that the gluon self-coupling differed from the form predicted by the $SU(3)$ gauge theory of the strong interactions would point to the existence of new color-related physics. This talk presents a model-independent effective Lagrangian analysis of possible non-standard contributions to color physics, and assesses the possibility of measuring the coefficients of the effective Lagrangian.

Suppose that some exotic color physics exists at an energy scale Λ. For instance, there might be new colored scalars or fermions with a mass of order Λ, such as squarks and gluinos (1), colored technihadrons (2), or fermions in non-fundamental representations of $SU(3)$. Or instead, perhaps the gluons and quarks are manifestly composite (3) when probed at a distance scale Λ^{-1}. Such non-standard physics would lead to new gluon self-interactions through virtual loops of heavy colored particles or through exchange of subcomponents.

A *complete* description of a given set of new phenomena would require a fundamental theory beyond the standard model. But at low energies $E \ll \Lambda$, where the underlying preon exchange or loops of new particles cannot be resolved, the new color physics causes multi-gluon contact interactions

---

[1]Presented by E.H. Simmons.

suppressed by inverse powers of Λ. These contact interactions are described by an effective Lagrangian

$$\mathcal{L}_{eff} = \mathcal{L}_{QCD} + \frac{1}{\Lambda^2}\sum_i C_i^{(6)}(\mu)O_i^{(6)}(\mu) + \frac{1}{\Lambda^4}\sum_i C_i^{(8)}(\mu)O_i^{(8)}(\mu) + O(\frac{1}{\Lambda^6})$$
(1)

that includes the conventional QCD Lagrangian plus non-renormalizable operators O_i that are constructed from gluon field strengths $G^{\mu\nu}$ or color-covariant derivatives $D^\mu = \partial^\mu - ig_s G^\mu$. The new operators obey the gauge and global symmetries of the standard model.

Our task is to identify the leading operators in this effective Lagrangian and determine which experiments are best able to detect their effects.

LEADING OPERATORS IN L_{EFF}

Since a non-renormalizable operator of dimension $(4+d)$ is suppressed by Λ^{-d}, the operators making the most visible contribution to physical processes will be those of lowest dimension.

The number of nonrenormalizable terms which arise at dimension 6 in the gluon sector is small. One can build only two gauge-invariant operators preserving C, P and T out of covariant derivatives and gluon field strengths (5):

$$O_1^{(6)} = g_s f_{abc} G_{a\nu}^\mu G_{b\lambda}^\nu G_{c\mu}^\lambda$$
(2)

$$O_2^{(6)} = \frac{1}{2} D^\mu G_{\mu\nu}^a D_\lambda G_a^{\lambda\nu}.$$
(3)

The triple gluon field strength term in Eq. 2, which we shall name G^3 for short, represents a true gluonic operator, contributing to three-gluon and four-gluon non-abelian vertices. The double gluon field strength operator in Eq. 3, which we will call $(DG)^2$, is not really gluonic in this sense. The classical equation of motion

$$D_\mu G_a^{\mu\nu} = -g_s \sum_{\text{flavors}} \bar{q}\gamma^\nu T_a q$$
(4)

relates its S-matrix elements to those of a color octet four-quark operator (17):

$$O_2^{(6)} \xrightarrow{EOM} \frac{g_s^2}{2} \sum_{\text{flavors}} (\bar{q}\gamma_\mu T_a q)(\bar{q}\gamma^\mu T_a q).$$
(5)

The two-field-strength operator thus affects parton processes involving external quarks rather than external gluons.

The list of CP-even gluon operators grows significantly at dimension eight. Classifying the operators according to the number of field strengths that they contain, we find one independent operator built from two field strengths and four covariant derivatives, two operators with three gluon field strengths and two derivatives, and a half-dozen operators containing four field strengths (6,9). Rather than listing all nine operators explicitly (for a list, see (6)) we merely mention that there are two situations in which dimension-8 operators may give noticeable effects. One of the two-field-strength operators,

$$O_3^{(8)} = g_s f_{abc} G^\mu_{a\nu} G^\nu_{b\lambda} D^2 G^\lambda_{c\mu}. \tag{6}$$

contributes at tree-level and order $1/\Lambda^4$ to the process $gg \to q\bar{q}$; it is the only $d=8$ gluonic operator to do so. The effect of this operator on angular distributions will feature in our discussion of $gg \to t\bar{t}$. In addition, the four-field-strength operators contribute significantly to the gluon four-point vertex and, hence, the process $gg \to gg$ which we will analyze shortly.

DIJET PRODUCTION

Having established our operator basis, we consider how best to detect the presence of non-standard gluon self-interactions. In principle, the operator G^3, being the lowest-dimension operator to affect multi-point gluon vertices, is the one to focus on. A logical beginning is to consider its effects on hadronic scattering at high-energy colliders like FNAL and the LHC. Because scattering at both colliders is dominated by gluon-gluon collisions, one expects non-standard gluon self-interactions to noticeably affect two-body scattering cross-sections. These cross-sections, in turn, dominate the well-measured (at FNAL) inclusive single-jet production cross-section. Hence, it appears that a strong limit on the coefficient $C_1^{(6)}/\Lambda^2$ should be forthcoming.

The leading contribution of the G^3 operator to the inclusive jet cross-section $p(\bar{p}) \to \text{jet} + X$ is expected to lie in its effect on the sub-process $gg \to gg$. Consider the Feynman diagrams contributing to the $gg \to gg$ scattering amplitude in pure QCD and those with one insertion of G^3 (i.e., one anomalous multi-gluon vertex per diagram). The ordinary QCD contribution to the scattering comes from squaring the sum of the QCD diagrams; the lowest-order ($\frac{1}{\Lambda^2}$) piece due to G^3 arises from the interference of the QCD and one-insertion diagrams. It has been shown (5), however, that the QCD amplitude is only in the [+ + ++] helicity channel and the one-insertion amplitude is purely in the orthogonal [+ + −−] and [+ − −−] channels. There is, consequently, no order $\frac{1}{\Lambda^2}$ contribution to $gg \to gg$. The leading effect of the G^3 operator arises at order $1/\Lambda^4$, from squaring the one-insertion diagrams and from interfering two-insertion diagrams with the QCD diagrams.

Where a lower-order effect is missing, one would hope to experimentally detect the remaining higher-order effect. This will be difficult. The leading contributions of the dimension-eight four-field-strength operators to $gg \to gg$ also arise at order $1/\Lambda^4$ when an amplitude with one insertion of a dimension-eight operator interferes with a QCD amplitude. Furthermore, the order $1/\Lambda^4$ contributions to gluon scattering of the operators G^3 and

$f_{eab}f_{ecd}G_a^{\mu\nu}G_b^{\lambda\rho}G_{c\mu\nu}G_{d\lambda\rho}$ are identical in form and similar in magnitude (6). Isolating the effects of G^3 in $gg \to gg$ does not appear possible.

The next most promising sub-processes appear to be those involving massless quarks as well as gluons, i.e. $gq \to gq$ and reactions related to it by crossing. Initial state gluons are still a possibility and G^3 can enter some diagrams through the three-gluon vertex. When the contribution of G^3 is calculated, however, the order $1/\Lambda^2$ piece vanishes. The leading effects are, again, order $1/\Lambda^4$ and compete with the effects of higher-dimension operators.

In summary: two-body scattering of massless partons does not put significant limits on non-standard gluon self-interactions involving the G^3 operator. Because the leading effects are of order $1/\Lambda^4$, there is competition from higher-dimension operators. Furthermore, the fact that gluons predominate at low x where the QCD background is greatest weakens the attainable bounds (6).

OTHER JET-PRODUCTION EXPERIMENTS

While the G^3 operator cannot be detected in $2 \to 2$ light parton scattering processes, there are several other options for studying non-standard strong interactions using massless final-state jets. As will become clear, none is fully satisfactory.

Although G^3 does not affect dijet production at tree level, the other dimension-six operator, $(DG)^2$, can. As noted earlier, this operator alters the propagator and coupling of internal gluons. At leading order its effects on scattering are equivalent to those of the four-quark operator $(\bar{\psi}\gamma^\mu T^a \psi)^2$. Consider, for example, the process $\mathbf{p\bar{p} \to jet + X}$, to which dijet production is the leading contributor. At FNAL, initial quarks play a larger role than initial gluons in scattering at large Bjorken x; thus high $p_\perp$ jets are more likely to originate from initial quarks. The inclusive jet cross-section is found to fall more slowly with $\hat{s}$ or $p_\perp$ when an insertion of the $(DG)^2$ operator is included than when only QCD is studied (5–7). This makes the transverse-momentum spectrum potentially sensitive to the presence of the $(DG)^2$ operator. Analyzing published CDF inclusive jet data (16) yields a lower bound of 2 TeV on the scale of new physics associated with this operator (7). This limit is useful – but because no external gluons are involved, it is only tangentially related to probing the structure of the gluon self-coupling.

At LEP, one source of 4-jet final states is the process wherein a Z decay to a quark/anti-quark pair is followed by radiation of a gluon that splits into a pair of gluons. The three-gluon vertex involved in this process can be affected by the presence of the G^3 operator. In (11,12) it was found that while most kinematic variables describing $Z \to 4j$ would reflect the presence of G^3 only in the overall *rate*, the dijet invariant mass distribution of the two most energetic jets changes *shape* when the G^3 operator is included. With 10pb^{-1} of data, it should be possible to set a limit of $\Lambda > 100$ GeV using this dijet invariant mass distribution; ten times the data would boost the limit to 175 GeV. The limiting factor is the energy at which LEP experiments are performed.

While the order $1/\Lambda^2$ contributions of the G^3 operator to dijet production vanish at tree level, the same is not true when larger numbers of jets are being produced. The very difference in helicity properties between the scattering amplitudes of pure QCD and those with one insertion of G^3 which keeps the $2 \to 2$ amplitudes from interfering provides a potential signal of the presence of G^3 in $2 \to 3$ processes (10). For example, if one considers $gg \to ggg$ when two of the outgoing gluons are nearly collinear one observes the following. Treating the nearby gluons as one effective gluon yields an approximate $2 \to 2$ process for which we know the helicity properties of scattering amplitudes with and without G^3. The pure QCD amplitude with its $[++++]$ helicity is symmetric under azimuthal rotations about the momentum vector of the effective gluon. The amplitude with an insertion of G^3 admits the $[++--]$ and $[+++-]$ helicities which allows the effective gluon to be linearly polarized, yielding azimuthal dependence. No limits on Λ have yet been suggested using this method; the limiting factor may be the experimental difficulty of studying 3-jet events in the near-collinear region.

THE TOP QUARK PRODUCTION CROSS-SECTION

Light jets having failed us, we turn to the possibility of studying anomalous gluon self-interactions through their effects on scattering involving heavy flavors. The two-body scattering process that both involves heavy fermions and benefits maximally from the high gluon luminosity at hadron colliders is $gg \to q\bar{q}$. We will find that the order $1/\Lambda^2$ contribution of G^3 to this process is proportional to m_q^2; hence the effect is greatest for top production. The top quark is also the easiest to tag since its leptonic decay channel can produce a high energy, isolated lepton in conjunction with a bottom quark. This distinctive signature cuts down on genuine backgrounds as well as false identifications. To the extent that $b\bar{b}$ and even $c\bar{c}$ final states can be cleanly identified, the signal for G^3 will be enhanced and our results can be applied to their study.

To study top quark production, we must enlarge our basis of higher-dimension operators. The gluonic operators described above are not the only higher-dimension operators that can affect top production. They also do not form a closed basis under one-loop renormalization, which we employ in running down from the scale of new physics to the top production threshold. The operators $O_1^{(6)}$ and $O_2^{(6)}$ do not mix with each other under the action of QCD at one-loop order. Instead, $O_1^{(6)}$ runs into itself and the chromomagnetic moment operator

$$O_0^{(6)} = \sum_{\text{flavors}} g_s m_q \bar{q} \sigma^{\mu\nu} T^a q G_{\mu\nu}^a. \tag{7}$$

Because the equations of motion relate $O_2^{(6)}$ to a color-octet four-quark operator, $O_2^{(6)}$ mixes at one loop with other four-quark operators

$$O_3^{(6)} = \frac{g_s^2}{2} \sum_{\text{flavors}} (\bar{q}\gamma_\mu \gamma^5 T_a q)(\bar{q}\gamma^\mu \gamma^5 T_a q) \tag{8}$$

$$O_4^{(6)} = \frac{g_s^2}{2} \sum_{\text{flavors}} (\bar{q}\gamma_\mu q)(\bar{q}\gamma^\mu q) \tag{9}$$

$$O_5^{(6)} = \frac{g_s^2}{2} \sum_{\text{flavors}} (\bar{q}\gamma_\mu \gamma^5 q)(\bar{q}\gamma^\mu \gamma^5 q) \tag{10}$$

The mixing matrices are given explicitly in (8,9).

Renormalization group evolution suppresses the coefficients $C_1^{(6)}$ and $C_2^{(6)}$. Given that one expects some new fundamental layer of physics to lie in the TeV regime, we will take $\Lambda = 2$ TeV. This ensures that our analysis will be valid over almost the entire energy range of present and anticipated hadron colliders. If the operator coefficients assume the values

$$(C_0^{(6)}, C_1^{(6)}, C_2^{(6)}, C_3^{(6)}, C_4^{(6)}, C_5^{(6)})(\Lambda) = (1, 1, 1, 0, 0, 0) \tag{11}$$

at a scale of 2 TeV, then they run down to the values

$$(0.7858, 0.7458, 0.8856, -0.0294, 0.0003, -0.0152) \tag{12}$$

at the top-antitop threshold (8).

The partonic sub-processes contributing to top quark pair-production at a hadronic collider are $gg \to t\bar{t}$ and $q\bar{q} \to t\bar{t}$. Both the G^3 and chromomagnetic moment operators contribute to the gluon fusion channel; the non-standard contribution to the squared matrix element is

$$\overline{\sum}' |\mathcal{A}(gg \to t\bar{t})|^2 = \frac{m_t^2}{\Lambda^2} \left[\frac{C_0^{(6)}(\frac{4}{3}\hat{s}^2 - 3\hat{t}\hat{u} - 3m_t^2\hat{s} + 3m_t^4) + \frac{9}{8}C_1^{(6)}(\hat{t}-\hat{u})^2}{(m_t^2 - \hat{t})(m_t^2 - \hat{u})} \right]$$
$$+ \frac{1}{\Lambda^4} \left[\frac{1}{6} C_0^{(6)2} \frac{m_t^2(14\hat{s}\hat{t}\hat{u} + m_t^2(31\hat{s}^2 - 36\hat{t}\hat{u}) - 50m_t^4\hat{s} + 36m_t^6)}{(m_t^2 - \hat{t})(m_t^2 - \hat{u})} \right. \tag{13}$$
$$\left. + \frac{9}{8} C_0^{(6)} C_1^{(6)} \frac{m_t^2 \hat{s}^3}{(m_t^2 - \hat{t})(m_t^2 - \hat{u})} + \frac{27}{4} C_1^{(6)2} (m_t^2 - \hat{t})(m_t^2 - \hat{u}) \right] + O\left(\frac{1}{\Lambda^6}\right).$$

where the bar over the Σ implies averaging (summing) over initial (final) spins and colors, while the prime indicates division by g_s^4. Notice that all of the nonrenormalizable operator terms except the last one are proportional to m_t^2. Because the last term is enhanced by a prefactor of 27/4 and increases quadratically with $\hat{s}$, well away from the $t\bar{t}$ threshold and over large regions of $C_1^{(6)}$ parameter space, the $O_1^{(6)}$ operator's squared amplitude is much larger than its interference with QCD.

One may question whether the $O(1/\Lambda^4)$ terms arising from dimension-8 gluon operators could be significant. The answer is generally no. The only dimension-8 operator that affects $gg \to t\bar{t}$ scattering at lowest order is $O_3^{(8)}$

$$\overline{\sum}'|\mathcal{A}(gg \to t\bar{t})|^2 = \cdots - \frac{3}{8}\frac{C_3^{(8)}}{\Lambda^4}\frac{m_t^2\hat{s}(\hat{t}-\hat{u})^2}{(m_t^2-\hat{t})(m_t^2-\hat{u})}. \tag{14}$$

This term has a smaller prefactor and increases more slowly with $\hat{s}$ than the term proportional to $(C_1^{(6)})^2$ and is unlikely to obscure any signal from $O_1^{(6)}$.

The $(DG)^2$, chromomagnetic moment, and four-quark operators make the following addition to the quark/anti-quark annihilation matrix element

$$\overline{\sum}' |\mathcal{A}(q\bar{q} \to t\bar{t})|^2 = \tag{15}$$

$$\frac{1}{9\hat{s}\Lambda^2}[4C_0^{(6)}m_t^2\hat{s} + C_2^{(6)}(\hat{t}^2+\hat{u}^2+4m_t^2\hat{s}-2m_t^4) + C_3^{(6)}\hat{s}(\hat{t}-\hat{u})]$$

$$+\frac{4}{9\Lambda^4}\Big[8{C_0^{(6)}}^2 m_t^2(\hat{t}\hat{u}+2m_t^2\hat{s}-m_t^4)/\hat{s} + 8C_0^{(6)}C_3^{(6)}m_t^2(\hat{t}-\hat{u})$$

$$+ 8C_0^{(6)}C_2^{(6)}m_t^2\hat{s} + ({C_2^{(6)}}^2 + \frac{1}{2}{C_4^{(6)}}^2)(\hat{t}^2+\hat{u}^2+4m_t^2\hat{s}-2m_t^4)$$

$$+ ({C_3^{(6)}}^2 + \frac{1}{2}{C_5^{(6)}}^2)(\hat{t}^2+\hat{u}^2-2m_t^4) + (2C_2^{(6)}C_3^{(6)}+C_4^{(6)}C_5^{(6)})\hat{s}(\hat{t}-\hat{u})\Big].$$

with the same conventions as before. Unlike Eq. 13 this expression contains no anomalously large order $1/\Lambda^4$ term. So we expect the effect of dimension-eight and higher operators upon $q\bar{q} \to t\bar{t}$ scattering to be small (8).

The squared amplitudes in Eqs. 13 and 15 enter the partonic cross section

$$\frac{d\sigma(ab\to t\bar{t})}{d\hat{t}} = \frac{\pi\alpha_s^2}{\hat{s}^2}\overline{\sum}'|\mathcal{A}(ab\to t\bar{t})|^2. \tag{16}$$

This is combined with distribution functions $f_{a/A}(x_a)$ and $f_{b/B}(x_b)$ specifying the probability of finding partons a and b inside hadrons A and B carrying momentum fractions x_a and x_b and summed over initial parton configurations. The resulting hadronic cross section

$$\frac{d^3\sigma}{dy_3 dy_4 dp_\perp}(AB \to t\bar{t}) = 2p_\perp \sum_{ab} x_a x_b f_{a/A}(x_a) f_{b/B}(x_b) \frac{d\sigma(ab\to t\bar{t})}{d\hat{t}}. \tag{17}$$

depends on the top and antitop rapidities y_3 and y_4 and their common transverse momentum $p_\perp$.

TOP QUARK PRODUCTION AND GLUON SELF-INTERACTIONS

We now compare the effects of various non-standard gluon interactions on the top quark production cross-section. This discussion will communicate the conclusions that can be drawn from the salient features of the kinematic distributions of the top quarks. Full details reside in (8).

We first examine the transverse momentum distribution obtained by integrating $d^3\sigma/dy_3 dy_4 dp_\perp$ over the rapidity range $-2.5 \le y_3, y_4 \le 2.5$. [2] The

---

[2]This convenient integration interval contains the bulk of the produced top quarks. Extending the range to $-6 \le y_3, y_4 \le 6$ does not alter our results.

resulting $p_\perp$ distribution of $t\bar{t}$ pairs produced at the LHC is plotted in Fig. 1, which shows curves for QCD and for the separate contributions of operators $O_0^{(6)}$, $O_1^{(6)}$ and $O_2^{(6)}$ with their respective $C_i(\Lambda)$ equal to 0.5. [3]

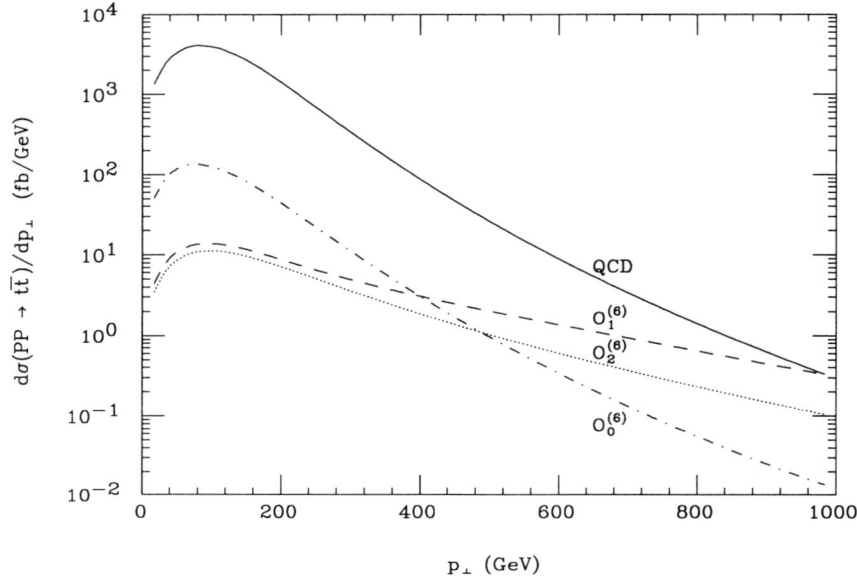

FIG. 1. $d\sigma(pp \to t\bar{t})/dp_\perp$ at the LHC with $\sqrt{s} = 14$ TeV. The solid curve represents pure QCD. The dot-dashed, dashed and dotted curves show the additional contributions when either $C_0^{(6)}(\Lambda)$, $C_1^{(6)}(\Lambda)$ or $C_2^{(6)}(\Lambda)$ is set to 0.5 with $\Lambda = 2$ TeV.

The $p_\perp$ dependence of the curves in Fig. 1 differentiates the dimension-6 operators from each other and from QCD terms in $\mathcal{L}_{eff}$. At high $p_\perp$, where the QCD background is lowest, the dimension-6 operator making the largest contribution to the rate of top quark production is G^3. Next in importance at large $p_\perp$ is $(DG)^2$; the chromomagnetic magnetic operator lags far behind. Placing a lower $p_\perp$ cut around 500 GeV, can eliminate most of the chromomagnetic moment operator's contribution in favor of that from G^3 and $(DG)^2$. In addition, the shapes of the curves for G^3 and $(DG)^2$ are noticeably different from one another and from the shape of the QCD curve; that of the chromomagnetic moment operator closely mimics QCD. Hence the G^3 operator should make the most visible contribution to $d\sigma/dp_\perp$ at the LHC.

A quantitative comparison of the QCD and effective lagrangian predictions for $d\sigma(pp \to t\bar{t})/dp_\perp$ at the LHC confirms this (8). To compare rates, we computed the ratio $R_\perp$ of the integrals of $d\sigma_{EFT}/dp_\perp$ and $d\sigma_{QCD}/dp_\perp$ over the momentum range 500 GeV $< p_\perp <$ 1000 GeV. We found that $R_\perp$ was

---

[3] We used the next-to-leading order parton distribution function set B of Harriman, Martin, Roberts and Stirling (18) evaluated at the $\mu = m_\perp \equiv \sqrt{m_t^2 + p_\perp^2}$.

fairly insensitive to $C_0^{(6)}$, and depended a few times more strongly on $C_1^{(6)}$ than on $C_2^{(6)}$. To compare shapes, we formed a χ^2 function for the difference in number of high-$p_\perp$ events predicted by QCD and $\mathcal{L}_{eff}$. Again, the strongest dependence was on $C_0^{(6)}$. Further, $R_\perp$ and the χ^2 function depend differently on $C_1^{(6)}$ and $C_2^{(6)}$, making the combined measurements even more powerful. We estimate that the LHC could set a limit $|C_1^{(6)}| < 0.5$.

The Tevatron analogues of the LHC differential cross-sections are shown in Fig. 2. The integrated cross-section for $t\bar{t}$ production is two orders of mag-

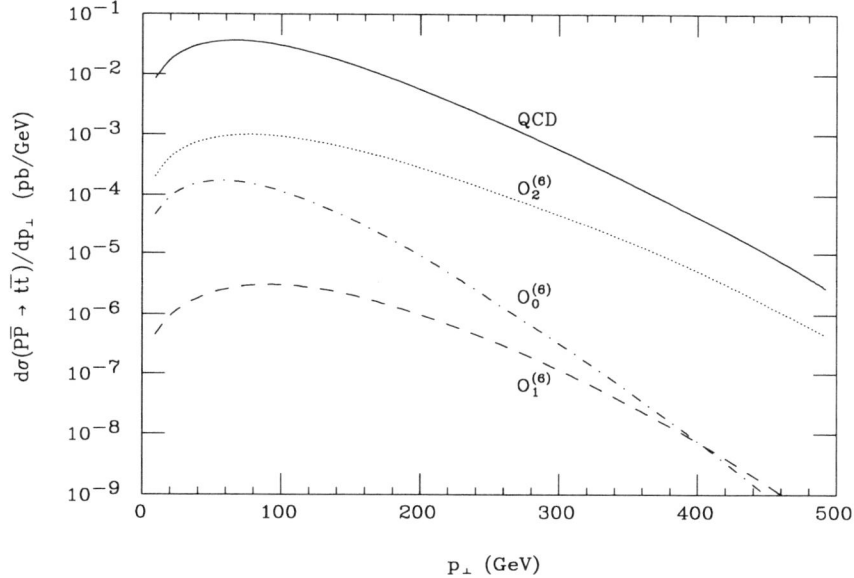

FIG. 2. $d\sigma(p\bar{p} \to t\bar{t})/dp_\perp$ at FNAL with $\sqrt{s} = 1.8$ TeV. The curves are labeled as in Fig. 1.

nitude lower at the Tevatron. And the relative importance of the dimension-6 terms in the effective Lagrangian depends upon $\sqrt{s}$ in a manner consistent with the parton content of the colliding hadrons. The most important dimension-six operators at Tevatron energies are the chromomagnetic moment operator $O_0^{(6)}$ and four-quark operators like $(DG)^2$; the effects of G^3 are (for equal values of the C_i at high energies) an order of magnitude smaller. Hence FNAL experiments are unlikely to find evidence for the G^3 operator in $t\bar{t}$ production. The effects of $O_0^{(6)}$ and $O_2^{(6)}$ may be visible in terms of enhanced production *rate*; again, the fact that the shape of the $O_0^{(6)}$ curve is identical to that of the QCD curve will make $O_0^{(6)}$ more difficult to detect.

We next study the angular distribution of the produced top quarks, $d\sigma(pp \to t\bar{t})/d\cos\theta^*$, where θ^* denotes the angle between the direction of the boost and that of the top quark in the parton center-of-mass frame. To enhance the signal we have imposed the cut $p_\perp \geq 500$ GeV. We also required the lab frame angle between the t or $\bar{t}$ and the beamline to exceed 25.4°; this ensures that the pseudorapidities of the decay products from high momentum tops will predominantly fall within $-2.5 \leq \eta \leq 2.5$, the approximate acceptance of planned LHC detectors.

The angular distribution is plotted in Fig. 3 for pure QCD and for QCD *plus* some of the O_i. The curves indicating the effects of the chromomagnetic

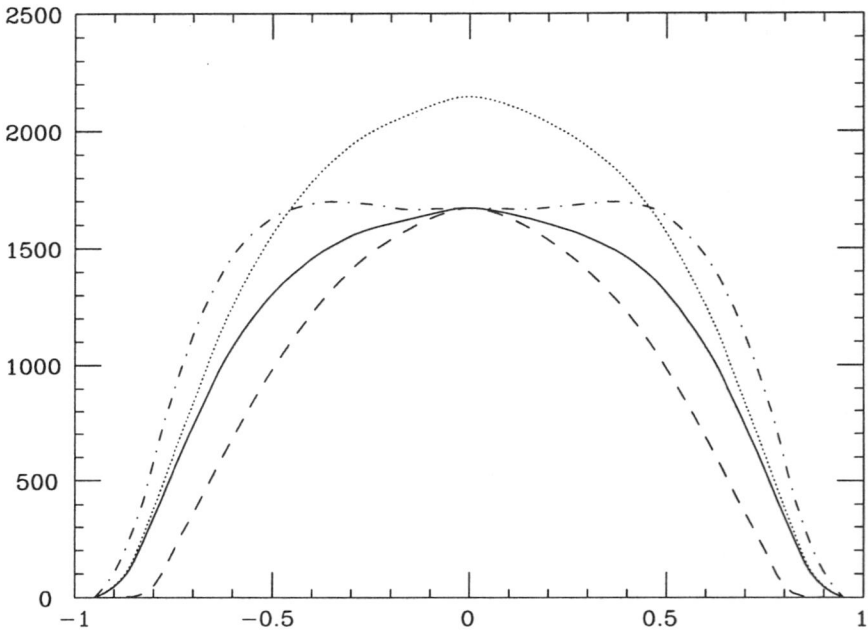

FIG. 3. $d\sigma(pp \to t\bar{t})/d\cos\theta^*$ at the LHC with $\sqrt{s} = 14$ TeV. The solid curve shows pure QCD. The dotted curve shows QCD plus $O_1^{(6)}$ with $C_1^{(6)}(2 \text{ TeV}) = 0.5$. The dashed (dot-dashed) curve shows QCD plus $O_3^{(8)}$ with $C_3^{(8)}(2 \text{ TeV}) = 0.5$ (-0.5).

moment operator are not included in the figure because they closely trace the QCD curve for $C_0^{(6)}(\Lambda) = \pm 0.5$. Likewise, the curves for $C_1^{(6)}(\Lambda) = -0.5$ and $C_2^{(6)}(\Lambda) = 0.5$ are nearly indistinguishable from the curve shown for $C_1^{(6)}(\Lambda) = 0.5$. The dimension-8 gluon operator $O_3^{(8)}$ induces deviations [4] from pure QCD which are clearly visible in $d\sigma/d\cos\theta^*$. This is quite interesting since the effect of $O_3^{(8)}$ upon the $t\bar{t}$ transverse momentum distribution was negligible. Indeed, we omitted the effects of $O_3^{(8)}$ from Fig. 1 and Fig. 2 since they would have been suppressed relative to the dimension-6 operators' effects by more than an order of magnitude.

---

[4] The coefficient $C_3^{(8)}$ was not evolved using the renormalization group but was instead simply fixed at its Λ scale value.

As in the analysis of the $p_\perp$ distributions, we distinguish between effects on the rate and the shape of the curves. In discussing rate, we compare the integral of a given curve (with respect to $\cos\theta^*$) to that of the QCD curve, denoting the ratio by R_{ang}. A curve's shape is compared with that of the QCD curve by forming the ratio (R_{rms}) of the respective root-mean-squared values of $\cos\theta^*$. For the curves arising when $C_0^{(6)}(\Lambda)$, $C_1^{(6)}(\Lambda)$, $C_2^{(6)}(\Lambda)$ or $C_3^{(8)}(\Lambda)$ is set equal to 0.5, we find $R_{\mathrm{ang}} = (1.03, 1.23, 1.46, 0.82)$ and $R_{\mathrm{rms}} = (0.999, 0.978, 0.991, 0.871)$. For analogous curves with the Λ scale coefficients set equal to -0.5, we find $R_{\mathrm{ang}} = (0.969, 1.23, 0.735, 1.18)$ and $R_{\mathrm{rms}} = (1.00, 0.978, 1.03, 1.08)$. The most striking implication is that the dimension-8 gluon operator alters the shape of the $t\bar{t}$ angular distribution more than any dimension-6 operator in $\mathcal{L}_{eff}$ for comparable values of the C_i. The magnetic moment operator's angular distribution is indistinguishable from that of pure QCD, while the distributions of the G^3 and $(DG)^2$ operators differ significantly from that of QCD in R_{ang} but not in R_{rms}.

Figure 4 summarizes the detectability of the operators we have studied. Each operator produces visible effects in a unique combination of experiments.

OPERATOR	EXPERIMENT				
	FNAL p_T	LHC p_T rate	p_T shape	angular rate	angular shape
$O_1^{(6)}$		✓	✓	✓	
$O_2^{(6)}$	✓	✓	✓	✓	
$O_0^{(6)}$	✓	✓			
$O_3^{(8)}$				✓	✓

FIG. 4. Experiments able to detect each type of non-standard gluon interaction.

CONCLUSIONS

Anomalous gluon self-interactions are elusive. Detecting them in dijet production is nearly impossible. Other measurements involving light jets are energy-limited or intrinsically difficult.

Heavy flavor production may offer the best hope of seeing non-standard gluon self-couplings. While only $t\bar{t}$ production is analyzed here, $b\bar{b}$ production should show similar effects. A strong signal will be provided by the shape of the transverse-momentum distribution of the produced heavy quarks; non-standard strong interactions can visibly affect the number of events at high transverse momentum. The angular distribution of the heavy fermions can also help discriminate among the effects of different higher-dimension operators.

Top-quark pair production at the LHC will test the three-gluon vertex well. The contribution of the G^3 operator to the transverse-momentum spectrum exceeds that of all other contact operators for similar values of their coefficients. For a scale of new physics $\Lambda = 2\text{TeV}$, LHC experiments should be able to set an upper bound of 0.5 on the coefficient of the G^3 operator. Using the more usual notation in which the coefficient C_i is set to 4π, the associated lower bound on Λ is of order 10 TeV. This compares well with the current lower bounds of order 1-2 TeV derived from FNAL data for both the 4-quark operator $(\bar{\psi}_L \gamma^\mu \psi_L)^2$ (16) and the $(DG)^2$ operator (7).

REFERENCES

1. P. Fayet, Nucl. Phys. B**90**, 104 (1975) and Phys. Lett. B**69**, 489 (1977); G.R. Farrar and P. Fayet, Phys. Lett. B**76**, 575 (1978); E. Witten, Nucl. Phys. B**188**, 513 (1981); S. Dimopoulos and H. Georgi, Nucl. Phys. B**193**, 150 (1981); N. Sakai, Z. Phys. C**11**, 153 (1981); L. Ibañez and G. Ross, Phys. Lett. B**105**, 439 (1981); R.K. Kaul, Phys. Lett. B**109**, 19 (1982); M. Dine, W. Fischler and M. Srednicki, Nucl. Phys. B**189**, 575 (1981); S. Dimopoulos and S. Raby, Nucl. Phys. B**192**, 353 (1981).
2. S. Weinberg, Phys. Rev. D**19**, 1277 (1979); L. Susskind, Phys. Rev. D**20**, 2619 (1979); E. Farhi and L. Susskind, Phys. Rep. **74**, 277 (1981).
3. See e.g. the review by M. Peskin in *Proceedings of the International Symposium on Lepton and Photon Interactions at High Energies, Bonn, 1981* edited by W. Pfeil (Physikalisches Institut, Universität Bonn, 1981), p. 880.
4. E. Eichten, K. Lane and M. Peskin, Phys. Rev. Lett. **50**, 811 (1983).
5. E.H. Simmons, Phys. Lett. B**226**, 132 (1989).
6. E.H. Simmons, Phys. Lett. B**246**, 471 (1990).
7. P. Cho and E.H. Simmons, Phys. Lett. B**323**, 401 (1994). hep-ph/9307232.
8. P. Cho and E.H. Simmons, Phys. Rev. D (1995). To appear. hep-ph/9408206.
9. A.Y. Mozorov, Sov. J. Nucl. Phys. **40**, 505 (1984); S. Narison and R. Tarrach, Phys. Lett. B**125**, 217 (1983).
10. L. Dixon and Y. Shadmi, Nucl. Phys. B**423**, 3 (1994).
11. H. Dreiner, A. Duff and D. Zeppenfeld, Phys. Lett. B**282**, 441 (1992).
12. A. Duff and D. Zeppenfeld, Z. Phys. C**53**, 529 (1992).
13. D. Atwood, A. Kagan and T. Rizzo (1994). hep-ph/9407408.
14. C. Hill and S. Parke, Phys. Rev. D**49**, 4454 (1994). hep-ph/9312324.
15. T. Rizzo, Phys. Rev. D**50**, 4478 (1994), hep-ph/9405391; T. Rizzo (1994). hep-ph/9407366.
16. CDF Collaboration, F. Abe et al., FERMILAB-PUB-91/231-E; Phys. Rev. Lett. **68**, 1104 (1992).
17. H.D. Politzer, Nucl. Phys. B**172**, 349 (1980).
18. P. Harriman, A. Martin, R. Roberts and J. Stirling, Phys. Rev. D**42**, 798 (1990).

Summary talk: Gauge Boson Self Interactions[1]

Ian Hinchliffe

Theoretical Physics Group
Lawrence Berkeley Laboratory
University of California
Berkeley, California 94720

A review is given of the theoretical expectations of the self couplings of gauge bosons and of the present experimental information on the couplings. The possibilities for future measurements are also discussed.

The electro-weak gauge bosons in the standard model of electroweak interactions interact with each other in a way that is fully described by the model. Deviations from the prescribed form cause the model to be non-renormalizable or, equivalently, to violate unitarity in high energy scattering (1). In this review talk, I shall present a personal perspective on the determination of, and expectations for, these couplings. I shall discuss the form of the deviations from the standard model and how they are parameterized and then discuss the expectations for the deviations in extensions to the standard model. I will review the current experimental information and the possible impact of future experiments.

Deviations from the standard model must be parameterized in some way that will still allow predictions for experimental quantities to be made. It is convenient to begin with the general form of the WWV coupling where V is either a Z boson or a photon (2).

$$L/(ig_v) = (W^{a\dagger}_{\mu\nu}W^{a\mu} - W^{a\dagger}_\mu W^{a\mu}_\nu)V^\nu g^V_1 + \kappa_V W^{a\dagger}_\mu W^a_\nu V^{\mu\nu}$$
$$+ \widetilde{\kappa_V} W^{a\dagger}_\mu W^a_\nu V_{\alpha\beta}\epsilon^{\mu\nu\alpha\beta} + \frac{\lambda_V}{M^2_W} W^{a\dagger}_{\rho\mu} W^{a\mu\nu} V^\rho_\nu - ig^V_5 \epsilon^{\mu\nu\rho\sigma}(W^{a\dagger}_\mu \overleftrightarrow{\partial_\rho} W^{a\dagger}_\nu)V_\sigma$$
$$+ \frac{\widetilde{\lambda_V}}{M^2_W} W^{a\dagger}_{\rho\mu} W^{a\mu}_\nu V_{\alpha\beta}\epsilon^{\rho\nu\alpha\beta} + ig^V_4 W^{a\dagger}_\mu W^a_\nu (\partial^\mu V^\nu + \partial^\nu V^\mu) \quad (1)$$

W^a_μ ($W^a_{\mu\nu}$) represents the W boson field (field strength) and V_μ (or $V_{\mu\nu}$) is that of the photon (γ) or Z boson. The $SU(2)$ index a will be dropped in what follows. Electromagnetic gauge invariance implies that $g^\gamma_5 = g^\gamma_4 = 0$. In the standard model, $\lambda_Z = \lambda_\gamma = g^\gamma_5 = g^Z_5 = g^A_4 = g^Z_4 = \widetilde{\kappa_V} = \widetilde{\lambda_V} = 0$,

---

[1] This work was supported by the Director, Office of Energy Research, Office of High Energy and Nuclear Physics, Division of High Energy Physics of the U.S. Department of Energy under Contract DE-AC03-76SF00098.

© 1996 American Institute of Physics

$\kappa_Z = \kappa_\gamma = g_1^Z = g_1^\gamma = 1$, $g_Z = e \cot\theta_W$ and $g_\gamma = e$. Radiative corrections can induce small changes in these values at higher order in perturbation theory. The terms $\tilde{\kappa}$, $\tilde{\lambda}$ and g_4 violate CP and are also zero at one loop in the standard model. Experimental constraints are often quoted in terms of λ and $\Delta\kappa = \kappa - 1$ which parameterize deviations from the standard model. The other possible self couplings are ZZZ, $ZZ\gamma$ and $Z\gamma\gamma$. In the standard model these are zero. They are severely constrained by electromagnetic gauge invariance and Bose symmetry and must vanish if all of the particles are on mass shell (2,3). I will phrase most of the following discussion in terms of κ_γ and λ_γ assuming that all the other couplings have the form given by the standard model. The arguments provided below can be extended to the other cases straightforwardly.

The standard, $SU(2) \times U(1)$ model, of electro-weak corrections has now been tested at the quantum (1-loop) level in experiments at LEP, SLC and elsewhere (4,5). In these radiative corrections, the gauge boson self interactions can appear in loop corrections to the W, Z and photon propagators. If all loops involving gauge boson self interactions are ignored, the agreement between theory and experiment is less good (6,7). Direct determination of these self interactions comes from direct observation of gauge boson pairs at the Tevatron or, eventually, at LEPII.

Extensions to the standard model can produce values of the parameters in Equation 1 that deviate from the standard model form. I will assume that whatever extensions exist, they must satisfy $SU(2) \times U(1)$ gauge invariance. A model that does not do this will be difficult to reconcile with current data[2]. It is convenient to distinguish two types of extensions to the standard model. First, there are models that, like the standard model, are renormalizable. In this case a finite number of new parameters is sufficient to fully describe the theory. Supersymmetric extensions of the standard model usually fall into this class. In models of this type the parameters in Equation 1 are modified by radiative (loop) corrections from the standard model values.

Second there are non-renormalizable theories. Such models have a mass scale Λ that appears in the coefficient of the higher dimension operators. For experiments that probe energy scales (E) less than Λ, the effects of these operators are suppressed by powers of (E/Λ). Although, such models contain, in principle, an infinite number of parameters, only a few of these will be relevant for experiment since the suppression will render the effects of most of them unobservable. The theory can then be regarded as an effective theory valid for $E < \Lambda$. At energies above Λ, the theory is replaced by a more fundamental one and the terms in the effective theory are computable in terms of the parameters of the more fundamental theory. This notion of an effective theory is a very useful one since it may be possible to severely constrain its form without knowing the full dynamics of the fundamental theory (9). The best example of this type is the theory that describes the interaction of pions

---

[2] For more discussion of this see the talk by Willenbrock at this meeting (8)

with each other at low energy. Introducing $U = exp(i\overline{\sigma}\cdot\overline{\pi}/f_\pi)$, where the vector $\overline{\pi}$ represents the $\pi^\pm, \pi^0$, the interactions are given by

$$tr(\partial_\mu U^\dagger \partial_\mu U) + O(\frac{1}{4\pi f_\pi})^2 \tag{2}$$

This Lagrangian well describes QCD, *i.e.* the dynamics of $\pi - \pi$ scattering, on energy scales less than a few hundred MeV. At higher energies the full dynamics of (non-perturbative) QCD, including the details or resonances is needed to fully describe the scattering. The low energy Lagrangian is determined by the symmetries of low energy QCD, *i.e.* the fact that the pions are the Goldstone bosons of spontaneously broken chiral symmetry.

If there is new dynamics on a mass scale of a few TeV, such as is the case in technicolor (10) models or models where there are strong interactions between longitudinally polarized W and Z bosons at high energy (11), the effects of this dynamics can be parameterized by adding terms to the standard model Lagrangian (12). These form of these terms is dictated by the requirement that they must not produce any effects that would invalidate the various standard model tests and they must be invariant under $SU(2) \times U(1)$. The form of the operators depends upon the particle content of the low energy effective theory. The theory must contain the quarks, leptons and gauge bosons; it may or may not contain Higgs scalars. If we assume that there are no light Higgs scalars then one can write 12 CP invariant operators of dimension 4 (13) or less. This lagrangian can be written as a gauged chiral model. In addition to the quark and lepton fields and the gauge boson fields, there is a field $\Sigma = exp(i\pi^a \tau^a / v)$ with $v = 246$ GeV. The field π^a provides the longitudinal degrees of freedom for the massive W and Z bosons. The kinetic energy for the gauge bosons is given by

$$\frac{v^2}{4}tr(D_\mu \Sigma^\dagger D^\mu \Sigma) - \frac{1}{2}W_{\mu\nu}W^{\mu\nu} - \frac{1}{2}B_{\mu\nu}B^{\mu\nu} \tag{3}$$

Here field $B_{\nu\mu}$ ($W_{\mu\nu}$) is the field strength of the $U(1)$ ($SU(2)$) part of the standard model. These terms also give the mass for the W and Z bosons and the photon. I will consider the effects of two of the additional operators

$$L_1 = -\frac{v^2}{\Lambda^2} 2ig\beta_1 tr(W_{\mu\nu} D^\mu \Sigma^\dagger D^\nu \Sigma)$$
$$L_2 = \frac{v^2}{\Lambda^2} g^2 \tan\theta_W \beta_2 (\Sigma B_{\mu\nu} \Sigma^\dagger W^{\mu\nu}) \tag{4}$$

These give a contribution to κ_γ

$$\Delta\kappa_\gamma = \frac{v^2}{\Lambda^2} g^2 (\beta_1 - \beta_2)$$
$$\lambda_\gamma = 0 \tag{5}$$

However the term L_2 also contributes to the two point function of the gauge bosons and is therefore constrained by measurements at LEP and elsewhere as I will now discuss.

Recall how tests of the standard model are carried out. The model is fully described in terms of a set of parameters which can be taken, to be the Fermi constant G_F, the fine structure constant α_{em}, the mass of the Z, the Higgs mass and the masses of the quarks and leptons. Taking these values as input, one computes the expected value of some experimentally observable quantity such as the cross section of $\nu-e$ scattering. This expected quantity has some error δ_{theory}, that arises from the uncertainties in the parameters and residual uncertainty arising from the the calculation having been carried out to some order in perturbation theory. This is then compared with an experimental measurement which has an error δ_{expt}. If the theory and experiment agree, the model is the tested with an accuracy that is the *larger* of δ_{theory} and δ_{expt}. A failure of the model is revealed when there are experimental results that disagree with theory by more than the *larger* of δ_{theory} and δ_{expt}. In a variant of the standard model, extra parameters appear and the values of these parameters can be adjusted to accommodate experimental values that the standard model fails to predict correctly.

The parameters appearing in equation 1 need to be related to physical quantities so that their values can be extracted from data. The general form of the WWV vertex for bosons of momenta p_1, p_2 and p_3 and polarization tensors ϵ_1^μ, ϵ_2^ν and ϵ_3^α depends upon the invariant mass of the three bosons viz. $\Gamma^{\mu\nu\alpha}(p_1^2, p_2^2, p_3^2)$. In the case of the $WW\gamma$ vertex, there is a physical point where all of the particles are on mass shell (static limit) *i.e.* $\Gamma^{\mu\nu\alpha}(M_W^2, M_W^2, 0)$. At this point the quantities appearing in equation 1 are related to physical properties of the W boson; κ_γ and λ_γ to the electric quadrapole moment (Q) and magnetic dipole moment (μ) of the W.

$$\mu = \frac{e}{2M_W}(1 + \kappa_\gamma + \lambda_\gamma) \tag{6}$$

$$Q = -\frac{e}{M_W^2}(\kappa_\gamma - \lambda_\gamma) \tag{7}$$

$$\tag{8}$$

However these static quantities are not sufficient to describe the general properties of $\Gamma^{\mu\nu\alpha}(p_1^2, p_2^2, p_3^2)$.

Consider the process $q\bar{q} \to W\gamma$; I will assume for simplicity that all of the parameters in the $WW\gamma$ vertex have the standard form except for κ_γ and λ_γ. There is a contribution for the Feynman diagram shown in figure which depends on $\Gamma^{\mu\nu\alpha}(s, M_W^2, 0)$ where $\sqrt{s}$ is the center of mass energy of the quark antiquark system. If κ_γ and λ_γ are taken to be constants, then this will result in a scattering amplitude of the form

$$A \sim a + b\sqrt{s}(\kappa_\gamma - 1 + \lambda_\gamma) + cs\lambda_\gamma \tag{9}$$

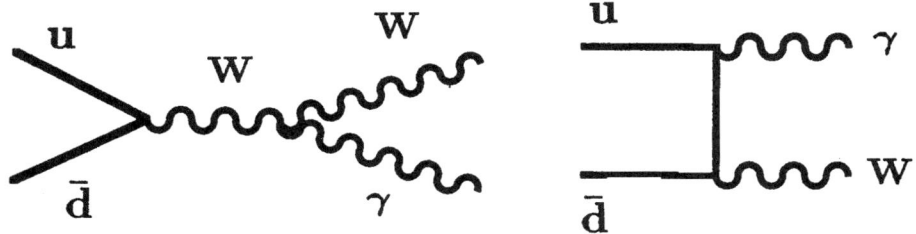

FIG. 1. Feynman diagrams showing the process $q\bar{q} \to W\gamma$

where a, b and c are independent of the center of mass energy ($\sqrt{s}$). This amplitude grows with s unless κ_γ and λ_γ have the standard model values of 1 and 0 respectively. This growth is a general feature of anomalous couplings. It is immediately clear that the sensitivity of an experiment to the anomalous couplings increases with the energy of the experiment and that a high energy experiment is more sensitive to λ_γ than to $\kappa_\gamma - 1$. Hence an $e^+e^- \to W^+W^-$ measurement at $\sqrt{s} \sim 500$ GeV can constrain λ_γ and $\kappa_\gamma - 1$ much more precisely than a measurement with comparable statistical power at $\sqrt{s} \sim 190$ GeV. Similarly in a hadron collider, the greatest sensitivity arises from the (few) events of largest energy.

This problem of unitarity violations can be avoided phenomenologically by the introduction of form factors (14) to damp the growth at large s *i.e.* $\lambda_\gamma \to \lambda_\gamma/(1+s/\Lambda^2)^{n_1}$ and $(\kappa_\gamma - 1) \to (\kappa_\gamma - 1)/(1+s/\Lambda^2)^{n_2}$ with $n_1, n_2 \geq 1$. It is conventional to use a dipole form factor, *i.e.* $n_2 = n_1 = 2$. An experiment measuring the $W\gamma$ production cross section can set a limit on $\Delta\kappa_\gamma$ and λ_γ given a value of n_1, n_2, and Λ. Note that for a given choice of n_1, n_2, and Λ, unitarity alone bounds λ_γ and $\kappa_\gamma - 1$. For $n_1 = n_2 = 2$, this bound is (15)

$$|\kappa_\gamma - 1| < 7.4 \text{ TeV}^2/\Lambda^2$$
$$|\lambda_\gamma| < 4.0 \text{ TeV}^2/\Lambda^2 \qquad (10)$$

An experiment that is not sensitive to values below these is not relevant.

Generally $\Gamma^{\mu\nu\alpha}(p_1^2, p_2^2, p_3^2)$ is not gauge invariant when computed beyond leading order in perturbation theory. This is directly related to the fact that it is not a physical quantity. As discussed in reference (16), it is possible to define a gauge invariant form by including some pieces of other corrections that would contribute at the same order in perturbation theory to a physical process. In the example of $q\bar{q} \to W\gamma$, a contribution of this type is shown in figure 2. It is convenient to quote the values of the physical quantities λ_γ and $\Delta\kappa_\gamma$ at the static limit as a measure of the expected size of the higher order corrections.

What values of anomalous couplings are to be expected in the standard model and its possible extensions? In the standard model the natural size

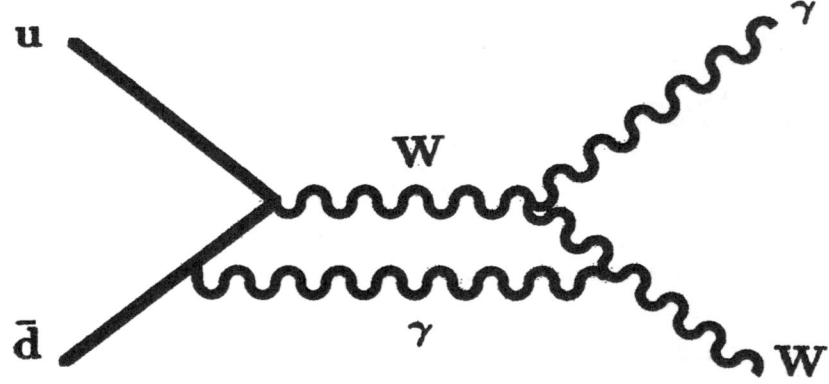

FIG. 2. An example of contribution to $q\bar{q} \to W\gamma$ which must be included along with the 1 loop corrections to the $WW\gamma$ vertex appearing in Figure

of $\kappa_\gamma - 1$ and λ_γ is α_{em}/π (17). For a top quark mass of 150 GeV and a Higgs mass of 100 GeV, $\lambda_\gamma = 0.006$ and $\kappa_\gamma + \lambda_\gamma - 1 = -0.0003$ (18). In the supersymmetric model the size of the corrections depends upon the masses of the supersymmetric particles. Note that the masses assumed must be consistent with other experimental constraints. For most of the values of the parameters, λ_γ is about 60% of its value in the standard model and $\kappa_\gamma + \lambda_\gamma - 1$ is about 5 times larger than its standard model value.

In extensions to the standard model where operators of the type in equation 4 are present, we need to estimate the size of $\beta_1, \beta_2,$ and Λ. Using the scale of new physics Λ to be 1 TeV we might expect $\delta\kappa_\gamma$ to be as large as 0.05 if $\beta_1 \sim \beta_2 \sim 1$ as would be expected if the new physics at scale Λ is strongly coupled. Other estimates yield values smaller than these (12). The term L_2 in equation 4 contributes to the gauge boson two point functions and in particular to the Peskin-Takeuchi (19) S parameter. Using the data from LEP, the constraint $|\beta_2| \lesssim 0.5$ (20) is obtained (again I have taken $\Lambda = 1$ TeV). Hence the contribution of L_2, to $\Delta\kappa_\gamma$ is restricted to be less than 0.013. The term L_1 is not directly constrained by LEP data. However since both β_1 and β_2 arise from the same (unknown) physics, it is to be expected that they will be of the same order of magnitude.

There have been observations of $W\gamma$, $Z\gamma$, WW and WZ final states at the Tevatron collider by both CDF (21) and D0 (22) that are reviewed at this meeting (23,24) The former constrains the $WW\gamma$ vertex while the latter constrains the $ZZ\gamma$ and $Z\gamma\gamma$ vertices and the last constrain WWZ and $WW\gamma$ vertices. The limits on κ_γ and λ_γ arising from observation of $W\gamma$ final states are shown in Figure 3. These limits use dipole form factors ($n_1 = n_2 = 2$) with $\Lambda = 1.5$ TeV. The limits are essentially unchanged if $\Lambda = 1$ TeV. The unitarity limits for $\Lambda = 1.5$ TeV are larger than the experimental constraints

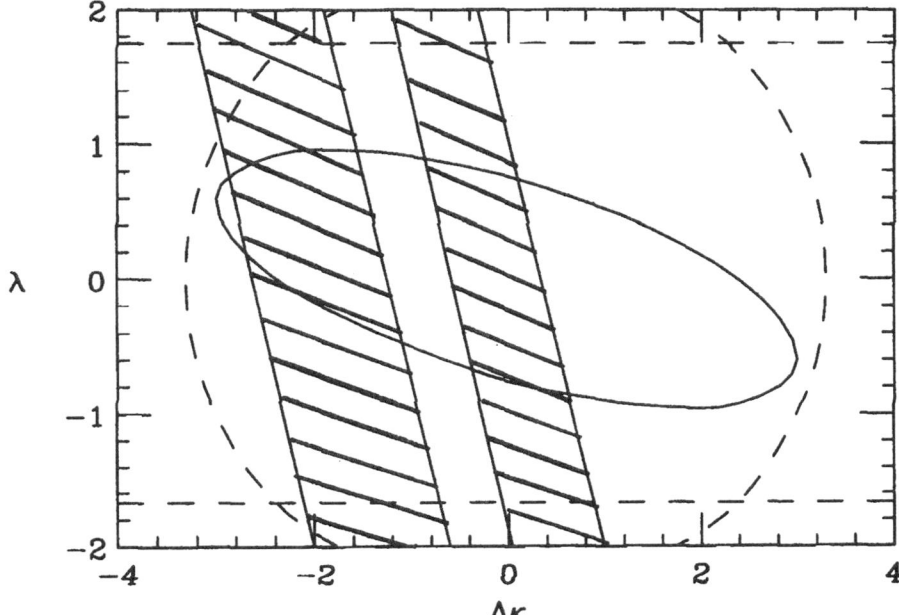

FIG. 3. The limits on $\Delta\kappa_\gamma$ and λ_γ from the D0 experiment (the area inside the oval region is allowed region) (22). The limits from CDF are similar (21). Also shown is the allowed region from the observation of $b \to s\gamma$ (the hatched area) from CLEO (28). The limits are shown at 95% confidence. The area outside the dashed circle is excluded by unitarity for the process $q\bar{q} \to W^+W^-$ with $n_1 = n_2 = 2$ and $\Lambda = 1.5$ TeV. The regions at the top and bottom of the figure bounded by the dashed horizontal lines are excluded by unitarity in $q\bar{q} \to W\gamma$

(see figure 3).

The limits on κ_Z and λ_Z arising from the observation of WW and WZ final states is similar to those on κ_γ and λ_γ (21). In the case of the $ZZ\gamma$ and $Z\gamma\gamma$ vertices, the limits are more sensitive to the assumed form factor behaviour of the vertices (3). This is due to the form of the vertex function, $\Gamma^{\mu\nu\alpha}(p_1^2, p_2^2, p_3^2)$, which must vanish when the particles are all on mass shell and therefore has powers of energy in the numerator. The form factors then introduced to prevent a unitarity violation must have $n \geq 3$. Constraints have also been placed on the $ZZ\gamma$ couplings by searching for events at LEP of the form $Z \to \gamma Z^*(\to \nu\bar{\nu})$ (27). These limits are comparable to those from CDF.

Note that the limits depend upon the ability to predict the event rates given the gauge boson self couplings requires an understanding of the QCD production process. This process is computed at next to leading order in α_{strong} and the resulting uncertainty should quite small (25). The angular distribution of the process $q\bar{q} \to W\gamma$ has a zero at a particular value of the scattering angle (26). This zero is not preserved by the higher order QCD corrections.

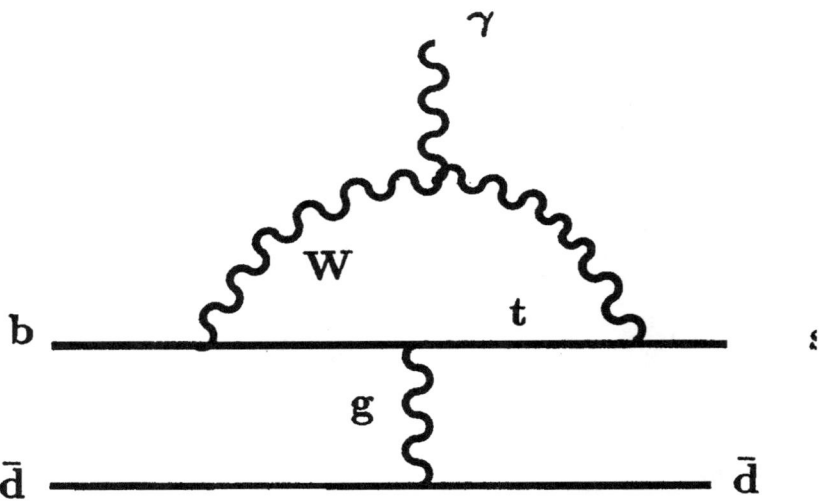

FIG. 4. A contribution to the process $B \to K\gamma$

The decay of a B meson to a photon and a strange meson, proceeds via loop effects. One relevant graph is given in Figure 4, where the $WW\gamma$ vertex is present. The experimental observation of this process (28) enables one to constrain κ_γ and λ_γ (29). The constraint is shown on Figure 3. Note that the constraint is less direct than that of CDF and D0. The interference between the graph shown in figure 4 and other graphs such as the one where the photon is radiated off the top quark, results in the odd shape for the allowed region. If there were other diagrams that could contribute to $b \to s\gamma$, such as would occur in a supersymmetric model, the constraint becomes a coupled limit involving the couplings of other particles (30).

I will end with a discussion of the prospects for future measurements. LEP II will be able to measure the $Z\gamma$ and WW and possibly the ZZ final state. Consequently it will probe the $WW\gamma$, $ZZ\gamma$, $Z\gamma\gamma$ and WWZ vertices. In the case of $WW\gamma$, the sensitivity of order 0.3 (0.5) to both λ and $\Delta\kappa$ at $\sqrt{s} = 192(176)$ GeV (32). This is approximately three times better than the current limits from the Tevatron. However these limits are based on ~ 15 pb^{-1} of data. They will improve by the end of the current when ~ 100 pb^{-1} will be available. If it is then possible to combine the CDF and D0 limits, they should fall by a factor of three or so. It seems reasonable to conclude therefore that any improvement that LEP II can provide over the Tevatron will be small.

There has been much discussion in the literature (20,31) and at this meeting of the extent to which the precision measurements of LEP imply that LEPII cannot see any effects of anomalous couplings. In order to address this question, possible models that differ from the standard model must be constructed so that they are consistent with LEP data and predictions for anomalous couplings or measurements at LEPII made. As discussed above, the LEP data constrain β_2 of Equation 4 sufficiently that the contribution of L_2 to anomalous couplings is too small to be seen at LEPII. The "natural" values of β_2 and β_1 should be roughly equal. In this case it is unlikely that LEPII (or the Tevatron) will see a positive effect. However, it might happen that $\beta_1 \gg \beta_2$. In QED, one can estimate the natural size of a process by assuming that the coefficient of the appropriate power of α_{em}/π is order one. Large coefficients such as π^2 that appears in the radiative corrections to Coulomb scattering (33) as well as ones that are less than one, such as the order $\alpha\pi$ correction to $g-2$ of the electron do occur.

The sensitivities of experiments discussed above are very far from the deviations from the standard model that can reasonably be expected. [3] Experiments at LHC (34,35) have greater sensitivity because of their greater energy. ATLAS expects a sensitivity of order $\Delta\kappa_\gamma \sim 0.04$ $\lambda_\gamma \sim 0.0025$ which is approaching values that are theoretically interesting (35). An e^+e^- collider with more energy than LEP will be more sensitive; at $\sqrt{s} = 500$ GeV (1.5 TeV) the sensitivities are λ_γ and $\Delta\kappa_\gamma$ are ~ 0.01 (~ 0.002) (36).

I am grateful to the members of the organizing committee, U. Baur, S. Errede and T. Müller for their work in making this conference such a success. The work was supported by the Director, Office of Energy Research, Office of High Energy Physics, Division of High Energy Physics of the U.S. Department of Energy under Contract DE–AC03–76SF00098. Accordingly, the U.S. Government retains a nonexclusive, royalty-free license to publish or reproduce the published form of this contribution, or allow others to do so, for U.S. Government purposes.

REFERENCES

1. H.H. Llewellyn Smith, Phys. Lett. **46B**, 233 (1973); S.D. Joglekar, Ann. Phys. (NY) **83**, 427 (1974); J.M. Cornwall, D.N. Levin and G. Tiktopolous, Phys. Rev. Lett. **30**, 1268 (1973).
2. K. Hagiwara, et al. Nucl. Phys. **B282**, 253 (1987).
3. U. Baur and E. Berger, Phys. Rev. **D47**, 4889 (1993).
4. D. Schaile these proceedings and CERN-PPE-94-162.
5. T. Takeuchi, these proceedings.
6. P. Gambini and A. Sirlin Phys. Rev. Lett. **73**, 621 (1994).
7. K. Hagiwara and S. Matsumoto, KEK-TH-375 (1994) and these proceedings.
8. S. Willenbrock, these proceeedings.
9. For a review, see e.g.H. Georgi, Ann. Rev. Nucl. and Part. Sci. **43**, 209 (1993).

---

[3] A participant asked me if this meant that theory obviated experiment.

10. For a review see, for example, K.D. Lane, BUHEP-94-26 (1994) and references therein.
11. M. Chanowitz, in Perspectives on High Energy Physics, Ed. G. Kane, World Scientific Publishing (1992).
12. C. Arst, M.N. Einhorn and J. Wudka, Nucl. Phys. **B433**, 41 (1995).
13. A Longhitano Nucl. Phys. **B188**, 118 (1981); T. Appelquist and C. Bernard, Phys. Rev. **D22**, 200 (1980).
14. U. Baur and D. Zeppenfeld, Nucl. Phys. **B308**, 127 (1988).
15. U. Baur and D. Zeppenfeld, Phys. Lett. **201B**, 383 (1988).
16. J.M. Cornwall, in Deeper Pathways in High Energy Physics, ed. B. Kursunoglu, A. Perlmutter and L. Scott, Plenum Press (1977), G. Degrassi and A. Sirlin Phys. Rev. **D46**, 3104 (1992).
17. W.A. Bardeen, R. Gastmans and B. Lautrup Nucl. Phys. **B46**, 319 (1982); K.J. Kim and Y.S. Tsai, Phys. Rev. **D12**, 3972 (1975).
18. E.N. Argyres et al., Nucl. Phys. **B391**, 23 (1993); J. Papavassiliou and K. Philoppides, Phys. Rev. **D48**, 4255 (1993).
19. M. Peskin and T. Takeuchi, Phys. Rev. Lett. **65**, 964 (1990).
20. S. Dawson and G. Valencia, BNL-60949 (1994).
21. F. Abe, et al. FNAL-PUB-95/036-E, Phys. Rev. Letters (submitted), FNAL-PUB-94/236-E Phys. Rev. Letters (submitted).
22. S. Abachi et al., Phys. Rev. Letters (submitted), J. Ellison in Proc. of 1994 DPF Meeting, Albuquerque, NM.
23. H. Aihara, these proceedings.
24. T. Fuess, these proceedings.
25. J. Ohnemus, these proceedings; U. Baur, T. Han, J. Ohnemus FSU-HEP-941010 (1994).
26. R.W. Brown, D. Sadhev and K.O. Mikaelian, Phys. Rev. **D20**, 1164 (1999); R.W. Brown, these proceedings.
27. P. Mattig, these proceedings; O. Adrianni et al., Phys. Lett. **B297**, 469 (1992); M. Acciarri, et al. Phys. Lett. **B345**, 609 (1995).
28. M.S. Alam et al., CLNS-94-1314 (1994); S. Playfer, these proceedings.
29. S. Chia Phys. Lett. **B240**, 467 (1990); K. Peterson Phys. Lett. **B282**, 207 (1992); T. Rizzo Phys. Lett. **B315**, 471 (1993) and X. He and B. McKellar Phys. Lett. **B320**, 165 (1994).
30. J. Hewett, SLAC-PUB 6521, Presented at SLAC Summer Inst. on Particle Physics, SLAC, Jul 6 - Aug 6, 1993.
31. P. Hernandez and F.J. Vegas, Phys. Lett. **B307**, 116 (1993); A. De Rujula, M.B. Gavela, P. Hernandez, and E. Masso Nucl. Phys. **B384**, 3 (1992); P.Hernandez, these proceedings.
32. Talks by G,. Gounaris and J.L. Knuer, LEPII workshop Jan 1995; R.L. Sekulin Phys. Lett. **B338**, 369 (1994); J. Hansen ALEPH-95-004.
33. J. Schwinger, Phys. Rev. **75**, 1912 (1949).
34. J. Womersley, these proceedings; ; CMS technical proposal CERN/LHCC/94-38.
35. ATLAS technical proposal CERN/LHCC/94-43.
36. T. Barklow, these proceedings, SLAC-PUB-6618 (1994).

B) PARALLEL TALKS

Rare Z^0 decays

P. Giacomelli

University of California, Riverside, CA 92521
and
PPE Division, CERN, CH-1211, Geneva 23

At the end of 1994 the four LEP experiments (ALEPH, DELPHI, L3 and OPAL) have each collected approximately 3.5 million hadronic Z^0 decays. These large samples allow to search for rare decays of the Z^0 boson at the level of 10^{-5} or better. A review of some of these searches is presented.

I. INTRODUCTION

The large samples of Z^0 events provided by the four LEP experiments allow to search for decay modes with rather small branching ratios. These rare decay modes could give hints of physics beyond the Standard Model. Indeed some of the searches described in the following have been inspired from predictions from theories beyond the SM, like SuperSymmetry or Compositeness. Previous searches at LEP have been described elsewhere, see for example (1).

It is also important to search, in a model independent way, for events with topologies which would differ from those expected by SM processes. This way one would retain sensitivity for new and unpredicted phenomena.

II. SEARCH FOR LEPTON FLAVOUR VIOLATING Z^0 DECAYS

Lepton flavour violation (LFV) is absolutely forbidden in the SM. The existence of massive neutrinos, or in theories beyond the SM, like SuperSymmetry, lepton flavour violation is allowed. Previous searches for LFV have given negative results (2). From the absence of the decay $\mu \to eee$ the following limit was set: $Br(Z^0 \to e\mu) < 7.4 \times 10^{-13}$ (3). Searches for neutrinoless τ decays lead to much less stringent limits on the conservation of τ lepton flavour.

At LEP energies the topologies of lepton flavour violating Z^0 decays would be quite clear and easily distinguishable from other processes. The events would have, in their final state, two back-to-back leptons of different families ($Z^0 \to e\mu$, $Z^0 \to e\tau$ and $Z^0 \to \mu\tau$), both with the beam momentum. Good energy and momentum resolution are a must for these type of analyses. The background contributions vary from channel to channel and are mainly given by $Z^0 \to \tau\tau$ or $Z^0 \to \mu\mu$ events. Results from the LEP experiments are published in several papers (4). To date the best 95% CL limits are: $Br(Z^0 \to$

$e\mu) < 0.17 \times 10^{-5}$, $Br(Z^0 \to e\tau) < 0.98 \times 10^{-5}$ and $Br(Z^0 \to \mu\tau) < 1.7 \times 10^{-5}$ (5).

III. $Z^0 \to \gamma\gamma\gamma$

In the SM the decay $Z^0 \to \gamma\gamma\gamma$ proceeds through fermion and W-loops and is strongly suppressed. The branching ratio is expected to be $\sim 5.4 \times 10^{-10}$ (6). An enhanced branching ratio would be an indication of new physics. Larger branching ratios are predicted in the context of composite Z models (7) and models assuming a light magnetic monopole coupling to the Z (8). The L3 collaboration performed a systematic study of 3 gamma final states. In the framework of the search for a composite Z, they have found 25 candidates; these are consistent with expectations from QED predictions. They expressed this negative result in terms of a 95% CL upper limit, $Br(Z^0 \to \gamma\gamma\gamma) < 1 \times 10^{-5}$. A composite Z might have scalar partners, X, which could be produced in $Z^0 \to \gamma X$, with the subsequent $X \to \gamma\gamma$ decay. Their results yield $Br(Z^0 \to \gamma X, X \to \gamma\gamma) < 0.4 - 1.3 \times 10^{-5}$ over the range $3 < m_X < 89$ GeV at 95% CL (9), as shown in figure 1.

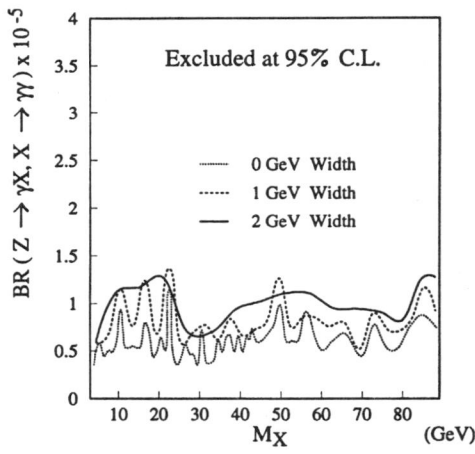

FIG. 1. Br ($Z^0 \to \gamma X$, $X \to \gamma\gamma$) limits vs. X mass (L3).

For the magnetic monopole analysis, they found 7 events, which are consistent with the (7.1±0.7) expected from QED. The branching ratio obtained is: $Br(Z^0 \to \gamma\gamma\gamma) < 0.8 \times 10^{-5}$. This limit can be interpreted in terms of a mass lower limit for a magnetic monopole: $m_{MM} \geq 510$ GeV (9).

IV. TOPOLOGICAL SEARCHES

Some particle search analyses, suggested by a specific theoretical model, may not be sensitive to new and unpredicted phenomena. New phenomena might manifest in the form of events with very peculiar topologies (acoplanar leptons, missing energy, single photons, monojets, etc.). It is therefore of interest to search for events with anomalous topologies. This can be efficiently done at an e^+e^- collider like LEP, due to the *cleanliness* of the events. The results of these searches can then be interpreted in the framework of different model predictions.

A. Low multiplicity events with anomalous topologies

The OPAL collaboration has searched for events with anomalous topologies using their low multiplicity data sample. They have *categorized* events by the number of cones containing charged tracks (N_c) and by the number of cones containing only calorimetric clusters and no charged tracks (N_n). After the various cuts no event survives. From this analysis they have extracted upper limits on $Br(Z^0 \to \eta\gamma)$, $Br(Z^0 \to \eta'\gamma)$, e^* and charged Higgs bosons (10).

B. Monojets

Monojet events are attractive since they are an almost background free topology for new particle searches. Monojet events could arise from light Higgs bosons, neutralinos, sneutrinos and more exotic phenomena. The ALEPH collaboration has looked for monojet events and found 3 such events in their sample of multihadronic events. Even though the number of observed events is in good agreement with expectations from SM background, coming from the process $e^+e^- \to \gamma^*\nu\bar{\nu}$ with $\gamma^* \to f\bar{f}$, their kinematical characteristics are not. Indeed the observed monojets have a rather high mass and a high p_t contrary to what is expected for events arising from the mentioned background process. The other LEP experiments, using comparable event samples, do not observe similar events. More data is needed to clarify this.

C. Search for particles with anomalous mass and/or charge

A search has been recently performed by the OPAL collaboration using the refined dE/dx information from the Jet chamber. The idea is rather simple: for each charged track, a cut is performed in the dE/dx vs. p/Q scatter plot, removing regions which are occupied by known particles. Then they look for particles with low ionization, like for example fractionally charged particles with $Q/e = \pm 2/3, 4/3$ and for Q=1 states with a large mass. From their negative results they have set limits of the order of 10^{-4} compared to $\sigma(e^+e^- \to \mu^+\mu^-)$ for particles with $Q/e=\pm 1, \pm 2/3, \pm 4/3, \pm 2$ (11), over the whole mass range accessible at LEP, as shown in figure 2.

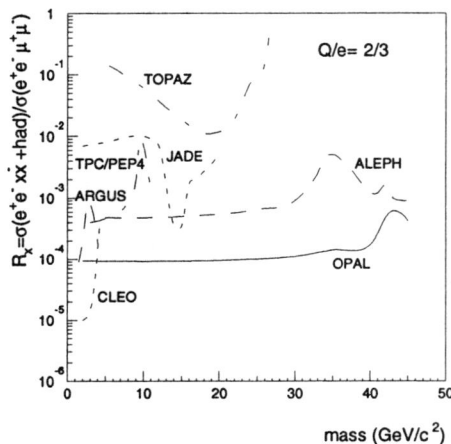

FIG. 2. Cross-section limit for particles with Q/e = 2/3 as a function of their mass (OPAL).

V. SEARCHES FOR LIGHT GLUINOS

Lower mass limits on relatively heavy gluinos have been set at hadronic colliders, $m_{\tilde{g}} > 141$ GeV (12). There is still disagreement on whether light gluinos, with mass of a few GeV or less, have been experimentally excluded or not (13). The OPAL collaboration has analysed their results on the search for particles of charge ±1 in the framework of light gluinos. These could in fact give rise to conglomerate states $(q\bar{q}\tilde{g})$ with charge ±1. If these conglomerate states have a long lifetime ($\tau > 10^{-7}$ s) they could be detected through their dE/dx deposit in the OPAL jet chamber. From the negative search they are able to exclude gluinos in the region $1.2 < m_{\tilde{g}} < 16.6$ GeV (11).

VI. SEARCHES FOR THE SCALAR TOP QUARK

The scalar top quark, $\tilde{t}$, (the SUSY partner of the t quark) could be lighter than the top quark itself. In fact, due to the high top quark mass, there could be significant mixing between the two supersymmetric top states, $\tilde{t}_L$ and $\tilde{t}_R$. The lightest stop state, $\tilde{t}_1$, would be: $\tilde{t}_1 = \tilde{t}_L \cos\theta_{mix} + \tilde{t}_R \sin\theta_{mix}$. The OPAL collaboration has searched for $\tilde{t}_1$ decays through these channels: $\tilde{t}_1 \to \tilde{\chi}_1^0 c$, $\tilde{t}_1 \to \tilde{\chi}_1^0 b f_1 \bar{f}_2$. No event survives the cuts, and they have set a mass lower limit of $m_{\tilde{t}_1} > 45.1$ GeV at the 95% CL (14), for a large set of values of θ_{mix}, as shown in figure 3.

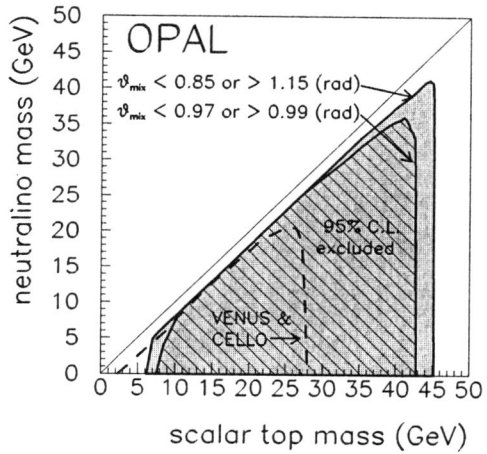

FIG. 3. Exclusion limits for a scalar top in the bidimensional plot neutralino mass vs. scalar top mass.

VII. CONCLUSIONS

Searches for rare Z^0 decays, not predicted by the SM, have so far given negative results. The upper limits on the branching ratios for various rare Z^0 decays are at the level of 10^{-5}. Other types of searches, like for fractionally charged particles and scalar top quarks have been briefly summarized.

REFERENCES

1. G. Giacomelli and P. Giacomelli, Riv. Nuovo Cimento **16** (1993) 1.
2. Particle Data Group, Phys. Rev. **D50** (1994).
3. Sindrum Collaboration, U. Bellgardt et al., Nucl. Phys. **B299** (1988) 1.
4. OPAL Collaboration, M.Z. Akrawy et al., Phys. Lett. **B254** (1991) 293.
   ALEPH Collaboration, D. Decamp et al., Phys. Rep. **216** (1992) 253.
   DELPHI Collaboration, P. Abreu et al., Phys. Lett. **B298** (1992) 247.
   L3 Collaboration, O. Adriani et al., Phys. Lett. **B316** (1993) 427.
5. OPAL Collaboration, R. Akers et al., CERN-PPE/95-032 (1995).
6. E. W. N. Glover and A. G. Morgan, Z. Phys. **C 60** (1993) 175.
7. F. Boudjema and F. Renard, CERN Report 89-08 vol 2. (1989) 185.
   M Baillargeon and F. Boudjema, CERN Report 92-04 (1992) 178.
8. A. De Rujula, CERN-TH 7273/94 (1994).
9. L3 Collaboration, M. Acciarri et al., Phys. Lett. **B345** (1995) 609.
10. OPAL Collaboration, R. Akers et al., PN151 (1994).
11. OPAL Collaboration, R. Akers et al., CERN-PPE/95-021 (1995).
12. CDF Collaboration, F. Abe et al., Phys. Rev. Lett. **69** (1992) 3439.
13. G. R. Farrar, RU-94-35 (1994).
14. OPAL Collaboration, R. Akers et al., Phys. Lett. **B337** (1994) 207.

The R_b Excess at LEP: Clue to New Physics at the TEVATRON?

Ernest Ma* and Daniel Ng†

*Department of Physics, University of California,
Riverside, California 92521, USA
†TRIUMF, 4004 Wesbrook Mall, Vancouver,
British Columbia, Canada V6T 2A3

> If the R_b excess at LEP is real, then any explanation in terms of renormalizable loop corrections leads to important new decay modes of the t quark and suppresses the $t \to bW$ branching ratio. In the two-Higgs-doublet model, the branching ratio of $Z \to b\bar{b}$ + a light boson which decays itself predominantly into $b\bar{b}$ is at least of order 10^{-4}.

INTRODUCTION

The awesome statistics of the four LEP collaborations have pinned down with great precision a host of measurable parameters in the standard model (1). The only quantity that shows a possible significant discrepancy with the theoretical prediction of the standard model is R_b which is defined as

$$R_b \equiv \frac{\Gamma(Z \to b\bar{b})}{\Gamma(Z \to \text{hadrons})}. \tag{1}$$

Using $m_t = 175$ GeV and $m_H = 300$ GeV, the standard model predicts that $R_b = 0.2158$, whereas LEP obtained $R_b = 0.2202 \pm 0.0020$ if the similarly defined R_c is assumed to be independent. If the latter is fixed at its standard-model value, then $R_b = 0.2192 \pm 0.0018$. In either case, the excess is about $2\% \pm 1\%$. If this is taken seriously, physics beyond the standard model is indicated.

TWO HIGGS DOUBLETS

The simplest extension of the standard model is to have two Higgs doublets instead of just one. The relevance of this model to R_b was studied in detail already a few years ago (2). To establish notation, let the two Higgs doublets be given by

$$\Phi_i = \begin{pmatrix} \phi_i^+ \\ \phi_i^0 \end{pmatrix} = \begin{bmatrix} \phi_i^+ \\ 2^{-1/2}(v_i + \eta_i + i\chi_i) \end{bmatrix}. \tag{2}$$

© 1996 American Institute of Physics

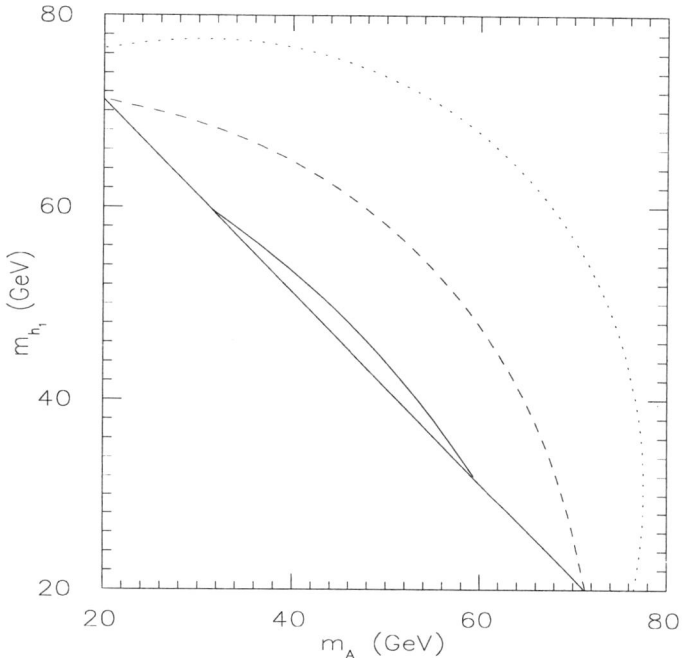

FIG. 1. $R_b = 0.2192$ (solid), 0.2174 (dashed) and 0.2164 (dotted) contours in the $m_{h_1} - m_A$ plane for $\alpha = 0$ and $\tan\beta = 70$. The straight line corresponds to $m_A + m_{h_1} = M_Z$. We have also assumed $m_{h^\pm} = m_{h_2} = 175$ GeV.

Let $\tan\beta \equiv v_2/v_1$, then

$$h^+ = \phi_1^+ \cos\beta - \phi_2^+ \sin\beta, \tag{3}$$
$$h_1 = \eta_1 \sin\alpha + \eta_2 \cos\alpha, \tag{4}$$
$$h_2 = \eta_1 \cos\alpha - \eta_2 \sin\alpha, \tag{5}$$
$$A = \chi_1 \cos\beta - \chi_2 \sin\beta. \tag{6}$$

Note that the $\bar{b}bA$ and $\bar{t}bh^+$ couplings involve the ratio $m_b \tan\beta/M_W$, hence they could be important for large values of $\tan\beta$. It was shown (2) that for $\tan\beta = 70 \simeq 2m_t/m_b$, the R_b excess peaks at about 4% near $m_A = m_{h_1} \simeq 40$ GeV for $\alpha = 0$. However, since $Z \to Ah_1$ is not observed, $m_A + m_{h_1} > M_Z$ is a necessary constraint. We show in Fig. 1 the contours in the $m_{h_1} - m_A$ plane for 3 values of R_b. It is clear that relatively light scalar bosons are required if the R_b excess is to be explained.

For $A(h_1)$ lighter than M_Z and having an enhanced coupling to $\bar{b}b$, the decay $Z \to b\bar{b} + A(h_1)$ becomes nonnegligible (3). As an illustration, we show in Fig. 2 the branching ratios of these two decays as functions of m_A with the constraint $m_A + m_{h_1} = M_Z + 10$ GeV so that a reasonable fit to the R_b excess

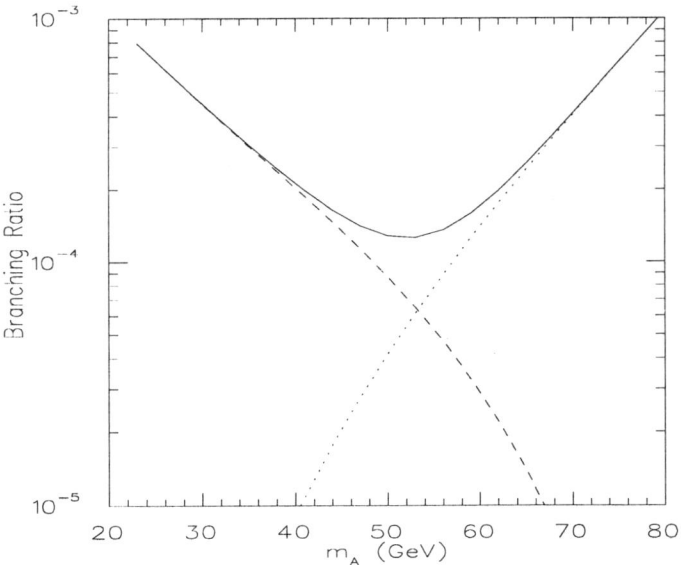

FIG. 2. The branching ratios, $Br(Z \to b\bar{b}A)$ (dashed) and $Br(Z \to b\bar{b}h_1)$ (dotted) and their sum (solid), as functions of m_A where we take $m_A = m_{h_1} = M_Z + 10$ GeV, $\tan\beta = 70$, $\alpha = 0$, and $m_{h^\pm} = m_{h_2} = 175$ GeV.

is obtained. It is seen that the sum of these two branching ratios is at least of order 10^{-4}. Once produced, A or h_1 decays predominantly into $b\bar{b}$ as well. Hence this scenario for explaining R_b can be tested at LEP if the sensitivity for identifying a $b\bar{b}$ pair as coming from A or h_1 in $b\bar{b}b\bar{b}$ final states can be pushed down below 10^{-4}.

MINIMAL SUPERSYMMETRIC STANDARD MODEL

In the Minimal Supersymmetric Standard Model (MSSM), there are two Higgs doublets, but their parameters are further constrained, hence the allowed region in the $m_{h_1} - m_A$ plane which gives a large enough R_b is further reduced by the experimental nonobservation of MSSM signals at LEP (4).

There is of course another possible contribution to R_b in the MSSM: the $Z \to b_L b_L$ vertex may be enhanced by the supersymmetric coupling of b_L to a scalar top quark and a chargino (5). In this case, both of the new particles must again be light, but now Z cannot decay into just one of these particles because of the assumed conservation of R parity, hence no further constraint is obtainable at LEP.

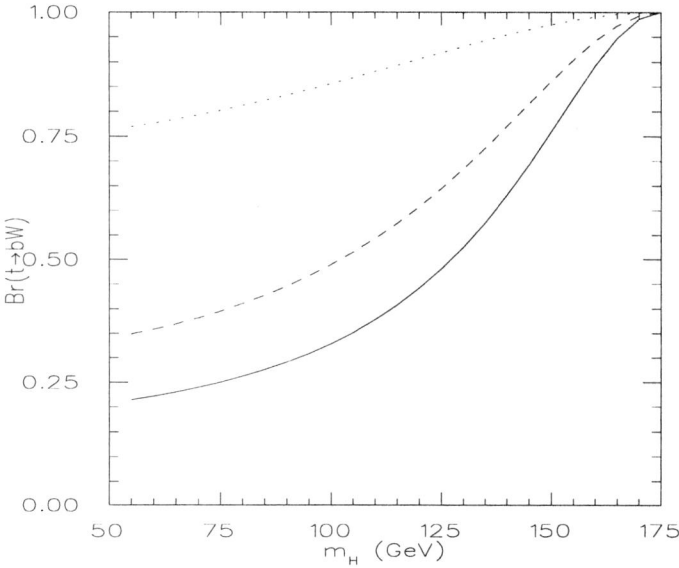

FIG. 3. The branching ratio $Br(t \to bW)$ as a function of m_h^+ for $\tan\beta = 70$ (solid), 50 (dashed), and 20 (dotted).

NECESSARY TOP DECAYS

Since b_L is involved in any enhanced coupling to light particles in explaining the R_b excess, its doublet partner t_L must necessarily have the same enhanced coupling to related particles. In the two-Higgs-doublet case, we must have an enhanced $\bar{t}bh^+$ coupling. Therefore, unless $m_{h^+} > m_t - m_b$, the branching ratio of $t \to bh^+$ will dominate over all others. In particular, the standard $t \to bW$ branching ratio will be seriously degraded. We show in Fig. 3 the branching ratio $Br(t \to bW)$ as a function of m_{h^+}. Large values of m_{h^+} are disfavored in this scenario because the splitting with A and h_1 would result in a large contribution to the ρ parameter (6). This poses a problem for top production at the TEVATRON because the number of observed top events is consistent with the assumption that top decays into bW 100% of the time. If that is not so, then top production must be enhanced by a large factor beyond that of the standard model. The two-Higgs-doublet model itself certainly does not provide for any such mechanism.

In the MSSM, if the R_b excess is attributed to a light scalar top quark and a light chargino, then we should look at the latter's doublet partner which is in general a linear combination of neutralino mass eigenstates. At least one of these, *i.e.* the Lightest Supersymmetric Particle (LSP), will be light enough to allow the top quark to decay into it and the scalar top. The ρ parameter also serves to disfavor large neutralino masses in this scenario. Hence the

$t \to bW$ branching ratio is again seriously degraded. Turning the argument around, this means that for every observed top event, there must be several others which correspond to the production of supersymmetric particles. If the R_b excess is really due to supersymmetry, top decay is the place to discover it!

CONCLUSION

If the R_b excess at LEP is real and we want to explain it in terms of renormalizable loop corrections, then light particles are unavoidable. However, these light particles may be produced also in Z decay such as in the two-Higgs-doublet case, where $Z \to b\bar{b} + A$ or h_1 is at least of order 10^{-4} in branching ratio. More importantly, there is necessarily a corresponding top decay into one of these light particles (such as the scalar top in the MSSM) and the other particle's doublet partner (the neutralino), which seriously degrades the $t \to bW$ branching ratio. Unless there is accompanying new physics which enhances the top production by a large factor at the TEVATRON, this generic explanation of the R_b excess in terms of light particles does not appear to be viable.

ACKNOWLEDGEMENT

The work of E.M. was supported in part by the U.S. Department of Energy under Contract No. DE-AT03-87ER40327. The work of D.N. was supported by the Natural Sciences and Engineering Research Council of Canada.

REFERENCES

1. The LEP Collaborations: ALEPH, DELPHI, L3, OPAL, and the LEP Electroweak Working Group, CERN Report No. CERN/PPE/94-187 (25 November 1994).
2. A. Denner, R. J. Guth, W. Hollik, and J. H. Kühn, Z. Phys. C51, 695 (1991).
3. A. Djouadi, P. M. Zerwas, and J. Zunft, Phys. Lett. B259, 175 (1991).
4. See for example A. Sopczak, CERN Report No. CERN/PPE/94-73 (9 May 1994).
5. G. Altarelli, R. Barbieri, and F. Caravaglios, Phys. Lett. B324, 357 (1993); J. D. Wells, C. Kolda, and G. L. Kane, Univ. of Michigan Report No. UM-TH-94-23 (July 1994); D. Garcia, R. A. Jimenez, and J. Sola, Univ. of Barcelona Reports Nos. UAB-FT-343,344 (September 1994).
6. A. K. Grant, Phys. Rev. D 51, 207 (1995).

The Measurement of Tri-Linear Gauge Boson Couplings at e^+e^- Colliders[1]

Gilles Couture*, Mikuláš Gintner† and Stephen Godfrey†

*Département de Physique, Université du Québec à Montréal
C.P. 8888, Succursale A, Montréal, Canada H3C 3P8
†Ottawa-Carleton Institute for Physics, Department of Physics,
Carleton University, Ottawa, Canada K1S 5B6

We describe a detailed study of the process $e^+e^- \to \ell\nu_\ell q\bar{q}$ and the measurement of tri-linear gauge boson couplings (TGV's) at LEP200 and at a 500 GeV and 1 TeV NLC. We included all tree level Feynman diagrams contributing to the four-fermion final states including gauge boson widths and non-resonance contributions. We employed a maximum likelihood analysis of a five dimensional differential cross section of angular distributions. This approach appears to offer an optimal strategy for measurement of TGV's. LEP200 will improve existing measurements of TGV's but not enough to see loop contributions of new physics. Measurements at the NLC will be roughly 2 orders of magnitude more precise which would probe the effects of new physics at the loop level.

INTRODUCTION

A driving force behind high energy physics is the search for physics beyond the Standard Model (SM). An approach receiving considerable attention generalizes the effects of new physics using effective Lagrangians ($\mathcal{L}_{eff}$) and tests for deviations from SM expectations. While the fermion gauge boson couplings have been measured to high precision by the LEP and SLC experiments the vector boson self-interactions have only just been experimentally verified by direct measurement (1). Because the standard model makes precise predictions for the TGV's, precision measurements constitute a stringent test of the gauge structure of the standard model (2). In this contribution we describe some recent work on precision measurements of tri-linear gauge boson couplings (TGV's) at e^+e^- colliders; LEP200 and 500 GeV and 1 TeV versions of the NLC. A more detailed account of this work is given in Ref. (3).

The processes $e^+e^- \to 4f$, including all tree level Feynman diagrams that contribute to the same final state, have been studied for a number of specific final states; $\ell\nu\ell'\bar{\nu}$, $\nu\bar{\nu}\mu^+\mu^-$, $q\bar{q}q'\bar{q}'$, and $\ell\nu q\bar{q}$. Our present work consists of a detailed study of the last process, $e^+e^- \to \ell\nu q\bar{q}$, and its sensitivity to

---

[1] Presented by S. Godfrey

TGV's. This process can be fully reconstructed and has the advantage of a higher branching ratio than the purely leptonic processes while avoiding the ambiguities and backgrounds associated with the purely hadronic modes.

There are three effective Lagrangians commonly used to describe TGV's which differ on the degree of generality assumed (2).

1. *Most General Parametrization.* The only assumptions here are Lorentz and $U(1)_{em}$ invariance. The parameters associated with the CP invariant operators are g_1^Z, κ_Z, κ_γ, λ_Z, and λ_γ. g_γ is always equal to 1 and in the SM at tree level $g_1^Z = \kappa_V = 1$ and $\lambda_V = 0$. Typically, radiative corrections from heavy particles will change κ_V by about 0.015 and λ_V by about 0.0025. The most robust limits come from associated $W\gamma$ and WZ production at the Tevatron (1); $-1.6 < \delta\kappa_\gamma < 1.8$, $|\lambda_\gamma| < 0.6$, $-8.6 < \delta\kappa_Z < 9.0$ and $|\lambda_Z| < 1.7$.

2. *Non-Linearly Realized Goldstone Bosons* or *The Chiral Lagrangian* includes custodial $SU(2)$ symmetry which is experimentally verified to a high degree of accuracy. The parameters of this Lagrangian are L_{9L}, L_{9R}, and L_{10}. L_{10} contributes to the gauge boson self energies where it is tightly constrained to $-1.1 \leq L_{10} \leq 1.5$ so we will not consider it further. $L_{9L,9R}$ are expected to be of order 1.

3. *Linearly Realized Goldstone Bosons* explicitly includes the Goldstone bosons.

Due to space limitations we only present results for the Chiral Lagrangian but note that the different Lagrangians can be mapped onto one another.

Calculations and Results

We studied the final states $\mu^{\pm}\nu_\mu q\bar{q}$ and $e^{\pm}\nu_e q\bar{q}$. In the first case ten diagrams contribute and in the second, 20 diagrams. We used the CALKUL helicity technique to calculate the amplitudes and integrated the resulting matrix elements using Monte Carlo methods to obtain the cross sections and distributions.

For our numerical results we used the following set of parameters: $\alpha = 1/128$, $\sin^2\theta_w = 0.23$, $M_Z = 91.187$, $\Gamma_Z = 2.49$, $M_W = 80.22$ and $\Gamma_W = 2.08$. We included the kinematic cut $170° > \theta > 10°$ where θ is the angle of the charged lepton, quark, or antiquark relative to the beam axis. We also took $E_{\ell,q,\bar{q}} > 10$ GeV. In some of our results we imposed that $|M_{(\ell\nu),(q\bar{q})} - M_W| \leq 10$ GeV where $M_{(\ell\nu),(q\bar{q})}$ is the invariant mass of the $\ell\nu$ ($q\bar{q}$) pair.

The object of the analysis is to maximize the sensitivity to anomalous gauge boson couplings. The longitudinal components of the W are most sensitive to anomalous couplings. W^-'s (W^+'s) produced parallel to the incoming e^- (e^+) are dominated by transverse W's while those in the backward direction have a large W_L content. To extract the W_L's from the W_T "background"

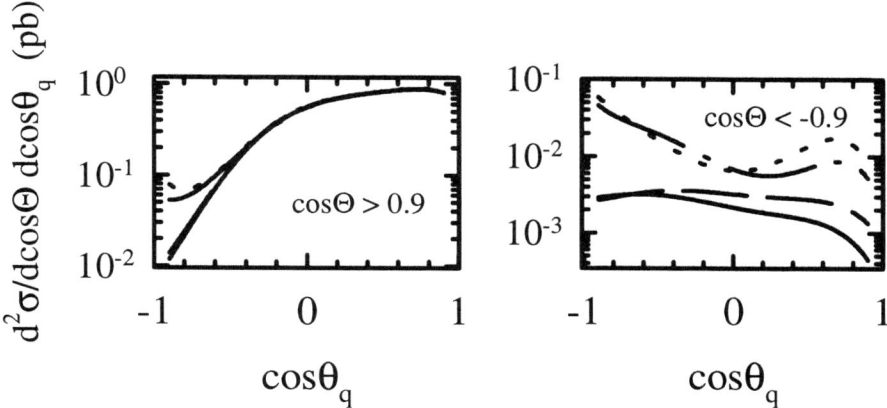

FIG. 1. Angular distributions of the outgoing quark in the rest frame of the W^- in the process $e^+e^- \to \mu^+\nu_\mu q\bar{q}$. The solid line is for the SM, the dashed line for $\kappa_Z = 1.1$, the dotted line for $\lambda_Z = 0.1$, and the dot-dashed line for $\lambda_\gamma = -0.1$.

we can use the W decay products as a polarimeter; W_T decay products peak at forward or backward angles while those of the W_L's peak about $\cos\theta = 0$ where θ is the angle of the decay product with respect to the W direction in the W rest frame. The changing mix of W_L and W_T is illustrated in Fig. 1 by the quark angular distributions for forward and backward W scattering angles (Θ). Additional information can be obtained by studying the decay product azimuthal distributions which are subject to more complicated interference.

To determine if anomalous couplings are measurable we use the maximum likelihood method applied to the differential cross section:

$$\frac{d^5\sigma}{d\cos\Theta\, d\cos\theta_q\, d\phi_q\, d\cos\theta_\ell\, d\phi_\ell}$$

where Θ is the scattering angle of the outgoing W's and $\theta_{q(\ell)}$ and $\phi_{q(\ell)}$ are the polar and azimuthal decay angles of the outgoing q (ℓ) in the W rest frame. We divided each of the angles into 4 bins. Summing over all 1024 bins and comparing the SM predictions to those for anomalous couplings we obtain the log likelihood function:

$$\ln \mathcal{L} = \sum_i [-r_i + r_i \ln(r_i) + \mu_i - r_i \ln(\mu_i)]$$

where r_i and μ_i are the measured and expected number of events respectively.

We show the 95% C.L. contours for the $L_{9L} - L_{9R}$ plane for LEP200 and 500 GeV and 1 TeV NLC options in Fig. 2.

One can obtain additional information from the processes studied here. Possibilities we have studied but have not included here for lack of space are:

1. *Initial state polarization.* The distributions are different for left and right handed initial state electrons mainly due to the contributions of the neutrino exchange diagram to the e_L^- mode but not the e_R^- mode. This results in different dependences on anomalous couplings adding to the measurement sensitivity.

2. *Single W production.* In the $e^+\nu q\bar{q}$ (or $e^-\bar{\nu}q\bar{q}$) final state, instead of imposing the kinematic cut that the final state e^+ (e^-) be observed we can impose the cut that it not be observed. In this case the cross section is dominated by the t-channel photon pole providing a mechanism for measuring the $WW\gamma$ vertex independent of the WWZ vertex.

3. *Off resonance production.* The deviations from the SM value cross sections can be dramatic off the W resonances. Although the off-shell contributions by themselves don't offer improvements to the on-shell W results, including the off resonance contributions can improve the overall sensitivities in certain cases.

SUMMARY

We have presented some results from a study of TGV's in the process $e^+e^- \to \ell\nu q\bar{q}$. We employed a maximum likelihood analysis of a five dimensional differential cross section of angular distributions. LEP200 will improve on existing measurements, but not sufficiently to observe deviations originating from loop contributions from heavy particles. On the other hand the NLC will be able to measure these couplings to better that a half of a percent which would be sensitive to radiative corrections to the TGV's.

ACKNOWLEDGMENTS

SG thanks the organizers of TGV95 for the invitation to attend a most enjoyable meeting and the Deans of Research and Science at Carleton University for financial support to attend the meeting. This research was supported in part by NSERC Canada and Les Fonds FCAR du Quebec.

REFERENCES

1. H. Aihara, these proceedings; F. Abe *et al.*, Phys. Rev. Lett. **74**, 1936 (1995); S. Errede, Proceedings of the *27th International Conference on High Energy Physics*, Glasgow, Scotland, July 1994.
2. For a recent review see H. Aihara *et al.*, to appear in *Electroweak Symmetry Breaking and Beyond the Standard Model*, eds. T. Barklow, S. Dawson, H. Haber and J. Siegrist (World Scientific).
3. M. Gintner, S. Godfrey, and G. Couture, Carleton report OCIP/C 95-3.

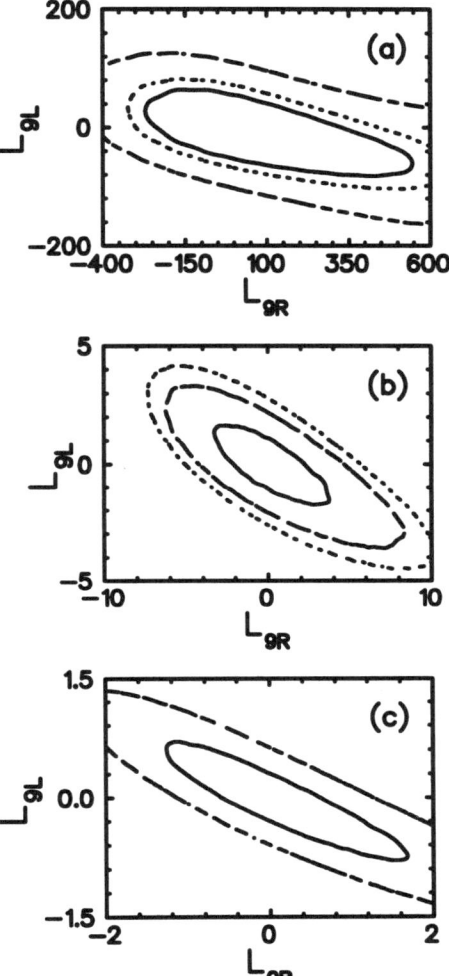

FIG. 2. 95% confidence level sensitivity contours for L_{9L} and L_{9R} at LEP200 and the 500 GeV and 1 TeV options of the NLC. In all cases the solid lines are the sensitivities obtained by combining the $e^\pm$ and $\mu^\pm$ modes, the dashed lines are for the μ^+ mode only, and the dotted line is for reduced luminosity and combining the 4 modes. (a) $\sqrt{s} = 175$ GeV: There are no cuts on $M_{\ell\nu}$ or $M_{q\bar{q}}$. The solid and dashed lines use L=500 pb^{-1} and the dotted line uses L=300 pb^{-1}. (b) 500 GeV: Include the cuts $|M_{\ell\nu(q\bar{q})} - M_W| < 10$ GeV. The solid and dashed lines use L=50 fb^{-1} and the dotted line uses L=10 fb^{-1}. (c) 1 TeV: Include the cuts $|M_{\ell\nu(q\bar{q})} - M_W| < 10$ GeV. The solid and dashed lines use L=200 fb^{-1}. Here the reduced luminosity contour of L=50 fb^{-1} lies on top of the μ^+ curve.

The Ward Identities of the Gauge Invariant Three Boson Vertices

Kostas Philippides

New York University, New York, New York 10003

We outline the pinch technique for constructing gauge invariant Green's functions in gauge theories and derive the Ward identities that must be satisfied by the gauge invariant three boson vertices of the standard model. They are generalizations of their tree level Ward identities and are shown to be crucial for the delicate gauge cancellations of the S-matrix.

THE PINCH TECHNIQUE

As is well known, the *off-shell* Green's functions of non Abelian gauge theories are in general gauge dependent as a result of the quantization procedure. For example in the t'Hooft covariant class of gauges R_ξ, the presence of the gauge fixing term and the Fadeev–Popov ghost term give rise to gauge boson, ghost, and unphysical scalar propagators that depend explicitly on the unphysical gauge parameters ξ_i, $i = g, \gamma, Z, W$. The gauge boson propagators are given by

$$\Delta_i^{\mu\nu}(q, \xi_i) = \frac{1}{q^2 - M_i^2}[g^{\mu\nu} - (1 - \xi_i)\frac{q^\mu q^\nu}{q^2 - \xi_i M_i^2}] \;, \qquad (1)$$

and the Green's functions of the theory are infected by the unphysical ξ_i parameters through the bosonic quantum loop corrections.

Although gauge invariance has been lost upon quantization, the effective quantum action still exhibits residual symmetries that lead to Ward identities (WI) between the gauge dependent Green's functions. In particular the BRST symmetry gives rise to complicated Slavnov–Taylor (ST) identities that make explicit reference to the gauge parameters and involve the ghosts' Green's functions while the background field gauge symmetry preserves the tree level WI but only for the background Green's functions.

In spite of the fact that the Green's functions of the theory are gauge dependent, the S-matrix is gauge independent (g.i.) order by order in perturbation theory, as a result of a subtle gauge cancellation between self–energy, vertex, and box graphs. Nevertheless the gauge dependence of the Green's functions is considered as an unhappy state of affairs for at least two reasons: i) In non–perturbative studies the infinite set of the SD equations is built out of the gauge dependent Green's functions; when casually truncated residual gauge dependences infest ostensibly g.i. quantities giving rise to meaningless approximations. ii) In perturbative studies form factors extracted from the usual vertices are gauge dependent, and thus void of any physical meaning, which makes the generalization of the familiar classical moments problematic at the quantum level.

The pinch technique, originally proposed by Cornwall (1), (2), has successfully addressed these issues. It exploits all the healthy properties of the S-matrix and allows the construction of modified g.i. n-point functions, through the order by order rearrangement of Feynman graphs, contributing to a certain physical and therefore g.i. amplitude.

The Green's functions constructed via the PT satisfy the following properties: i) They are gauge parameter independent and ii) gauge fixing procedure independent as has been shown by explicit calculations in a wide variety of gauges (R_ξ, Unitary, Axial, R_{ξ_Q}) (2), (3), (4), (5). ii) They are process independent (6) iii) Running couplings defined directly from the PT self energies $\widehat{\Pi}$, obey the RG equations (2), (3), and the form factors extracted from the PT vertices $\widehat{\Gamma}$, are well behaved (8), (9) iv) Finally they satisfy naive QED-like WI which are just their respective tree level classical WI (7), (8), (9), (10), (11).

Although in principle the PT can be applied to any order, so far explicit results have been obtained only up to one loop.

THE WARD IDENTITIES

After the PT rearrangement has been completed, the amplitude we consider has been reorganized into individually g.i. loop structures (Green's functions) connected by gauge dependent tree level propagators. In other words, the PT algorithm only cancels all gauge dependences originating from the tree-level propagators appearing *inside* the loops, but a residual gauge dependence, stemming from boson propagators *outside* of loops, survives at the end of the pinching process.

When the external currents are conserved any residual gauge dependence will automatically cancel in the R_ξ gauges. However this is not the case in the axial gauges and when the currents are not conserved this remaining gauge dependence will persist even in the R_ξ gauges.

The cancellation of this remaining gauge dependence from the S-matrix becomes possible due to a set of WI satisfied by the g.i. Green's functions. One can actually derive these WI *without* any detailed knowledge of the algorithm which gives rise to the g.i. Green's functions. All one needs to assume is that such an algorithm exists (in our case the PT algorithm), and demand the complete gauge independence of the S-matrix. So, once the g.i. Green's functions have been constructed, one should examine whether or not they actually satisfy the required WI, as a self-consistency check.

It is instructive to illustrate the derivation of the WI for the simple case of the g.i. W propagator. We consider the one-loop S-matrix element of the charged current process $e^- + \nu_e \to \nu_e + e^-$ and apply the PT rules in the context of the R_ξ gauges. After the application of the PT, the part of the S-matrix $\widehat{T}_1(q^2)$, which only depends on the momentum transfer q^2, will consist of four amplitudes each one corresponding to the relevant g.i. 2-point function of the W and its associated unphysical scalar ϕ. We use the current relation $q_\mu J^\mu_{W^\pm} = iM_W J_{\phi^\pm}$ in order to pull out a common factor $q_\alpha J^\alpha_{W^+} q_\beta J^\beta_{W^-}$ from all these amplitudes and then employ the standard identity

$$\Delta^{\mu\nu}_W(q,\xi_W) = U^{\mu\nu}_W(q) - \frac{q^\mu q^\nu}{M_W^2}\Delta_\phi(q,\xi_W) , \qquad (2)$$

in order to isolate all remaining gauge dependences into scalar propagators of the ϕ, which is given by

$$\Delta_\phi(q,\xi_W) = \frac{-1}{q^2 - \xi_W M_W^2} . \qquad (3)$$

$U^{\mu\nu}_W(q)$ is the g.i. propagator of the W at tree level, namely the propagator in the unitary gauge. After this last step one observes that the remaining gauge dependences will arise from terms containing either one or two Δ_ϕ. The requirement that $\widehat{T}_1(q^2)$ is ξ_W-independent, gives rise to two independent equations; the first enforces the cancellation of the terms with only one Δ_ϕ factor, whereas the second enforces the cancellation of the terms with a $\Delta_\phi \Delta_\phi$ factor. It then follows that

$$q^\mu \widehat{\Pi}^W_{\mu\nu}(q) \mp iM_W \widehat{\Pi}^\pm_\nu(q) = 0 \qquad (4)$$

$$q^\mu \widehat{\Pi}^\pm_\mu(q) \pm iM_W \widehat{\Pi}^\phi(q) = 0 \qquad (5)$$

$$q^\mu q^\nu \widehat{\Pi}^W_{\mu\nu}(q) - M_W^2 \widehat{\Pi}^\phi(q) = 0 \qquad (6)$$

Along the same line of reasoning similar WI can be obtained for the Z and χ self energies. For the photon and the gluon, the relevant WI are obtained by repeating the above arguments in an axial gauge.

We now turn to our main objective, namely the derivation of the WI for the g.i three boson vertices (TBV). In order to construct a g.i. n-point function in the context of the S-matrix PT one has to employ an $n \to n$ process. We choose to use fermions as external test particles and the process $e^-(n) + \nu(\ell) + e^-(r) \to e^-(\hat{n}) + e^-(\hat{\ell}) + \nu(\hat{r})$ where $q = n - \hat{n}$, $p_1 = \ell - \hat{\ell}$, $p_2 = r - \hat{r}$, are the momentum transfers at the corresponding fermion lines; they represent the incoming momenta of each of the bosons, merging in the TBV. Following the PT prescription we identify the pinch parts and allot them to the relevant graphs ending up with g.i. "blobs" (Green's functions) connected by gauge dependent tree level bosonic propagators. Concentrating on the amplitudes from which an overall factor containing the three external fermionic currents can be pulled out, we use the identity of Eq.(2) and its Z, χ analogue to isolate all remaining gauge dependences in the form of propagators of the unphysical scalars. In this case there are many gauge dependent terms displaying characteristic structures that depend on the kind and number of scalar propagators they contain, and the momenta carried by these propagators. Clearly if two such terms differ in any of the above three aspects they are linearly independent and the cofactor in front of them must vanish individually. We are thus led to a number of conditions, which enforce the remaining gauge cancellations and give rise to a tower of WI satisfied by the three boson vertices and relating them to the boson self-energies. In particular the final cancellation of the $\Delta_\chi(q, \xi_Z)$ terms leads to

$$q^\mu \widehat{\Gamma}_{\mu\alpha\beta}^{ZW^-W^+} + iM_Z \widehat{\Gamma}_{\alpha\beta}^{\chi W^-W^+} = gc \left[\widehat{\Pi}_{\alpha\beta}^W(p_1) - \widehat{\Pi}_{\alpha\beta}^W(p_2) \right] , \qquad (7)$$

while the cancellation of the $\Delta_\phi(p_1, \xi_W)$ and $\Delta_\phi(p_2, \xi_W)$ requires respectively

$$p_1^\alpha \widehat{\Gamma}_{\mu\alpha\beta}^{VW^-W^+} + iM_W \widehat{\Gamma}_{\mu\beta}^{V\phi^-W^+} = g_V \left[\widehat{\Pi}_{\mu\beta}^W(p_2) - \widehat{\Pi}_{\mu\beta}^V(q) - \frac{g_{V'}}{g_V} \widehat{\Pi}_{\mu\beta}^{VV'}(q) \right] , \qquad (8)$$

$$p_2^\beta \widehat{\Gamma}_{\mu\alpha\beta}^{VW^-W^+} + iM_W \widehat{\Gamma}_{\mu\alpha}^{VW^-\phi^+} = g_V \left[\widehat{\Pi}_{\mu\alpha}^V(q) + \frac{g_{V'}}{g_V} \widehat{\Pi}_{\mu\alpha}^{VV'}(q) - \widehat{\Pi}_{\mu\alpha}^W(p_1) \right] , \qquad (9)$$

where $V = \gamma, Z$, $g_\gamma = gs$, $g_Z = gc$, and $s^2 = 1 - c^2$ is an abbreviation for $\sin^2 \theta_W$. To derive the analogue of Eq.(7) for the $\widehat{\Gamma}_{\mu\alpha\beta}^{\gamma W^-W^+}$ vertex we have to work in an axial gauge, in which case we obtain

$$q^\mu \widehat{\Gamma}_{\mu\alpha\beta}^{\gamma W^-W^+} = gs \left[\widehat{\Pi}_{\alpha\beta}^W(p_1) - \widehat{\Pi}_{\alpha\beta}^W(p_2) \right] . \qquad (10)$$

Finally WI where the g.i. TBV are contracted with two or three momenta can be easily derived, by demanding the cancellation of gauge dependences stemming from the terms with more than one unphysical scalar propagators e.g. $\Delta_\chi(q, \xi_Z) \Delta_\phi(p_1, \xi_W)$ etc. .

All of the above WI for the PT Green's functions have been explicitly proven up to one loop order, for some cases in more than one gauge. We emphasize that they constitute an extension, to higher orders, of the tree level WI stemming from the classical gauge invariance, and in this respect they are QED-like. These same WI also allow a consistent g.i. truncation procedure for the SD equations of the theory. In contrast to the ST identities, they are simple, manifestly g.i. and make no reference to ghost Green's functions . Since they are just the tree level WI, they coincide with the WI satisfied by the background Green's functions of the background field method (BFM) but contrary to the latter they are not restricted to background Green's functions only, they relate g.i. quantities and can be obtained within any gauge fixing procedure.

The derivation of the WI presented above, displays in a very transparent way the mechanism responsible for the gauge cancellations of the S-matrix. The amplitudes of an S-matrix reorganize themselves through pinching; this suffices to cancel all gauge dependences inside loops and makes possible the definition of g.i. Green's functions. The Green's functions thus constructed satisfy their tree level WI, which enforce the elimination of all remaining gauge dependences that were initially present outside of the loops. This implies that, the gauge dependences originating inside the loops cancel independently from the ones residing outside of loops, which points towards an interesting property of the S-matrix already known to hold true in QED. Namely, one could freely choose *two different* ξ-parameters ξ_l and ξ_t, to gauge fix the bosonic propagators that appear inside and outside of the loops respectively. Although this dual choice of gauges seems quite arbitrary and lacks any field theoretical justification at a formal level, the WI guarantee that the S-matrix will still be gauge invariant and independent of both ξ_t and ξ_l.

In closing, we point out that the WI of Eqs.(7,10) can find immediate application, in incorporating, the full one loop W-width effects in the tree level cross sections of processes involving the production or decay of on-shell Ws, in a way consistent with gauge invariance.

ACKNOWLEDGMENTS

I would like to thank J. M. Cornwall, and J. Papavassiliou for useful discussions, and E. Karagiannis for his kind hospitality during my visit in Los Angeles.

REFERENCES

1. J. M. Cornwall, in *Proceedings of the 1981 French-American Seminar on Theoretical Aspects of Quantum Chromodynamics*, Marseille, France, 1981, edited by J. W. Dash (Centre de Physique Théorique, Marseille, 1982).
2. J. M. Cornwall, Phys. Rev. D 26 (1982) 1453.
3. G. Degrassi and A. Sirlin , Phys. Rev. D 46 (1992) 3104
4. J. Papavassiliou and A. Sirlin, Phys. Rev. D 50,5951 (1994).
5. J. Papavassiliou, Phys. Rev. D 51 (1995) 856.
6. N. Jay Watson, CPT-94-P-3133, hep-ph/9412319.
7. J. M. Cornwall and J. Papavassiliou , Phys. Rev. D 40 (1989) 3474.
8. J. Papavassiliou, Phys. Rev. D 41 (1990) 3179.
9. J. Papavassiliou and K. Philippides, Phys. Rev. D 48 (1993) 4255.
10. J. Papavassiliou, Phys. Rev. D 50 (1994) 5958.
11. J. Papavassiliou and K. Philippides, NYU-TH-95/03/03, hep-ph/9503377.

One-Loop Effects of a Heavy Higgs Boson: a Functional Approach[1]

Stefan Dittmaier* and Carsten Grosse-Knetter[†]

*University of Bielefeld, Bielefeld, Germany
[†]University of Montreal, Montreal, Canada

We integrate out the Higgs boson in the electroweak standard model at one loop, assuming that it is very heavy. We construct a low-energy effective Lagrangian, which parametrizes the one-loop effects of the heavy Higgs boson at $\mathcal{O}(M_H^0)$. Instead of applying conventional diagrammatical techniques, we integrate out the Higgs boson directly in the path integral.

INTRODUCTION

Effective Lagrangians are used in order to describe the low-energy effects of heavy particles. An effective Lagrangian contains only light particles, and the heavy particles' effects are parametrized in terms of effective interactions of the light ones.

An effective Lagrangian can be constructed from the underlying theory by integrating out the heavy particles. This can be done in two different ways:

- *The diagrammatical method:* One calculates the relevant Feynman graphs with heavy particles and matches the full theory to the effective one.

- *The functional method:* One integrates out the the heavy fields directly in the path integral.

Here we focus on the functional method which turns out to be much more elegant and simpler than the diagrammatical one.

As a phenomenologically important application we consider the electroweak standard model (SM) provided that $M_H \gg M_W, E$. We integrate out the Higgs boson and determine the formal limit $M_H \to \infty$ of the SM, i.e. all contributions of the Higgs boson to the resulting effective Lagrangian at $\mathcal{O}(M_H^0)$ (which includes $\log M_H$-terms).

In these proceedings we can sketch our method and our results only very briefly. The reader who is interested in a more detailed presentation is referred to the original articles Refs. (1,2).

---

[1]Talk presented by C. Grosse-Knetter

INTEGRATING OUT THE HIGGS FIELD

We briefly describe the basic concepts of our method to integrate out heavy fields in the path integral.

The background-field method

The SM Lagrangian contains terms cubic and quartic in the Higgs field. Thus, the integral over the Higgs field is not of Gaussian type. However, this problem can be circumvented by applying the background-field method (BFM) (3,4), where each field is split into a (classical) background field and a quantum field, such that the functional integral is only performed over the latter. The background fields correspond at the diagrammatical level to tree lines while the quantum fields correspond to lines in loops. Thus, at one loop it is sufficient to consider only the part of the Lagrangian which is quadratic in the quantum fields. Then the heavy quantum field can be integrated out by Gaussian integration.

The Stueckelberg formalism

Another advantage of the use of the BFM is that different gauges may be chosen for the quantum and background fields, respectively (3,4). For our purposes it is useful to choose a generalized R_ξ-gauge (4) for the quantum fields (such that the quantized Lagrangian is still invariant under gauge transformations of the background fields (3,4)) but the unitary gauge for the background fields. This aim can be achieved by applying a generalized (1,2,5) Stueckelberg transformation (6,7) to the (background and quantum) vector fields, which removes the background Goldstone fields from the Lagrangian. After all calculations are done, this transformation is inverted in order to reintroduce the background Goldstone fields and to obtain a manifestly gauge-invariant result.

Diagonalization of the Lagrangian

The one-loop Lagrangian of the SM (i.e. the part quadratic in the quantum fields) contains terms linear in the quantum Higgs field H. After Gaussian integration these would yield terms with inverse operators acting on the quantum fields. However, one can apply appropriate shifts to the quantum fields, such that these linear terms are removed while the Higgs-independent part of the Lagrangian remains unaffected (1,5,8). The resulting Lagrangian can then be written in the symbolic form

$$\mathcal{L}^{1-\text{loop}} = -\frac{1}{2} H \tilde{\Delta}_H(x, \partial_x) H + \mathcal{L}^{1-\text{loop}}\big|_{H=0}, \qquad (1)$$

where the operator $\tilde{\Delta}_H$, which depends on the background fields, also contains contributions from the terms originally linear in the quantum Higgs field H owing to the shifts.

$1/M_H$-expansion

Next the Gaussian integration over the quantum Higgs field can be performed. The resulting functional determinant can be parametrized in terms of an effective Lagrangian (1,9)

$$\mathcal{L}_{\text{eff}} = \frac{i}{2} \int \frac{d^4p}{(2\pi)^4} \log\left(\tilde{\Delta}_H(x, \partial_x + ip)\right). \qquad (2)$$

Then one can perform a Taylor expansion of the expression $\tilde{\Delta}_H(x, \partial_x + ip)$ around $\tilde{\Delta}_H(x, ip)$ (derivative expansion) followed by a Taylor expansion of the logarithm. These expansions yield one-loop vacuum integrals of the type

$$\int d^4p \frac{p_{\mu_1} \cdots p_{\mu_{2k}}}{(p^2 - M_H^2)^l (p^2 - \xi M_i^2)^m}, \qquad M_i = M_W, M_Z. \qquad (3)$$

These are $\mathcal{O}(M_H^0)$ or higher only for $4 + 2(k - l - m) \geq 0$. Thus, in the two above-mentioned Taylor expansions only a finite number of terms contribute to the effective Lagrangian $\mathcal{L}_{\text{eff}}$ at $\mathcal{O}(M_H^0)$.

Elimination of the background Higgs field

After the integration over the quantum Higgs field H, the effective Lagrangian still contains the background Higgs field $\hat{H}$. The latter corresponds to Higgs tree lines, and thus can easily be eliminated by a propagator expansion Diagrammatically this means that the $\hat{H}$ propagator shrinks to a point rendering such (sub-)graphs irreducible, which contain background Higgs lines only. Equivalently, the background Higgs-field can be eliminated by applying the classical equations of motion which are valid at tree level. Before eliminating the field $\hat{H}$, the Higgs sector has to be renormalized by adding the Higgs-dependent part of the counterterm Lagrangian.

THE HEAVY-HIGGS LIMIT OF THE STANDARD MODEL

Proceeding as explained above, we find the formal limit of the SM for $M_H \to \infty$ – i.e. the Lagrangian which contains the non-decoupling ($\mathcal{O}(M_H^0)$) effects of the Higgs boson – at one loop (2):

$$\left. \mathcal{L}_{\text{SM}}^{1-\text{loop}} \right|_{M_H \to \infty} = \mathcal{L}_{\text{GNLSM}}^{1-\text{loop}} + \mathcal{L}_{\text{eff}}. \qquad (4)$$

In eq. 4 $\mathcal{L}_{GNLSM}^{1-loop}$ is the one-loop Lagrangian of the gauged nonlinear σ-model (GNLSM) (10), which is obtained from the SM Lagrangian by simply omitting the Higgs field in the unitary-gauge. $\mathcal{L}_{\text{eff}}$ is the effective Lagrangian generated by integrating out the Higgs field and parametrizes the one-loop effects of the heavy Higgs field. Omitting terms which do not contribute to the S-matrix we find (2)

$$\begin{aligned}\mathcal{L}_{\text{eff}}^{\text{S-matrix}} =& \frac{1}{16\pi^2}\frac{3}{8}\left(\Delta_{M_{\text{H}}} + \frac{5}{6}\right)\frac{g_1^2}{g_2^2}M_W^2\left(\text{tr}\{T\hat{V}_\mu\}\right)^2 \\ &- \frac{1}{16\pi^2}\frac{1}{24}\left(\Delta_{M_{\text{H}}} + \frac{5}{6}\right)g_1 g_2 \hat{B}_{\mu\nu}\,\text{tr}\{T\hat{W}^{\mu\nu}\} \\ &+ \frac{1}{16\pi^2}\frac{1}{48}\left(\Delta_{M_{\text{H}}} + \frac{17}{6}\right)ig_1 \hat{B}_{\mu\nu}\,\text{tr}\{T[\hat{V}^\mu,\hat{V}^\nu]\} \\ &- \frac{1}{16\pi^2}\frac{1}{24}\left(\Delta_{M_{\text{H}}} + \frac{17}{6}\right)ig_2\,\text{tr}\{\hat{W}_{\mu\nu}[\hat{V}^\mu,\hat{V}^\nu]\} \\ &- \frac{1}{16\pi^2}\frac{1}{12}\left(\Delta_{M_{\text{H}}} + \frac{17}{6}\right)\left(\text{tr}\{\hat{V}_\mu\hat{V}_\nu\}\right)^2 \\ &- \frac{1}{16\pi^2}\frac{1}{24}\left(\Delta_{M_{\text{H}}} + \frac{79}{3} - \frac{27\pi}{2\sqrt{3}}\right)\left(\text{tr}\{\hat{V}_\mu\hat{V}^\mu\}\right)^2 + \mathcal{O}(M_{\text{H}}^{-2}) \quad (5)\end{aligned}$$

with

$$\Delta_{M_{\text{H}}} = \Delta - \log\left(\frac{M_{\text{H}}^2}{\mu^2}\right), \qquad \Delta = \frac{2}{4-D} - \gamma_E + \log(4\pi), \tag{6}$$

where D is the space-time dimension in dimensional regularization, γ_E is Euler's constant, and μ is the reference mass.

In eq. 5 we have used the notation of Refs. (2,10,11), which we specify here only for the case of the U-gauge, where the background Goldstone fields are absent, and thus the physical content of the terms in $\mathcal{L}_{\text{eff}}$ is most obvious:

$$\hat{V}^\mu = -\frac{i}{2}\left(g_2\hat{W}_i^\mu\tau_i + g_1\hat{B}^\mu\tau_3\right), \qquad T = \tau_3, \qquad \hat{W}^{\mu\nu} = \frac{1}{2}\hat{W}_i^{\mu\nu}\tau_i, \tag{7}$$

where $\hat{B}^\mu$ and $\hat{W}^\mu$ are the $U(1)$ and $SU(2)$ gauge fields, respectivly, g_1 and g_2 are the corresponding gauge couplings and the τ_i are the Pauli matrices. The hats over the fields indicate that these are background fields.

The result of our functional calculation agrees with the result of the diagrammatical calculation in Ref. (11).

The first two terms in eq. 5 contain vector-boson two-point (and higher) functions, the next two three-point (and higher) functions and the last two four-point functions. Thus, the first two parametrize the effects of the heavy Higgs boson to LEP 1 physics, the next two the effects to LEP 2 physics and the last two those to LHC physics.

$\mathcal{L}_{\text{eff}}$ (eq. 5) does not contain custodial $SU(2)$ violating terms of dimension 4, although there are 7 such terms which are by naive power counting expected to be generated when integrating out the Higgs field (10,11). As shown in Ref. (2), within our functional calculation it is obvious that these terms vanish (while in a diagrammatical calculation (11) their absence seems to be an accidental cancellation).

The effective interaction terms in eq. 5 have logarithmically divergent coefficients. Owing to the renormalizability of the SM, these UV-divergences cancel against the logarithmically divergent one-loop contributions from the GNLSM Lagrangian in eq. 4 (10). Since logarithmic divergences and $\log M_H$-terms in $\mathcal{L}_{\text{eff}}$ always occur in the combination Δ_{M_H} (eq. 6), the $\log M_H$-terms in the SM coincide with the logarithmically divergent terms in the GNLSM, as assumed in Ref. (10). However, in addition eq. 5 contains finite and constant differences between the SM and the GNLSM at one loop.

CONCLUSION

Our purpose with this project is twofold: On the one hand, we integrate out the Higgs boson in the electroweak standard model at one loop. We parametrize the non-decoupling (i.e. $\mathcal{O}(M_H^0)$) effects of this field in terms of a low-energy effective Lagrangian. On the other hand, we have developed a functional method to integrate out heavy fields directly in the path integral. This method can be applied to integrate out any kind of non-decoupling heavy field and also be generalized to a decoupling scenario.

Our method is an alternative to the conventional diagrammatical techniques and turns out to be a huge simplification. While in a diagrammatical calculation various Green functions (i.e. very many Feynman diagrams) have to be calculated and the effective Lagrangian has to be determined indirectly by matching the full theory to the effective one (11), in a functional calculation the effective Lagrangian is generated *directly* by integrating out the heavy fields. The use of the BFM and of the Stueckelberg formalism automatically ensures gauge invariance of the result.

REFERENCES

1. S. Dittmaier and C. Grosse-Knetter, BI-TP 95/01, hep-ph/9501285
2. S. Dittmaier and C. Grosse-Knetter, BI-TP 95/10, in preparation
3. L.F. Abbott, Nucl. Phys. **B185**, 189 (1981) and references therein;
   L.F. Abbott, M.T. Grisaru and R.K. Schaefer, Nucl. Phys. **B229**, 372 (1983)
4. A. Denner, S. Dittmaier and G. Weiglein, Phys. Lett. **B333**, 420 (1994); BI-TP 94/50, hep-ph/9410338, to appear in Nucl. Phys. B
5. O. Cheyette, Nucl. Phys. **B297**, 183 (1988)
6. E.C.G. Stueckelberg, Helv. Phys. Acta **11**, 299 (1938); **30**, 209 (1956);
   T. Kunimasa and T. Goto Prog. Theor. Phys. **37**, 425 (1967)
7. B.W. Lee and J. Zinn-Justin, Phys. Rev. **D5**, 3155 (1972);
   C. Grosse-Knetter and R. Kögerler, Phys. Rev. **D48**, 2865 (1993)
8. J. Gasser and H. Leutwyler, Ann. Phys. (NY) **158**, 142 (1984)
9. L.-H. Chan, Phys. Rev. Lett. **54**, 1222 (1985); **57**, 1199 (1986)
10. A.C. Longhitano, Nucl. Phys. **B188**, 118 (1981)
11. M.J. Herrero, E. Ruiz Morales, Nucl. Phys. **B418**, 413 (1994); **B437**, 319 (1995)

Search For W Boson Pair Production in Dilepton Decay Modes at DØ

Hossein Johari

Department of Physics
Northeastern University, Boston, MA 02115

For the DØ Collaboration

> The results of a search for W boson pair production in $p\bar{p}$ collisions at $\sqrt{s} = 1.8$ TeV with subsequent decay to dilepton (e^+e^-, $e^\pm\mu^\mp$, and $\mu^+\mu^-$) channels are presented. For an integrated luminosity of 14 pb^{-1}, one event is observed with an expected background of 0.56 ± 0.13 events. Limits on the anomalous coupling parameters, assuming $\Delta\kappa_\gamma = \Delta\kappa_Z$ and $\lambda_\gamma = \lambda_Z$, are $-2.6 < \Delta\kappa < 2.8$ and $-2.2 < \lambda < 2.2$ at 95% confidence level.

INTRODUCTION

This report describes a search for $p\bar{p} \to WW + X$ with subsequent decays to ee, $e\mu$, and $\mu\mu$ channels. The data used in this analysis were taken with the DØ detector during 1992-93 Tevatron collider run at Fermilab.

The W pair production includes the $WW\gamma$ and WWZ vertices that can be described with several coupling parameters (1). In this analysis, four coupling parameters κ_V and λ_V, where $V = \gamma, Z$ denotes the coupling to the photon or Z boson, are studied. The SM corresponds to the $\kappa_V = 1$ ($\Delta\kappa_V \equiv \kappa_V - 1 = 0$) and $\lambda_V = 0$. It is assumed that $\Delta\kappa_\gamma = \Delta\kappa_Z$ and $\lambda_\gamma = \lambda_Z$. Unitarity violation in the theory is avoided by defining the coupling parameters as form factors with a scale, Λ, (e.g. $\lambda(\hat{s}) = \lambda/(1 + \hat{s}/\Lambda^2)^2$) (2). The constraint $\Lambda \leq (\frac{6.88}{(\kappa-1)^2 + 2\lambda^2})^{1/4}$ TeV (3) is used to determine the maximum value of Λ allowed by unitarity. Limits on the coupling parameters are set by comparing the measured cross section for W boson pair production to the predicted non-SM values.

DØ DETECTOR

The DØ detector (4) consists of three major components, the tracking system, calorimeter, and muon system. Electrons and photons are detected using a hermetic, compensating, uranium-liquid argon calorimeter system with an energy resolution of $15\%\sqrt{E(\text{GeV})}$. The resolution for the transverse component of missing energy is 1.1 GeV $+ 0.02(\sum E_T)$, where $\sum E_T$ is the scalar sum of traverse energy deposited in the calorimeter. The central and forward drift chambers are used to identify charged tracks for $|\eta| \leq 3.2$. There is no central magnetic field. Muons are identified with three layers of proportional drift tubes, one inside and two outside of the magnetized iron toroids. The muon momentum resolution is $\sigma(1/p) = 0.18(p-2)/p^2 \oplus 0.008$ (p in GeV/c).

PARTICLE IDENTIFICATION

Electrons are identified as an isolated energy cluster in the calorimeter with an associated track in the central tracking detector. It is required that the longitudinal and transverse shower shape to be consistent with test beam results. A criterion on ionization (dE/dx), measured in the central and forward tracking detectors, is imposed to reduce backgrounds from photon conversions and hadronic showers. Electrons are required to be within a fiducial region of $|\eta| \leq 2.5$. Muons are required to be isolated, to have energy deposition in the calorimeter corresponding to at least that of a minimum ionizing particle, to point to the primary interaction vertex, and to have $|\eta| \leq 1.7$. For the $\mu\mu$ channel, a requirement that the muons have timing consistent with the beam crossing is added to remove cosmic ray muons.

EVENT SELECTION

For the ee channel, two electrons are required, each with $E_T \geq 20$ GeV. The $\not{E}_T$ is required to be greater than 20 GeV. The $Z \to ee$ background is reduced by removing events where the dielectron invariant mass is between 77 and 105 GeV/c^2. The background from $Z \to \tau\tau$ as well as ee with the energy of one or both electrons mismeasured is reduced by requiring $20° \leq \Delta\phi(p_T^e, \not{E}_T) \leq 160°$ for the lower energy electron if $\not{E}_T \leq 50$ GeV, where $\Delta\phi(p_T^e, \not{E}_T)$ is the angle between the lower energy electron and $\not{E}_T$. In order to suppress background from $t\bar{t}$ production, a cut on the transverse energy of the recoiling system against WW, $\vec{E}_T^{had} \equiv -(\vec{E}_T^{l1} + \vec{E}_T^{l2} + \vec{\not{E}}_T)$, is introduced. $|\vec{E}_T^{had}|$ is required to be less than 40 GeV in all channels. For WW events, non-zero values of $|\vec{E}_T^{had}|$ are due to gluon radiation and detector resolution. For $t\bar{t}$ events, the most significant contribution is the b-quark jets from the t-quark decays. The efficiency of this selection criterion for SM W boson pair production events is $0.95^{+0.01}_{-0.04}$ and decreases slightly with increasing W boson pair invariant mass. The integrated luminosity in the ee channel is 13.9 ± 1.7 pb^{-1}. One event survives these selection requirements.

For the $e\mu$ channel, a muon with $p_T \geq 15$ GeV/c and an electron with $E_T \geq 20$ GeV are required. $\not{E}_T$ is required to be ≥ 20 GeV. The background from $Z \to \tau\tau$ and $b\bar{b}$ are reduced by requiring that $20° \leq \Delta\phi(p_T^\mu, \not{E}_T) \leq 160°$ if $\not{E}_T \leq 50$ GeV, where $\Delta\phi(p_T^\mu, \not{E}_T)$ is the angle between the muon and $\not{E}_T$. After applying $|\vec{E}_T^{had}| \leq 40$ GeV no events survive in a data sample corresponding to an integrated luminosity of 13.5 ± 1.6 pb^{-1}.

For the $\mu\mu$ channel two muons are required, one with $p_T \geq 20$ GeV/c and another with $p_T \geq 15$ GeV/c. The background from $Z \to \mu\mu$ events is

Background	$e\mu$	ee	$\mu\mu$
$Z \to ee$ or $\mu\mu$	—	0.02 ± 0.01	0.066 ± 0.026
$Z \to \tau\tau$	0.11 ± 0.05	$< 10^{-3}$	$< 10^{-3}$
Drell-Yan dileptons	—	$< 10^{-3}$	$< 10^{-3}$
$W\gamma$	0.04 ± 0.03	0.02 ± 0.01	—
QCD	0.07 ± 0.07	0.15 ± 0.08	$< 10^{-3}$
$t\bar{t}$	0.04 ± 0.02	0.03 ± 0.01	0.009 ± 0.003
Total	0.26 ± 0.10	0.22 ± 0.08	0.075 ± 0.026

TABLE 1. Summary of backgrounds to $WW \to e\mu$, $WW \to ee$ and $WW \to \mu\mu$. The units are expected number of background events in the data sample. The uncertainties include both statistical and systematic contributions.

reduced by requiring that the $\not{E}_T$ projected on the dimuon bisector in the transverse plane be greater than 30 GeV. This selection requirement is less sensitive to the momentum resolution of the muons than a dimuon invariant mass cut. The $Z \to \mu\mu$ background is further reduced by requiring that $\Delta\phi(p_T^\mu, \not{E}_T) \leq 170°$ for the higher p_T muon. $|\vec{E}_T^{had}| \leq 40$ GeV is also applied to reduce the background from the $t\bar{t}$ production. No events survive these selection requirements in a data sample corresponding to an integrated luminosity of 11.8 ± 1.4 pb^{-1}.

EFFICIENCY AND BACKGROUND

The detection efficiency for W boson pair production events is estimated using the Monte Carlo provided by the authors of reference (2) with a parameterized detector simulation (5) and the PYTHIA (6) event generator followed by GEANT detector simulation program (DØGEANT) (7) and the DØ data reconstruction program DØRECO. The two above methods give consistent results for the SM case. The former method is used to determine the $\Delta\kappa$ and λ dependence of the detection efficiency of the W pair production events and is used in the determination of the limits on the W boson pair production cross section.

The overall detection efficiency for SM $WW \to e\mu$ is 0.092 ± 0.010. For the $\mu\mu$ channel it is 0.033 ± 0.003. For the ee channel the overall detection efficiency is 0.094 ± 0.008.

The backgrounds due to physics processes are estimated using the PYTHIA and ISAJET (8) Monte Carlo event generators followed by the DØGEANT detector simulation. The backgrounds due to one or two jets mis-identified as electrons, $b\bar{b}$, and $c\bar{c}$ are estimated using the data. The background estimates are summarized in Table 1.

LIMITS ON THE COUPLING PARAMETERS

The 95% confidence upper limit on the W boson pair production cross section is calculated based on the Poisson-distributed numbers of events convoluted with Gaussian uncertainties for the detection efficiencies, background and luminosity. The 95% confidence limit on the W boson pair production cross section is 91 pb and corresponds to limits on the coupling parameters of $-2.6 < \Delta\kappa < 2.8$ $(\lambda = 0)$ and $-2.2 < \lambda < 2.2$ $(\Delta\kappa = 0)$ (5). Figure 1 shows the contour of the coupling limits with the form factor scale $\Lambda = 900$ GeV.

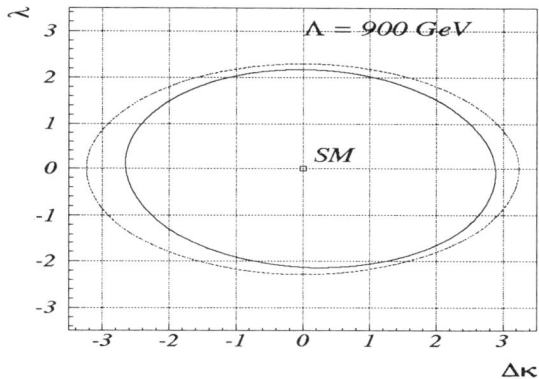

FIG. 1. 95% limits on the anomalous couplings, λ and $\Delta\kappa$, assuming $\Delta\kappa_\gamma = \Delta\kappa_Z$ and $\lambda_\gamma = \lambda_Z$. The dotted contour is the unitarity limit for the form factor scale $\Lambda = 900$ GeV.

SUMMARY

A search for W pair production events with dilepton decays in $p\bar{p}$ collisions at $\sqrt{s} = 1.8$ TeV is presented. In 14 pb^{-1} of data, one event is found with an expected background of 0.56 ± 0.13 events. From the Standard Model, 0.46 ± 0.08 events are expected. The limits on the anomalous coupling parameters are $-2.6 < \Delta\kappa < 2.8$ $(\lambda = 0)$ and $-2.2 < \lambda < 2.2$ $(\Delta\kappa = 0)$, using the form factor scale $\Lambda = 900$ GeV.

REFERENCES

1. K. Hagiwara et al, Nucl. Phys. B **282**, 253, (1987).
2. U. Baur and D. Zeppenfeld, Phys. Lett. B **201**, 383, (1988).

3. K. Hagiwara et al, Phys. Rev. D **41**, 2113, (1990).
4. S. Abachi et al., Nucl. Instr. Meth. A **324**, 53, (1993).
5. H. Johari, Ph.D. thesis, Northeastern University, 1995 (unpublished).
6. T. Sjöstrand, "PYTHIA 5.6 and Jetset 7.3 Physics and Manual", CERN-TH.6488/92, (1992).
7. F. Carminati *et al.*, "GEANT Users Guide", CERN Program Library, December 1991 (unpublished). DØGEANT is the DØ adaptation of GEANT program.
8. F. Paige and S. Protopopescu, BNL Report BNL38034, 1986 (unpublished), release V6.49.

Search For $WW \to l^+l^- + X$ at $\sqrt{s} = 1.8$ TeV

Duncan L. Carlsmith* and Liqun Zhang*

Department of Physics
University of Wisconsin at Madison
Madison, Wisconsin 53705
* Representing CDF collaboration

A search for the production of WW in $\bar{p}p$ collisions at $\sqrt{s} = 1.8$ TeV using the Collider Detector at Fermilab (CDF) is presented. In a 45 pb^{-1} data sample, we find 2 candidates from the semileptonic process $W^+W^- \to l^+l^-\nu\bar{\nu} + X$. Electroweak WW production is expected to produce 1.33±0.48 events while $0.89^{+0.42}_{-0.35}$ events are expected from background processes. We measure the WW cross section $\sigma(\bar{p}p \to WW)$ and set a 95% confidence level upper limit.

INTRODUCTION

Observation of WW production in $\bar{p}p$ collisions provides an opportunity to check the standard model prescription of WWγ and WWZ couplings (1). Anomalous couplings, or new sources such as top or heavy Higgs decays, may result in an enhanced WW production cross section (2–4). We have searched for events corresponding to the decay channel $W^+W^- \to l^+l^-\nu\bar{\nu} + X$ in a 45 pb^{-1} data sample. Although the WW production cross section is only about 10 pb, we can still use this limited data to measure the WW cross section $\sigma(\bar{p}p \to WW)$ and set an upper limit.

In this paper we summary the CDF $W^+W^- \to l\bar{l}$ search during 1992-93. A total of 19.3 pb^{-1} of data was collected.

EVENT SELECTION AND ACCEPTANCE

The CDF is described in detail in Ref. (5). In this analysis, electrons are required to have transverse energy $E_T > 20$ GeV, and muons transverse momentum $P_T > 20$ GeV/c. Other lepton selection cuts are described in detail in Ref. (4).

The total detection efficiency can be decomposed into several factors:

$$\epsilon_{total} = \epsilon_{geom.P_T} \epsilon_{ID} \epsilon_{Isol} \epsilon_{event} \epsilon_{0-jet} \epsilon_{trigger} \qquad (1)$$

The acceptance due to geometrical and transverse momentum cuts, denoted by $\epsilon_{geom.P_T}$, is the fraction of WW events for which both leptons are inside

TABLE 1. Dilepton detection efficiency. The various efficiencies are explained in the text

	$\epsilon_{geom.P_T}$	ϵ_{ID}	ϵ_{Isol}	ϵ_{event}	ϵ_{0-jet}	$\epsilon_{trigger}$	ϵ_{total}
CE-CE	5.1	76.6	97.0	55.0	50.9	99.3	1.05
CE-PE	2.4	54.3	99.0	54.1	52.8	98.4	0.36
CE-MU	10.7	80.1	96.5	72.1	51.2	97.2	2.97
CE-MI	1.5	75.8	95.3	75.4	56.7	91.6	0.42
PE-MU	2.4	56.8	99.0	67.1	55.2	93.8	0.47
PE-MI	0.3	53.8	96.4	73.6	48.7	76.9	0.04
MU-MU	3.8	83.7	97.7	57.4	47.1	94.9	0.79
MU-MI	0.9	79.2	97.4	58.9	49.0	87.2	0.17
Total(%)	27.1	77.0	97.0	65.7	51.1	96.2	6.27

the fiducial region and pass the P_T cut. The branching ratio has been taken into account in it. The detailed description of the lepton identification cut can be found in Ref. (4). This fraction has been determined by using the ISAJET version V6_43 Monte Carlo (6) to generate WW events and the fast detector simulation QFL. The lepton identification efficiency, ϵ_{ID}, includes the combined effect of all lepton identification cuts. The detailed description of the identification efficiencies can be found elsewhere (4). All the leptons are required to pass the calorimeter isolation cut - the fractional transverse energy in a cone of radius 0.4 in the η-ϕ plane, excluding the lepton energy must be less than 0.1.. Here η is the pseudo-rapidity and ϕ is the azimuthal angle.

To discriminate against a $t\bar{t} \to l\bar{l}b\bar{b}$, we remove events with any jet with uncorrected transverse energy exceeding 10 GeV. Gluon radiation accompanying electroweak continuum WW implies an efficiency ϵ_{0-jet} of approximately 50 percent.

To discriminate against backgrounds from $Z \to \tau\tau$, $b\bar{b}$ production, the Drell-Yan process and lepton misidentification, events are required to have a minimal missing transverse energy $\not{E}_T$ exceeding 25 GeV. To further reduce these backgrounds, we reject events where $\not{E}_T$ points along the direction of one of the leptons (4). For events with $\not{E}_T < 50$ GeV, we require $\Delta\phi(\vec{\not{E}}_T,$ lepton) $> 20^0$, where $\Delta\phi$ is the azimuthal angle between the direction of $\not{E}_T$ and the direction of the nearest lepton. Also in $WW \to ee(\mu\mu)$ channel, we remove events with 75 GeV/$c^2 < M_{l\bar{l}} < 105$ GeV/c^2. This cut, together with $\not{E}_T$ cut, will effectively reduce the Drell-Yan background. The efficiency of these three cuts is denoted by ϵ_{event}. The trigger efficiency $\epsilon_{trigger}$ has been determined from the data. In Table 1 we summarize all the results.

The uncertainties in the efficiency calculation come from the lepton identification cuts(6%), lepton isolation cuts(2%), structure functions(2%) and Monte Carlo statistics(3%) (4). The 10% uncertainty of the jet energy scale affects the accuracy of $\not{E}_T$, as well as the jet multiplicity. We find 4% uncertainty by changing the jet energy scale within ±10%.

To check the ISAJET prescription for gluon radiation, we compare the jet

TABLE 2. Various backgrounds to WW search.

	without 0-jet cut	All cuts
$t\bar{t}$	2.91	0.14±0.02
$Z \to \tau\tau$	0.98	0.07±0.02
$b\bar{b}$	0.51	0.16±0.07
Drell-Yan	0.65	$0.00^{+0.21}_{-0.00}$
Fake	1.03	0.52±0.35
Total background	6.08	$0.89^{+0.42}_{-0.35}$

multiplicity from ISAJET to that given by perturbation calculations at parton level for associated production of jets (1, 7). The ratio of events with no jets to events with less than two jets within a rapidity range of $|\eta| < 2.4$ is 0.65 for the ISAJET Monte Carlo and 0.75 according to the perturbation calculation. Since multiple jet production is relatively rare, we can assume that this ratio is approximately equal to ϵ_{0-jet} when estimating the uncertainty. We take the difference (16%) as the systematic uncertainty in ϵ_{0-jet} due to QCD radiation modeling.

BACKGROUND

As we mentioned early, $t\bar{t}, b\bar{b}$, the Drell-Yan process, $Z \to \tau\tau$ and fake leptons can mimic the WW signature. We use Monte Carlo programs, as well as data samples, to study these processes, applying the selection criteria for WW. The results are summarized in Table 2

A Monte Carlo study shows that only 3.9% of top events have no jet with energy exceeding 10 GeV (4). In $t\bar{t} \to b\bar{b} + l\bar{l}$ channel, a total of 0.14±0.02 events are expected in our sample for a top mass $M_{top} = 170$ GeV/c^2.

The background from $Z \to l\bar{l}$ is estimated using $Z \to$ ee events in which the two leptons are replaced with two τ's which are then allowed to decay to leptons. After all cuts, 0.07±0.02 are expected in this data sample.

Heavy flavor backgrounds have been investigated using ISAJET and CLEO Monte Carlo programs. We expect 0.16±0.07 $b\bar{b}$ events from this background.

After Z mass window cut, a total of 0.65 events from the Drell-Yan process are expected before jet multiplicity cut. The Z data is then filtered to find jet multiplicity in this data sample. All of events surviving this cut have at least one jet with $E_T > 10$ GeV. We therefore estimate this background to be $0.00^{+0.21}_{-0.00}$.

Events from QCD multijet or W+jets processes, with at least one misidentified lepton, can mimic the WW lepton signature and are referred to as 'fake dilepton' background. 0.52±0.35 events are expected from this background. The fake rate is dominated by W+single jet events where the single jet is misidentified.

Other backgrounds, such as WZ and ZZ, $Z \to b\bar{b}$, are found negligible.

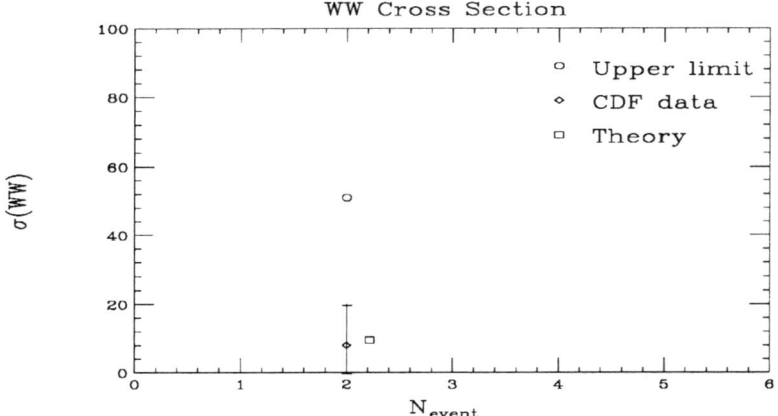

FIG. 1. $\sigma(\bar{p}p \to W^+W^- + X)$ measured in CDF. Experiment measurement and upper limit. See text for detail.

RESULTS

The expected number of WW events can be written as

$$N_{expected} = \sigma(\bar{p}p \to WW + X)\epsilon_{total} BR\mathcal{L} \quad (2)$$

where $\mathcal{L} = 45 \pm 4.5 \; pb^{-1}$ is the integrated luminosity, $BR = \frac{4}{81}$ is four times the branching fraction for $WW \to ee + X$, and $\sigma(\bar{p}p \to WW + X)$ is the production cross section. An order of α_s calculation (8) indicates a continuum standard model production cross section of 9.5 pb with 30% theoretical uncertainty. Therefore we expect a signal of 1.33±0.48 events with a total background of $0.89^{+0.42}_{-0.35}$ events.

Now we can measure the WW cross section $\sigma(\bar{p}p \to WW)$, although the statistics is very limited. Given $N_{event} = 1.11$ after the background substraction, we have

$$\sigma(\bar{p}p \to WW) = 7.9^{+11.4}_{-7.9}(stat) \pm 2.2(sys) pb \quad (3)$$

Where the statistical fluctuation is based on the Poisson distribution for three events. The uncertainties from ϵ_{total} and $\mathcal{L}$ are 18% and 10%, respectively. Figure 1 shows this result.

The upper limit of WW production cross section can be written as

$$\sigma(\bar{p}p \to WW) = \frac{N^{upperlimit}}{\epsilon_{total} BR\mathcal{L}} \quad (4)$$

With 95% confidence level, the background subtracted upper limit is found to be $N^{upperlimit} = 5.5$ if two events are observed. We conclude that the upper limit of continuum WW production cross section is

$$\sigma(\bar{p}p \to WW) = 39.5 pb \quad (5)$$

DISCUSSION

The CDF preliminary result of WW production is consistent with the standard model prediction. Although the statistics is very limited, based on our studies, the anomalous couplings are unlikely to happen.

In the near future, a total integrated luminosity 100 pb^{-1} is anticipated by CDF. The increase in luminosity will permit a further decrease in the cross section limit. Further improvements could be achieved with the inclusion of events with a single jet. Clearly this limit will be improved when CDF collects more data in the future.

REFERENCES

1. V. Barger, T. Han, D. Zeppenfeld and J. Ohnemus, Phys. Rev. D **41**, 2782 (1990), and references therein.
2. K. Hagiwara, J. Woodside, D. Zeppenfeld, Phys.Rev. D **41**, 2113 (1990).
3. B. W. Lee, C. Quigg, and H. Thacker, Phys.Rev. D **16**, 1519 (1977).
4. CDF Collaboration, F. Abe et al., Phys. Rev. D **50**, 2966 (1994), and references therein.
5. CDF Collaboration, F. Abe et al., Nucl. Instrum. Methods A **271**, 387 (1988).
6. F. Paige and S.D. Protopopescu, BNL Report No. BNL, 38034, 1986.
7. U. Baur, E. W. N. Glover, and J. J. van der Bij, Nucl. Phys. **B318**, 106 (1989).
8. J. Ohnemus, Phys.Rev. D **44**, 1403 (1991).
9. Review of Particle Properties, Phys.Rev. D **45**, (1992).

Search for Anomalous $ZZ\gamma$ and $Z\gamma\gamma$ Couplings with DØ

Greg Landsberg
(for the DØ Collaboration[†])

Fermi National Accelerator Laboratory, MS #122, Batavia, IL 60510

A direct test of the Standard Model by searching for anomalous $ZZ\gamma$ and $Z\gamma\gamma$ couplings is presented. We analyze $p\bar{p} \to \ell\ell\gamma + X$, ($\ell = e, \mu$) events at $\sqrt{s} = 1.8$ TeV with the DØ detector at the Fermilab Tevatron Collider. A fit to the transverse energy spectrum of the photon in the signal events, based on the data set corresponding to an integrated luminosity of 13.9 pb^{-1} (13.3 pb^{-1}) for the electron (muon) channel, yields the following 95% CL limits on the anomalous CP-conserving $ZZ\gamma$ couplings: $-1.9 < h^Z_{30} < 1.8$ ($h^Z_{40} = 0$), and $-0.5 < h^Z_{40} < 0.5$ ($h^Z_{30} = 0$), for a form-factor scale $\Lambda = 500$ GeV. Limits on the $Z\gamma\gamma$ couplings and CP-violating couplings are also discussed.

Direct measurement of the $ZV\gamma$ ($V = Z, \gamma$) trilinear gauge boson couplings is possible by studying $Z\gamma$ production in $p\bar{p}$ collisions at the Tevatron ($\sqrt{s} = 1.8$ TeV). The most general Lorentz and gauge invariant $ZV\gamma$ vertex is described by four coupling parameters, h^V_i, ($i = 1...4$) (1). Combinations of the CP-conserving (CP-violating) parameters h^V_3 and h^V_4 (h^V_1 and h^V_2) correspond to the electric (magnetic) dipole and magnetic (electric) quadrupole transition moments of the $ZV\gamma$ vertex. In the Standard Model (SM), all the $ZV\gamma$ couplings vanish at the tree level. Non-zero (i.e. *anomalous*) values of the h^V_i couplings result in an increase of the $Z\gamma$ production cross section and change the kinematic distribution of the final state particles (2). Partial wave unitarity of the general $f\bar{f} \to Z\gamma$ process restricts the $ZV\gamma$ couplings uniquely to their vanishing SM values at asymptotically high energies (3). Therefore, the coupling parameters have to be modified by form-factors $h^V_i = h^V_{i0}/(1 + \hat{s}/\Lambda^2)^n$, where $\hat{s}$ is the square of the invariant mass of the $Z\gamma$ system, Λ is the form-factor scale, and h^V_{i0} are coupling values at the low energy limit ($\hat{s} \approx 0$) (2). Following Ref. (2) we assume $n = 3$ for $h^V_{1,3}$ and $n = 4$ for $h^V_{2,4}$. Such a choice yields the same asymptotic energy behavior for all the couplings. Unlike $W\gamma$ production where the form-factor effects do not play a crucial role, the Λ-dependent effects cannot be ignored in $Z\gamma$ production at Tevatron energies. This is due to the higher power of $\hat{s}$ in the vertex function, a direct consequence of the additional Bose-Einstein symmetry of the $ZV\gamma$ vertices (2).

This paper describes a measurement (4) of the $ZV\gamma$ couplings using $p\bar{p} \to \ell\ell\gamma + X$ ($\ell = e, \mu$) events. Similar measurements were recently performed by

CDF (5) and L3 (6). The data used was observed with the DØ detector during the 1992–1993 run, corresponding to an integrated luminosity of $13.9\pm1.7\text{pb}^{-1}$ ($13.3\pm1.6\text{pb}^{-1}$) for the electron (muon) data. Since the initial state radiation and radiative Z decays can also contribute to the $\ell\ell\gamma$ final state, one would expect a non-zero signal even for the SM (i.e., vanishing) $ZV\gamma$ couplings.

The DØ detector (7) consists of three main systems: (i) uranium-liquid argon sampling calorimeter with coverage in pseudorapidity (η) for $|\eta| \leq 4.4$. and energy resolution of $15\%/\sqrt{E/\text{GeV}}$ for electrons and $50\%/\sqrt{E/\text{GeV}}$ for isolated pions (8); (ii) central and forward drift chambers with a combined coverage of $|\eta| \leq 3.2$; and (iii) muon system providing coverage for $|\eta| \leq 3.3$ and momentum resolution for central ($|\eta| < 1.0$) muons of $\delta(1/p)/(1/p) = 0.18(p-2)/p \oplus 0.008p$ (p in GeV/c).

$Z\gamma$ candidates are selected by searching for events containing two isolated electrons (muons) with high transverse energy E_T (transverse momentum p_T), and an isolated photon. The $ee\gamma$ sample is selected from a trigger requiring two isolated EM clusters, each with $E_T \geq 20$ GeV. An electron cluster is required to be within the fiducial region of the calorimeter ($|\eta| \leq 1.1$ in the central calorimeter (CC), or $1.5 \leq |\eta| \leq 2.5$ in the end calorimeters (EC)). Offline electron identification requirements are: (i) the ratio of the EM energy to the total shower energy must be > 0.9; (ii) the lateral and longitudinal shower shape must be consistent with an electron shower (9); (iii) the isolation variable of the cluster (I) must be < 0.1, where I is defined as $I = [E_{\text{tot}}(0.4) - E_{\text{EM}}(0.2)]/E_{\text{EM}}(0.2)$, $E_{\text{tot}}(0.4)$ is the total shower energy inside a cone defined by $\mathcal{R} = \sqrt{(\Delta\eta)^2 + (\Delta\phi)^2} = 0.4$, and $E_{\text{EM}}(0.2)$ is the EM energy inside a cone of $\mathcal{R} = 0.2$; (iv) at least one of the two electron clusters must have a matching track in the drift chambers; and (v) $E_T > 25$ GeV for both electrons.

The $\mu\mu\gamma$ sample is selected from a trigger requiring an EM cluster with $E_T > 7$ GeV and a muon track with $p_T > 5$ GeV/c. A muon track is required to have $|\eta| \leq 1.0$ and must have: (i) hits in the inner drift-tube layer; (ii) a good overall track fit; (iii) bend view impact parameter < 22 cm; (iv) a matching track in the central drift chambers; and (v) minimum energy deposition of 1 GeV in the calorimeter along the muon path. The muon must be isolated from a nearby jet ($\mathcal{R}_{\mu-\text{jet}} > 0.5$). At least one of the muon tracks is required to traverse a minimum length of magnetized iron ($\int Bdl > 1.9$ Tm); it is also required that $p_T^{\mu_1} > 15$ GeV/c and $p_T^{\mu_2} > 8$ GeV/c.

The requirements for photon identification are common to both electron and muon samples. We require a photon transverse energy $E_T^\gamma > 10$ GeV and the same quality cuts as those on the electron, except that there must be no track pointing toward the calorimeter cluster. Additionally, we require that the separation between a photon and both leptons be $\Delta\mathcal{R}_{\ell\gamma} > 0.7$. This cut suppresses the contribution of the radiative $Z \to \ell\ell\gamma$ decays (2). The above selection criteria yield four $ee\gamma$ and two $\mu\mu\gamma$ candidates. Figure 1 shows the E_T^γ distribution for these events. Three $ee\gamma$ and both $\mu\mu\gamma$ candidates have a three body invariant mass close to that of the Z and low separation between the photon and one of the leptons, consistent with the interpretation of these events as radiative $Z \to \ell\ell \to \ell\ell\gamma$ decays. The remaining candidate in electron

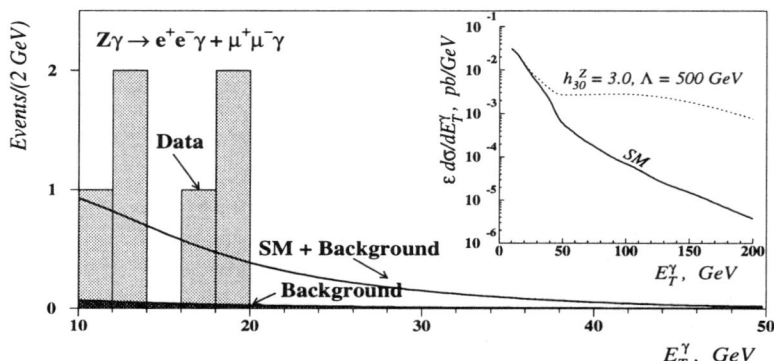

FIG. 1. Photon transverse energy spectrum for $ee\gamma$ and $\mu\mu\gamma$ events. The shadowed bars correspond to data, the hatched curve represents the total background, and the solid line shows the sum of the SM predictions and the background. The insert shows $d\sigma/dE_T^\gamma$ folded with the efficiencies for SM and anomalous ($h_{30}^Z = 3.0$) couplings.

channel has a dielectron mass compatible with that of the Z and a photon well separated from the leptons, an event topology typical for direct $Z\gamma$ production in which a photon is radiated from one of the interacting partons (2).

The estimated background includes contributions from (i) Z + jet(s) production where one of the jets fakes a photon or an electron (the latter case corresponds to the $ee\gamma$ signature if additionally one of the electrons from the $Z \to ee$ decay is not detected in a tracking chamber); (ii) QCD multijet production with jets being misidentified as electrons or photons; (iii) $\tau\tau\gamma$ production followed by decay of each τ to $\ell\bar{\nu}_\ell\nu_\tau$.

We estimate the QCD background from data using the probability, $P(\text{jet} \to e/\gamma)$, for a jet to be misidentified as an electron/photon. This probability is determined by measuring the fraction of non-leading jets in samples of QCD multijet events that pass our electron/photon identification cuts, and takes into account a 0.25 ± 0.25 fraction of direct photon events in the multijet sample (10). We find the probabilities $P(\text{jet} \to e/\gamma)$ to be $\sim 10^{-3}$ in the typical E_T ranges for the electrons and photons of between 10 and 50 GeV. We find the background from Z + jet(s) and QCD multijet events in the electron channel by applying misidentification probabilities to the jet E_T spectrum of the inclusive ee + jet(s) and $e\gamma$ + jet(s) data. The background is 0.43 ± 0.06 events. For the muon channel the QCD background is estimated by applying the misidentification probability to the inclusive $\mu\mu$ + jet(s) spectrum. The estimation of the QCD background from data in the muon case also accounts for cosmic ray background. The combined background from QCD multijet and cosmic ray events in the muon channel is found to be 0.02 ± 0.01 events.

The $\tau\tau\gamma$ background is estimated using the ISAJET Monte Carlo event generator (11) followed by a full simulation of the DØ detector, resulting in 0.004 ± 0.002 events for $ee\gamma$ and 0.03 ± 0.01 events for $\mu\mu\gamma$ channels.

Subtracting the estimated backgrounds from the observed number of events, the signal is $3.57^{+3.15}_{-1.91} \pm 0.06$ for the $ee\gamma$ channel and $1.95^{+2.62}_{-1.29} \pm 0.01$ for the $\mu\mu\gamma$

channel, where the first and dominant uncertainty is due to Poisson statistics, and the second is due to the systematic error of the background estimate.

The acceptance of the DØ detector for the $ee\gamma$ and $\mu\mu\gamma$ final states was studied using the leading order event generator of Baur and Berger (2) followed by a fast detector simulation program which takes into account effects of the electromagnetic and missing transverse energy resolutions, muon momentum resolution, variations in position of the vertex along the beam-axis, and trigger and offline efficiencies. These efficiencies are estimated using $Z \to ee$ data for the electron channel. The muon trigger efficiency is estimated from the $e\mu$ data selected using non-muon triggers. The offline efficiency for the muon channel is calculated based on $e\mu$ and $Z \to \mu\mu$ samples. The trigger efficiency for $ee\gamma$ is 0.98 ± 0.01 while the efficiency of offline dielectron identification is 0.64 ± 0.02 in the CC and 0.56 ± 0.03 in the EC. For the muon channel the trigger efficiency is $0.94^{+0.06}_{-0.09}$, and the offline dimuon identification efficiency is 0.54 ± 0.04. The photon efficiency depends slightly on E_T^γ due to the calorimeter cluster shape algorithm and the isolation cut, and accounts for loss of the photon due to a random track overlap (which results in misidentification of the photon as an electron) and the photon conversion into an e^+e^- pair before the outermost tracking chamber. The average photon efficiency is 0.53 ± 0.05. The geometrical acceptance for the electron (muon) channel is 53% (20%) for the SM case and increases slightly for non-zero anomalous couplings. The overall efficiency for the electron (muon) channel for SM couplings is 0.17 ± 0.02 (0.06 ± 0.01). The MRSD$-'$ (12) set of structure functions is used in the calculations. The uncertainties due to the choice of structure function (6%, as determined by variation of the results for different sets) are included in the systematic error of the Monte Carlo calculation. The effect of higher order QCD corrections are accounted for by multiplying the rates by a constant factor $k = 1.34$ (2).

The observed number of events is compared with the SM expectation using the estimated efficiency and acceptance. We expect the signal in the e and μ channels for SM couplings to be: $S_{ee\gamma}^{SM} = 2.7 \pm 0.3$ (sys) ± 0.3 (lum) and $S_{\mu\mu\gamma}^{SM} = 2.2 \pm 0.4$ (sys) ± 0.3 (lum) events, where the first error is due to the uncertainty in the Monte Carlo modelling, and the second reflects the uncertainty in the integrated luminosity calculation. Our observed signal agrees within errors with the SM prediction for both channels.

To set limits on the anomalous coupling parameters, we fit the observed E_T spectrum of the photon (E_T^γ) with the Monte Carlo predictions plus the estimated background, combining the information in the spectrum shape and the event rate. The fit is performed for the $ee\gamma$ and $\mu\mu\gamma$ samples, using a binned likelihood method (13), including constraints to account for our understanding of luminosity and efficiency uncertainties. Because the contribution of the anomalous couplings is concentrated in the high E_T^γ region, the differential distribution $d\sigma/dE_T^\gamma$ is more sensitive to the anomalous couplings than a total cross section (see insert in Fig. 1 and Ref. (2)). To optimize the sensitivity of the experiment for the low statistics, we assume Poisson statistics for each E_T^γ bin and use the maximum likelihood method to fit the experimental data.

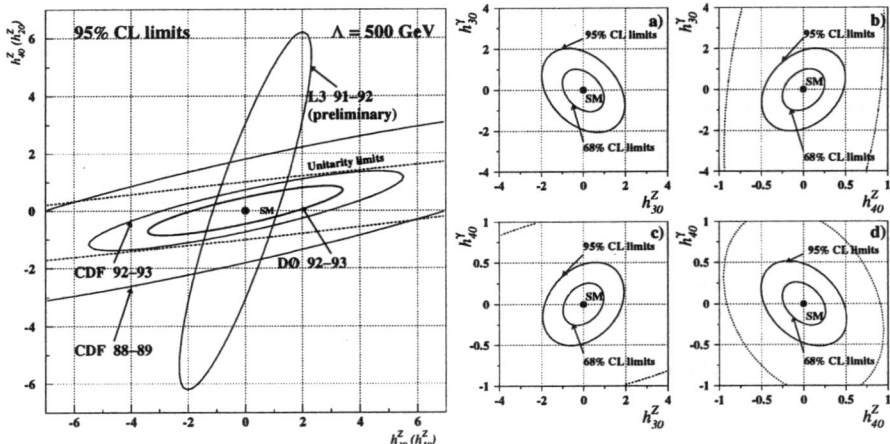

FIG. 2. Comparison of the limits on the correlated CP-conserving (CP-violating) anomalous $ZZ\gamma$ coupling parameters h_{30}^Z and h_{40}^Z (h_{10}^Z and h_{20}^Z). The solid ellipses represent 95% CL exclusion contours for DØ, CDF (5), and L3 (6) experiments. The dashed curve shows limits from partial wave unitarity for $\Lambda = 500$ GeV.

FIG. 3. Limits on the weakly correlated CP-conserving pairs of anomalous $ZV\gamma$ couplings: a) $(h_{30}^Z, h_{30}^\gamma)$, b) $(h_{40}^Z, h_{30}^\gamma)$, c) $(h_{30}^Z, h_{40}^\gamma)$, and d) $(h_{40}^Z, h_{40}^\gamma)$. The solid ellipses represent 68% and 95% CL exclusion contours. Dashed curves show limits from partial wave unitarity for $\Lambda = 500$ GeV.

To exploit the fact that anomalous coupling contributions lead to an excess of events at high transverse energy of the photon, a high-E_T^γ bin, in which we observe no events is explicitly used in the histogram (13). The results were cross-checked using an unbinned likelihood fit which yields similar results.

Figure 1 shows the observed E_T^γ spectrum with the SM prediction plus the estimated background for the $e+\mu$ combined sample. The 95% confidence level (CL) limit contour for anomalous coupling parameters h_{30}^Z and h_{40}^Z (h_{10}^Z and h_{20}^Z) is shown in Fig. 2. Also shown in this figure are the recent CDF (5) and L3 (6) limits. A form-factor scale $\Lambda = 500$ GeV is used for the calculations of the limits and partial wave unitarity constraints. We obtain the following 95% CL limits for the CP-conserving $ZZ\gamma$ and $Z\gamma\gamma$ couplings (in the assumption that all couplings except one are at the SM values, i.e. zeros):

$$-1.9 < h_{30}^Z < 1.8; \quad -0.5 < h_{40}^Z < 0.5$$
$$-1.9 < h_{30}^\gamma < 1.9; \quad -0.5 < h_{40}^\gamma < 0.5$$

The correlated limits for pairs of couplings (h_{30}^V, h_{40}^V) are less stringent due to the strong interference between these couplings:

$$-3.3 < h_{30}^Z < 3.3; \quad -0.9 < h_{40}^Z < 0.9$$
$$-3.5 < h_{30}^\gamma < 3.5; \quad -0.9 < h_{40}^\gamma < 0.9$$

Limits on the CP-violating $ZV\gamma$ couplings are numerically the same as those for the CP-conserving couplings. The limits on the h_{20}^Z, h_{40}^Z, and h_{i0}^γ couplings are currently the most stringent.

Global limits on the anomalous couplings (i.e., limits independent of the values of other couplings) are close to the correlated limits for (h_{30}^V, h_{40}^V) and (h_{10}^V, h_{20}^V) pairs, since other possible combinations of couplings interfere with each other only at the level of 10%. This is illustrated in Fig. 3, which shows the limits for pairs of couplings of the same CP-parity (couplings with different CP-parity do not interfere with each other).

We also study the form-factor scale dependence of the results. The chosen value of the scale $\Lambda = 500$ GeV is close to the sensitivity limit of this experiment for the $h_{20,40}^V$ couplings: for larger values of the scale partial wave unitarity is violated for certain values of anomalous couplings allowed at 95% CL by this measurement.

The luminosity expected in the current Tevatron run may enable us to improve the limits on the anomalous couplings by a factor of ~ 2. Another way to improve the sensitivity toward the couplings is to include the $\nu\bar{\nu}$ decay channel of the Z in the above analysis. This study is in progress.

I would like to thank my DØ colleagues for a lot of help and efforts they have put into this analysis. I am grateful to U. Baur for providing us with the $Z\gamma$ Monte Carlo program and for many helpful discussions.

REFERENCES

† See, e.g. (7) for the list of the DØ Collaboration.

1. K. Hagiwara et al., Nucl. Phys. **B282**, 253 (1987).
2. U. Baur and E.L. Berger, Phys. Rev. **D47**, 4889 (1993).
3. K. Hagiwara and D. Zeppenfeld, Nucl. Phys. **B274**, 1 (1986); U. Baur and D. Zeppenfeld, Phys. Lett. **201B**, 383 (1988).
4. S. Abachi et al. (DØ Collaboration), preprint Fermilab–Pub–95/042–E, Phys. Rev. Lett. (in print).
5. F. Abe et al. (CDF Collaboration), preprint Fermilab–Pub–94/304–E, Phys. Rev. Lett. (in print); preprint Fermilab–Pub–94/244–E, Phys. Rev. **D** (in print).
6. M. Acciarri et al. (L3 Collaboration), preprint CERN–PPE/94–216, Phys. Lett. **B** (in print).
7. S. Abachi et al. (DØ Collaboration), Nucl. Instrum. Methods **A338**, 185 (1994).
8. S. Abachi et al. (DØ Collaboration), Nucl. Instrum. Methods **A324**, 53 (1993); H. Aihara et al. (DØ Collaboration), Nucl. Instrum. Methods **A325**, 393 (1993).
9. M. Narain (DØ Collaboration), "Proceedings of the APS Division of Particles and Fields Conference", Fermilab (1992), eds. R. Raja and J. Yoh; R. Engelmann et al., Nucl. Instrum. Methods **A216**, 45 (1983).
10. S. Fahey (DØ Collaboration), to appear in the "Proceedings of the APS Division of Particles and Fields Conference", Albuquerque, (1994).
11. F. Paige and S. Protopopescu, BNL Report no. BNL–38034, 1986 (unpublished), release V6.49.
12. A.D. Martin, R.G. Roberts and W.J. Stirling, Phys. Lett. **306B** 145 (1993); Erratum, Phys. Lett. **309B** 492 (1993).
13. G. Landsberg, in Proc. Workshop on Physics at Current Accelerators and the Supercollider, ANL–HEP–CP–93–92, 303 (1992); Ph.D. Dissertation, SUNY at Stony Brook (1994), unpublished.

CDF Results on $Z\gamma$ Production

Robert G. Wagner,
for the CDF Collaboration

*High Energy Physics Division
Argonne National Laboratory*[1]
Argonne, Illinois 60439

We discuss final results on $Z\gamma$ production observed by the Collider Detector at Fermilab (CDF) from the Tevatron Collider Run 1a and preliminary results from 36pb^{-1} of integrated luminosity from Run 1b. Both the Run 1a and 1b results are consistent with expectations from the Standard Model.

INTRODUCTION

Since the Z is not coupled to the photon within the framework of the Standard Model, production of photons in association with Zs beyond that predicted by initial state radiation and radiative Z decay would indicate a composite Z and/or additional bosons beyond those of the Standard Model. A summary of $Z\gamma$ production at the Tevatron Collider can be found in the plenary talk at this conference by Aihara (1). In this paper, we review the analysis of $Z\gamma$ production by the CDF experiment at the Tevatron, present the final limits on $Z\gamma$ couplings and transition multipole moments from Run 1a data, show kinematic distributions from the first 36 pb^{-1} of data from Run 1b, and finally discuss two unusual $e^+e^-\gamma$ events observed during the Run 1b data taking period.

EXPERIMENTAL ANALYSIS OF $Z\gamma$ CANDIDATES

From samples of inclusive electrons (pseudorapidity,$| \eta | < 1.0$) and muons ($| \eta | < 0.6$), Z candidates were selected by requiring an isolated charged lepton with $E_T > 20$ GeV. The electron (identified by a cluster in the central EM calorimeter (CEM)) or muon (identified by a "stub" track in the central muon drift chambers (CMU)) was required to have a matching track reconstructed in the central tracking chamber (CTC). Because Zs are readily identified by the presence of two high-p_T, opposite signed charged leptons, less stringent requirements were imposed on the second lepton. For the electron channel, the second lepton was required to have $E_T > 20$ GeV if in the CEM, $E_T > 15$ GeV if in the plug EM calorimeter, and $E_T > 10$ GeV if in the forward

---

[1] Work supported by the Department of Energy under Contract W-31-109-Eng-38

EM calorimeter. For electrons in the CEM, additional identification of the EM cluster as an electron was provided by shower shape measurements and track/cluster matching using the shower max detector (CES) embedded in the central EM calorimeter. Finally, a Z mass window selection was imposed: 70 GeV/c^2 $\leq M(e^+e^-) \leq$ 110 GeV/c^2. In the muon channel, the second lepton was required to have $p_T >$ 20 GeV/c, to have a calorimeter energy deposit consistent with a minimum ionizing particle, and to lie within the $|\eta|<$ 1.2 region of the CTC. The Z mass window selection imposed on the muon pair was 65 GeV/c$^2 \leq M(\mu^+\mu^-) \leq$ 115 GeV/c^2.

Photon candidates were identified in the Z sample by requiring an isolated CEM energy cluster with $E_T >$ 7 GeV separated from the closest charged lepton by $\Delta R_{\ell\gamma} = \sqrt{\Delta\eta^2 + \Delta\phi^2} >$ 0.7. Because the Z identification is quite clean, the *crucial* aspect of the experiment is to reduce and understand the background to the photon signal that arises mainly from neutral QCD jets faking a photon signal. To this end, the photon coverage was restricted to the CEM detector region where there exist several well-understood tools for photon identification:

- The hadronic calorimeter to electromagnetic calorimeter energy ratio was required to be consistent with a shower arising from a purely electromagnetic origin.

- The EM cluster was required to be isolated in energy in the CEM.

- No tracks reconstructed in the CTC were allowed to point to the CEM photon cluster; and the cluster was required to be isolated in tracking by demanding the transverse momentum sum of tracks within a $\Delta R = 0.4$ cone be less than 2 GeV/c.

- The shower profile in the CES was used to demand a CES/CEM profile consistency, a CES signal indicative of a single photon shower, and the absence of additional CES clusters indicating the possible presence of multiple photons from π°s and/or ηs.

An independent sample of inclusive jet events recorded by an $E_T >$ 16 GeV central photon trigger was used to estimate the number of fake photons remaining in the final $Z\gamma$ sample.

Beginning in Run 1a with 133,805 inclusive electron events and 83,051 inclusive muon events, the Z selection criteria resulted in 1237 electron channel Zs and 507 muon channel Zs. Imposition of the photon selections gave a final sample of 4 $Z\gamma$ candidates in each of the two channels. The background estimated from the sample of inclusive jet events passing the central photon trigger was 0.4 ± 0.1(stat) ± 0.2(syst) events in the electron channel and 0.10 ± 0.03(stat) ± 0.04(syst) events in the muon channel; giving 0.5 ± 0.2 events for the combined sample.

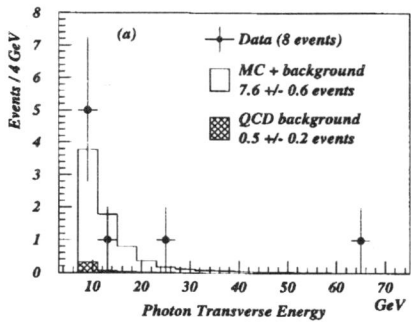 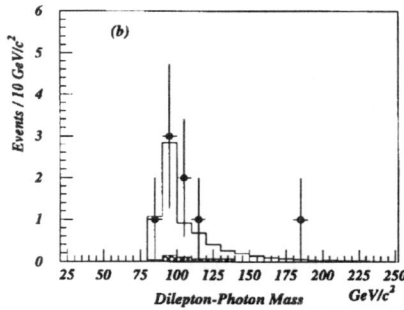

FIG. 1. Final kinematic distributions for $Z\gamma$ events from the Tevatron Collider Run 1a. a) The photon transverse energy distribution for the electron and muon channels combined. b) The dilepton-photon mass distribution.

RESULTS FROM RUN 1A

In figure 1, we show the photon transverse energy and the dilepton-photon mass distributions. Also shown is the expectation for the Standard Model (derived from a Monte Carlo plus parametrized detector simulation) plus background. The Standard Model reproduces the data adequately, but we note the presence of one event in the muon channel with a photon $E_T \sim 64$ GeV and a three body mass of ~ 188 GeV/c^2. The combined electron plus muon total of 8 events with a background estimate of 0.5 ± 0.2 events implies $[\sigma \cdot B(Z+\gamma)]_{\rm exp} = 5.1 \pm 1.9({\rm stat}) \pm 0.3({\rm syst})$ pb. This is to be compared with the Standard Model prediction of $[\sigma \cdot B(Z+\gamma)]_{\rm SM} = 5.2 \pm 0.6({\rm stat} \oplus {\rm syst})$ pb.

Anomalous $Z\gamma$ couplings are assumed to be regulated by generalized dipole form factors that are functions of the initial $q\bar{q}$ center of momentum energy, $\sqrt{\hat{s}}$, and the energy scale, Λ_Z, at which the new interactions become manifest (2,3). The four possible couplings for each of two vertices ($ZZ\gamma$ and $Z\gamma\gamma$) are denoted by $h_{i0}^{V=Z,\gamma}$ which represent the low energy, ($\sqrt{\hat{s}} \to 0$) limit for the couplings. Since the experimental limits determined by CDF on the couplings reach the unitarity limit for $\Lambda_Z \simeq 500$ GeV, the scale parameter has been fixed at this value in determining limits on couplings. Limits were obtained in pairwise fashion setting all other couplings to zero. A log-likelihood fit was performed to the photon E_T spectrum parametrized by the couplings. Details of the fitting procedure are found in refs. (3,4). Limits for the CP conserving (violating) combination h_{30}^Z, h_{40}^Z (h_{10}^Z, h_{20}^Z) are shown in figure 2a. The couplings are related to the transition electromagnetic multipole moments (see ref. (3)). Figure 2b shows the limit contour for the magnetic quadrupole (q_Z^m) versus electric dipole (δ_Z^*). Using instead the CP violating couplings, h_{10}^Z, h_{20}^Z, results in an identical contour for the electric quadrupole (q_Z^e) versus magnetic dipole (g_Z^*) moments.

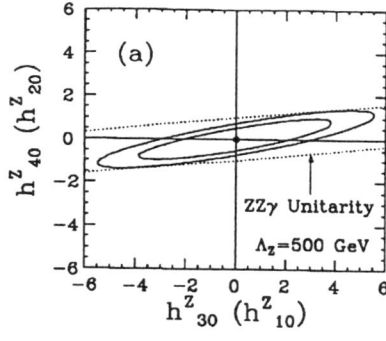 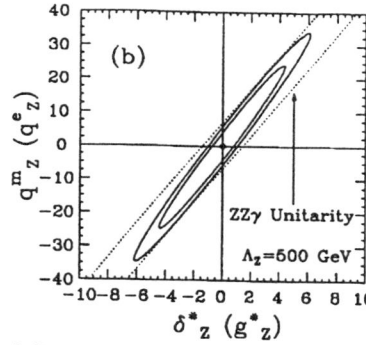

FIG. 2. Final 68% and 95% confidence level limit contours from Tevatron Collider Run 1a. a) The limit contours for CP conserving (violating) anomalous $ZZ\gamma$ couplings. Essentially identical limits are obtained for the $Z\gamma\gamma$ coupling. b) Limit contours for transition electromagnetic multipole moments. Symbols are explained in the text.

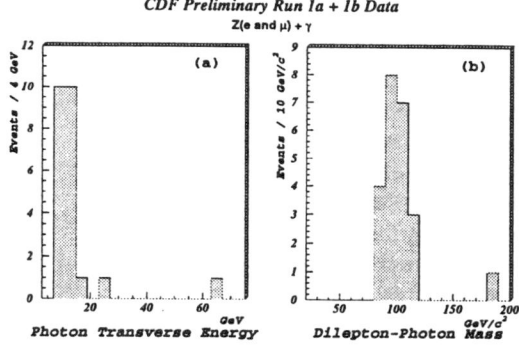

FIG. 3. Preliminary kinematic distributions for $Z\gamma$ events from the combined Tevatron Collider Run 1a and 1b data. a) The photon transverse energy. b) The dilepton-photon mass.

PRELIMINARY RESULTS FROM RUN 1B

We have to date acquired for Run 1b an integrated luminosity of 36pb^{-1}. The combined Run 1a and 1b photon transverse energy and dilepton-photon mass distributions are shown in figure 3. As we have not yet obtained acceptances and efficiencies for the run 1b data, we have not included any Standard Model expectation plots. However, comparison of figure 3 with figure 1 leads to the tentative conclusion that our consistency with Standard Model expectations continues in the larger data sample.

We have, however, recently recorded two unusual events in Run 1b. The first event is very clean and passes all $Z\gamma$ selection requirements *except* the Z mass window requirement. It has $M(e^+e^-) \sim 255\,\text{GeV}/c^2$ and $M(e^+e^-\gamma) \sim$

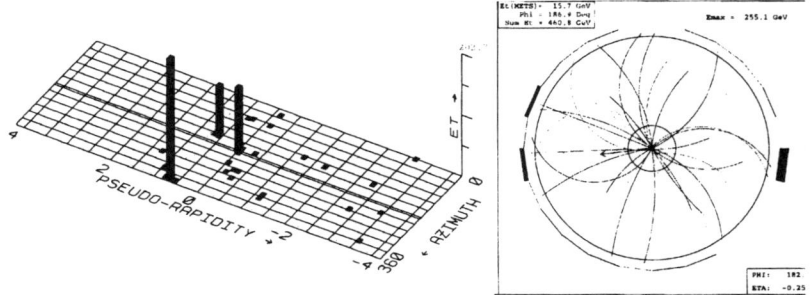

FIG. 4. Transverse energy "lego" and tracking chamber plots for an $e^+e^-\gamma$ event with a very high E_T photon. The photon is the highest energy cluster near azimuthal angle 360°. The e^+e^- pair reconstruct to within 5 GeV of the Z mass.

274 GeV/c^2. The second event shown in figure 4 was recorded after processing had been completed for the preliminary data set shown in figure 3. It is a $Z\gamma$ candidate with $M(e^+e^-) \sim 86$ GeV/c^2, $E_T^\gamma \sim 195$ GeV, and $M(e^+e^-\gamma) \sim 417$ GeV/c^2. The probability of producing an event with a three body mass of at least this value was estimated using a $Z\gamma$ Standard Model Monte Carlo including Drell-Yan contributions and is found to be at the level of about 0.05%.

CONCLUSIONS

We have discussed the analysis of $Z\gamma$ production as observed by the CDF experiment at the Tevatron Collider during Runs 1a and 1b. Final results from Run 1a indicate good agreement with the Standard Model. Results analyzed to date from Run 1b tentatively confirm this agreement for a larger integrated luminosity. We have observed at least two unusual $e^+e^-\gamma$ events recently in Run 1b. No conclusions can be drawn at this time about the events. The 100pb^{-1} data sample expected to be obtained from the whole of Run 1b should either provide additional events of this type or confirm the events as low probability fluctuations of the Standard Model.

REFERENCES

1. See contribution of H. Aihara to these proceedings.
2. U. Baur and E. Berger, Phys. Rev. D **47**, 4889 (1993).
3. F. Abe et al., Phys. Rev. Lett. 74, 1941 (1995).
4. F. Abe et al., Fermilab Report No. FERMILAB-Pub-94/244-E, submitted to Phys. Rev. D (1994).

Results on $W\gamma$ Production at CDF

Dirk Neuberger

University of California, Los Angeles [1]
Los Angeles, California 90024

CDF has collected $\approx 20\ pb^{-1}$ data in $p\bar{p}$ collisions at $\sqrt{s} = 1.8\ TeV$ during the Tevatron Run 1a (1992 - 1993). We discuss final results on $W\gamma$ production from Run 1a and present preliminary results based on $\approx 35\ pb^{-1}$ data from the ongoing Run 1b. No deviation from Standard Model predictions were found in either data set. In Run 1a we set limits on CP-conserving anomalous $W\gamma$ couplings, which are numerically identical with the CP-violating couplings: $-2.3 < \Delta\kappa < 2.2$ for $\lambda = 0$ and $-0.7 < \lambda < 0.7$ for $\Delta\kappa = 0$ (all at 95% C.L.).

INTRODUCTION

In the framework of the Standard Model the weak bosons W and Z are assumed to be fundamental particles. Many theories, however, view these bosons as composite rather than pointlike entities (Ref. (1)). Such models allow non-Standard Model couplings among electroweak bosons, which can be studied in $p\bar{p}$ collisions at $\sqrt{s} = 1.8\ TeV$ at the Fermilab Tevatron Collider.

In the case of $W\gamma$ production, limits on anomalous couplings between a photon and a W boson can be set by comparing the measured excess of events with photons at high transverse energy to Standard Model predictions (Ref. (2)). Compositeness of W bosons would reveal itself in anomalous electromagnetic moments, which are proportional to the couplings:

$$\begin{aligned}
\mu_W &= \tfrac{e}{2M_W}(2 + \Delta\kappa + \lambda) &= \tfrac{e}{2M_W} g_W & \quad \text{Magnetic Dipole Moment} \\
Q_W^e &= -\tfrac{e}{M_W^2}(1 + \Delta\kappa - \lambda) &= -\tfrac{e}{M_W^2} q_W^e & \quad \text{Electric Quadrupole Moment} \\
d_W &= \tfrac{e}{2M_W}(\tilde{\kappa} + \tilde{\lambda}) &= \tfrac{e}{2M_W}\delta_W & \quad \text{Electric Dipole Moment} \\
Q_W^m &= -\tfrac{e}{M_W^2}(\tilde{\kappa} - \tilde{\lambda}) &= -\tfrac{e}{M_W^2} q_W^m & \quad \text{Magnetic Quadrupole Moment}
\end{aligned}$$

The coupling constants $\Delta\kappa, \lambda, \tilde{\kappa}$ and $\tilde{\lambda}$ vanish identically in the Standard Model at tree level. The observation of non-zero values would indicate compositeness of the W boson or yet unknown interactions of substructures.

In the following, we discuss final results on $W\gamma$ production from $\approx 20\ pb^{-1}$ of electron and muon data, which were taken during the Tevatron Run 1a (1992 - 1993). In addition, we report on preliminary results based on $\approx 35\ pb^{-1}$ electron and muon data from the ongoing Tevatron Run 1b, which started late 1993.

---

[1] Supported by the Department of Energy under Contract DE-FG03-91ER-40662.

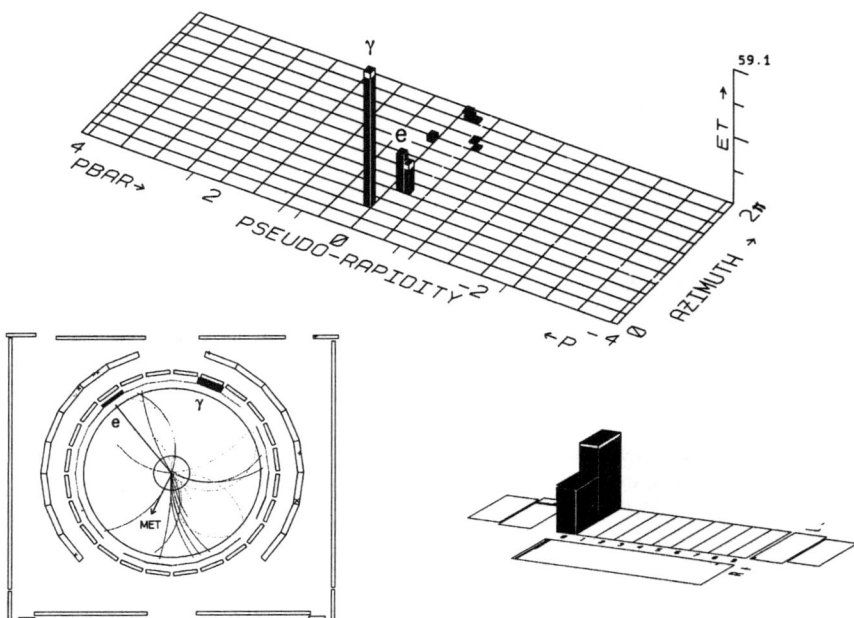

FIG. 1. Event display of a $W\gamma$ event with an energetic photon of $E_T = 59\,GeV$. a.) "Lego" plot showing the energy deposition of the W electron and the photon in the central calorimeter. b.) View of the central tracking chamber showing the track of the W electron, the trackless photon response and the direction of the missing E_T vector, c.) central strip chamber response of the photon.

SEARCH FOR $W\gamma$ CANDIDATES

The $W\gamma$ data sample used for the analysis was taken from a sample of W bosons where the decay electron (muon) was found in the central region of the detector. A detailed description of the CDF detector is given in Ref. (4).

Using Run 1a+b data, we found $\approx 39,000$ W events in the electron channel by requiring a central ($|\eta| \leq 1.05$) and isolated electron (positron) with transverse energy $E_T \geq 20\,GeV$. In the muon channel, a total of $\approx 17,000$ W events were selected by requiring an isolated track with $P_T \geq 20$ GeV/c matched to a track stub in the central muon chambers ($|\eta| \leq 0.6$). In both channels, the missing transverse energy had to be greater than $20\,GeV$.

The samples were searched for isolated central photons ($|\eta| < 1.1$) with a transverse energy of $E_T > 7\,GeV$. A more detailed discussion of the photon selection is presented in these proceedings, Ref. (3) (see also Ref. (2)). By applying all photon selection criteria, we defined a final electron (muon) sample of 18 (7) and 43 (19) $W\gamma$ candidates in Run 1a and 1b, respectively. For illustration we show in Fig. 1 the display of the $W\gamma$ event with the highest

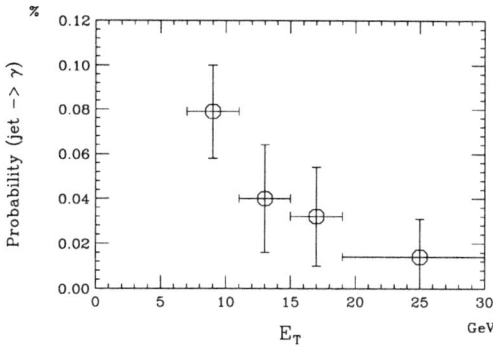

FIG. 2. Probability that a jet is mistakenly identified as a photon.

photon E_T found.

The main background to these events is due to W + jet production. A hadronic jet may fragment into a single, isolated meson that decays to multiple photons. Minor background contributions stem from $Z\gamma$ events where only one of the Z decay leptons was identified, and from $(W \to \nu\tau) + \gamma$ production. The $Z\gamma$ contributions were suppressed by vetoing events with a second isolated charged particle above $P_T = 10$ GeV/c that, together with the detected lepton, forms an invariant mass consistent with the Z mass. Using Monte Carlo simulations (Ref. (5)), we estimated the non-QCD contributions for Run 1a to be 2.2 ± 0.3.

The jet fragmentation background was estimated by measuring the probability for a jet to be misidentified as an isolated photon in a large inclusive jet ('QCD') sample. Since the QCD sample also contains single photons from Compton-type processes that do not come from jet fragmentation, we statistically removed such photons using a shower shape analysis (Ref. (6)). Fig. 2 shows the resulting E_T dependent probability for a jet to fake a photon. Folded with the jet distribution from the electron and muon $W\gamma$ sample, we estimated the overall jet fragmentation background in Run 1a to $6.5 \pm 1.2_{stat} \pm 1.6_{sys}$ events.

Using identical photon selection criteria, the background for Run 1b is expected to approximately scale with luminosity, resulting in ≈ 13 background events out of 62 $W\gamma$ candidates. More detailed studies are presently under way.

RESULTS FROM THE TEVATRON RUNS 1A AND 1B

In Fig. 3a. we compare, for Run 1a, the measured photon transverse energy and transverse cluster mass (Ref. (7)) to the Standard Model Monte Carlo prediction (Ref. (5)) including the overall background estimate. We find good agreement. In Fig. 3b. we show the same quantities for the

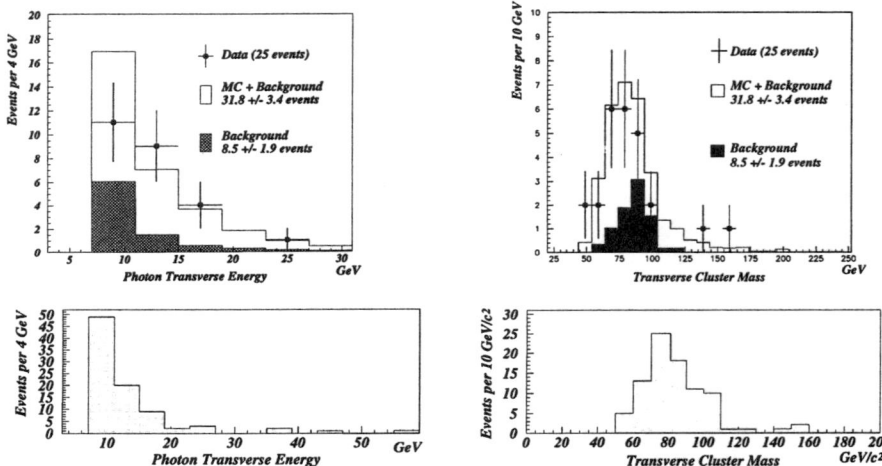

FIG. 3. a.) Measured photon E_T and transverse cluster mass spectrum compared to Standard Model Monte Carlo predictions plus background estimate for Run 1a. b.) Measured photon E_T and transverse cluster mass distributions for Run 1b.

combined Run 1a+b data set. These distributions also agree well with a luminosity-scaled Monte Carlo plus background prediction. No events with unusually high tranverse photon energy or high transverse cluster mass were found. In fact, the Monte Carlo predicts about 20% (9%) of all events to have transverse cluster masses, M_{TC}, higher than $100\,GeV/c^2$ ($120\,GeV/c^2$). In the combined data set we find 15 (4) events, i.e. 24% (6.5%), above $M_{TC} = 100\ GeV/c^2$ ($M_{TC} = 120\ GeV/c^2$), in good agreement with the Standard Model prediction.

Radiative events, where the photon is radiated off a quark or charged W decay lepton, cannot be distinguished from $W\gamma$ production involving a W-photon vertex (Ref. (8)). Because radiative contributions are negligible above transverse cluster masses $M_{TC} \approx 100\,GeV/c^2$, our measurement is first - statistically significant - evidence of the presence of $WW\gamma$ gauge boson couplings.

To extract limits on anomalous couplings we performed a log-likelihood fit to the photon E_T distribution from Run 1a (Fig. 3a.). Fig. 4a. shows the confidence level contours in $\Delta\kappa$ versus λ, with the CP-violating constants $\tilde\kappa$ and $\tilde\lambda$ set to zero. We obtain direct limits on CP-conserving $WW\gamma$ anomalous couplings of $-2.3 < \Delta\kappa < 2.2$ for $\lambda = 0$ and $-0.7 < \lambda < 0.7$ for $\Delta\kappa = 0$ at 95% C.L. These limits are a factor of three more stringent than those from previous measurements. As discussed in the introduction, these results translate into limits on the electromagnetic moments (Fig. 4b.). At 95% C.L. we obtain $0.8 < g_W < 3.1$ for $q_W^e = 1$ and $-0.6 < q_W^e < 2.7$ for $g_W = 2$. The measured limits for the CP-violating couplings $\tilde\kappa$ and $\tilde\lambda$ are numerically identical with

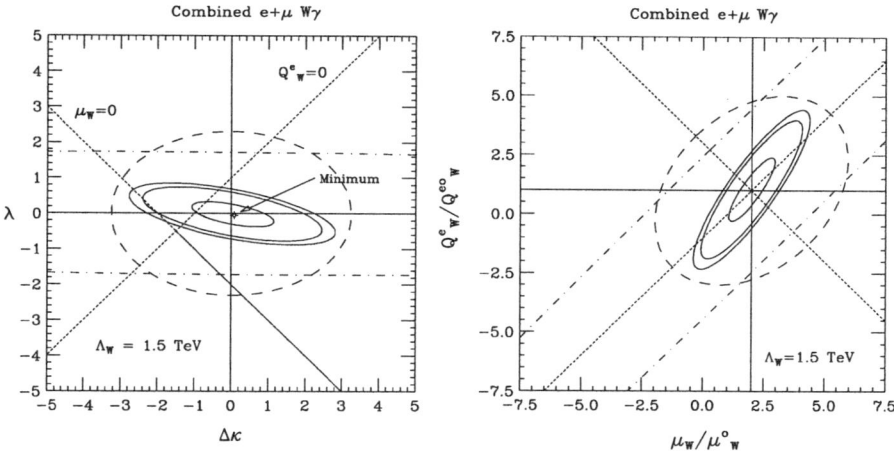

FIG. 4. Final limit contours at 68%, 90% and 95% C.L. from Run 1a for a.) the CP- conserving couplings, b.) the magnetic dipole and electric quadrupole moments. The dashed (dash-dotted) curves are the unitarity limits derived for WW ($W\gamma$) couplings.

the above limits on CP-conserving couplings.

Because the $WW\gamma$ couplings are restricted by unitarity to their Standard Model values at high energies, they have to be described by momentum dependent form factors. Beyond an energy scale Λ_W at which new physics becomes important, the form factors converge rapidly to zero. In our analysis, we chose a form factor scale of $1.5\,TeV$. However, the experimental results are insensitive to the choice of a scale $\Lambda_W > 0.3\,TeV$.

Further improved limits based on the high statistics data samples from the ongoing Run 1b will be available in the near future.

CONCLUSIONS

We have summarized results of the measurement of $W\gamma$ couplings based on the Tevatron Collider Run 1a data sample. Even when almost tripling the amount of data by adding the $W\gamma$ events from a preliminary analysis of the ongoing Run 1b, we find no deviation from the Standard Model that might indicate compositeness of the W boson. With expected $100\,pb^{-1}$ data in Run 1b, we will be able to quantify this statement further and significantly increase our sensitivity to anomalous couplings.

REFERENCES

1. F.M. Renard, Nucl. Phys. B **196**, 93 (1982); R. Barbieri, H. Harari and M. Leurer, Phys. Lett. B **141**, 455 (1985); J.P. Eboli, A.V. Olinto, Phys. Rev. D **38**, 3461 (1988).
2. F. Abe *et al.*, Phys. Rev. Lett. 74, 1936 (1995).
3. See contributions of R. G. Wagner to these proceedings.
4. F. Abe *et al.*, Nucl. Instrum. Methods A **271**, 387 (1988).
5. We used the Baur Monte Carlo plus a parameterized CDF detector simulation.
6. F. Abe *et al.*, Phys. Rev. Lett. 71, 679 (1993).
7. The transverse cluster mass of the $W\gamma$ system is defined as

$$\left\{\left[\left(M_{\ell\gamma}^2 + |\vec{P}_T^\gamma + \vec{P}_T^\ell|^2\right)^{\frac{1}{2}} + |\vec{P}_T^{\nu_\ell}|\right]^2 - |\vec{P}_T^\gamma + \vec{P}_T^\ell + \vec{P}_T^{\nu_\ell}|^2\right\}^{\frac{1}{2}},$$

where $M_{\ell\gamma}$ is the invariant mass of the lepton-photon system and c = 1; see also J. Cortes, K. Hagiwara, and F. Herzog, Nucl. Phys. B**278**, 27 (1986).
8. U. Baur and E. L. Berger, Phys. Rev. D **41**, 1476 (1990).

Towards Probing the $WW\gamma$ Vertex at HERA

André Schöning

Deutsches Elektronen-Synchrotron DESY
Notkestr. 85, D-22603 Hamburg
H1-Collaboration

The processes $ep \to eWX$ (W production) and $ep \to \nu\gamma X$ (Radiative Charged Current) which provide a test of possible anomalous $WW\gamma$ couplings at HERA processes are discussed. For each process a selection of candidate events based on a simulation is described which is sensitive to the triple gauge boson vertex. These selections are applied to 1994 data recorded by the H1 detector. One possible candidate for W production and three radiative CC events are found in agreement with the Standard Model.

INTRODUCTION

In 1994 the H1 experiment at the electron–proton collider HERA collected an integrated luminosity of about $4\,pb^{-1}$. HERA was operated with $27.5\,GeV$ positrons (electrons) and $820\,GeV$ protons. The large center of mass energy of $\sqrt{s} = 300\,GeV$ allows the study of electroweak processes which are sensitive to the triple gauge boson vertex (TGV). Measurements of the TGV are a crucial test of the Standard Model (SM). Anomalous coupling constants $\Delta\kappa = \kappa - 1$ and λ are defined for the $WW\gamma$ and WWZ vertex which do not violate discrete symmetries (1) and are expected to be zero in the SM.

HERA is sensitive to the $WW\gamma$ vertex via two processes. In W production ($ep \to eWX$) the $WW\gamma$ vertex is formed by a W boson emitted from a quark of the proton and a photon emitted from the incoming lepton (fig. 1a). The process $ep \to \nu WX$ in which the W boson is emitted from the lepton has a much smaller cross section and will not be considered here.

In radiative Charged Current (CC) events ($ep \to \nu\gamma X$) the photon can be emitted from the exchanged W boson and thus forming the $WW\gamma$ vertex (fig. 1b). Sensitivity to the TGV is expected for photons with high transverse momenta (p_t) as discussed in (2).

Both processes which are at HERA energies mainly sensitive to the coupling constant κ are discussed in the next sections.

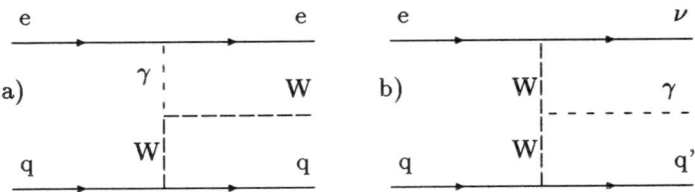

FIG. 1. *Diagrams with a $WW\gamma$ vertex at HERA. a) W production. b) Radiative Charged Current.*

W PRODUCTION

The W production process at HERA is discussed in (1), (3), (4). The expected total cross section of W^+ and W^- production at HERA is about $1\,pb$ in the SM. 60% of the W bosons are produced in photoproduction (γp: $Q^2 < 4\;GeV^2$), 40% are observed in Deep Inelastic Scattering (DIS: $Q^2 > 4\;GeV^2$).

The main contribution is W emission from the initial quark. The contribution of the $WW\gamma$ vertex depends strongly on the coupling constant κ (see table 1).

In this paper only leptonic W decay modes are considered ($W \to e\nu_e, \mu\nu_\mu$). The W has an average momentum of about $100\,GeV$ and is predominantly produced close to the proton beam direction. These processes have a clear signature because of the high transverse momentum leptons and substantial missing transverse momentum due to the undetected neutrinos.

Selection Cuts

For a selection of leptonic W decays we require:

$$p_t^{e,\mu} > 20\,GeV \quad \wedge \quad p_t^\nu = p_t^{miss} > 20\,GeV \tag{1}$$

The first requirement provides a clear lepton identification, the second requirement suppresses background from neutral current DIS and γp processes

TABLE 1. *Cross sections for W production with leptonic decay $[B = Br(W \to e\nu, \mu\nu]$ for different couplings κ before and after acceptance cuts.*

coupling κ	$B\sigma(W^+)/fb$	accepted (rel.)	$B\sigma(W^-)/fb$	accepted (rel.)
0	73	53 (74%)	84	70 (76%)
1	104	75 (73%)	102	84 (76%)
2	159	117 (75%)	132	107 (76%)
10	1510	1120 (74%)	750	570 (74%)

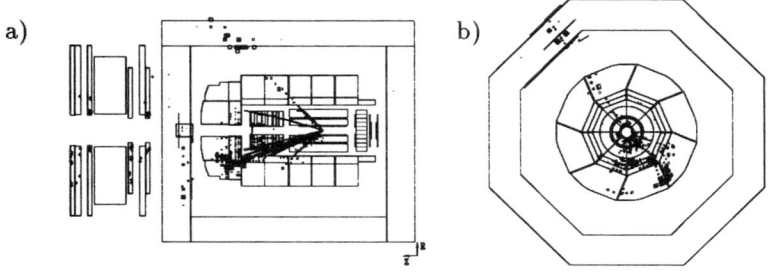

FIG. 2. Display of the high p_t muon event : a) R - z view , b) R - phi view.

which can have missing transverse momenta only due to energy fluctations.

The expected cross section for W^+ and W^- production for different values of κ is given in table 1 before and after applying these cuts. Combining both decay channels leads to an expected 1σ limit of $\kappa = 1^{+0.6}_{-1.0}$ for an integrated luminosity of 100 pb^{-1}.

A Candidate Event

In the data sample collected in 1994 an event with a high transverse momentum muon ($p_t^\mu = 23^{+7}_{-6}\ GeV$) was observed (fig. 2). The muon is of positive charge and it recoils against the visible hadronic system with $p_t^{had} = 42 \pm 4\ GeV$. The event has a missing transverse momentum of about 20 GeV (5).

A possible explanation is the production of a W^+ which decays into muon and neutrino. In this interpretation the W^+ has a transverse momentum of 42 GeV and is produced into the extreme forward direction ($\theta = 7°$). The expected SM cross section for the production of high p_t W^+ bosons with subsequent decay into muons is $\sigma(p_t^W > 40\ GeV) = 7\ fb$, which leads to a probability of 0.03 for this explanation.

Another possible explanation is a large fluctuation of a γp event. A high p_t muon could be faked by a jet which fluctuates into a single particle or by an isolated muon from a semileptonic decay in heavy quark production. The cross section for photoproduced high transverse momentum jets is $\sigma(p_t^W > 40 GeV) < 10 pb$. Taking into account the disfavoured fluctuations one obtains a probability $< 10^{-3}$ for this explanation.

RADIATIVE CHARGED CURRENT EVENTS

In CC DIS processes photons can be emitted from the incoming lepton, from the quark line or from the exchanged W boson (diagram 1b). The

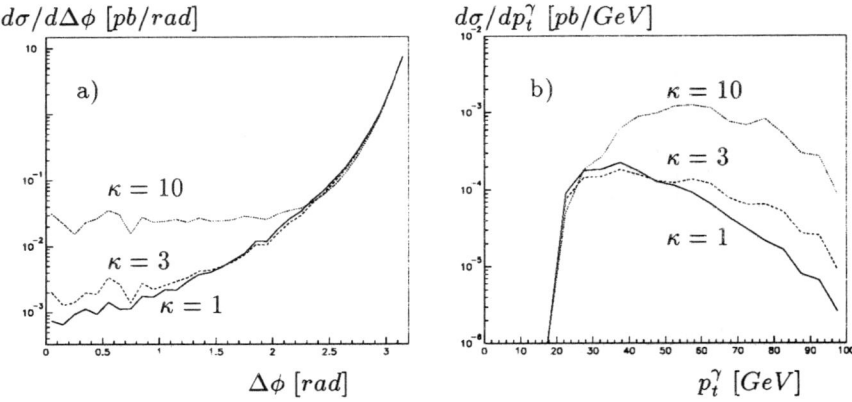

FIG. 3. Expected sensitivity to the $WW\gamma$ vertex in radiative CC from MC (2) a) $d\sigma/d\Delta\phi^{\nu,had}$ and b) $d\sigma/dp_t^\gamma$ for $\Delta\phi^{\nu,had} < 1.5$ for different κ.

largest contribution is due to leptonic Initial State Radiation (ISR). Radiative photons are preferentially emitted collinear to the quarks and electrons. The photons are thus expected close to the beam direction or close to the hadronic final state and have an $1/k$ energy distribution. In case of photon emission from the W boson isolated photons with high energies are expected.

Event Selection

Starting from CC events (6) with a missing transverse momentum $> 25\,GeV$, electromagnetic clusters in the central part of the detector with a minimum energy of $5\,GeV$ were selected. Isolation criteria were applied to suppress possible background from the hadronic final state: $R^{had,\gamma} = (\Delta\eta^{had,\gamma\,2} + \Delta\phi^{had,\gamma\,2})^{1/2} > 0.5$ and $\theta_\gamma > 0.5$. Here $\Delta\eta^{had,\gamma}$, $\Delta\phi^{had,\gamma}$ are differences in pseudorapidity and azimuthal angle between photon and hadronic system. θ_γ is the polar angle of the photon w.r.t the proton beam direction. About 5% of the CC events with $p_t^\nu > 25\,GeV$ fulfill these requirements.

The sensitivity to the $WW\gamma$ vertex properties is shown in fig. 3 for different values of the coupling constant κ. Fig. 3a shows the azimuthal difference between neutrino and hadronic system $\Delta\phi^{\nu,had} = |\phi^\nu - \phi^{had}|$ as derived from MC events selected as stated above.

For $\Delta\phi^{\nu,had} < 1.5$ in fig. 3b the p_t^γ distribution is shown again for different values of κ. One can conclude that the $\Delta\phi^{\nu,had}$ distribution as well as the transverse momentum of the photons are sensitive to an anomalous coupling κ. Sensitivity to $\kappa = 10$ is expected for an integrated luminosity of $20\,pb^{-1}$.

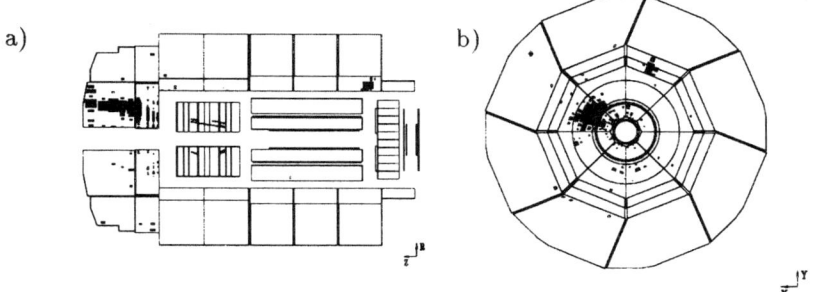

FIG. 4. Display of the radiative CC candidate with hightest transverse momentum of the photon ($E^\gamma = 16\,GeV, p_t^\gamma = 11\,GeV$); a) R - z view , b) R - phi view.

Candidates

Applying the photon selection to the CC candidates with $p_t^{miss} > 25\,GeV$ in the 1994 data sample yields three events. This is in agreement with the theoretical expectation of 2.4 for SM couplings. Fig. 4 shows the most prominent candidate with a photon energy of $16\,GeV$ ($p_t^\gamma = 11\,GeV$). The missing transverse momentum is $36\,GeV$.

CONCLUSION

At HERA two processes are sensitive to the coupling of the $WW\gamma$ vertex: W production and radiative charged current events. For each process a selection of candidate events was discussed. Expected sensitivities to the $WW\gamma$ coupling were derived. W production is the most promising process. The expected sensitivity for an integrated luminosity of $100\,pb^{-1}$ is $\kappa = 1^{+0.6}_{-1.0}$ (1σ limit).

Event selections performed on 1994 data yield one possible candidate for W production with subsequent decay into a muon and three radiative CC events.

For the 1995 data taking period an integrated luminosity at least three times higher than in 1994 is expected which will give the possibility to derive first limits on the $WW\gamma$ vertex.

REFERENCES

1. U. Baur and D. Zeppenfeld Nucl. Phys. B **325**, 253 (1989).
2. T. Helbig and H. Spiesberger Nucl. Phys. B **373**, 73 (1992).
3. E. Gabrielli Mod. Phys. Lett. A **1**, 465 (1986).
4. U. Baur, J. A. M. Vermaseren and D. Zeppenfeld Nucl. Phys. B **375**, 3 (1992).
5. H1 Collaboration, T. Ahmed et al., DESY preprint **94-248**, (1994).
6. H1 Collaboration, T. Ahmed et al., Phys. Lett. B **324**, 241(1994).

Limits on Rare B Decays
$B \to \mu^+\mu^- K^\pm$ and $B \to \mu^+\mu^- K^*$

Carol Anway-Wiese [1]

Physics Department, University California, Los Angeles
405 Hilgard Ave, Los Angeles, California 90024-1547, USA, caw@fnald.fnal.gov

---

We report on a search for flavor-changing neutral current decays of B mesons into $\mu\mu K^\pm$ and $\mu\mu K^*$ using data obtained in the Collider Detector at Fermilab (CDF) 1992-1993 data taking run. To reduce the amount of background in our data we use precise tracking information from the CDF silicon vertex detector to pinpoint the location of the decay vertex of the B candidate, and accept only events which have a large decay time. We compare this data to a B meson signal obtained in a similar fashion, but where the muon pairs originate from ψ decays, and calculate the relative branching ratios. In the absence of any indication of flavor-changing neutral current decays we set an upper limits of BR($B \to \mu\mu K^\pm$) $< 3.5 \times 10^{-5}$, and BR($B \to \mu\mu K^*$) $< 5.1 \times 10^{-5}$ at 90% confidence level, which are consistent with Standard Model expectations but leave little room for non-standard physics.

INTRODUCTION

Rare B decays provide us a way to test the Standard Model against possible effects of different form factors, anomalous magnetic moment of the $W^\pm$, and charged Higgs. Several theorists have predicted the rate of these decays. Differences in the form factors used in these calculations give relatively small uncertainties, but deviations from Standard Model physics can dramatically increase the expected rate. For a top quark mass of 150 GeV/c^2, A. Ali (1) predicts non-resonant branching ratios of BR($B \to \mu\mu K^\pm$) $= 4.4 \times 10^{-7}$ and BR($B \to \mu\mu K^*$) $= 2.3 \times 10^{-6}$ using the hadronic matrix elements of Isgur and Wise (2). G. Baillie (3) has also used heavy quark effective theory to calculate the ratio of the decay rate in a portion of the non-resonant dimuon mass spectrum to the decay rate of $B \to \psi K, \psi \to \mu\mu$. He finds the ratio of the decay rates to be 2×10^{-3} for each decay mode at a top quark mass of 170 GeV/c^2. The portion of the non-resonant dimuon mass spectrum he uses goes from $\hat{s} \equiv M(\mu\mu)^2/M_B^2 = 0.35$ to 0.48 and 0.50 to $\hat{s}_{max} = (M_B - M_K)^2/M_B^2$, similar to the region we use here. We also use the non-resonant theoretical differential decay rate as a function of dimuon mass from Isgur and Wise

---

[1] Representing the CDF Collaboration MS 318, Fermilab, PO Box 500 Batavia, IL 60510

to extrapolate our results from the small dimuon mass region to the overall non-resonant region.

DATA AND METHOD

At CDF (4) we have accurate momentum resolution in the central tracking chamber (CTC), further improved by using vertex position information from the silicon vertex detector (SVX) (5). In this analysis we accept only pairs of muons which have traversed the SVX, the CTC, and have muon stubs in the central muon chambers (up to $\eta = 0.6$). Both muons must have $Pt(\mu) > 2.0$ GeV/c, and one $Pt(\mu) > 2.8$ GeV/c.

We select muon pairs with an invariant mass between 2.8 and 4.5 GeV/c^2, assign the $K^\pm$ and $\pi^\pm$ masses to tracks in the central tracking chamber and use a secondary vertex fit to help reconstruct candidate B's. Tight cuts on the vertex fit quality and the transverse proper B candidate decay time (0.1 mm) help reduce combinatoric background. Background from hadronic punch-through is largely reduced by requiring the B candidate to carry the majority of the momentum in a cone. The transverse momentum cuts on the events are as follows: For $B \to \mu\mu K^\pm$, $Pt(B) > 5.0$ GeV/c, $Pt(K) > 1.0$ GeV/c. For $B \to \mu\mu K^*$, $Pt(B) > 6.0$ GeV/c, $Pt(K^*) > 2.0$ GeV/c. We assign dimuons to the ψ resonance if their invariant mass falls between 3.017 and 3.177 GeV/c^2, and to the *(partial)* non-resonant region if their mass falls in the range 3.3 to 3.6 or 3.8 to 4.5 GeV/c^2.

$B \to \mu\mu K$ RESULTS

After making the cuts above, we see 57 ± 10 $B \to \psi K^\pm, \psi \to \mu\mu$ events above background. We compare this to 10 $B \to \mu\mu K^\pm$ candidate events in our signal region, which is consistent with 13.0 ± 2.5 background events, as estimated from the sidebands. Using $BR(B \to \psi K^\pm, \psi \to \mu\mu) = 6.5 \pm 1.0 \times 10^{-5}$ (6) we can compare the number of ψ events to the number of non-resonant events to calculate the branching ratio limits according to the method of reference (7). We extrapolate to the overall dimuon mass region by multiplying the partial branching ratios by 4.4, as calculated using Monte Carlo and the theoretical models of Isgur and Wise (2,3).

$$\frac{BR(B \to \mu\mu K, partial)}{BR(B \to \psi K, \psi \to \mu\mu)} < 0.12 \text{ at } 90\% \text{ CL}, \quad < 0.15 \text{ at } 95\% \text{ CL}$$

$$\frac{BR(B \to \mu\mu K)}{BR(B \to \psi K, \psi \to \mu\mu)} < 0.50 \text{ at } 90\% \text{ CL}, \quad < 0.66 \text{ at } 95\% \text{ CL}$$

$$BR(B \to \mu\mu K, partial) < 0.8 \times 10^{-5} \text{ at } 90\% \text{ CL}, \; < 1.0 \times 10^{-5} \text{ at } 95\% \text{ CL}$$

$$BR(B \to \mu\mu K) < 3.5 \times 10^{-5} \text{ at } 90\% \text{ CL}, \; < 4.4 \times 10^{-5} \text{ at } 95\% \text{ CL}$$

$B \to \mu\mu K^*$ RESULTS

We see 33.5 ± 6.7 $B \to \psi K^*, \psi \to \mu\mu$ events above background. We compare this to 7 $B \to \mu\mu K^*$ candidate events in our signal region, which is consistent with 7.5 ± 1.9 background events. Using BR($B \to \psi K^*, \psi \to \mu\mu$) $= 7.76 \pm 2.41 \times 10^{-5}$ (8) we can compare the number of ψ events to the number of non-resonant events to calculate the branching ratio limits. We extrapolate to the overall dimuon mass region by multiplying the partial branching ratios by 3.75.

$$\frac{\mathrm{BR}(B \to \mu\mu K^*, partial)}{\mathrm{BR}(B \to \psi K^*, \psi \to \mu\mu)} < 0.21 \text{ at } 90\% \text{ CL}, \quad < 0.26 \text{ at } 95\% \text{ CL}$$

$$\frac{\mathrm{BR}(B \to \mu\mu K^*)}{\mathrm{BR}(B \to \psi K^*, \psi \to \mu\mu)} < 0.77 \text{ at } 90\% \text{ CL}, \quad < 0.96 \text{ at } 95\% \text{ CL}$$

BR($B \to \mu\mu K^*$, partial) $< 1.3 \times 10^{-5}$ at 90% CL, $< 1.7 \times 10^{-5}$ at 95% CL

BR($B \to \mu\mu K^*$) $< 5.1 \times 10^{-5}$ at 90% CL, $< 6.4 \times 10^{-5}$ at 95% CL

ACKNOWLEDGEMENTS

We thank Grant Baillie, UCLA, for his help in the theory behind the RareB Monte Carlo event generator and for many valuable discussions on the subject of rare B decays.

We thank the Fermilab staff and the technical staffs of the participating institutions for their vital contributions. This work was supported by the U.S. Department of Energy and National Science Foundation; the Italian Istituto Nazionale di Fisica Nucleare; the Ministry of Education, Science and Culture of Japan; the Natural Sciences and Engineering Research Council of Canada; the National Science Council of the Republic of China; the A. P. Sloan Foundation; and the Alexander von Humboldt-Stiftung.

REFERENCES

1. A. Ali, Phys. Lett. B 264 (1991) 447, Erratum, B 274 (1992) 526.
2. N. Isgur and M. B. Wise, Phys. Lett. B 232 (1989) 113; and Phys. Lett. B 237 (1990) 527.
3. G. Baillie, UCLA/93/TEP/26, hep-ph 9307369 (1993).
4. F. Abe et al. NIM A 271 (1988) 387, and references therein.
5. B. Barnett et al. NIM A 315 (1992) 125.
6. T. Browder, K. Honscheid, S. Playfer, CLNS 93/1261 (1994).
7. O. Helene, NIM 212 (1983) 319; and G. Zech, NIM A 277 (1989) 608.
8. Review of Particle Properties, Phys. Rev. D 45 (1992).

WGZ95-Participants

	Name	Affiliation
1	Aihara, Hiroaki	LBL
2	Anway-Wiese, Carol	UCLA
3	Arisaka, Katsushi	UCLA
4	Arzt, Christopher	Caltech
5	Barklow, Timothy L.	SLAC
6	Baur, Ulrich	SUNY
7	Berger, Edmond	ANL
8	Bonushkin, Yuri	UCLA
9	Brown, Robert W.	Case West. U.
10	Busenitz, Jerome	U. Alabama
11	Byers, Nina	UCLA
12	Carithers, Bill	LBL
13	Carlsmith, Duncan	U. Wisc.
14	Chang, Lay Nam	Virginia Tech.
15	Chen, Liang Ping	LBL
16	Chivukula, Sekhar	Boston U.
17	Choudhary, Brajesh	Fermilab
18	Chrisman, David	UCLA
19	Cline, David	UCLA
20	Cornwall, Mike	UCLA
21	Cousins, Robert	UCLA
22	Van Dalen, Gordon	UCR
23	Diehl, H. Thomas	Fermilab
24	Ellison, John	UCR
25	Eno, Sarah	U. Maryland
26	Errede, Steven	U. Illinois
27	Fabbri, Fabrizio	INFN-Bologna
28	Fahland, Tom	Brown U.
29	Falk, Adam F.	John Hopkins
30	Fuess, Theresa A.	ANL
31	Fujita, Hiroyuki	Case West. U.
32	Giacomelli, Paolo	UCR
33	Glenn, Steven	UCD
34	Godfrey, Stephen	Carleton U.
35	Grosse-Knetter, Carsten	U. Montreal
36	Hagiwara, Kaoru	KEK
37	Han, Tao	UCD
38	Hauser, Jay	UCLA
39	Hernandez, M. Pilar	Harvard
40	Hinchliffe, Ian	LBL
41	Johari, Hossein	FNAL
42	Johnson, Rolland P.	DOE
43	Kelly, Michael	U. Notre Dame
44	Kinnunen, Ritva	CERN
45	Klem, Dan	LLNL

WGZ95-Participants

	Name	Affiliation
46	Kolonko, Jim	UCLA
47	Lahanas, Athanasios B.	U. Athens
48	Landsberg, Greg	Stony Brook
49	Lareneta, Melinda	UCLA
50	Layter, John	UCR
51	Lane, Ken	Boston U.
52	Lecomte, Pierre	CERN
53	Lee, Clarence	Caltech
54	Ligeti, Zoltan	Caltech
55	Lindgren, Michael	UCLA
56	Ma, Ernest	UCR
57	Mattig, Peter	U. Bonn
58	Mery, Pierre	U. Mediterranee
59	Mo, Luke	Virgina Poly.
60	Muller, Thomas	UCLA
61	Miyamoto, Akiya	KEK
62	Neuberger, Dirk	UCLA
63	Newman, Harvey B.	Caltech
64	Ohnemus, James	UCD
65	Papavassiliou, Joannis	NYU
66	Peccei, Roberto	UCLA
67	Philippides, Kostas	NYU
68	Playfer, Stephen	Syracuse U.
69	Rajpoot, Subhash	CSULB
70	Riles, Keith	U. Michigan
71	Schaile, Dorothee	CERN
72	Schoning, Andre	DESY
73	Simmons, Elizabeth	Boston U.
74	Spadafora, Anthony	LBL
75	Takeuchi, Tatsu	Fermilab
76	Tkabladze, Avto	IHEP Moscow
77	Valencia, German E.	Iowa St. U.
78	Wagner, Bob	ANL
79	Walczak, Roman	Oxford U.
80	Wendt, Christopher	U. of Wisc.
81	Willenbrock, Scott	U. Illinois
82	Womersley, John	Fermilab
83	Wudka, Jose	UCR
84	Yasuda, Takahiro	Northeastern U.
85	Zeppenfeld, Dieter	U. Wisc.
86	Zhang, Liqun	U. Wisc.
87	Zhang, Xinmin	Iowa St. Univ.
88	Zhu, Renyuan	Caltech

AUTHOR INDEX

A
Aihara, H., 72
Anway-Wiese, C., 404

B
Barklow, T. L., 307
Brown, R. W., 261
Busenitz, J., 273

C
Carlsmith, D. L., 377
Chivukula, R. S., 239
Cho, P., 323
Couture, G., 357

D
Dittmaier, S., 367

F
Falk, A. F., 136
Fuess, T. A., 84

G
Giacomelli, P., 347
Gintner, M., 357
Godfrey, S., 209, 357
Grosse-Knetter, C., 367

H
Hagiwara, K., 182
Han, T., 224
Hernández, P., 250
Hinchliffe, I., 335

J
Johari, H., 372

L
Lahanas, A. B., 110
Landsberg, G., 382

M
Ma, E., 352
Mättig, P., 148

N
Neuberger, D., 393
Ng, D., 352

O
Ohnemus, J., 60

P
Papavassiliou, J., 98
Philippides, K., 362
Playfer, S. M., 124

S
Schaile, D., 33
Schöning, A., 399
Simmons, E. H., 323

T
Takeuchi, T., 20

V

Valencia, G., 160

W

Wagner for the CDF Collaboration, R. G., 388
Walczak, R., 198

Wendt, C., 285
Willenbrock, S., 3
Womersley, J., 299
Wudka, J., 171

Z

Zeppenfeld, D., 46
Zhang, L., 377

AIP Conference Proceedings

		L.C. Number	ISBN
No. 110	Hadron Substructure in Nuclear Physics (Indiana University, 1983)	84-70165	0-88318-309-9
No. 111	Production and Neutralization of Negative Ions and Beams (3rd Int'l Symposium) (Brookhaven, NY, 1983)	84-70379	0-88318-310-2
No. 112	Particles and Fields – 1983 (APS/DPF, Blacksburg, VA)	84-70378	0-88318-311-0
No. 113	Experimental Meson Spectroscopy – 1983 (7th International Conference, Brookhaven, NY)	84-70910	0-88318-312-9
No. 114	Low Energy Tests of Conservation Laws in Particle Physics (Blacksburg, VA, 1983)	84-71157	0-88318-313-7
No. 115	High Energy Transients in Astrophysics (Santa Cruz, CA, 1983)	84-71205	0-88318-314-5
No. 116	Problems in Unification and Supergravity (La Jolla Institute, 1983)	84-71246	0-88318-315-3
No. 117	Polarized Proton Ion Sources (TRIUMF, Vancouver, 1983)	84-71235	0-88318-316-1
No. 118	Free Electron Generation of Extreme Ultraviolet Coherent Radiation (Brookhaven/OSA, 1983)	84-71539	0-88318-317-X
No. 119	Laser Techniques in the Extreme Ultraviolet (OSA, Boulder, CO, 1984)	84-72128	0-88318-318-8
No. 120	Optical Effects in Amorphous Semiconductors (Snowbird, UT, 1984)	84-72419	0-88318-319-6
No. 121	High Energy e^+e^- Interactions (Vanderbilt, 1984)	84-72632	0-88318-320-X
No. 122	The Physics of VLSI (Xerox, Palo Alto, CA, 1984)	84-72729	0-88318-321-8
No. 123	Intersections Between Particle and Nuclear Physics (Steamboat Springs, CO, 1984)	84-72790	0-88318-322-6
No. 124	Neutron-Nucleus Collisions: A Probe of Nuclear Structure (Burr Oak State Park, 1984)	84-73216	0-88318-323-4
No. 125	Capture Gamma-Ray Spectroscopy and Related Topics – 1984 (Int'l Symposium, Knoxville, TN)	84-73303	0-88318-324-2
No. 126	Solar Neutrinos and Neutrino Astronomy (Homestake, 1984)	84-63143	0-88318-325-0

No. 127	Physics of High Energy Particle Accelerators (BNL/SUNY Summer School, 1983)	85-70057	0-88318-326-9
No. 128	Nuclear Physics with Stored, Cooled Beams (McCormick's Creek State Park, IN, 1984)	85-71167	0-88318-327-7
No. 129	Radiofrequency Plasma Heating (Sixth Topical Conference) (Callaway Gardens, GA, 1985)	85-48027	0-88318-328-5
No. 130	Laser Acceleration of Particles (Malibu, CA, 1985)	85-48028	0-88318-329-3
No. 131	Workshop on Polarized ^{3}He Beams and Targets (Princeton, NJ, 1984)	85-48026	0-88318-330-7
No. 132	Hadron Spectroscopy – 1985 (International Conference, Univ. of Maryland)	85-72537	0-88318-331-5
No. 133	Hadronic Probes and Nuclear Interactions (Arizona State University, 1985)	85-72638	0-88318-332-3
No. 134	The State of High Energy Physics (BNL/SUNY Summer School, 1983)	85-73170	0-88318-333-1
No. 135	Energy Sources: Conservation and Renewables (APS, Washington, DC, 1985)	85-73019	0-88318-334-X
No. 136	Atomic Theory Workshop on Relativistic and QED Effects in Heavy Atoms (Gaithersburg, MD, 1985)	85-73790	0-88318-335-8
No. 137	Polymer-Flow Interaction (La Jolla Institute, 1985)	85-73915	0-88318-336-6
No. 138	Frontiers in Electronic Materials and Processing (Houston, TX, 1985)	86-70108	0-88318-337-4
No. 139	High-Current, High-Brightness, and High-Duty Factor Ion Injectors (La Jolla Institute, 1985)	86-70245	0-88318-338-2
No. 140	Boron-Rich Solids (Albuquerque, NM, 1985)	86-70246	0-88318-339-0
No. 141	Gamma-Ray Bursts (Stanford, CA, 1984)	86-70761	0-88318-340-4
No. 142	Nuclear Structure at High Spin, Excitation, and Momentum Transfer (Indiana University, 1985)	86-70837	0-88318-341-2
No. 143	Mexican School of Particles and Fields (Oaxtepec, México, 1984)	86-81187	0-88318-342-0
No. 144	Magnetospheric Phenomena in Astrophysics (Los Alamos, NM, 1984)	86-71149	0-88318-343-9
No. 145	Polarized Beams at SSC & Polarized Antiprotons (Ann Arbor, MI & Bodega Bay, CA, 1985)	86-71343	0-88318-344-7

No. 146	Advances in Laser Science—I (Dallas, TX, 1985)	86-71536	0-88318-345-5
No. 147	Short Wavelength Coherent Radiation: Generation and Applications (Monterey, CA, 1986)	86-71674	0-88318-346-3
No. 148	Space Colonization: Technology and The Liberal Arts (Geneva, NY, 1985)	86-71675	0-88318-347-1
No. 149	Physics and Chemistry of Protective Coatings (Universal City, CA, 1985)	86-72019	0-88318-348-X
No. 150	Intersections Between Particle and Nuclear Physics (Lake Louise, Canada, 1986)	86-72018	0-88318-349-8
No. 151	Neural Networks for Computing (Snowbird, UT, 1986)	86-72481	0-88318-351-X
No. 152	Heavy Ion Inertial Fusion (Washington, DC, 1986)	86-73185	0-88318-352-8
No. 153	Physics of Particle Accelerators (SLAC Summer School, 1985) (Fermilab Summer School, 1984)	87-70103	0-88318-353-6
No. 154	Physics and Chemistry of Porous Media—II (Ridgefield, CT, 1986)	83-73640	0-88318-354-4
No. 155	The Galactic Center: Proceedings of the Symposium Honoring C. H. Townes (Berkeley, CA, 1986)	86-73186	0-88318-355-2
No. 156	Advanced Accelerator Concepts (Madison, WI, 1986)	87-70635	0-88318-358-0
No. 157	Stability of Amorphous Silicon Alloy Materials and Devices (Palo Alto, CA, 1987)	87-70990	0-88318-359-9
No. 158	Production and Neutralization of Negative Ions and Beams (Brookhaven, NY, 1986)	87-71695	0-88318-358-7
No. 159	Applications of Radio-Frequency Power to Plasma: Seventh Topical Conference (Kissimmee, FL, 1987)	87-71812	0-88318-359-5
No. 160	Advances in Laser Science—II (Seattle, WA, 1986)	87-71962	0-88318-360-9
No. 161	Electron Scattering in Nuclear and Particle Science: In Commemoration of the 35th Anniversary of the Lyman-Hanson-Scott Experiment (Urbana, IL, 1986)	87-72403	0-88318-361-7
No. 162	Few-Body Systems and Multiparticle Dynamics (Crystal City, VA, 1987)	87-72594	0-88318-362-5

No. 163	Pion–Nucleus Physics: Future Directions and New Facilities at LAMPF (Los Alamos, NM, 1987)	87-72961	0-88318-363-3
No. 164	Nuclei Far from Stability: Fifth International Conference (Rosseau Lake, ON, 1987)	87-73214	0-88318-364-1
No. 165	Thin Film Processing and Characterization of High-Temperature Superconductors (Anaheim, CA, 1987)	87-73420	0-88318-365-X
No. 166	Photovoltaic Safety (Denver, CO, 1988)	88-42854	0-88318-366-8
No. 167	Deposition and Growth: Limits for Microelectronics (Anaheim, CA, 1987)	88-71432	0-88318-367-6
No. 168	Atomic Processes in Plasmas (Santa Fe, NM, 1987)	88-71273	0-88318-368-4
No. 169	Modern Physics in America: A Michelson-Morley Centennial Symposium (Cleveland, OH, 1987)	88-71348	0-88318-369-2
No. 170	Nuclear Spectroscopy of Astrophysical Sources (Washington, DC, 1987)	88-71625	0-88318-370-6
No. 171	Vacuum Design of Advanced and Compact Synchrotron Light Sources (Upton, NY, 1988)	88-71824	0-88318-371-4
No. 172	Advances in Laser Science—III: Proceedings of the International Laser Science Conference (Atlantic City, NJ, 1987)	88-71879	0-88318-372-2
No. 173	Cooperative Networks in Physics Education (Oaxtepec, Mexico, 1987)	88-72091	0-88318-373-0
No. 174	Radio Wave Scattering in the Interstellar Medium (San Diego, CA, 1988)	88-72092	0-88318-374-9
No. 175	Non-neutral Plasma Physics (Washington, DC, 1988)	88-72275	0-88318-375-7
No. 176	Intersections Between Particle and Nuclear Physics (Third International Conference) (Rockport, ME, 1988)	88-62535	0-88318-376-5
No. 177	Linear Accelerator and Beam Optics Codes (La Jolla, CA, 1988)	88-46074	0-88318-377-3
No. 178	Nuclear Arms Technologies in the 1990s (Washington, DC, 1988)	88-83262	0-88318-378-1
No. 179	The Michelson Era in American Science: 1870–1930 (Cleveland, OH, 1987)	88-83369	0-88318-379-X
No. 180	Frontiers in Science: International Symposium (Urbana, IL, 1987)	88-83526	0-88318-380-3

No. 181	Muon-Catalyzed Fusion (Sanibel Island, FL, 1988)	88-83636	0-88318-381-1
No. 182	High T_c Superconducting Thin Films, Devices, and Applications (Atlanta, GA, 1988)	88-03947	0-88318-382-X
No. 183	Cosmic Abundances of Matter (Minneapolis, MN, 1988)	89-80147	0-88318-383-8
No. 184	Physics of Particle Accelerators (Ithaca, NY, 1988)	89-83575	0-88318-384-6
No. 185	Glueballs, Hybrids, and Exotic Hadrons (Upton, NY, 1988)	89-83513	0-88318-385-4
No. 186	High-Energy Radiation Background in Space (Sanibel Island, FL, 1987)	89-83833	0-88318-386-2
No. 187	High-Energy Spin Physics (Minneapolis, MN, 1988)	89-83948	0-88318-387-0
No. 188	International Symposium on Electron Beam Ion Sources and their Applications (Upton, NY, 1988)	89-84343	0-88318-388-9
No. 189	Relativistic, Quantum Electrodynamic, and Weak Interaction Effects in Atoms (Santa Barbara, CA, 1988)	89-84431	0-88318-389-7
No. 190	Radio-frequency Power in Plasmas (Irvine, CA, 1989)	89-45805	0-88318-397-8
No. 191	Advances in Laser Science—IV (Atlanta, GA, 1988)	89-85595	0-88318-391-9
No. 192	Vacuum Mechatronics (First International Workshop) (Santa Barbara, CA, 1989)	89-45905	0-88318-394-3
No. 193	Advanced Accelerator Concepts (Lake Arrowhead, CA, 1989)	89-45914	0-88318-393-5
No. 194	Quantum Fluids and Solids—1989 (Gainesville, FL, 1989)	89-81079	0-88318-395-1
No. 195	Dense Z-Pinches (Laguna Beach, CA, 1989)	89-46212	0-88318-396-X
No. 196	Heavy Quark Physics (Ithaca, NY, 1989)	89-81583	0-88318-644-6
No. 197	Drops and Bubbles (Monterey, CA, 1988)	89-46360	0-88318-392-7
No. 198	Astrophysics in Antarctica (Newark, DE, 1989)	89-46421	0-88318-398-6
No. 199	Surface Conditioning of Vacuum Systems (Los Angeles, CA, 1989)	89-82542	0-88318-756-6
No. 200	High T_c Superconducting Thin Films: Processing, Characterization, and Applications (Boston, MA, 1989)	90-80006	0-88318-759-0

No. 201	QED Structure Functions (Ann Arbor, MI, 1989)	90-80229	0-88318-671-3
No. 202	NASA Workshop on Physics From a Lunar Base (Stanford, CA, 1989)	90-55073	0-88318-646-2
No. 203	Particle Astrophysics: The NASA Cosmic Ray Program for the 1990s and Beyond (Greenbelt, MD, 1989)	90-55077	0-88318-763-9
No. 204	Aspects of Electron-Molecule Scattering and Photoionization (New Haven, CT, 1989)	90-55175	0-88318-764-7
No. 205	The Physics of Electronic and Atomic Collisions (XVI International Conference) (New York, NY, 1989)	90-53183	0-88318-390-0
No. 206	Atomic Processes in Plasmas (Gaithersburg, MD, 1989)	90-55265	0-88318-769-8
No. 207	Astrophysics from the Moon (Annapolis, MD, 1990)	90-55582	0-88318-770-1
No. 208	Current Topics in Shock Waves (Bethlehem, PA, 1989)	90-55617	0-88318-776-0
No. 209	Computing for High Luminosity and High Intensity Facilities (Santa Fe, NM, 1990)	90-55634	0-88318-786-8
No. 210	Production and Neutralization of Negative Ions and Beams (Brookhaven, NY, 1990)	90-55316	0-88318-786-8
No. 211	High-Energy Astrophysics in the 21st Century (Taos, NM, 1989)	90-55644	0-88318-803-1
No. 212	Accelerator Instrumentation (Brookhaven, NY, 1989)	90-55838	0-88318-645-4
No. 213	Frontiers in Condensed Matter Theory (New York, NY, 1989)	90-6421	0-88318-771-X 0-88318-772-8 (pbk.)
No. 214	Beam Dynamics Issues of High-Luminosity Asymmetric Collider Rings (Berkeley, CA, 1990)	90-55857	0-88318-767-1
No. 215	X-Ray and Inner-Shell Processes (Knoxville, TN, 1990)	90-84700	0-88318-790-6
No. 216	Spectral Line Shapes, Vol. 6 (Austin, TX, 1990)	90-06278	0-88318-791-4
No. 217	Space Nuclear Power Systems (Albuquerque, NM, 1991)	90-56220	0-88318-838-4
No. 218	Positron Beams for Solids and Surfaces (London, Canada, 1990)	90-56407	0-88318-842-2

No. 219	Superconductivity and Its Applications (Buffalo, NY, 1990)	91-55020	0-88318-835-X
No. 220	High Energy Gamma-Ray Astronomy (Ann Arbor, MI, 1990)	91-70876	0-88318-812-0
No. 221	Particle Production Near Threshold (Nashville, IN, 1990)	91-55134	0-88318-829-5
No. 222	After the First Three Minutes (College Park, MD, 1990)	91-55214	0-88318-828-7
No. 223	Polarized Collider Workshop (University Park, PA, 1990)	91-71303	0-88318-826-0
No. 224	LAMPF Workshop on (π, K) Physics (Los Alamos, NM, 1990)	91-71304	0-88318-825-2
No. 225	Half Collision Resonance Phenomena in Molecules (Caracas, Venezuela, 1990)	91-55210	0-88318-840-6
No. 226	The Living Cell in Four Dimensions (Gif sur Yvette, France, 1990)	91-55209	0-88318-794-9
No. 227	Advanced Processing and Characterization Technologies (Clearwater, FL, 1991)	91-55194	0-88318-910-0
No. 228	Anomalous Nuclear Effects in Deuterium/ Solid Systems (Provo, UT, 1990)	91-55245	0-88318-833-3
No. 229	Accelerator Instrumentation (Batavia, IL, 1990)	91-55347	0-88318-832-1
No. 230	Nonlinear Dynamics and Particle Acceleration (Tsukuba, Japan, 1990)	91-55348	0-88318-824-4
No. 231	Boron-Rich Solids (Albuquerque, NM, 1990)	91-53024	0-88318-793-4
No. 232	Gamma-Ray Line Astrophysics (Paris-Saclay, France, 1990)	91-55492	0-88318-875-9
No. 233	Atomic Physics 12 (Ann Arbor, MI, 1990)	91-55595	088318-811-2
No. 234	Amorphous Silicon Materials and Solar Cells (Denver, CO, 1991)	91-55575	088318-831-7
No. 235	Physics and Chemistry of MCT and Novel IR Detector Materials (San Francisco, CA, 1990)	91-55493	0-88318-931-3
No. 236	Vacuum Design of Synchrotron Light Sources (Argonne, IL, 1990)	91-55527	0-88318-873-2
No. 237	Kent M. Terwilliger Memorial Symposium (Ann Arbor, MI, 1989)	91-55576	0-88318-788-4

No. 238	Capture Gamma-Ray Spectroscopy (Pacific Grove, CA, 1990)	91-57923	0-88318-830-9
No. 239	Advances in Biomolecular Simulations (Obernai, France, 1991)	91-58106	0-88318-940-2
No. 240	Joint Soviet-American Workshop on the Physics of Semiconductor Lasers (Leningrad, USSR, 1991)	91-58537	0-88318-936-4
No. 241	Scanned Probe Microscopy (Santa Barbara, CA, 1991)	91-76758	0-88318-816-3
No. 242	Strong, Weak, and Electromagnetic Interactions in Nuclei, Atoms, and Astrophysics: A Workshop in Honor of Stewart D. Bloom's Retirement (Livermore, CA, 1991)	91-76876	0-88318-943-7
No. 243	Intersections Between Particle and Nuclear Physics (Tucson, AZ, 1991)	91-77580	0-88318-950-X
No. 244	Radio Frequency Power in Plasmas (Charleston, SC, 1991)	91-77853	0-88318-937-2
No. 245	Basic Space Science (Bangalore, India, 1991)	91-78379	0-88318-951-8
No. 246	Space Nuclear Power Systems (Albuquerque, NM, 1992)	91-58793	1-56396-027-3 1-56396-026-5 (pbk.)
No. 247	Global Warming: Physics and Facts (Washington, DC, 1991)	91-78423	0-88318-932-1
No. 248	Computer-Aided Statistical Physics (Taipei, Taiwan, 1991)	91-78378	0-88318-942-9
No. 249	The Physics of Particle Accelerators (Upton, NY, 1989, 1990)	92-52843	0-88318-789-2
No. 250	Towards a Unified Picture of Nuclear Dynamics (Nikko, Japan, 1991)	92-70143	0-88318-951-8
No. 251	Superconductivity and its Applications (Buffalo, NY, 1991)	92-52726	1-56396-016-8
No. 252	Accelerator Instrumentation (Newport News, VA, 1991)	92-70356	0-88318-934-8
No. 253	High-Brightness Beams for Advanced Accelerator Applications (College Park, MD, 1991)	92-52705	0-88318-947-X
No. 254	Testing the AGN Paradigm (College Park, MD, 1991)	92-52780	1-56396-009-5

No. 255	Advanced Beam Dynamics Workshop on Effects of Errors in Accelerators, Their Diagnosis and Corrections (Corpus Christi, TX, 1991)	92-52842	1-56396-006-0
No. 256	Slow Dynamics in Condensed Matter (Fukuoka, Japan, 1991)	92-53120	0-88318-938-0
No. 257	Atomic Processes in Plasmas (Portland, ME, 1991)	91-08105	0-88318-939-9
No. 258	Synchrotron Radiation and Dynamic Phenomena (Grenoble, France, 1991)	92-53790	1-56396-008-7
No. 259	Future Directions in Nuclear Physics with 4π Gamma Detection Systems of the New Generation (Strasbourg, France, 1991)	92-53222	0-88318-952-6
No. 260	Computational Quantum Physics (Nashville, TN, 1991)	92-71777	0-88318-933-X
No. 261	Rare and Exclusive B&K Decays and Novel Flavor Factories (Santa Monica, CA, 1991)	92-71873	1-56396-055-9
No. 262	Molecular Electronics—Science and Technology (St. Thomas, Virgin Islands, 1991)	92-72210	1-56396-041-9
No. 263	Stress-Induced Phenomena in Metallization: First International Workshop (Ithaca, NY, 1991)	92-72292	1-56396-082-6
No. 264	Particle Acceleration in Cosmic Plasmas (Newark, DE, 1991)	92-73316	0-88318-948-8
No. 265	Gamma-Ray Bursts (Huntsville, AL, 1991)	92-73456	1-56396-018-4
No. 266	Group Theory in Physics (Cocoyoc, Morelos, Mexico, 1991)	92-73457	1-56396-101-6
No. 267	Electromechanical Coupling of the Solar Atmosphere (Capri, Italy, 1991)	92-82717	1-56396-110-5
No. 268	Photovoltaic Advanced Research & Development Project (Denver, CO, 1992)	92-74159	1-56396-056-7
No. 269	CEBAF 1992 Summer Workshop (Newport News, VA, 1992)	92-75403	1-56396-067-2
No. 270	Time Reversal—The Arthur Rich Memorial Symposium (Ann Arbor, MI, 1991)	92-83852	1-56396-105-9
No. 271	Tenth Symposium Space Nuclear Power and Propulsion (Vols. I–III) (Albuquerque, NM, 1993)	92-75162	1-56396-137-7 (set)

No. 272	Proceedings of the XXVI International Conference on High Energy Physics (Vols. I and II) (Dallas, TX, 1992)	93-70412	1-56396-127-X (set)
No. 273	Superconductivity and Its Applications (Buffalo, NY, 1992)	93-70502	1-56396-189-X
No. 274	VIth International Conference on the Physics of Highly Charged Ions (Manhattan, KS, 1992)	93-70577	1-56396-102-4
No. 275	Atomic Physics 13 (Munich, Germany, 1992)	93-70826	1-56396-057-5
No. 276	Very High Energy Cosmic-Ray Interactions: VIIth International Symposium (Ann Arbor, MI, 1992)	93-71342	1-56396-038-9
No. 277	The World at Risk: Natural Hazards and Climate Change (Cambridge, MA, 1992)	93-71333	1-56396-066-4
No. 278	Back to the Galaxy (College Park, MD, 1992)	93-71543	1-56396-227-6
No. 279	Advanced Accelerator Concepts (Port Jefferson, NY, 1992)	93-71773	1-56396-191-1
No. 280	Compton Gamma-Ray Observatory (St. Louis, MO, 1992)	93-71830	1-56396-104-0
No. 281	Accelerator Instrumentation Fourth Annual Workshop (Berkeley, CA, 1992)	93-072110	1-56396-190-3
No. 282	Quantum 1/f Noise & Other Low Frequency Fluctuations in Electronic Devices (St. Louis, MO, 1992)	93-072366	1-56396-252-7
No. 283	Earth and Space Science Information Systems (Pasadena, CA, 1992)	93-072360	1-56396-094-X
No. 284	US-Japan Workshop on Ion Temperature Gradient-Driven Turbulent Transport (Austin, TX, 1993)	93-72460	1-56396-221-7
No. 285	Noise in Physical Systems and 1/f Fluctuations (St. Louis, MO, 1993)	93-72575	1-56396-270-5
No. 286	Ordering Disorder: Prospect and Retrospect in Condensed Matter Physics: Proceedings of the Indo-U.S. Workshop (Hyderabad, India, 1993)	93-072549	1-56396-255-1
No. 287	Production and Neutralization of Negative Ions and Beams: Sixth International Symposium (Upton, NY, 1992)	93-72821	1-56396-103-2

No. 288	Laser Ablation: Mechanismas and Applications-II: Second International Conference (Knoxville, TN, 1993)	93-73040	1-56396-226-8
No. 289	Radio Frequency Power in Plasmas: Tenth Topical Conference (Boston, MA, 1993)	93-72964	1-56396-264-0
No. 290	Laser Spectroscopy: XIth International Conference (Hot Springs, VA, 1993)	93-73050	1-56396-262-4
No. 291	Prairie View Summer Science Academy (Prairie View, TX, 1992)	93-73081	1-56396-133-4
No. 292	Stability of Particle Motion in Storage Rings (Upton, NY, 1992)	93-73534	1-56396-225-X
No. 293	Polarized Ion Sources and Polarized Gas Targets (Madison, WI, 1993)	93-74102	1-56396-220-9
No. 294	High-Energy Solar Phenomena A New Era of Spacecraft Measurements (Waterville Valley, NH, 1993)	93-74147	1-56396-291-8
No. 295	The Physics of Electronic and Atomic Collisions: XVIII International Conference (Aarhus, Denmark, 1993)	93-74103	1-56396-290-X
No. 296	The Chaos Paradigm: Developments an Applications in Engineering and Science (Mystic, CT, 1993)	93-74146	1-56396-254-3
No. 297	Computational Accelerator Physics (Los Alamos, NM, 1993)	93-74205	1-56396-222-5
No. 298	Ultrafast Reaction Dynamics and Solvent Effects (Royaumont, France, 1993)	93-074354	1-56396-280-2
No. 299	Dense Z-Pinches: Third International Conference (London, 1993)	93-074569	1-56396-297-7
No. 300	Discovery of Weak Neutral Currents: The Weak Interaction Before and After (Santa Monica, CA, 1993)	94-70515	1-56396-306-X
No. 301	Eleventh Symposium Space Nuclear Power and Propulsion (3 Vols.) (Albuquerque, NM, 1994)	92-75162	1-56396-305-1 (Set) 156396-301-9 (pbk. set)
No. 302	Lepton and Photon Interactions/ XVI International Symposium (Ithaca, NY, 1993)	94-70079	1-56396-106-7

No. 303	Slow Positron Beam Techniques for Solids and Surfaces Fifth International Workshop (Jackson Hole, WY 1992)	94-71036	1-56396-267-5
No. 304	The Second Compton Symposium (College Park, MD, 1993)	94-70742	1-56396-261-6
No. 305	Stress-Induced Phenomena in Metallization Second International Workshop (Austin, TX, 1993)	94-70650	1-56396-251-9
No. 306	12th NREL Photovoltaic Program Review (Denver, CO, 1993)	94-70748	1-56396-315-9
No. 307	Gamma-Ray Bursts Second Workshop (Huntsville, AL 1993)	94-71317	1-56396-336-1
No. 308	The Evolution of X-Ray Binaries (College Park, MD 1993)	94-76853	1-56396-329-9
No. 309	High-Pressure Science and Technology—1993 (Colorado Springs, CO 1993)	93-72821	1-56396-219-5 (Set)
No. 310	Analysis of Interplanetary Dust (Houston, TX 1993)	94-71292	1-56396-341-8
No. 311	Physics of High Energy Particles in Toroidal Systems (Irvine, CA 1993)	94-72098	1-56396-364-7
No. 312	Molecules and Grains in Space (Mont Sainte-Odile, France 1993)	94-72615	1-56396-355-8
No. 313	The Soft X-Ray Cosmos ROSAT Science Symposium (College Park, MD 1993)	94-72499	1-56396-327-2
No. 314	Advances in Plasma Physics Thomas H. Stix Symposium (Princeton, NJ 1992)	94-72721	1-56396-372-8
No. 315	Orbit Correction and Analysis in Circular Accelerators (Upton, NY 1993)	94-72257	1-56396-373-6
No. 316	Thirteenth International Conference on Thermoelectrics (Kansas City, Missouri 1994)	95-75634	1-56396-444-9
No. 317	Fifth Mexican School of Particles and Fields (Guanajuato, Mexico 1992)	94-72720	1-56396-378-7
No. 318	Laser Interaction and Related Plasma Phenomena 11th International Workshop (Monterey, CA 1993)	94-78097	1-56396-324-8

No. 319	Beam Instrumentation Workshop (Santa Fe, NM 1993)	94-78279	1-56396-389-2
No. 320	Basic Space Science (Lagos, Nigeria 1993)	94-79350	1-56396-328-0
No. 321	The First NREL Conference on Thermophotovoltaic Generation of Electricity (Copper Mountain, CO 1994)	94-72792	1-56396-353-1
No. 322	Atomic Processes in Plasmas Ninth APS Topical Conference (San Antonio, TX)	94-72923	1-56396-411-2
No. 323	Atomic Physics 14 Fourteenth International Conference on Atomic Physics (Boulder, CO 1994)	94-73219	1-56396-348-5
No. 324	Twelfth Symposium on Space Nuclear Power and Propulsion (Albuquerque, NM 1995)	94-73603	1-56396-427-9
No. 325	Conference on NASA Centers for Commercial Development of Space (Albuquerque, NM 1995)	94-73604	1-56396-431-7
No. 326	Accelerator Physics at the Superconducting Super Collider (Dallas, TX 1992-1993)	94-73609	1-56396-354-X
No. 327	Nuclei in the Cosmos III Third International Symposium on Nuclear Astrophysics (Assergi, Italy 1994)	95-75492	1-56396-436-8
No. 328	Spectral Line Shapes, Volume 8 12th ICSLS (Toronto, Canada 1994)	94-74309	1-56396-326-4
No. 329	Resonance Ionization Spectroscopy 1994 Seventh International Symposium (Bernkastel-Kues, Germany 1994)	95-75077	1-56396-437-6
No. 330	E.C.C.C. 1 Computational Chemistry F.E.C.S. Conference (Nancy, France 1994)	95-75843	1-56396-457-0
No. 331	Non-Neutral Plasma Physics II (Berkeley, CA 1994)	95-79630	1-56396-441-4
No. 332	X-Ray Lasers 1994 Fourth International Colloquium (Williamsburg, VA 1994)	95-76067	1-56396-375-2
No. 333	Beam Instrumentation Workshop (Vancouver, B. C., Canada 1994)	95-79635	1-56396-352-3
No. 334	Few-Body Problems in Physics (Williamsburg, VA 1994)	95-76481	1-56396-325-6

No. 335	Advanced Accelerator Concepts (Fontana, WI 1994)	95-78225	1-56396-476-7 (Set) 1-56396-474-0 (Book) 1-56396-475-9 (CD-Rom)
No. 336	Dark Matter (College Park, MD 1994)	95-76538	1-56396-438-4
No. 337	Pulsed RF Sources for Linear Colliders (Montauk, NY 1994)	95-76814	1-56396-408-2
No. 338	Intersections Between Particle and Nuclear Physics 5th Conference (St. Petersburg, FL 1994)	95-77076	1-56396-335-3
No. 339	Polarization Phenomena in Nuclear Physics Eighth International Symposium (Bloomington, IN 1994)	95-77216	1-56396-482-1
No. 340	Strangeness in Hadronic Matter (Tucson, AZ 1995)	95-77477	1-56396-489-9
No. 341	Volatiles in the Earth and Solar System (Pasadena, CA 1994)	95-77911	1-56396-409-0
No. 342	CAM -94 Physics Meeting (Cacun, Mexico 1994)	95-77851	1-56396-491-0
No. 343	High Energy Spin Physics Eleventh International Symposium (Bloomington, IN 1994)	95-78431	1-56396-374-4
No. 344	Nonlinear Dynamics in Particle Accelerators: Theory and Experiments (Arcidosso, Italy 1994)	95-78135	1-56396-446-5
No. 345	International Conference on Plasma Physics ICPP 1994 (Foz do Iguaçu, Brazil 1994)	95-78438	1-56396-496-1
No. 346	International Conference on Accelerator-Driven Transmutation Technologies and Applications (Las Vegas, NV 1994)	95-78691	1-56396-505-4
No. 347	Atomic Collisions: A Symposium in Honor of Christopher Bottcher (1945-1993) (Oak Ridge, TN 1994)	95-78689	1-56396-322-1
No. 350	International Symposium on Vector Boson Self-Interactions (Los Angeles, CA, 1995)	95-79865	1-56396-520-8